ENERGY METHODS IN STRESS ANALYSIS

with an introduction to finite element techniques

ELLIS HORWOOD SERIES IN ENGINEERING SCIENCE

Editors: J. M. ALEXANDER, Professor of Applied Mechanics, and
J. MUNRO, Reader in Civil Engineering Systems,
Imperial College of Science and Technology, University of London

Engineering disciplines have been strongly characterised by immense expansion in the range of applicable knowledge in recent years. The days when a professional engineer could base the scientific part of his career on his undergraduate training have gone, and it is now accepted that he must keep abreast of the developing technology of his chosen field by a continual process of updating and retraining. This can be achieved in a variety of ways, but is greatly facilitated by the availability of authoritative texts.

Similarly, the changing climate of engineering practice requires a continual re-appraisal of undergraduate teaching so that the embryo engineer may be properly introduced to the fundamental concepts of his technology and at the same time acquire sufficient flexibility and breadth of understanding so that he can respond to the ever changing demands of his vocation.

Thus the Ellis Horwood Engineering Science Series has two objectives, of satisfying the requirements of post-graduate and mid-career education and of providing clear and modern texts for more basic undergraduate topics in the fields of civil and mechanical engineering. It is furthermore the editors' intention to include English translations of outstanding texts originally written in other languages, thereby introducing works of international merit to English language audiences.

In these ways it is hoped that the series will render a valuable service to practising engineers, students and teachers, by providing noteworthy additions to technological literature.

Published and Forthcoming Books:

ADVANCED MANUFACTURING PROCESSES TECHNOLOGY
J. M. ALEXANDER, R. C. BREWER, Imperial College of Science and Technology, University of London, J. R. CROOKALL, Cranfield Institute of Technology

VIBRATION ANALYSIS AND CONTROL SYSTEM DYNAMICS
CHRISTOPHER BEARDS, Imperial College of Science and Technology, University of London

COMPUTER AIDED DESIGN FOR ENGINEERS
C. B. BESANT, Imperial College of Science and Technology, University of London

BASIC LUBRICATION THEORY 2nd Edition
ALASTAIR CAMERON, Imperial College of Science and Technology, University of London

ADVANCED MECHANICS OF MATERIALS 2nd Edition
Sir HUGH FORD, F.R.S. and J. M. ALEXANDER, Imperial College of Science and Technology, University of London

TECHNIQUES OF FINITE ELEMENT ANALYSIS
BRUCE M. IRONS, University of Calgary, and S. AHMAD, Bangladesh University of Engineering and Technology, Dacca

ENERGY METHODS IN STRESS ANALYSIS: With an Introduction to Finite Element Techniques
T. H. RICHARDS, University of Aston, Birmingham

ENERGY METHODS IN STRESS ANALYSIS

with an introduction to finite element techniques

T. H. RICHARDS

Senior Lecturer, Department of Mechanical Engineering, University of Aston, Birmingham

ELLIS HORWOOD LIMITED

Publisher • Chichester

Halsted Press: a division of

JOHN WILEY & SONS INC.

New York • London • Sydney • Toronto

First published in 1977 by

ELLIS HORWOOD LIMITED PUBLISHER
Coll House, Westergate, Chichester, Sussex, England

Distributors:

Australia, New Zealand, South-east Asia:
JOHN WILEY & SONS AUSTRALASIA PTY LIMITED
1-7 Waterloo Road, North Ryde, N.S.W. 2113, Australia

Canada:
JOHN WILEY & SONS CANADA LIMITED
22 Worcester Road, Rexdale, Ontario, Canada

Europe, Africa:
JOHN WILEY & SONS LIMITED
Baffins Lane, Chichester, Sussex, England

North and South America and the rest of the world:
HALSTED PRESS a division of
JOHN WILEY & SONS INC.
605 Third Avenue, New York, N.Y. 10016, U.S.A.

Library of Congress Cataloging in Publication Data
Richards, T. H., Energy Methods in Stress Analysis
1. Mechanics. Applied. 2. Finite element method.
I. Title.
TA350.R55 620.1'04 76-29647
ISBN 0-470-98960-2 (Halsted Press)
ISBN 85312-048-X (Ellis Horwood) Library Edition
ISBN 85312-050-1 (Ellis Horwood) Student Edition

The publisher's colophon is reproduced from James Gillison's drawing of the ancient Market Cross, Chichester

Set in 10 point Press Roman by Speedpress Repro Services,
14 St. John's Precinct, Hebburn, Tyne & Wear

Printed in Great Britain by Butler & Tanner Ltd, Frome and London

Dedication

To my wife Frances and children Mark, Louise and David who put up with my pre-occupation with energies, variations and things over a long period

Table of Contents

Preface

Mechanics has developed as a branch of rational thought from very early times; it is the oldest of the physical sciences and its principles, formulated to describe mechanical behaviour in nature, form the basis of engineering calculations for a vast range of devices and structures. In applying these principles to mechanical design, an essential activity is the formulation of conceptual models, or idealizations, of real situations to which one's current stock of analytic techniques may be applied: the greater the precision one seeks in the correlation between actual and predicted behaviour, the more sophisticated does the model need to be. Unfortunately, this increased sophistication almost always incurs increased complexity in analysis and it is this which ultimately places a practical limit on the implementation of advanced methods to design. However, the developments which have taken place in the capability of computing devices in recent years have made possible the implementation of advanced methods yielding levels of precision which previously could not have been contemplated. From the point of view both of formulating theoretical models for, and the practical determination of numerical solutions to, engineering problems, the energy (or variational) principles of mechanics hold a position of central importance. In this book, we are concerned with developing these principles in their application to equilibrium stress and deformation analysis: the subject matter is developed in a careful manner from the treatment of discrete rigid body/spring systems onto bar and more sophisticated continuum stressing situations culminating in an introduction to the finite element method.

The book is directed at a senior undergraduate and first year graduate level of work so that the usual basic experience of statics and 'mechanics of materials' is taken for granted. However, in teaching advanced engineering students over a number of years, the author has learnt that a frequent cause of difficulty with some advanced theory is an imperfect recollection of elementary principles, studied quite some time previously, with a consequent difficulty in relating these principles to the current situation. This seems to be especially true of the energy principles of mechanics which often receive only a small amount of attention in undergraduate courses. Accordingly the subject matter is developed from basic principles to a good level in a reasonably self-contained manner so as to facilitate this process of reference back from advanced to elementary ideas. It is intended to be useful for individual study or as supplementary reading for a course. The presentation is such that the first two chapters alone could form the basis for a series of lectures to undergraduate mechanical or structural engineers; the remainder is written for the first year graduate student or practicing engineer who has cause to take up advanced methods in his professional work.

The principles of virtual work and virtual complementary work are concerned with describing the static equilibrium of a system by comparing the relative merits of a number of candidate solutions and identifying the true one by requiring that some property of the system shall attain a stationary value. Whilst the overtones of economy in nature implicit in this approach have considerable plausibility; it, and the concomitant notions of virtual displacements and virtual forces, are a source of unease to engineers who are usually more familiar with the cause and effect philosophy of vector mechanics. The earlier sections of the book are consequently devoted to overcoming this unease by providing a very full account of the energy principles in relation to discrete systems; in this way the underlying philosophy can be absorbed without the distractions of the higher level of mathematics required when continuum systems are discussed. It is only when these basic concepts have been thoroughly understood as a result of reflection and practice in treating relatively simple systems that one can approach more complex situations with confidence.

Chapter 1 is concerned only with discrete or lumped spring and rigid body systems. It begins with a brief review of some very basic notions in mechanics and continues with a careful discussion of system configuration in which constrained and generalised coordinates are introduced. For the purposes of contrast with the subsequently developed energy methods, the principles of vector statics are reviewed and it is shown that the solution to any problem in the statics of deformable bodies involves three essential ingredients (a) the consideration of equilibrium of forces, (b) the requirement of geometric fit, or compatibility of the displacements and (c) some force-deformation characteristic for the elements of the system. Work, energy, the idea of a virtual displacement, the principles of virtual work and stationary potential energy are then described and applied to a few simple situations. In a similar way, we then discuss and apply the ideas of complementary work and energy, statically independent forces and the principles of virtual complementary work and complementary energy. The presentation is such as to highlight the complementary nature of two characteristic approaches to problems in structural mechanics: on the one hand, the virtual work principle leads to the stiffness method of analysis whilst on the other the virtual complementary work principle leads to the flexibility method. We conclude with a discussion of the stability of equilibrium. The author considers this chapter to be of crucial importance in that all the essentials for a variational approach to mechanics problems are introduced; the remainder of the book is concerned with the refinements necessary to facilitate tackling problems of greater complexity.

In Chapter 2, we apply the previously established principles to structures comprised of bar-type elements. We study the characteristics of individual elements under conditions of axial load, torsion, bending and transverse shear and consider the idealised assemblages of pin jointed and rigid jointed frames. By means of these relatively straightforward systems we illustrate the freedom of choice one has for virtual displacements and virtual forces and show how the former can be used to evaluate the degree of statical indeterminancy of a given structure. The principles of virtual work and virtual complementary work are applied to such situations as will highlight their scope and limitations without undue computational effort and once more to formulate the alternative stiffness or flexibility methods of structural analysis.

Solid continuum mechanics is the body of analytic techniques by means of which a high level of precision has been attained in the stress analysis of many complex systems: Chapter 3 is concerned with the foundations of this topic. Whether one bases this level of treatment on vector or variational principles, it is necessary to clarify the concepts of a continuum, the state of stress and strain and to establish a constitutive relation between stress and strain for the continuum material: the earlier parts of the chapter are concerned with these matters. The interpretation of equilibrium and compatibility conditions at a continuum level are discussed and, with these foundations firmly lain down, the formulation of the general problem of elastic stress analysis follows. The chapter concludes with an account of some specific classes of problem.

When variational principles are applied to continuum problems, the appropriate mathematical tool is the calculus of variations. Chapter 4 is concerned with an account of its principles which is by no means exhaustive but which the author feels is adequate for the purposes of this book. Its central purpose is to show that when the behaviour of a continuous system is in accordance with some minimum principle, the procedures of the variational calculus yield differential equations together with appropriate boundary conditions describing the behaviour. The subject matter is essential preparation for Chapter 6.

The so-called direct methods of the calculus of variations provide powerful practical techniques for obtaining numerical solutions to very complex stressing problems of engineering significance. They comprise a body of approximation methods for treating continuum problems. Several such methods have been developed over the years, but in Chapter 5 we restrict ourselves to a discussion of the more widely used ones due to Rayleigh, Ritz, Kantorovitch, Galerkin, Euler and Trefftz. These methods are of importance in their own right, but they also form a valuable preparation for an understanding of the more modern finite element procedure.

Chapter 6 is concerned with formulating the variational principles of mechanics for a solid continuum. Here the ideas and methods of Chapters 3 and 4 are brought together in a generalisation of the procedures of Chapters 1 and 2 and Reissner's principle is introduced. We see how the relevant equations of equilibrium and compatibility can be deduced for an 'exact' theory and, by way of specific classes of problem, how all the relevant equations for an 'approximate' theory can be deduced in a systematic and consistent manner. A designer cannot be content with merely producing an adequate approximate theory to describe his current problem, he has to produce quantitative data from which suitable proportions may be determined for structural and machine parts. Then, since even an approximate theory may yield differential equations which are difficult to handle, the direct methods of the calculus of variations can be extremely valuable. Chapter 7 is a collection of substantial examples in which the methods of Chapter 5 are applied to a variety of situations of practical interest. The examples show that the effect of these approaches is to descretise a continuum model, that is the behaviour of a system with an infinite number of degrees of freedom is approximated by one having only a finite number. Under these circumstances the virtual work principle is concerned with determining a compatible configuration for which equilibrium is satisfied only in some average sense whilst the virtual complementary work principle is concerned with equilibriant stress systems for which compatibility is satisfied only in some average sense. Considerable emphasis is placed on clarifying these aspects of the work since confidence in applying and interpreting the results of these methods only comes from appreciating the physical implications of the various mathematical operations carried out.

The finite element method is introduced in Chapter 8. Emphasis is placed on the so-called displacement method and a conscious effort is made to relate the method to the material in the earlier sections of the book. Following a discussion on generating displacement models, a number of classes of problem are treated showing that the characteristic feature of the method is the generation of stiffness equilibrium equations in which the basic unknowns are displacements. The torsion problem, formulated in terms of the Prandtl stress function, is treated to provide an example of a complementary energy approach whilst the chapter is concluded by a brief introduction to the more sophisticated hybrid and mixed formulations which have a number of important applications.

Some useful results from mathematics are collected together in the Appendix.

The variational methods of mechanics are amongst the most beautiful of man's intellectual achievements and at the same time provide some of his most useful analytic tools. Their range of application stretches well beyond that covered here but to keep the book to a moderate length a number of interesting topics have had to be put aside. Thus, to add a satisfactory treatment of shell problems, plasticity or the stability and dynamics of elastic systems would require substantially more space; on the other hand, to replace the introduction to finite elements by, say, an account of stability or dynamics is largely a matter of personal preference. The object has been to produce a reasonably self contained developing account of the use of energy methods in determining the displacements and stresses in a mechanical system in a state of stable equilibrium. It is my hope that, with the pointers given in the Introduction, the reader who wishes to proceed to the literature dealing with the more specialist areas of this work will feel equipped so to do.

T. H. Richards,
The University of Aston,
Birmingham.

Author's Introduction

In spite of the advanced stage to which Engineering Science has progressed, it is not yet capable of describing the behaviour of real engineering systems in every particular. Consequently, an essential element in the process of designing such systems is the generation of conceptual models (idealisations) of real situations to which one's current stock of analytic and numerical techniques may be applied. Depending on the requirements of the problem under consideration and the current state of one's capability for solution, models at various levels of sophistication may be contemplated and calculations based on these models will yield predictions at various levels of precision. Although one may hope that such calculations will be adequate for the purpose at hand, it is often the case (and *usually* so for advanced products) that the development of a satisfactory design requires laboratory work and in-service experience to confirm the adequacy of some theoretical model or to overcome some problem for which no theoretical treatment is contemplated.

In the design of a given product, one may need to refer to several branches of engineering science. Yet, whatever function the device must perform, it will need to be structurally sound. Now it is very obvious that if a product actually breaks it certainly will not have been adequate, but malfunction may also occur due to excessive distortion under load or to spurious response to some environmental disturbance such as temperature or vibration. Evidently stress and deformation analysis is central to design and in this book we are concerned with showing what a powerful aid to this work is offered by the variational principles of mechanics. The aim is to develop the subject in a self contained coherent manner from elementary beginnings to advanced applications. To this end, we first consider one of the most basic models of a mechanical system, a lumped spring/rigid body arrangement, and acquire an appreciation of the underlying notions of the variational approach without the distractions of the higher level of mathematics necessary for continuum situations. In this way, we gain a 'feel' for virtual displacements (variations to a configuration), virtual forces (variations to equilibrium states) and a sense of nature working with sublime economy of effort to achieve some end. In preparation for a more sophisticated treatment of systems with distributed mechanical properties, we proceed to clarify the basic ideas of continuum mechanics and to examine the variational calculus, which is the essential mathematical tool for manipulating energy principles applied to continua. With this ground work consolidated, it is possible fully to appreciate the elegance and practical utility of the variational approach and to benefit from the insight it provides. In formulating, by the vector approach, an approximate theory say for beams or plates, it is not always clear just what mathematical terms are significant and what are not: in contrast, we perceive that the variational approach yields a consistent statement of the boundary value problem automatically. Further, in the event of even the approximate theory being too difficult to handle analytically, a whole range of numerical processes, culminating in the finite element method, is available to us. Except for a brief reference to stability in Section 1.9, we shall be concerned with problems in *equilibrium* stress analysis; a consideration of large deformations, buckling, dynamics and plasticity or viscoelaticity would require far more space than we can afford. However, to provide some indication of how the energy methods may be applied to problem areas beyond those of immediate interest in this book, we shall devote the remainder of these introductory pages to a very brief discussion of just three beam problems. We shall discover that whilst the equilibrium configuration of a finite degree of freedom system can be described by a set of simultaneous algebraic equations,

$$[K]\{q\} = \{Q\} \, , \tag{I.1}$$

an analogous treatment of the buckling or free vibration of such systems leads to the **eigenvalue** problem

$$([K] - \lambda[M])\{q\} = 0 \, . \tag{I.2}$$

Suppose a beam, simply supported at both ends, carries a distributed load of intensity $p(x)$ per unit length over its entire length l. As we shall see in Chapter 1, the equilibrium configuration of a mechanical system may be identified by means of the principle of stationary total potential energy; then, in the case of this beam, the deflexion curve $w(x)$ is identified by

$$\delta V = 0 \tag{I.3}$$

where

$$V = \int_0^l \frac{EI}{2}(w'')^2 \, dx - \int_0^l pw dx \dagger \, , \tag{I.4}$$

The appropriate mathematical tool for handling equation (I.3) when (I.4) is substituted into it is the Calculus of Variations (Chapter 4). If we follow its rules, we find that equation (I.3) leads to

$$\frac{d^2}{dx^2}\left(EI \frac{d^2w}{dx^2}\right) = p \tag{I.5}$$

together with appropriate boundary conditions; equation (I.5) is the familiar differential equation of equilibrium for a beam. If the bending stiffness EI is constant, solving equation (I.5) is a straightforward matter, but if EI is a function of position along the beam, the solution may be awkward to achieve. An alternative approach, which has increasing attraction as the systems of interest become more complex, is to seek an *approximate* solution by *assuming* a representation for the deflexion curve in the form of a series of prescribed functions $\phi(x)$.

† See Chapters 2 and 6.

Then,

$$w(x) = \sum_{i=1}^{n} q_i\phi_i(x) \quad , \tag{I.6}$$

where the q's are parameters to be determined. Substituting from equation (I.6) into (I.4),

$$V = \frac{1}{2}\sum_i \sum_j q_i q_j \int_0^l EI\phi_i''\phi_j''\mathrm{d}x - \sum_i q_i \int_0^l p\phi_i\mathrm{d}x \ . \tag{I.7}$$

Denoting

$$k_{ij} = \int_0^l EI\phi_i''\phi_j''\mathrm{d}x = k_{ji} \tag{I.8}$$

and

$$Q_i = \int_0^l p\phi_i\mathrm{d}x \quad , \tag{I.9}$$

equation (I.7) becomes

$$V = \frac{1}{2}\sum_i \sum_j k_{ij}q_i q_j - \sum_i Q_i q_i \ . \tag{I.10}$$

The reader who is familiar with matrix algebra† will recognize this as

$$V = \tfrac{1}{2}\{q\}^t[K]\{q\} - \{q\}^t\{Q\} \ . \tag{I.11}$$

Here, $[K]$ is a matrix of the **stiffness coefficients** k_{ij}, $\{q\}$ is a vector of **generalised coordinates** and $\{Q\}$ is a vector of **generalised forces** corresponding to the $\{q\}$ (see Chapter 1): we observe from equation (I.8) that $[K]$ is symmetric.

Equation (I.11) expresses the typical form of the potential energy expression for a finite degree of freedom system and we see that the energy approach has furnished, in a rational manner, the appropriate mechanical characteristic $[K]$ and disturbance $\{Q\}$ for a substitute discrete system whose behaviour approximates that of the real beam. (We will not dwell on the considerations affecting the choice of the ϕ's here, they are discussed in the main body of the text). $\delta V = 0$, applied to equation (I.11) now ensures satisfaction of beam equilibrium only in some average way summarised in

$$[K]\{q\} = \{Q\} \ . \tag{I.12}$$

Whilst the above beam takes up its bowed shape, its ends will travel towards each other by some small amount Δ. As this movement occurs, the moment arms of the support reactions change very slightly, but in the usual theory, where the deflexion is small, this effect is negligible. If, on the other hand, the beam also supports an axial thrust T, as shown in Fig. I.1, the bow in the beam causes T to have a moment arm (equal to the deflexion) and hence a bending action as well as the ordinary stress = force divided by area effect. This feature of an alteration in the action of a load as the deflexion occurs is a simple example of

† The reader who is not familiar with this notation will find a summary in the Appendix.

non-linear behaviour rooted in structural geometric effects rather than a non-linear stress-strain law for the material.

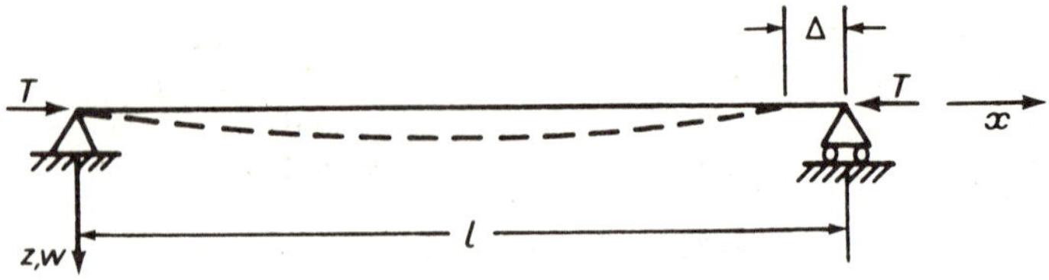

Fig. I.1

To apply the energy method to this problem, it is only necessary to take into account the work done by T in travelling through the distance Δ.† Then, the total potential energy of the system is

$$V = \int_0^l \frac{EI}{2}(w'')^2 \mathrm{d}x - \int_0^l pw\mathrm{d}x - T\Delta \ . \qquad \text{(I.12)}$$

We are concerned with bending actions as opposed to direct compression here so that the final length of the bowed beam is still l. Then,

$$l = \int_{x=0}^{x=(l-\Delta)} \mathrm{d}s \ . \qquad \text{(I.13)}$$

Now
$$\mathrm{d}s = \left[1 + \left(\frac{\mathrm{d}w}{\mathrm{d}x}\right)^2\right]^{1/2} \mathrm{d}x \qquad \text{(I.14)}$$

and, if we suppose $\mathrm{d}w/\mathrm{d}x$ to be small, expanding by the binomial theorem and ignoring terms in $(\mathrm{d}w/\mathrm{d}x)^4$ etc, yields

$$\mathrm{d}s \doteqdot \left[1 + \frac{1}{2}\left(\frac{\mathrm{d}w}{\mathrm{d}x}\right)^2\right] \mathrm{d}x \ . \qquad \text{(I.15)}$$

Substituting from equation (I.15) into (I.13), we have

$$\Delta = \frac{1}{2}\int_0^l (w')^2 \mathrm{d}x \ . \qquad \text{(I.16)}$$

For equilibrium, $\delta V = 0$ and applying the formal processes of the calculus of variations to equation (I.12), we have the Euler-Lagrange equation for a uniform **beam column** as

$$EI\frac{\mathrm{d}^4 w}{\mathrm{d}x^4} + T\frac{\mathrm{d}^2 w}{\mathrm{d}x^2} = p \ . \qquad \text{(I.17)}$$

† Notice that Δ arises from the bow in the beam alone and not from direct compression. We have ignored this latter effect since we are interested in the buckling problem and it can be shown that the value of the buckling load is not affected by the value of the direct compression energy.

If the beam supports the end thrust alone, $p = 0$ and equation (I.17) becomes

$$EI\frac{d^4w}{dx^4} + T\frac{d^2w}{dx^2} = 0 \ . \tag{I.18}$$

Equation (I.18) gives non-trivial solutions for w only for certain discrete values T_{crit} of the end thrust. These critical values of T are the **eigenvalues** of the problem and the corresponding solutions for w are the **eigenfunctions.** Equation (I.18) is a **buckling equation** and the smallest value of T_{crit} is the **buckling load.**

If we assume a solution for $w(x)$ in the form of a series of prescribed functions $\phi(x)$,

$$w(x) = \sum_{i=1}^{n} q_i\phi_i(x) \ , \tag{I.19}$$

with the q_i parameters to be determined, the potential energy becomes

$$V = \frac{1}{2}\sum_i\sum_j q_iq_j\left(\int_0^l EI\phi_i''\phi_j''\mathrm{d}x\right) - \sum_i q_i\left(\int_0^l p\phi_i\,\mathrm{d}x\right) - \frac{1}{2}\sum_i\sum_j q_iq_j\left(\int_0^l T\phi_i'\phi_j'\mathrm{d}x\right) \ . \tag{I.20}$$

Denoting

$$k_{ij} = \int_0^l EI\phi_i''\phi_j''\mathrm{d}x = k_{ji} \ , \quad s_{ij} = \int_0^l T\phi_i'\phi_j'\mathrm{d}x = s_{ji} \ ,$$

$$Q_i = \int_0^l p\phi_i\mathrm{d}x \ , \tag{I.21}$$

k_{ij} is the usual stiffness coefficient with respect to q_i and q_j implicit in the assumed deflexion shape, s_{ij} is called the **geometric stiffness** coefficient and Q_i is a generalised force. Then,

$$V = \frac{1}{2}\sum_i\sum_j (k_{ij} - s_{ij})q_iq_j - \sum Q_iq_i \tag{I.22}$$

or

$$V = V(q\text{'s}) \ .$$

Now, we shall see in Section 1.9 that the equilibrium configuration ceases to be stable when the second variation $\delta^2 V$ ceases to be positive definite. For the present example, this implies that:

$$|[K] - [S]| = 0 \ . \tag{I.23}$$

This is an algebraic eigenvalue problem: expanding the determinant yields a polynomial in T, the roots of which are its critical values T_{crit}. The smallest value of T_{crit} is the **buckling load.** If we denote $T_{crit} = \lambda T$, where λ is a parameter, equation (I.23) can be written in the standard form

$$|[K] - \lambda[S]| = 0 \ . \tag{I.24}$$

Equations (I.23) and (I.24) correspond to systems with a finite number of degrees of freedom and we see, once more, how an energy approach provides a convenient and rational means for calculating the appropriate parameters $[K]$ and $[S]$ for the discretised continuum.

It is often necessary to compute the response of a structure to dynamic loading: when a structure is set in motion, its elements experience accelerations so that inertia forces are called into play. According to **D'Alembert's principle,** a dynamics problem may be converted into a statics one by taking these inertia forces into account so that the principle of virtual work† may be employed. If the displacements are considered to be prescribed at two instants of time t_1 and t_2, so that any virtual displacements must vanish then, integrating the virtual work equation with respect to time, with the inertia forces incorporated into the body force term, leads to **Hamilton's principle.** This may be regarded as *the* basic principle of dynamics and states that:

Of all the geometrically possible motions which a system may execute, the true one is that which renders

$$\delta A = \delta \int_{t_1}^{t_2} L \, dt = 0 . \qquad (I.25)$$

The quantity A is called the **Action,** whilst $L = T - U - \Omega$ is the **Langrangian** function. T is the kinetic energy, U the strain energy and Ω the potential energy of the applied loads.

Application of the formal processes of the calculus of variations to equation (I.25) for a continuum leads to partial differential equations of motion and appropriate boundary conditions. Alternatively, we may seek an approximate solution after the fashion we now employ in a further study of the beam.

Suppose the beam is excited by a distributed load $p(x, t)$, which is now a function of time and position, and that the deflected shape taken up can be approximated by

$$w(x, t) = \sum_{i=1}^{n} q_i(t)\phi_i(x) . \qquad (I.26)$$

The ϕ's are prescribed bent shapes for the beam and so are functions of position x alone satisfying the boundary conditions, whilst the q's, which are functions of t, are the amplitudes of the constituent shapes making up $w(x, t)$.

If the mass per unit length of the beam is m, the kinetic energy is

$$T = \frac{1}{2}\int_0^l m(\dot{w})^2 dx = \frac{1}{2}\sum_i \sum_j \dot{q}_i\dot{q}_j \int_0^l m\, \phi_i \phi_j \, dx . \qquad (I.27)$$

Denoting $$m_{ij} = \int_0^l m\, \phi_i \phi_j \, dx = m_{ji} , \qquad (I.28)$$

† See Chapter 1.

as elements of the mass matrix $[M]^\dagger$,

$$T = \tfrac{1}{2}\{\dot{q}\}^t[M]\{\dot{q}\} \quad . \tag{I.29}$$

The strain energy is given by

$$U = \tfrac{1}{2}\{q\}^t[K]\{q\} \quad , \tag{I.30}$$

where $$k_{ij} = \int_0^l EI\phi_i''\phi_j''\mathrm{d}x$$

as before and $$\Omega = -\{q\}^t\{Q\} \quad ,$$

with $$Q_i(t) = \int_0^l p\phi_i\mathrm{d}x \quad . \tag{I.31}$$

Hamilton's principle now becomes

$$\delta\left[\int_{t_1}^{t_2}\left(\frac{1}{2}\{\dot{q}\}^t[M]\{\dot{q}\} - \frac{1}{2}\{q\}^t[K]\{q\} + \{q\}^t\{Q\}\right)\mathrm{d}t\right] = 0 \tag{I.32}$$

or $$\int_{t_1}^{t_2}(\{\delta\dot{q}\}^t[M]\{\dot{q}\} - \{\delta q\}^t[K]\{q\} + \{\delta q\}^t\{Q\})\mathrm{d}t = 0 \tag{I.32a}$$

where we have taken advantage of the symmetry in $[M]$ and $[K]$. Integrating the first term by parts,††

$$\text{First term} = |\{\delta q\}^t[M]\{\dot{q}\}|_{t_1}^{t_2} - \int_{t_1}^{t_2}\{\delta q\}^t[M]\{\ddot{q}\}\mathrm{d}t \quad . \tag{I.33}$$

But $\{\delta q\} = 0$ at t_1 and t_2 so that equation (I.32) becomes

$$\int_{t_1}^{t_2}\{\delta q\}^t([M]\{\ddot{q}\} + [K]\{q\} - \{Q\})\mathrm{d}t = 0 \quad . \tag{I.34}$$

At any instant, the $\{\delta q\}$ are arbitrary so that the motion is described by

$$[M]\{\ddot{q}\} + [K]\{q\} = \{Q\} \quad . \tag{I.35}$$

The energy method has again proved to be a very convenient means of generating the significant form of the excitation force, $\{Q\}$, and the relevant system parameters $[M]$ and $[K]$.

† Notice that $[M]$ is not merely a diagonal matrix as it would be for a simple lumped parameter system.

†† Alternatively, one could use **Lagrange's Equations**

$$\frac{\mathrm{d}}{\mathrm{d}t}\left(\frac{\partial L}{\partial q_i}\right) - \frac{\partial L}{\partial q_i} = Q_i$$

which also follow from Hamilton's principle.

When the beam is in a state of **Free Vibration,** $\{Q\} = \{0\}$ and $\{q\}$ is a harmonic function of time. Then,

$$w(x, t) = \sum_{i=1}^{n} \hat{q} \sin \omega t \, . \, \phi(x) \qquad \text{(I.36)}$$

Substitution in equation (I.35) yields

$$([K] - \lambda[M])\{\hat{q}\} = \{0\} \ , \qquad \text{(I.37)}$$

where $\lambda = \omega^2$. Non-trivial solutions $\{\hat{q}\}$ are obtained only when

$$|[K] - \lambda[M]| = 0 \ . \qquad \text{(I.38)}$$

The roots $\lambda_1, \lambda_2 \ldots\ldots \lambda_n$ of this polynomial correspond to the **Natural frequencies** of vibration of the discretised system whilst the corresponding $\{\hat{q}_1\}, \{\hat{q}_2\} \ldots$ are its **Natural modes.**

This book is concerned with the application of the variational principles of mechanics to the solution of problems in elastic stress analysis. In these introductory pages, the aim has been to give an impressionist view of the topics covered and to provide a few pointers as to how they may be extended to treat other problem areas which are significant in practical design. It is hoped that the student who has read this text and wishes to proceed to the literature dealing with these more specialist fields will feel equipped so to do.

Further reading

Bishop, R.E.D. *and* Johnson, D.C.	*The Mechanics of Vibration,* Cambridge University Press (1960).
Bishop, R.E.D., Gladwell, G.M.L. *and* Michaelson, S.	*The Matrix Analysis of Vibration,* Cambridge University Press (1965).
Bleich, F.	*Buckling Strength of Metal Structures,* McGraw Hill (1952).
Cox, H.L.	*The Buckling of Plates and Shells,* Pergamon Press (1961).
Dym, C.L. *and* Shames, I.H.	*Solid Mechanics: A Variational Approach,* McGraw Hill (1973).
Flugge, W.	*Stresses in Shells,* Springer Verlag (1960).
Lanczos, C.	*The Variational Principles of Mechanics,* University of Toronto Press, 4th Edn. (1970).
Langhaar, H.L.	*Energy Methods in Applied Mechanics,* John Wiley (1962).
Love, A.E.H.	*A Treatise on the Mathematical Theory of Elasticity,* Cambridge University Press, 4th Edn. (1927).
Meirovitch, L.	*Analytical Methods in Vibrations,* McGraw Hill (1970).
Novozhilov, V.V. (Trans. P.G. Lowe)	*The Theory of Thin Shells,* P. Noordhoff Ltd (1959).

Przemieniecki, J.S. *Theory of Matrix Structural Analysis,* McGraw Hill (1968).

Rayleigh, J.W.S. *The Theory of Sound,* Dover Publications (1945).

Rubinstein, M.F. *Structural Systems – Statics, Dynamics and Stability,* Prentice-Hall (1970).

Temple, G. *and* Bickley, W.G. *Rayleigh's Principle and its Applications to Engineering,* Dover Publications (1956).

Timoshenko, S. *and* Gere, J.M. *Theory of Elastic Stability,* McGraw Hill (1961).

Timoshenko, S. *and* Woinowsky-Kreiger, S. *Theory of Plates and Shells,* McGraw Hill (1959).

Timoshenko, S., Young, D.H. *and* Weaver, W. *Vibration Problems in Engineering,* John Wiley, 4th Edn. (1974).

Washizu, K. *Variational Methods in Elasticity and Plasticity,* 2nd Edition, Pergamon Press (1973)

Ziegler, H. *Principles of Structural Stability,* Blaisdell Publishers (1968).

Chapter 1

Basic Principles of Statics

1.1 Introduction

Mechanics is the oldest of the physical sciences, having attracted the attention of philosophers since before the birth of Christ. Aristotle and Archimedes had discussed the equilibrium of the lever as early as the third century B.C., but very little further progress was made until late in the sixteenth century. Then, Stevinus made apparent the vector nature of a force and laid the foundations of statics; Galileo advanced the subject and laid the foundations of dynamics. It seems that Aristotle, Stevinus and Galileo regarded the work of a force as a significant quantity and used it in formulating conditions of equilibrium: the beginnings of the virtual work method are thus to be found very early indeed. By the early years of the eighteenth century, considerable progress had been made: Newton had propounded his laws of motion, thus codifying and generalising Galileo's ideas. Leibnitz had established what is now called the law of kinetic energy and John Bernoulli had formulated the principle of virtual work in the form used at the present time. From this period, developments in mathematics facilitated the further rapid advance of the subject and this occurred along two distinct paths. In **vectorial mechanics**, force and momentum are regarded as fundamental quantities characterising the behaviour of mechanical systems: the rate of change of momentum of a system is a measure of the action of a force in accordance with Newton's laws. In **analytical mechanics**, the work of a force and the kinetic energy of the system are the fundamental quantities describing the motion: the behaviour is in accordance with the principle of virtual work generalised for dynamical situations by D'Alembert's principle.

It is valid to approach a problem in mechanics by either the vector or analytical method. However, as the systems under consideration become increasingly complex, so the advantages of the work methods in simplicity of treatment and economy of effort become more and more significant. We may see the nature of these advantages as follows. Newton's laws were originally proposed for particles moving freely: by visualising engineering systems to be comprised of assemblages of particles, we can evidently apply Newton's laws to each particle in turn, but clearly, we must take account of *all* the forces acting on each one. When systems of particles are constrained to move relative to each other in some way, taking account of the constraint forces can become cumbersome. In contrast, the energy principles are applied to the system as a whole without the need to identify particles. Indeed, if certain particles are constrained to move together as a rigid body, the corresponding constraint forces do no work and so do not enter into the discussion at all. We shall see in due course that this is a significant feature whether we are obtaining numerical solutions for discrete systems or, on a higher level of sophistication, formulating approximate theories for continuum problems. The second great advantage of energy methods derives from the scalar nature of work: this gives us considerable freedom in choosing the most convenient set of coordinates with which to describe the system under consideration.

In this chapter, we review some of the basic concepts of vector statics, establish the various energy principles for discrete systems and contrast their applications in a number of examples. The remainder of the book is then concerned with further development of the energy principles and their applications to stress analysis.

1.2 Mechanical systems

On an engineering scale, a body is viewed as matter continuously occupying a region of space bounded by a surface. However for many analytic purposes it has been found convenient to introduce the notion of a material point, or particle, which has mass but no volume; then, a body may be regarded as a collection or system of particles distributed within a bounding surface. Since machines and structures are assemblages of bodies, we may ultimately regard any mechanical device as constituting a system of particles. The subject of mechanics is concerned with determining the response of the system to actions across its bounding surface. At a higher level of analysis, it proves necessary to introduce the notion of a **continuum** representation of a solid: **continuum mechanics**† provides the means of describing the response of a continuum to the various actions which may be impressed on it.

It usually happens that the information sought regarding a given system requires us to identify sub-collections of its particles (or sub-regions of the continuum) by drawing bounding surfaces around them. Then, we would have cause to discuss the responses of these newly generated sub-systems to actions across *their* boundaries. Thus, the frame of Fig. 1.2.1(a) is an assemblage of bars each of which is a collection of particles. At various stages in a stress analysis of the frame, we would find it necessary to carry out various isolations into sub-systems by means of the dotted surfaces of Figs 1.2.1(b), (c) and (d).

† The idea of a continuum is discussed further in Chapter 5.

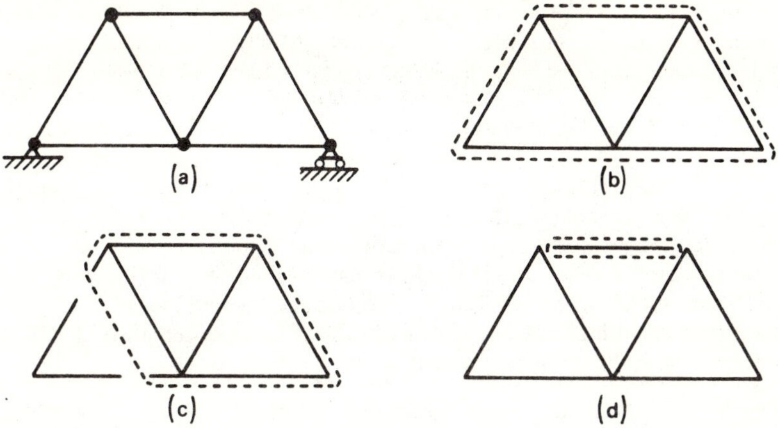

Fig. 1.2.1

The process of representing on paper the isolated systems and sub-systems together with the actions across their bounding surface is the familiar one of drawing free body diagrams.

As part of a mechanical analysis, we shall clearly have cause to describe the geometry of any mechanical arrangement: we speak of the **configuration** of a system as being the simultaneous positions taken up by all its constituent particles. To define the configuration, we need a frame of reference against which the relative positions of all the particles are measured. A set of numbers which locates the system with respect to the frame of reference comprises a set of coordinates. Thus, if the two particles of Fig. 1.2.2 constitute a system, its configuration is completely defined relative to the arbitrarily chosen reference frame by the Cartesian coordinates x_1, y_1, z_1 and x_2, y_2, z_2

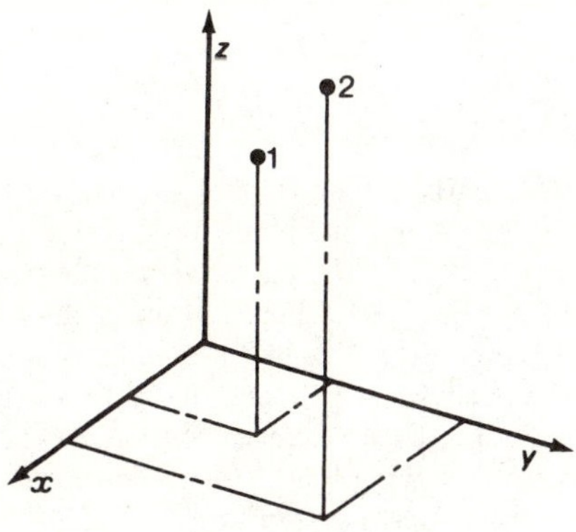

Fig. 1.2.2

Any body may experience actions from outside itself so that we come to the idea of determining the response of a system of particles to actions across the bounding surface which isolates the system. In mechanics, the actions of interest are forces and, since they have magnitude and direction, they are vector quantities. More precisely, forces are *inter*actions: thus, in Fig. 1.2.3(a), the attraction between the satellite and the centre of the earth is shown by $\mathbf{F}_1$ and $\mathbf{F}_2$; in Fig. 1.2.3(b), $\mathbf{F}_1$ and $\mathbf{F}_2$ represent the interaction at the point of contact between the heavy block and the table whilst Fig. 1.2.3(c) shows the interaction $\mathbf{F}_1$, $\mathbf{F}_2$ across an *imaginary* cut in a spring supporting a mass. In the latter case $\mathbf{F}_3$ shows the action of some further system on m. In each instance, the dotted line identifies the system of interest and the interactions are seen to cross the bounding surface. Newton's third law of motion states that $\mathbf{F}_1$ and $\mathbf{F}_2$ are always equal and opposite and have the same line of action.

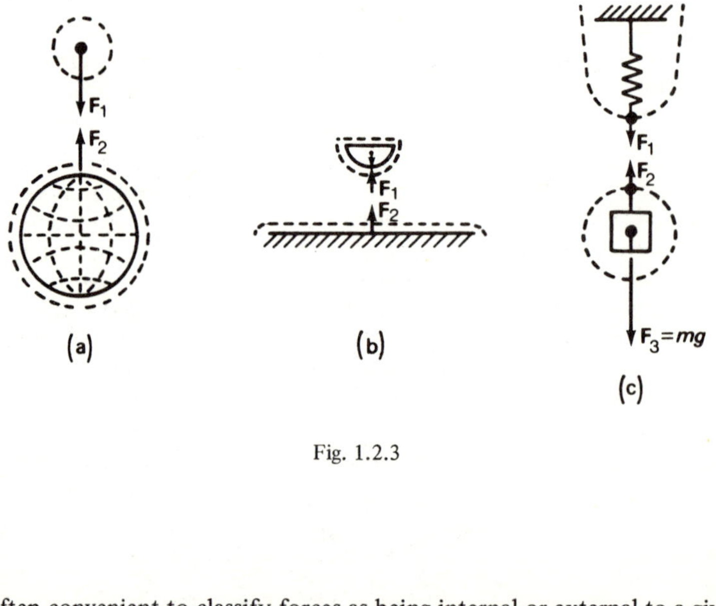

Fig. 1.2.3

It is often convenient to classify forces as being internal or external to a given system. In Fig. 1.2.4(a), the dotted line identifies a system comprised of four particles. We see that forces $\mathbf{F}_1$, $\mathbf{F}_2$, $\mathbf{F}_3$, $\mathbf{F}_4$ act *across* the bounding surface and so represent the actions *on* the system due to interactions with other systems external to it. Forces $\mathbf{F}_{12}$ etc., do not cross the bounding surface and represent interactions between the particles comprising the system itself. With respect to the chosen system, $\mathbf{F}_1$, $\mathbf{F}_2$, $\mathbf{F}_3$, $\mathbf{F}_4$ are *external forces* whilst $\mathbf{F}_{12}$, $\mathbf{F}_{23}$ etc., are *internal* forces. It is worth observing that if the same four particles are bounded in a different way to define the two systems of Fig. 1.2.4(b), then forces $\mathbf{F}_{12}$, $\mathbf{F}_{13}$, $\mathbf{F}_{14}$, $\mathbf{F}_{21}$, $\mathbf{F}_{31}$, $\mathbf{F}_{41}$ become external forces relative to each of the new systems.

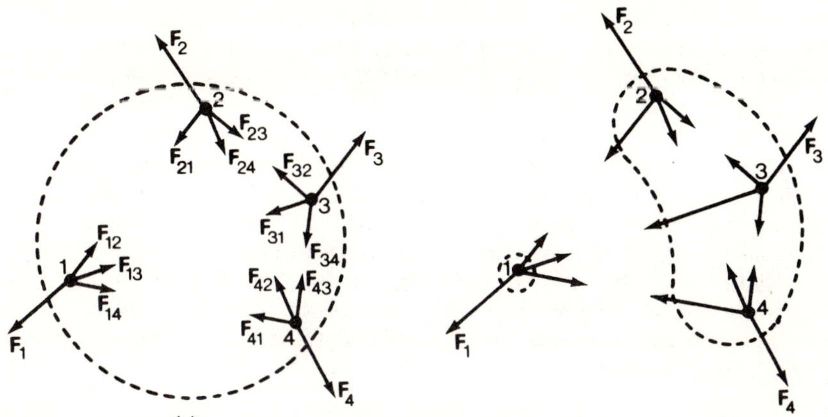

(a) Fig. 1.2.4 (b)

The overall constructed shape of a system 'at rest' constitutes its basic or datum configuration: then a fundamental problem in mechanics is to determine the displaced configuration of the system when it is subjected to some disturbance. This displacement may comprise an overall movement together with some change in shape or deformation: if the disturbance is a function of time so also will be the response. In the statical situations of interest in this book, we have a steady disturbance which causes the system to take up a new **equilibrium** configuration. In engineering analysis, we usually take the datum position for granted and calculations relate to changes or displacements from this datum. This is the case in stress analysis, and it then proves convenient to regard the displacements themselves as defining the current configuration.

1.3 System configuration and coordinates

Suppose we have a mechanical system 'as assembled' to consist of just two independent particles lying on a plane (two billiard balls on a table perhaps) and that we choose the rectangular reference frame of Fig. 1.3.1 to locate the system. The datum configuration is completely defined by the four quantities x_1, y_1, x_2, y_2.

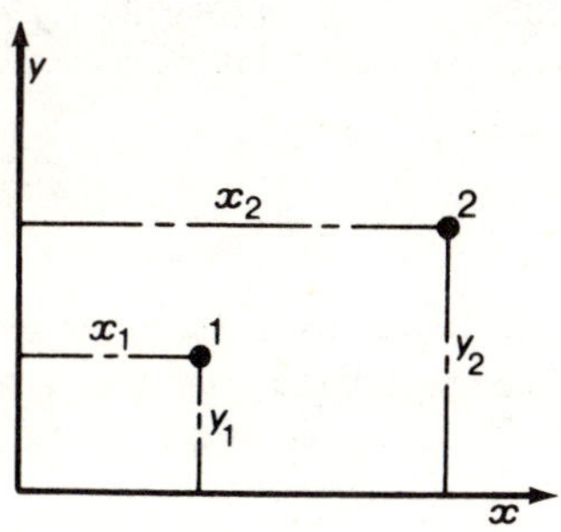

Fig. 1.3.1

It is possible to move each particle in any direction in the plane so that each of the four quantities $x_1 \ldots\ldots y_2$ can be changed independently of each other: each new set of values for $x_1 \ldots\ldots y_2$ then defines a possible configuration for the system. If we regard the configuration of Fig. 1.3.1 as a datum, then a displacement of the system occurs when any one or all of the u_1, v_1, u_2, v_2 of Fig. 1.3.2 have non-zero values. If we take the datum position for granted, then u_1, v_1, u_2, v_2 can serve as coordinates to define the displaced configuration instead of $x_1 + u_1$, $y_1 + v_1$, $x_2 + u_2$, $y_2 + v_2$. It is worth observing that whilst in this example $u_1 \ldots\ldots v_2$ have directions corresponding to the 'global' or 'system' coordinates xOy, it is not essential for this to be so. Indeed, it may be particularly convenient to describe the displacements of an element of a complex system relative to some coordinate system local to the element: such situations are discussed in Chapter 8.

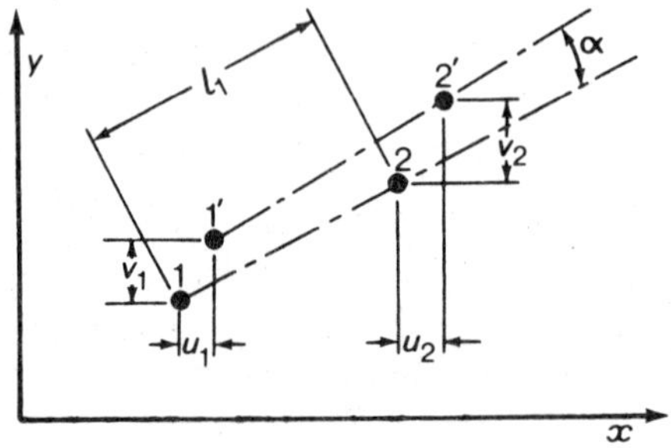

Fig. 1.3.2

If the arrangement is modified by joining the particles by a rigid (inextensible) link in the datum position, we see immediately that if x_1 and y_1 are temporarily held fixed we cannot alter x_2 without altering y_2 since the length of the link, l_1, must remain unchanged. We can thus say that u_1, v_1, u_2, v_2 are no longer independent coordinates for the system of two particles, but are constrained to satisfy the auxilliary condition (or **equation of constraint**).

$$[(x_2 + u_2) - (x_1 + u_1)]^2 + [(y_2 + v_2) - (y_1 + v_1)]^2 = l_1^2 . \tag{1.3.1}$$

We can readily visualise further possible constraints: thus, suppose that particle 2 is also connected to a permanently fixed point P by a second rigid link l_2 as shown in Fig. 1.3.3. Then the u_1, v_1, u_2, v_2 are also related by a second equation of constraint

$$[x_3 - (x_2 + u_2)]^2 + [y_3 - (y_2 + v_2)]^2 = l_2^2 . \tag{1.3.2}$$

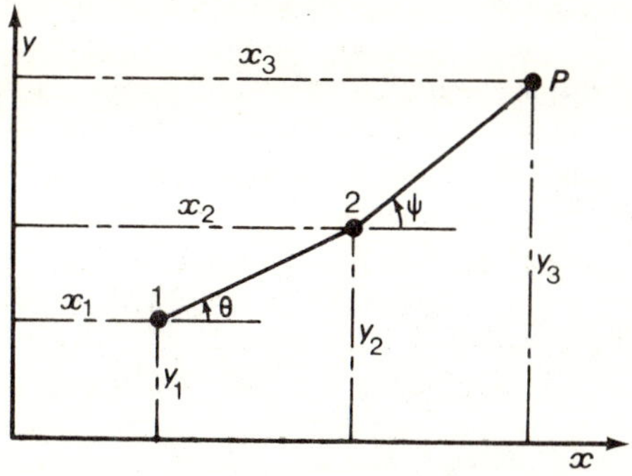

Fig. 1.3.3

Equations 1.3.1 and 1.3.2 show relatively complicated relationships between the constrained coordinates. Suppose however that the displacements are so small that quantities such as u_1^2 or $2u_1u_2$, etc., are negligible compared with such as $2x_1u_1$, then equations (1.3.1) and (1.3.2) simplify considerably to yield

$$u_1 + v_1 \tan\theta - u_2 - v_2 \tan\theta = 0 \tag{1.3.3}$$

$$u_2 + v_2 \tan\psi = 0\ . \tag{1.3.4}$$

Here, since the datum position is taken for granted, $\tan\theta$ and $\tan\psi$ are definite fixed numbers and we see that the constraints imply two *linear* relations between the four coordinates u_1, v_1, u_2, v_2.

The system of Fig. 1.3.1 is said to have four **degrees of freedom** since to define any new configuration requires a specification of all four u_1, v_1, u_2, v_2†, each of which can be altered independently of the others to define new configurations. In Fig. 1.3.2, the least number of quantities required to define the configuration unambiguously is three: we may use u_1, v_1, u_2 or u_1, u_2, v_1 or u_1, v_1, α, etc. This least number equals the number of degrees of freedom of the system and the quantities themselves are called its **independent** or **generalised coordinates**: each of them can be changed independently of the others to define new configurations without violating any constraints. In general, if m arbitrarily chosen coordinates defining a configuration are related by n equations of constraint, then the number of degrees of freedom possessed by the system is

$$d = m - n\ . \tag{1.3.5}$$

† The datum position being taken for granted.

Thus, corresponding to the u_1, v_1, u_2, v_2 of the system of Fig. 1.3.3, $m = 4$, $n = 2$ so that $d = 4 - 2 = 2$. In equations (1.3.1) and (1.3.2) [or (1.3.3) and (1.3.4) for small displacements], any two of $u_1 \ldots\ldots v_2$ can be regarded as independent. It is common to use the symbol q to denote generalised coordinates, then if we choose v_1 and v_2 as independent and denote $v_1 = q_1$, $v_2 = q_2$, we have

$$\begin{aligned} u_1 &= -q_1 \tan\theta + q_2(\tan\theta - \tan\psi) \\ v_1 &= q_1 \\ u_2 &= -q_2 \tan\psi \\ v_2 &= q_2 \;. \end{aligned} \tag{1.3.6}$$

Matrix algebra provides a means of handling such transformations systematically (see Appendix). Then

$$\begin{Bmatrix} u_1 \\ v_1 \\ u_2 \\ v_2 \end{Bmatrix} = \begin{bmatrix} -\tan\theta & (\tan\theta - \tan\psi) \\ 1 & 0 \\ 0 & -\tan\psi \\ 0 & 1 \end{bmatrix} \begin{Bmatrix} q_1 \\ q_2 \end{Bmatrix}$$

or

$$\{u\} = [C]\,\{q\} \;. \tag{1.3.6a}$$

Here, $[C]$ is called a **connexion** or **constraint** or **compatibility matrix**, whilst $\{u\}$ and $\{q\}$ are displacement vectors having components u_1, v_1, u_2, v_2 in the constrained coordinate system and q_1, q_2 in the generalised coordinate system respectively.

More generally, if there is a system of N particles in three dimensional space located by $3N$ coordinates $x_i{}^0$, $y_i{}^0$, $z_i{}^0$, which are related by m equations of constraint, the number of degress of freedom is $n = 3N - m$ and we can write for the datum configuration

$$\begin{aligned} x_1{}^0 &= f_1{}^0(q_1{}^0, q_2{}^0, \ldots\ldots q_n{}^0) \\ y_1{}^0 &= g_1{}^0(q_1{}^0, q_2{}^0, \ldots\ldots q_n{}^0) \\ &\vdots \\ z_n{}^0 &= h_n{}^0(q_1{}^0, q_2{}^0, \ldots\ldots q_n{}^0) \end{aligned}$$

where the functions $f_1{}^0$, etc., are prescribed.

If the system suffers an infinitesimal displacement $\Delta q_j^0, j = 1, 2 \ldots\ldots n$ from its datum position, the corresponding changes in the coordinates of a selected particle s will be

$$\Delta x_s^0 = \sum_{j=1}^{n} \frac{\partial f_s^0}{\partial q_j^0} \Delta q_j^0 \text{ , etc.} \tag{1.3.8}$$

Evidently, Δx_s^0 is *linearly* related to the Δq_j^0.

If we represent the set $\Delta x_1^0, \Delta x_1^0 \ldots\ldots \Delta z_N^0$ by the $3N \times 1$ matrix (or $3N$ dimensional vector) $\{u\}$, the set $\Delta q_1^0, \Delta q_2^0 \ldots\ldots \Delta q_n^0$ by the n dimensional vector $\{q\}$ and represent the coefficients $\partial f_i^0 / \partial q_j^0$ by the elements C_{ij} of a $3N \times n$ rectangular matrix $[C]$, equations (1.3.8) can be written as

$$\{u\} = [C]\{q\} \text{ .} \tag{1.3.9}$$

This is a generalisation of the coordinate transformation (1.3.6a) of the example above.

Equations (1.3.8) or (1.3.9) show that the total change in the position of a selected particle due to simultaneous infinitesimal changes in the generalised coordinates is the sum of the changes in position caused when each generalised coordinate is changed individually. This is an example of the *principle of superposition of small effects* which has far reaching significance in stress analysis and other problem areas.

So far we have discussed the configuration of systems comprised of a finite number of discrete particles. When we introduced the rigid links of the earlier examples we were not interested in them as such, because they were merely tangible devices for providing constraints: in fact we were only really concerned that there be strong forces capable of maintaining the geometric constraints conceived. Suppose now we identify a system comprising the links as well as the original particles of Fig. 1.3.3: then we have an assemblage of an infinite number of particles and it would appear that we need an infinite number of discrete coordinates to describe the configuration. However, if we recall the supplementary information of rigidity of the links, we see immediately that we still only need two generalised coordinates to define the configuration: should we wish to locate a special particle on a link, a simple supplementary calculation suffices. Pursuing this argument further, we find that a rigid body in space has six degrees of freedom. Thus, to locate three non-collinear particles in the body requires nine rectangular coordinates however, the condition of rigidity implies three equations of constraint between them so that the body has $9 - 3 = 6$ degrees of freedom.

The idea of using 'discrete' generalised coordinates for an overall description of the configuration of a continuous system will prove very valuable in later discussions even when the system comprises deformable bodies. Three examples will serve to illustrate.

The uniform bar of Fig. 1.3.4 is stretched by force P_1. In the context of mechanics of materials theory, the bar is in a state of uniform strain so that a typical particle originally at distance x from the fixed end travels axially a distance $u = (x/l)u_1$. The displacement u_1 of the right hand end thus characterises the deformation of the bar and can be used as a discrete generalised coordinate for it.

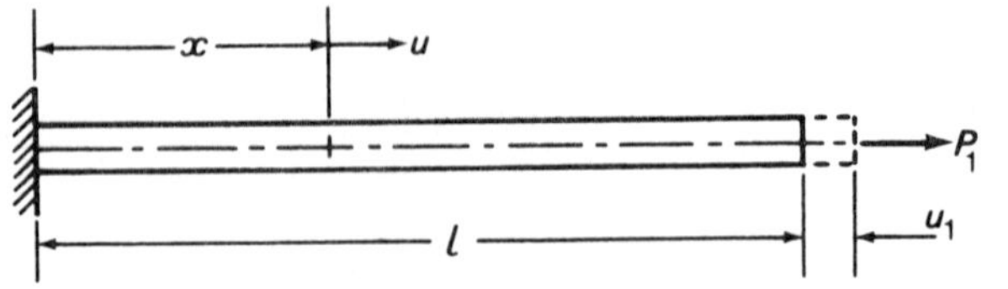

Fig. 1.3.4

The cantilever beam of Fig. 1.3.5 supports a transverse force and a couple at its end. Here, provided we 'take for granted' the supplementary information of engineering beam theory, which gives the deformed *shape,* the discrete coordinates v_1 and θ_1 characterise the entire configuration for the specified type of loading. In fact

$$v = (3x^2/l^2 - 2x^3/l^3)v_1 + (-x^2/l^2 + x^3/l^3)l\theta_1 .$$

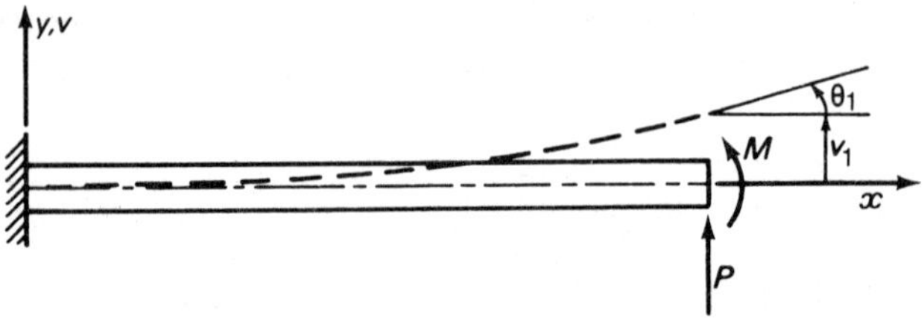

Fig. 1.3.5

Finally, discrete coordinates $u_1 \ldots\ldots u_6$ relating to nodes A and B of the completely flexible portal frame of Fig. 1.3.6, characterise the deformed shape taken up under the actions of the nodal forces $F_1 \ldots\ldots F_6$.

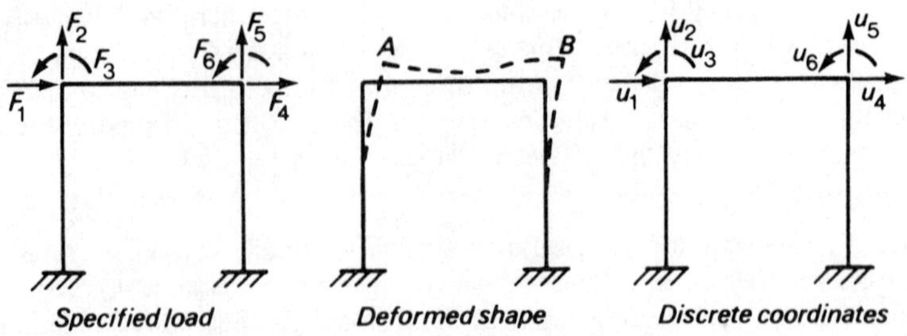

Fig. 1.3.6

In the more complex situations discussed later, we will see how to effect an approximate solution by *assuming* some continuous distribution of displacement or deformed shape for the system. Thus, if the bar of Fig. 1.3.4 is tapered we know that the state of the strain varies along the length so that the axial displacement at a typical section is not given by $u = (x/l)u_1$. However, if we *assume* this to be the deformed shape and treat u_1 as an as yet unknown parameter we can obtain an approximate solution by, say, the principle of virtual work. As we shall see in due course, this procedure yields a value for u_1 which is the 'best' one within the limitations of our *assumed* displacement field; it serves as a generalised coordinate.

The preceding discussion showed that when discrete coordinates are used to describe the configuration of a continuous system, supplementary information is required if we are to provide data relating to points away from the nodes. A convenient alternative procedure is to use **coordinate functions**, or **distributed coordinates**, which give a continuous distribution of information directly. The theory of Fourier Series shows that, subject to certain restrictions, a given function $f(x)$ may be represented over the range $(-l, l)$ in the form of a trigonometric series

$$f(x) = a_0 + \sum_{n=1}^{\infty} a_n \cos\frac{n\pi x}{l} + \sum_{n=1}^{\infty} b_n \sin\frac{n\pi x}{l} \qquad (1.3.10)$$

where the coefficients a_0, a_n, b_n are determined by certain rules. More generally, it may be possible to expand the function in a series of other non-trigonometric functions, say

$$f(x) = \sum_{n=1}^{\infty} a_n \phi_n(x) \ . \qquad (1.3.11)$$

Here, the $\phi_n(x)$ are known and, in certain circumstances, only a finite number of terms may be required.

A beam is a continuously deformable body which can take up an infinite number of bent shapes depending on the load. In beam theory interest centres on the deflexion curve taken up by the centre line of the beam and this will be defined by some function of position along the length. If $w = f(x)$ is the deflexion function for the simply supported beam of Fig. 1.3.7, it can be expanded as a half range sine series over $0 < x < l$ so that

$$w = f(x) = \sum_{n=1}^{\infty} b_n \sin\frac{n\pi x}{l} \ . \qquad (1.3.12)$$

Fig. 1.3.7

On purely geometric grounds the values of the b_n's can be adjusted independently of each other and each new set of values will describe new deflexion shapes for the beam so that they constitute generalised coordinates for the system.† A distributed load of intensity $p(x)$ per unit length may also be represented in a similar way so that

$$p(x) = \sum_{n=1}^{\infty} b_n' \sin \frac{n\pi x}{l} . \tag{1.3.13}$$

The relation between b_n and b_n' is analogous to that between u_i and F_i in the discrete coordinate system.

EXAMPLE 1.3.1

Three particles are attached to a straight rigid link as shown in Fig. 1.3.8. Show that the system has three degrees of freedom and determine the $[C]$ matrix for small displacements from the position given. Adopt u_3, v_3, $\Delta\theta$ as the generalised coordinates q_1, q_2, q_3.

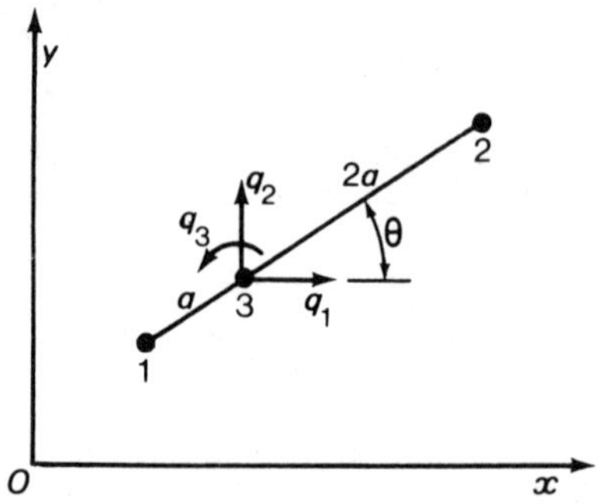

Fig. 1.3.8

Six coordinates $x_1^0, \ldots\ldots y_3^0$ are required to locate the three particles. However, the rigidity of the link provides three equations of constraint relating them. Number of degrees of freedom is thus 3.

$$x_1^0 = x_3^0 - a\cos\theta , \qquad y_1^0 = y_3^0 - a\sin\theta$$

$$x_2^0 = x_3^0 + 2a\cos\theta , \qquad y_2^0 = y_3^0 + 2a\sin\theta .$$

Then, with $\Delta x_3^0 = u_3 = q_1$, $\Delta y_3^0 = v_3 = q_2$, $\Delta\theta = q_3$,

$$\Delta x_1^0 = u_1 = \Delta x_3^0 + a\sin\theta\Delta\theta = q_1 + a\sin\theta\, q_3$$

$$\Delta y_1^0 = v_1 = \Delta y_3^0 - a\cos\theta\Delta\theta = q_2 - a\cos\theta\, q_3$$

$$\Delta x_2^0 = u_2 = \Delta x_3^0 - 2a\sin\theta\Delta\theta = q_1 - 2a\sin\theta\, q_3$$

$$\Delta y_2^0 = v_2 = \Delta y_3^0 + 2a\cos\theta\Delta\theta = q_2 + 2a\cos\theta\, q_3$$

† See also examples (7.2.1) and (7.2.2).

$$\begin{Bmatrix} u_1 \\ v_1 \\ u_2 \\ v_2 \\ u_3 \\ v_3 \end{Bmatrix} = \begin{bmatrix} 1 & 0 & a \sin \theta \\ 0 & 1 & -a \sin \theta \\ 1 & 0 & -2a \sin \theta \\ 0 & 1 & 2a \cos \theta \\ 1 & 0 & 0 \\ 0 & 1 & 0 \end{bmatrix} \begin{Bmatrix} q_1 \\ q_2 \\ q_3 \end{Bmatrix}$$

EXAMPLE 1.3.2

In the portal frame of Fig. 1.3.9, the nodal displacements due to axial elongations of the bars are negligible compared with those due to bending. Determine the number of degrees of freedom and the $[C]$ matrix for small displacements.

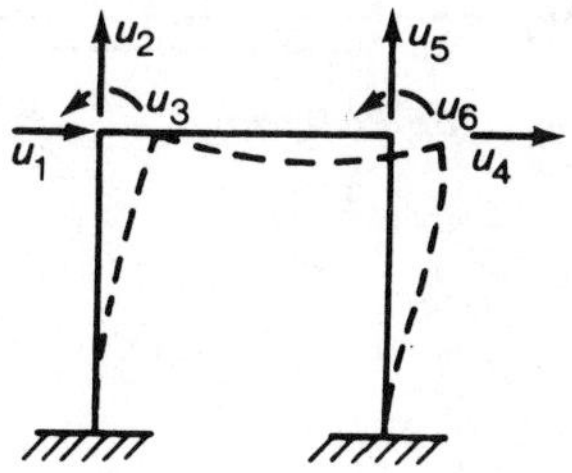

Fig. 1.3.9

Since axial displacements are negligible the equations of constant are three in number:

$$u_2 = 0\,, \quad u_5 = 0\,, \quad u_4 - u_1 = 0$$

Number of $\mathrm{d}.o.f = 6 - 3 = 3$.

Choosing generalised coordinates to be

$$u_1 = q_1\,, \quad u_3 = q_2\,, \quad u_6 = q_3$$

$$\begin{Bmatrix} u_1 \\ u_2 \\ u_3 \\ u_4 \\ u_5 \\ u_6 \end{Bmatrix} = \begin{bmatrix} 1 & 0 & 0 \\ 0 & 0 & 0 \\ 0 & 1 & 0 \\ 1 & 0 & 0 \\ 0 & 0 & 0 \\ 0 & 0 & 1 \end{bmatrix} \begin{Bmatrix} q_1 \\ q_2 \\ q_3 \end{Bmatrix} .$$

Problems 1.3

1.3.1 Show that the system of Fig. 1.3.10 has two degrees of freedom if the rollers turn without slipping. Deduce the relation $\{u\} = [C]\{q\}$ when the generalised coordinates are chosen (a) $q_1 = u_1, q_2 = u_2$, (b) $q_1 = u_4$, $q_2 = u_5$, (c) $q_1 = u_3, q_2 = u_6$.

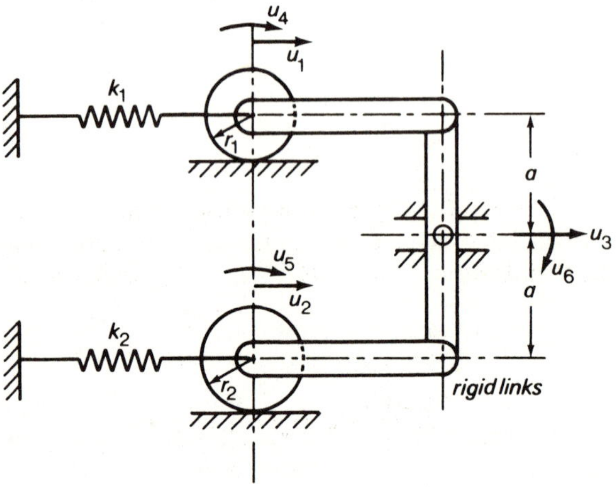

Fig. 1.3.10

Ans:

$$[C_a]^t = \begin{bmatrix} 1 & 0 & \frac{1}{2} & \frac{1}{r_1} & 0 & \frac{1}{2a} \\ 0 & 1 & \frac{1}{2} & 0 & \frac{1}{r_2} & -\frac{1}{2a} \end{bmatrix}$$

1.3.2 If the longitudinal extensions of the bars are negligible in the frame of Fig. 1.3.11, show that

$$[C]^t = \begin{bmatrix} 1 & 1 & 0 & \tan\theta & 0 & 0 \\ 0 & 0 & 0 & 0 & 1 & 0 \\ 0 & 0 & 0 & 0 & 0 & 1 \end{bmatrix}$$

when $\{u\} = [C]\{q\}$ and $q_1 = u_1, q_2 = u_5, q_3 = u_6$.

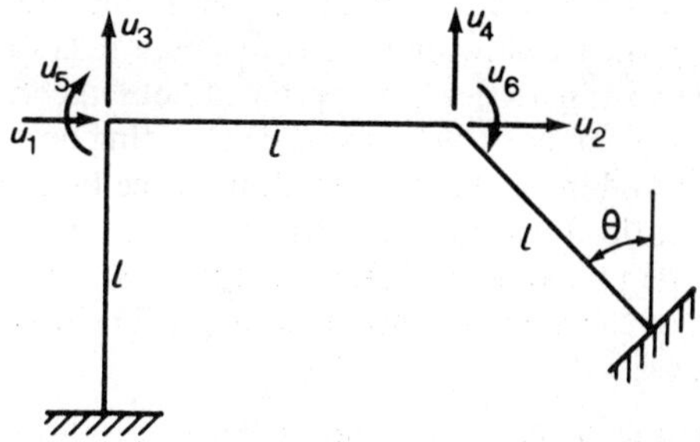

Fig. 1.3.11

1.3.3 Fig. 1.3.12(a) shows the deflexions at three stations on a cantilever of variable section supporting a complicated distribution of load. If the deflexion curve is to be represented by

$$w(x) = \sum_{i=1}^{3} p_i \phi_i(x)$$

where the $\phi_i(x)$ are normal modes with the amplitudes shown in Fig. 1.3.12 (b), (c), (d), show that the values of the distributed coordinates are $p_1 = -1.222$, $p_2 = -0.833$, $p_3 = 0.889$.

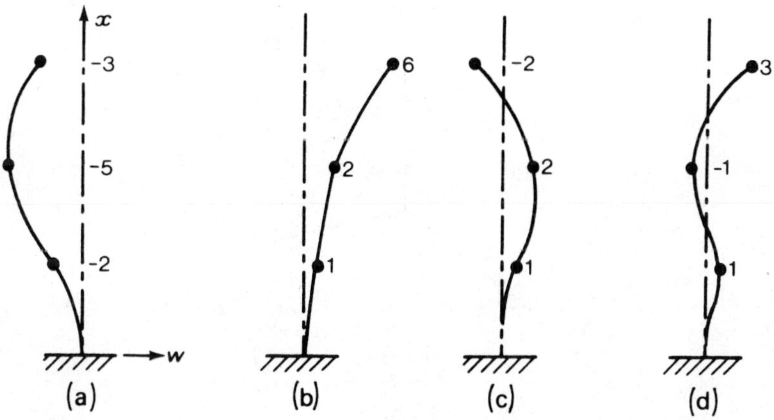

Fig. 1.3.12

1.4 Elements of vector statics

A basic problem in statics is the determination of the equilibrium configuration taken up by a mechanical system under the action of some prescribed loading: alternatively, we may require to find the forces necessary to maintain some specified deformed shape. In establishing the rules which describe the response of mechanical systems to some disturbance, it is an important feature of vector mechanics that such rules follow directly from those relating to particles. Whilst this pre-occupation with particles is a significant help in constructing the theories of vector mechanics of systems, it is also the source of the practical difficulties encountered in calculations relating to the more complex systems of interest in design. In this section we briefly review the elements of vector statics and apɒly them to some problems, with a view to presenting a contrast with the approach of the energy methods which will be developed in subsequent portions of this book.

According to Newton, the necessary and sufficient condition that a single particle shall be in equilibrium is that the resultant force acting on it shall be zero. Then, for the particle of Fig. 1.4.1, since the resultant force **R** is given by

$$\mathbf{R} = \text{vector sum of } (\mathbf{F}_1, \mathbf{F}_2, \mathbf{F}_3, \mathbf{F}_4) \tag{1.4.1}$$

or

$$\mathbf{R} = \sum_{i=1}^{4} \mathbf{F}_i \ ,$$

we have for equilibrium that

$$\mathbf{R} = \mathbf{O} \ . \tag{1.4.2}$$

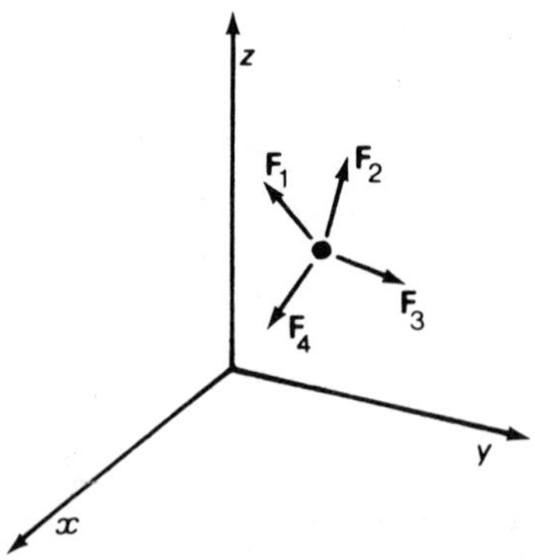

Fig. 1.4.1

For numerical work it is convenient to express the forces in terms of their components, then with **i**, **j**, **k** as unit vectors in the x, y, z directions respectively, (see Appendix), combining equations (1.4.1) and (1.4.2) we have

$$\mathbf{R} = R_x\mathbf{i} + R_y\mathbf{j} + R_z\mathbf{k} = \mathbf{O}$$

$$\text{or } \mathbf{O} = (F_{1x} + F_{2x} + F_{3x} + F_{4x})\mathbf{i} + (F_{1y} + F_{2y} + F_{3y} + F_{4y})\mathbf{j} + (F_{1z} + F_{2z} + F_{3z} + F_{4z})\mathbf{k} \ . \tag{1.4.2a}$$

The single vector equation of equilibrium may thus be replaced by three scalar ones

$$
\begin{aligned}
R_x &= F_{1x} + F_{2x} + F_{3x} + F_{4x} &= \Sigma F_x &= 0 \\
R_y &= F_{1y} + F_{2y} + F_{3y} + F_{4y} &= \Sigma F_y &= 0 \\
R_z &= F_{1z} + F_{2z} + F_{3z} + F_{4z} &= \Sigma F_z &= 0\ . \qquad (1.4.3)
\end{aligned}
$$

Suppose now that we have a system of n particles located by position vectors $\mathbf{r}_1\ \mathbf{r}_2 \ldots\ldots \mathbf{r}_n$ having a common origin, or base point, O and that resultant forces $\mathbf{R}_1, \mathbf{R}_2, \ldots\ldots \mathbf{R}_n$ act on them. The total force on the system is then defined to be

$$\mathbf{R} = \text{vector sum of } (\mathbf{R}_1, \mathbf{R}_2, \ldots\ldots \mathbf{R}_n)$$

or

$$\mathbf{R} = \sum_{i=1}^{n} \mathbf{R}_i\ . \qquad (1.4.4)$$

The moment, or torque, of force $\mathbf{R}_i$ about the origin O is defined by the vector product

$$\mathbf{M}_i = \mathbf{r}_i \times \mathbf{R}_i\ . \qquad (1.4.5)$$

Accordingly $\mathbf{M}_i$ itself is also a vector: it has its initial point at O, is normal to the plane containing $\mathbf{r}_i$ and $\mathbf{R}_i$ and has a sense given by the right hand screw rule as shown in Fig. 1.4.2 (see also Appendix).

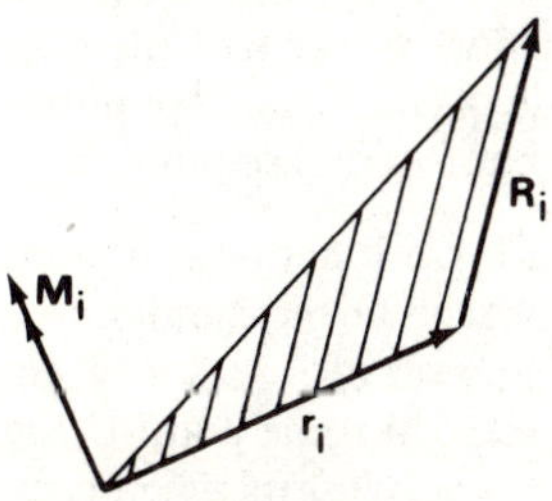

Fig. 1.4.2

The total moment about O of all the forces on the system is then given by

$$\mathbf{M} = \text{vector sum of } (\mathbf{M}_1, \mathbf{M}_2, \ldots\ldots \mathbf{M}_n)$$

or

$$\mathbf{M} = \sum_{i=1}^{n} \mathbf{r}_i \times \mathbf{R}_i\ . \qquad (1.4.6)$$

It is common to represent moment vectors with double arrow heads on drawings, as in Fig. 1.4.2.

These definitions show that the total force **R** does not depend upon the choice of base point O whereas the total moment **M** does so depend. When two force systems have the same value of **R** and the same value of **M** for a given base point, they are said to be statically equivalent or **equipollent**.

A particularly important special case of equation (1.4.6) occurs when there are only two forces $\mathbf{R}_1$ and $\mathbf{R}_2$, equal in magnitude, parallel and opposite in sense as in Fig. 1.4.3. Now,

$$\mathbf{M} = \mathbf{r}_1 \times \mathbf{R}_1 + \mathbf{r}_2 \times \mathbf{R}_2 = \mathbf{r}_1 \times \mathbf{R}_1 + (\mathbf{r}_1 + \mathbf{a})\mathbf{R}_2$$

or

$$\mathbf{M} = \mathbf{r}_1 \times (\mathbf{R}_1 + \mathbf{R}_2) + \mathbf{a} \times \mathbf{R}_2 \quad ,$$

but $\mathbf{R}_2 = -\mathbf{R}_1 = \mathbf{R}$ say, so that

$$\mathbf{M} = \mathbf{a} \times \mathbf{R} \ . \tag{1.4.7}$$

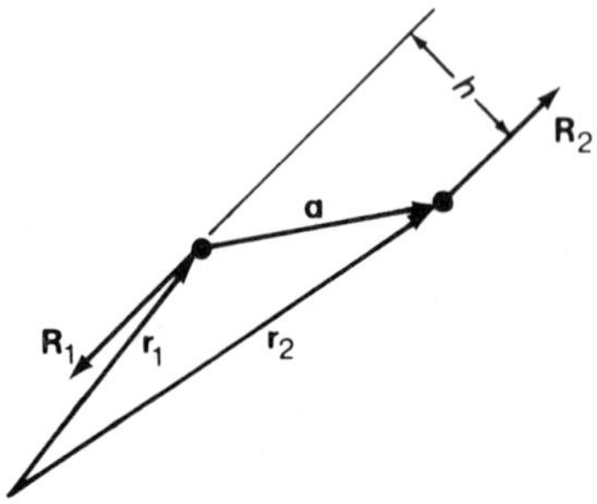

Fig. 1.4.3

Furthermore, if h is the perpendicular distance between the forces, the definition of vector product in the Appendix shows that the magnitude of $\mathbf{M}$ is $M = Rh$. Such a pair of forces constitute a *couple* and we notice that the moment of a couple is independent of the choice of base point O.

The above results show that a force $\mathbf{R}$ acting at a point $\mathbf{r}$ is statically equivalent to a second force $\mathbf{R}$, acting through O and parallel to the original $\mathbf{R}$ *together with* a couple $\mathbf{r} \times \mathbf{R}$. It follows that a set of forces may, for statical purposes, be reduced to a single resultant force at some point O together with a couple having an axis through O. This result is very useful when discussing the *overall* behaviour of a system of particles, however if we wish to describe the *deformation* of the system we must take into account the actual *distribution* of forces within it.

A system of particles is in equilibrium only if each one is in equilibrium individually. This statement has implications for the total force and moment. Referring to Fig. 1.2.4, a typical particle experiences internal and external forces: let $\mathbf{R}_i$ be the resultant external force on particle i and $\mathbf{R}_{i1}, \mathbf{R}_{i2}, \ldots\ldots$ be the internal forces due to the other particles. Then for equilibrium

$$\mathbf{R}_i + \mathbf{R}_{i1} + \mathbf{R}_{i2} \ldots\ldots = \mathbf{0} \ . \tag{1.4.8}$$

Adding all such forces on the system and observing that $\mathbf{R}_{ij} = -\mathbf{R}_{ji}$ for equality of action and reaction, we find that for equilibrium of the system as a whole,

$$\mathbf{R} = \sum_{i=1}^{n} \mathbf{R}_i = \mathbf{0} \tag{1.4.9}$$

That is, the total *external* force on the system must be zero. The resultant moment about O of all the forces acting on the i'th particle is

$$\mathbf{M}_i = \mathbf{r}_i \times (\mathbf{R}_i + \mathbf{R}_{i1} + \mathbf{R}_{i2} + \ldots\ldots) . \qquad (1.4.10)$$

By virtue of equation (1.4.8), $\mathbf{M}_i$ is zero for each particle, so that if we add all such moments and again observe that the internal forces occur in self cancelling pairs having the same line of action, equilibrium of the system requires

$$\mathbf{M} = \mathbf{M}_i = \sum_{i=1}^{n} \mathbf{r}_i \times \mathbf{R}_i = \mathbf{0} ; \qquad (1.4.11)$$

that is, the total moment of all the *external* forces on the system must be zero.

The vector equations (1.4.9) and (1.4.11) are the necessary conditions for equilibrium, but for numerical work it is convenient to express them in terms of their scalar components. Alternatively, then, there are six scalar equations relating the external forces and couples acting on systems of particles in equilibrium:

$$\Sigma R_x = 0, \quad \Sigma R_y = 0, \quad \Sigma R_z = 0$$
$$\Sigma M_x = 0, \quad \Sigma M_y = 0, \quad \Sigma M_z = 0 . \qquad (1.4.12)$$

Here, ΣR_x is the algebraic sum of the external forces in the x direction, ΣM_x is the algebraic sum of the moments of the external forces about the x axis, etc.

We now illustrate the application of these principles to two systems.

EXAMPLE 1.4.1

Two light rigid links are pinned together at P and to a rigid support at O as shown in Fig. 1.4.4. If the forces are applied to the *pins* at P and Q, determine the equilibrium configuration if friction forces at the pins are negligible. Take the datum position to correspond to both links along Ox.

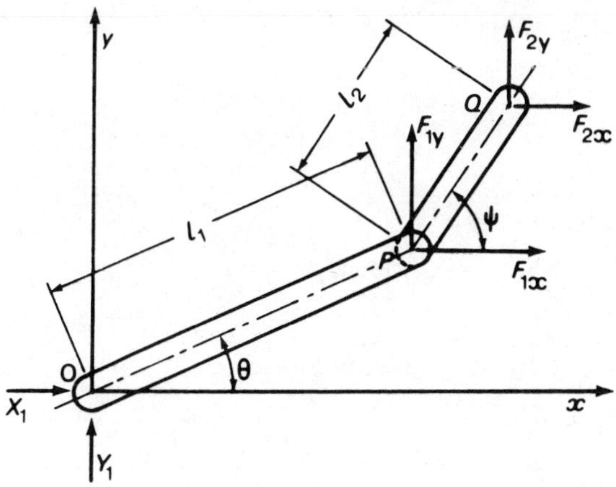

Fig. 1.4.4

Since the links are rigid, the reasoning of Section 1.3 shows that the system has two degrees of freedom. We choose θ and ψ as generalised coordinates.

If we first isolate the system of two links and three pins *as a whole* and consider its equilibrium we must first draw in the actions X_1 and Y_1 of the support *on* the system in Fig. 1.4.4. Then,using conditions (1.4.12) in the plane,

$$\Sigma F_x = X_1 + F_{1x} + F_{2x} = 0 \quad \text{(a)}$$

$$\Sigma F_y = Y_1 + F_{1y} + F_{2y} = 0 \quad \text{(b)}$$

$$\Sigma M_0 = F_{1y}l_1 \cos\theta - F_{1x}l_1 \sin\theta + F_{2y}(l_1 \cos\theta + l_2 \cos\psi) - F_{2x}(l_1 \sin\theta + l_2 \sin\psi) = 0 . \quad \text{(c)}$$

Evidently, these considerations alone are not sufficient to yield θ and ψ since X_1 and Y_1 are also unknown. To proceed, we must isolate various sub-systems and consider *their* equilibrium. To effect a solution, we must recall that

(i) Action and reaction are equal and opposite.

(ii) Forces which were *internal* to the system of Fig. 1.4.4 become *external* to the various sub-systems of Fig. 1.4 5.

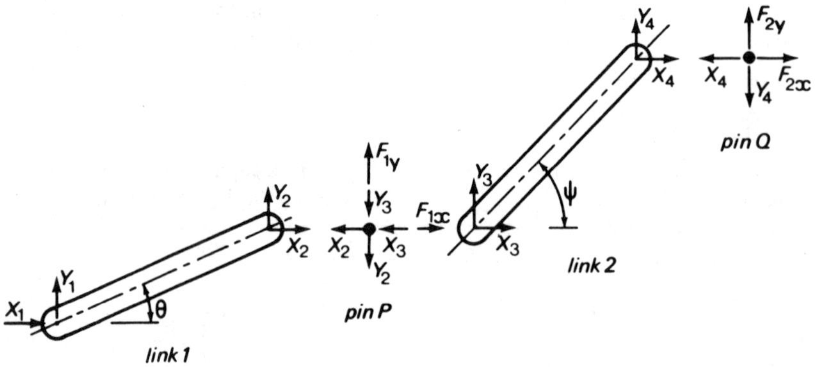

Fig. 1.4.5

Equilibrium of link 1

$$\Sigma F_x = X_1 + X_2 = 0 \quad \text{(d)}$$

$$\Sigma F_y = Y_1 + Y_2 = 0 \quad \text{(e)}$$

$$\Sigma M_o = Y_2 l_1 \cos\theta - X_2 l_1 \sin\theta = 0 . \quad \text{(f)}$$

Equilibrium of pin P

$$\Sigma F_x = F_{1x} - X_2 - X_3 = 0 \quad \text{(g)}$$

$$\Sigma F_y = F_{1y} - Y_2 - Y_3 = 0 . \quad \text{(h)}$$

Equilibrium of link 2

$$\Sigma F_x = X_3 + X_4 = 0 \qquad \text{(j)}$$

$$\Sigma F_y = Y_3 + Y_4 = 0 \qquad \text{(k)}$$

$$\Sigma M_p = Y_4 l_2 \cos\psi - X_4 l_2 \sin\psi = 0 \;. \qquad \text{(l)}$$

Equilibrium of pin Q

$$\Sigma F_x = F_{2x} - X_4 = 0 \qquad \text{(m)}$$

$$\Sigma F_y = F_{2y} - Y_4 = 0 \;. \qquad \text{(n)}$$

From (m) and (n) in (l)

$$\psi = \tan^{-1}(F_{2y}/F_{2x}) \;. \qquad \text{(p)}$$

From (m) and (n) in (j) and (k) in (g) and (h) in (f)

$$\theta = \tan^{-1}[(F_{1y} + F_{2y})/(F_{1x} + F_{2x})] \;. \qquad \text{(q)}$$

This simple example highlights the important features of the vector approach.
(i) All relevant forces, i.e. forces to maintain the geometric constraints are found as part of the solution process;
(ii) If we only wanted the equilibrium configuration we could not avoid explicitly taken the constraint forces into account.

EXAMPLE 1.4.2

Fig. 1.4.6 shows a Hooke's joint coupling two shafts lying in the xOy plane. Given the input torque $\mathbf{M}_1$, we require the load torque $\mathbf{M}_2$ necessary to maintain equilibrium *in the configuration shown*. Draw the free body diagrams required to effect a solution.

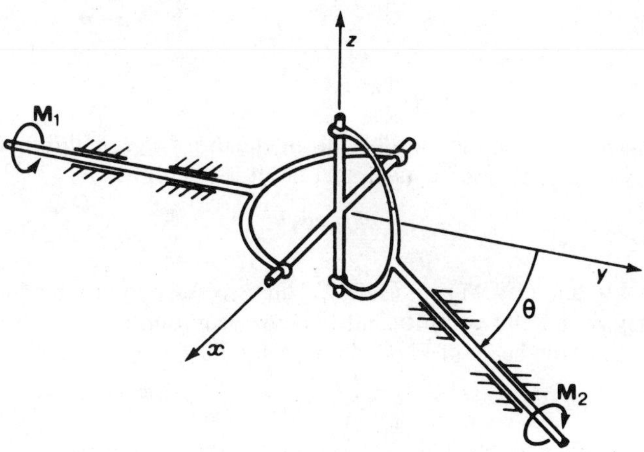

Fig. 1.4.6

We assume that the friction is negligible and that the bearings are short enough to provide point supports. It is found that an overall isolation of the shaft-cross-shaft system is not adequate for the analysis and that we need to consider the sub-systems of Fig. 1.4.7.

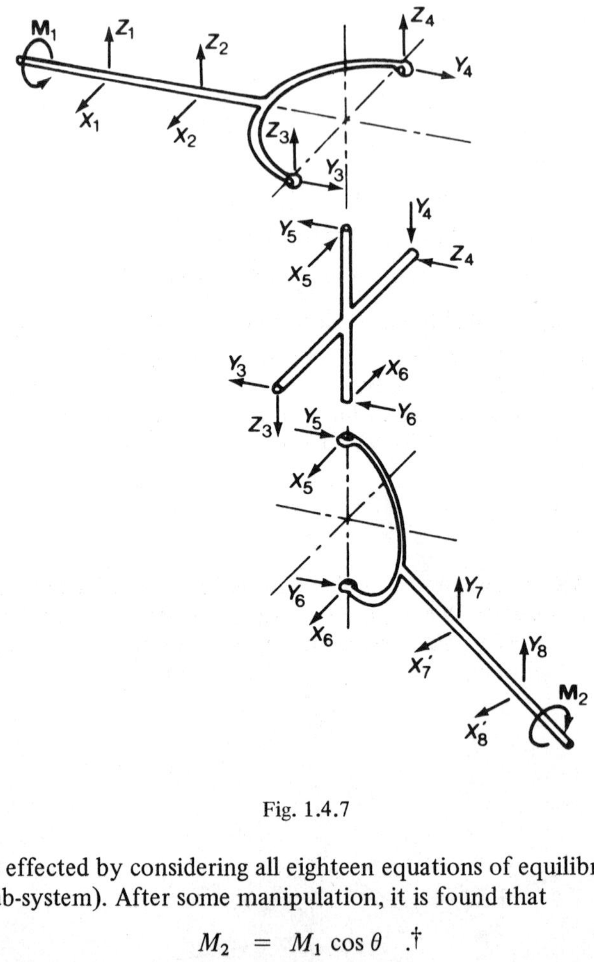

Fig. 1.4.7

Solution is effected by considering all eighteen equations of equilibrium (six for each sub-system). After some manipulation, it is found that

$$M_2 = M_1 \cos\theta \ .\dagger$$

Equations (1.4.9) and (1.4.11) [or (1.4.12)] are necessary conditions for equilibrium but they are not always sufficient to allow a solution for the configuration. Thus, suppose that the beam of Fig. 1.4.8 is in equilibrium under a plane loading arrangement.

† Good insight will often allow short cuts in a solution: the reader should reflect on this example. However, the intention here is to contrast a formal application of vector methods with a later application of the virtual work principle. See Example 1.6.2.

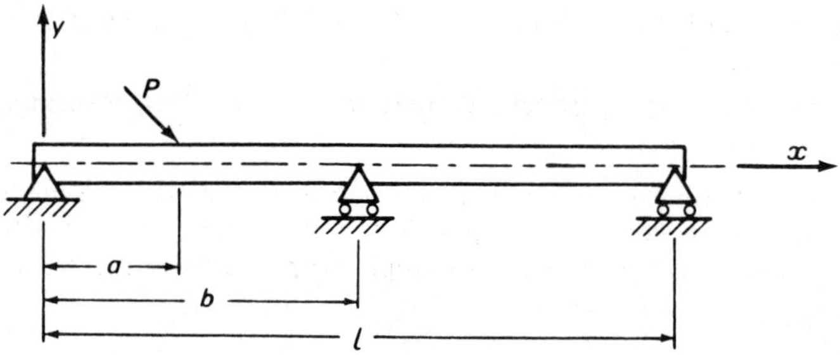

Fig. 1.4.8

Isolating the beam as a free body, we have

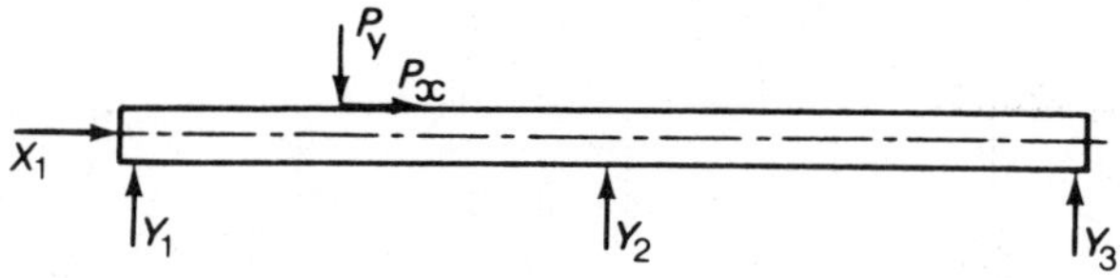

Fig. 1.4.9

For equilibrium

$$\Sigma F_x = P_x + X_1 = 0 \quad \text{(a)}$$

$$\Sigma F_y = Y_1 + Y_2 + Y_3 - P_y = 0 \quad \text{(b)}$$

$$\Sigma M_{\text{pin}} = Y_2 b + Y_3 l - P_y a = 0 \;. \quad \text{(c)}$$

Equilibrium conditions lead to only three equations relating the forces on the beam, whereas there are four unknown reactions to be found. Such systems are **statically indeterminate**: regarding P_x and P_y as prescribed forces, considerations of equilibrium alone allows an infinite number of combinations of values for Y_1, Y_2, Y_3. In particular, one of the vertical supports is said to be **redundant** since if it were removed, equilibrium could still be maintained and the beam would not move like a mechanism: from a purely statical point of view it is immaterial which of Y_1, Y_2 and Y_3 is regarded as the redundant reaction. In a structural system consisting of a number of inter-connected members, statical indeterminary may be internal to the system as well as at its boundaries: then, some internal members may also be removed without upsetting equilibrium. To effect a solution to such problems, a rigid body idealisation of the individual members is no longer adequate, and it is necessary to take their deformation into account.

All but the most trivial problems in stress analysis are concerned with statically indeterminate situations and we find that the solution process always involves the manipulation of three essential ingredients.

(i) Satisfaction of equilibrium requirements.

(ii) Satisfaction of the conditions of geometric fit (compatibility) in the deformed configuration.

(iii) The use of some (experimentally determined) force-deformation relations appropriate to the constituent parts of the system.

As we shall see in due course, the way in which these essential considerations are accommodated becomes very sophisticated in the more complex situations. However, it will be useful to treat two simple examples to illustrate the reasoning involved.

EXAMPLE 1.4.3

The particle in Fig. 1.4.10 (a) and (b) is in equilibrium under the action of P_x, P_y and the four linear springs having stiffnesses k_1, k_2, k_3, k_4. If in the no-load, or datum, position there is just no force in each spring, determine the *small* displacements, u and v, of the particle.

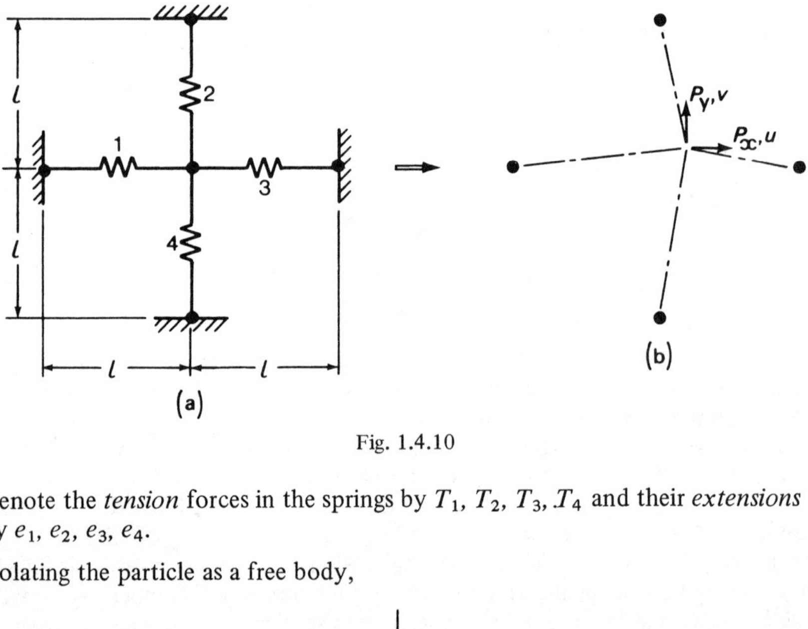

Fig. 1.4.10

Denote the *tension* forces in the springs by T_1, T_2, T_3, T_4 and their *extensions* by e_1, e_2, e_3, e_4.

Isolating the particle as a free body,

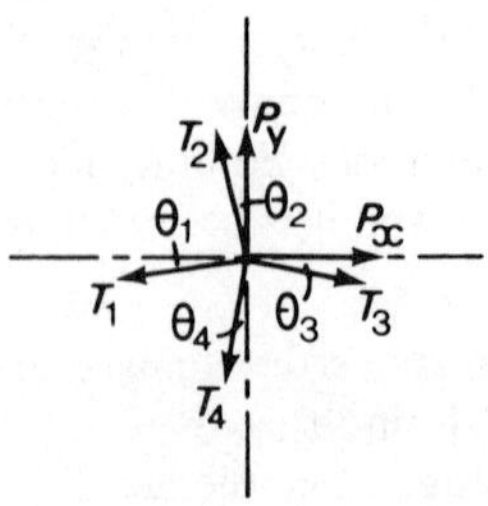

Fig. 1.4.10(c)

For equilibrium

$$\Sigma F_x = P_x - T_1 \cos\theta_1 - T_2 \sin\theta_2 + T_3 \cos\theta_3 - T_4 \sin\theta_4 \quad \text{(a)}$$

$$\Sigma F_y = P_y - T_1 \sin\theta_1 + T_2 \cos\theta_2 - T_3 \sin\theta_3 - T_4 \cos\theta_4 \ . \quad \text{(b)}$$

Since only two equations relate four unknown forces, the system is statically indeterminate.

For compatibility of displacements and spring elongations,

$$e_1 = [\sqrt{(l+u)^2 + v^2} - l] \quad \text{(c)}$$

$$e_2 = [\sqrt{(l-v)^2 + u^2} - l] \quad \text{(d)}$$

$$e_3 = [\sqrt{(l-u)^2 + v^2} - l] \quad \text{(e)}$$

$$e_4 = [\sqrt{(l+v)^2 + u^2} - l] \ . \quad \text{(f)}$$

Force deformation relations

$$T_1 = k_1 e_1, \quad T_2 = k_2 e_2, \quad T_3 = k_3 e_3, \quad T_4 = k_4 e_4 \ . \quad \text{(g)}$$

Now, if the displacements are small, $\sin\theta \doteqdot 0$† and $\cos\theta \doteqdot 1$, then the equations of equilibrium reduce to

$$P_x + T_3 - T_1 = 0 \quad \text{(h)}$$

$$P_y + T_2 - T_4 = 0 \quad \text{(j)}$$

which refer to the datum configuration.

Re-writing equation (c)

$$e_1 = l[\sqrt{1 + 2u/l + (u^2 + v^2)/l^2} - 1] \ .$$

For small displacements, the squared terms are negligible compared with $2u/l$ and since by the binomial theorem

$$\sqrt{1 + 2u/l} = [1 + \tfrac{1}{2}(2u/l) + \ldots\ldots]$$

$$e_1 = u \ . \quad \text{(k)}$$

Similarly $e_2 = -v, \quad e_3 = -u, \quad e_4 = v \ .$ (l)

The use of such approximations is central to the small displacement theory of stress analysis.

Altogether, then, we have ten equations relating the ten unknowns $T_1, \ldots\ldots,$ $e_1 \ldots\ldots, u, v$ and so they can be found. Indeed

$$u = P_x/(k_1 + k_3) \quad \text{(m)}$$

$$v = P_y/(k_2 + k_4) \ . \quad \text{(n)}$$

† If we use the usual approximation $\sin\theta \doteqdot \theta$, we will be left with first order small terms which are negligible compared with the finite ones. In equations (a) and (b) the net result is the same.

To complete the analysis, the forces in the members are given by

$$T_1 = P_x k_1/(k_1 + k_3), \qquad T_2 = -P_y k_2/(k_2 + k_4) \qquad \text{(p)}$$

$$T_3 = -P_x k_3/(k_1 + k_3), \qquad T_4 = P_y k_4/(k_2 + k_4)\ . \qquad \text{(q)}$$

The minus signs indicate that springs 2 and 3 are in compression as we would have expected.

EXAMPLE 1.4.4

A light rigid bar is in equilibrium in the arrangement of Fig. 1.4.11(a). Determine the equilibrium configuration assuming displacements are small.

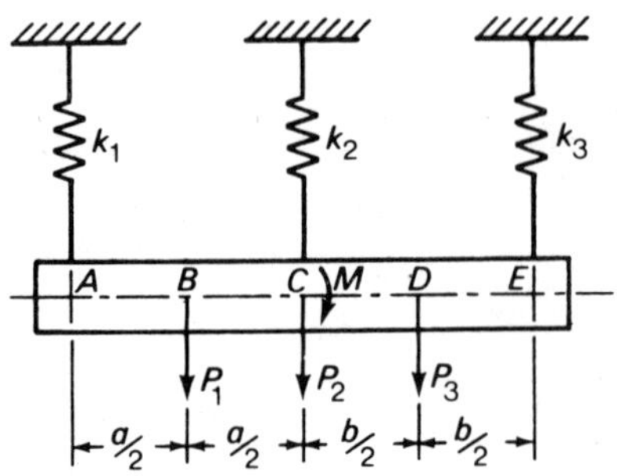

Fig. 1.4.11(a)

Denote the spring tensions and extensions by T_1 . . . . . , e_1 . . . . . Then isolating the bar as a free body we have Fig. 1.4.11(b):

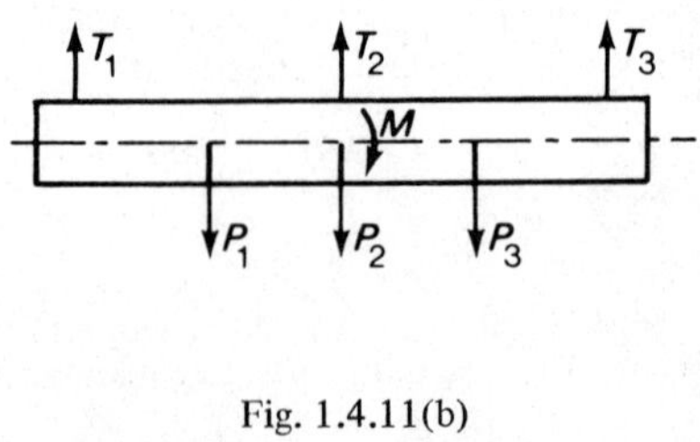

Fig. 1.4.11(b)

For equilibrium:

$$\Sigma F_x = \text{O identically}$$

$$\Sigma F_y = T_1 + T_2 + T_3 - P_1 - P_2 - P_3 = 0 \quad \text{(a)}$$

$$\Sigma M_{\text{centre}} = -M - T_1 a - P_3 b/2 + P_1 a/2 + T_3 b = 0 \ . \quad \text{(b)}$$

Thus, the system is statically indeterminate.

For compatibility: since sideways movement is not contemplated, the bar has two degrees of freedom. We choose u and θ of Fig. 1.4.11(c) as generalised coordinates

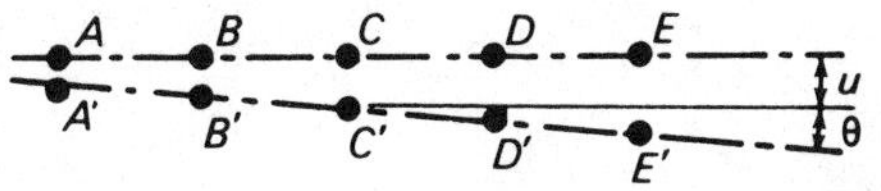

Fig. 1.4.11(c)

Compatibility requires

$$e_1 = u - a\theta \ , \quad e_2 = u \ , \quad e_3 = u + b\theta \ . \quad \text{(c)}$$

Force-deformation relations: for Hooke's law springs,

$$T_1 = k_1 e_1 \ , \quad T_2 = k_2 e_2 \ , \quad T_3 = k_3 e_3 \ . \quad \text{(d)}$$

Thus, we have eight equations from which $T_1 \ldots\ldots e_3$, u, θ may be found. In fact, using (c) and (d) in (a) and (b), we find that expressing the equilibrium equations in terms of the unknown u and θ yields

$$(k_1 + k_2 + k_3)u - (ak_1 - bk_3)\theta = P_1 + P_2 + P_3 \quad \text{(e)}$$

$$-(ak_1 - bk_3)u + (a^2 k_1 + b^2 k_3)\theta = \frac{P_3 b - P_1 a}{2} + M \ . \quad \text{(f)}$$

Looking ahead to more general developments later, if we denote $u = q_1$, $\theta = q_2$, we may write (e) and (f) compactly as

$$\begin{aligned} K_{11}q_1 + K_{12}q_2 &= Q_1 \\ K_{21}q_1 + K_{22}q_2 &= Q_2 \end{aligned} \quad \text{(g)}$$

or

$$\begin{bmatrix} K_{11} & K_{12} \\ K_{21} & K_{22} \end{bmatrix} \begin{Bmatrix} q_1 \\ q_2 \end{Bmatrix} = \begin{Bmatrix} Q_1 \\ Q_2 \end{Bmatrix} \quad \text{(h)}$$

or

$$[K]\{q\} = \{Q\} \ . \quad \text{(j)}$$

We shall see in due course that this is the characteristic form of the **stiffness equilibrium equation** for finite degree of freedom systems: $[K]$ is called the **stiffness matrix** of the assembled system. In this case we have actually calculated that $K_{12} = K_{21}$ so that the stiffness matrix is symmetric. This is not a chance occurrence, it is always true for a linear elastic system as will be shown later.

Solving equations (e) and (f) or (j), and using equations (c) in (d), the spring forces may be found and the stress analysis is complete. Indeed, if $P_1 = P_2 = P_3 = P/3, M = 0, a = b$ and $k_1 = k_2 = k_3$, equation (f) shows that $\theta = 0$. Then $u = P/3k$ and $T_1 = T_2 = T_3 = P/3$. It is only under these special circumstances that the springs carry an equal share of the load. In general, the distribution of force 'into the system' depends on its stiffness characteristics.

Problems 1.4

1.4.1 The trapdoor of Fig. 1.4.12 is hinged at A and B and propped at D. Determine the reactions.

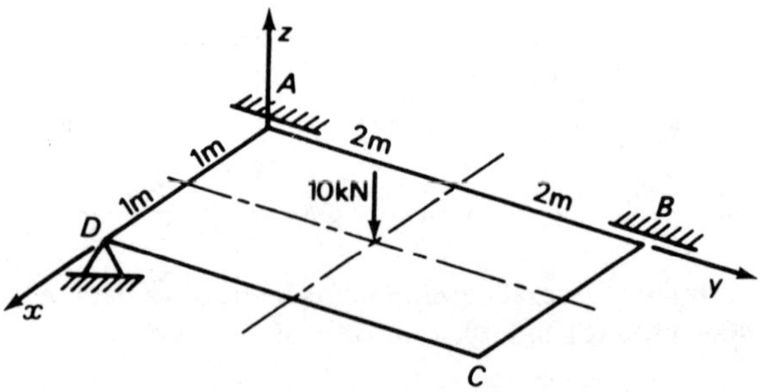

Fig. 1.4.12

Ans: $R_A = 0\,kN, \quad R_B = 5\,kN, \quad R_D = 5\,kN$

1.4.2 Determine the thrust in the rod and the tensions in the wires of the system of Fig. 1.4.13.

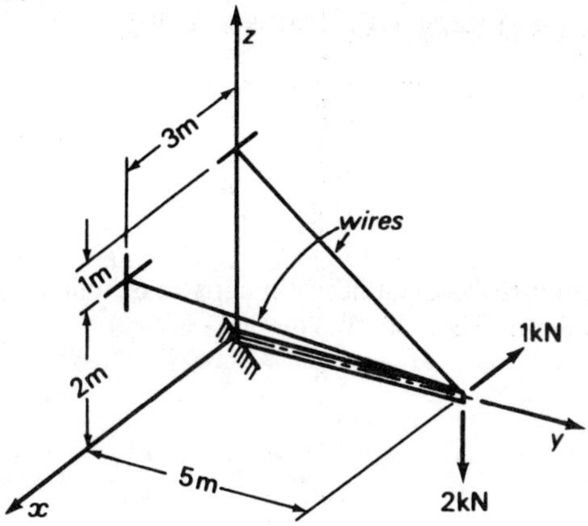

Fig. 1.4.13

Ans: $T_1 = \sqrt{43}/3\ kN$, $T_2 = \sqrt{41}/12\ kN$, $T_{rod} = 17/12\ kN$

1.4.3 Deduce the stiffness equilibrium equations for the system of Fig. 1.4.14, when displacements are small.

What can you conclude when $k_1 = k_2 = k_3 = k_4$?

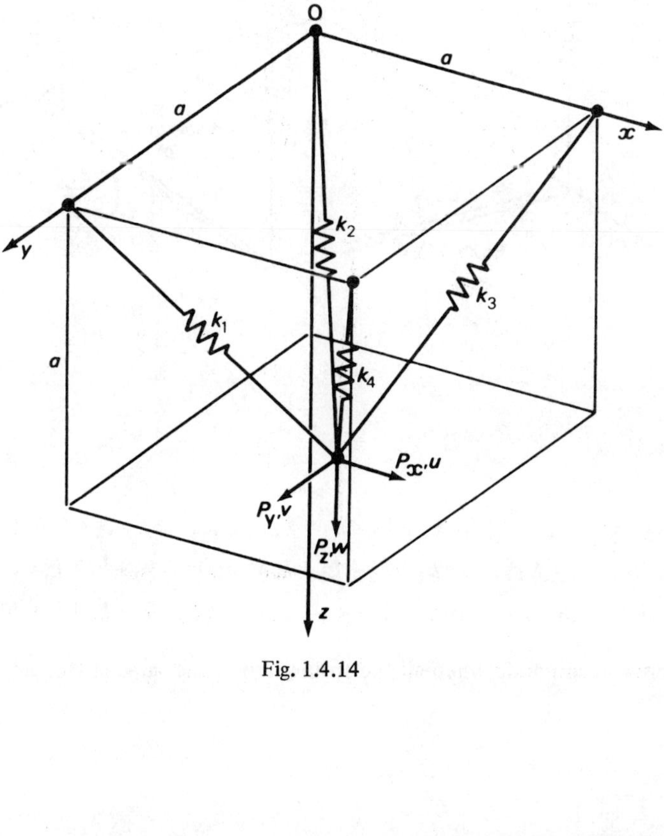

Fig. 1.4.14

Ans:

$$\frac{1}{6}\begin{bmatrix}(k_1+k_2+k_3+k_4) & (-k_1+k_2-k_3+k_4) & 2(k_1+k_2-k_3-k_4)\\ & (k_1+k_2+k_3+k_4) & 2(-k_1+k_2+k_3-k_4)\\ \text{Symm} & & 4(k_1+k_2+k_3+k_4)\end{bmatrix}\begin{Bmatrix}u\\v\\w\end{Bmatrix}=\begin{Bmatrix}P_x\\P_y\\P_z\end{Bmatrix}$$

1.4.4 Deduce the stiffness equilibrium equations, when displacements are small, for the system of Fig. 1.4.15. The plate is rigid.

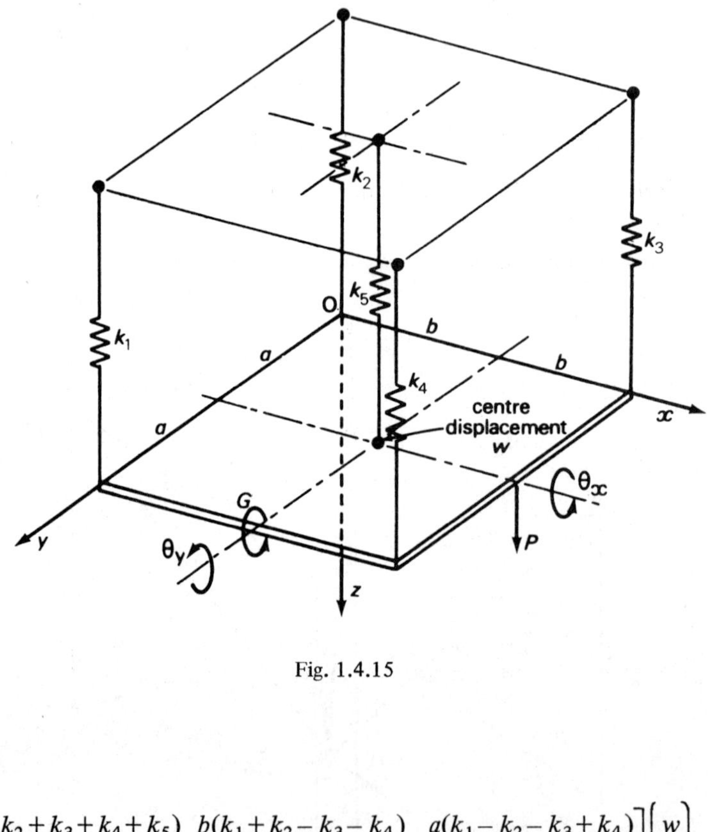

Fig. 1.4.15

Ans:

$$\begin{bmatrix}(k_1+k_2+k_3+k_4+k_5) & b(k_1+k_2-k_3-k_4) & a(k_1-k_2-k_3+k_4)\\ b(k_1+k_2-k_3-k_4) & b^2(k_1+k_2+k_3+k_4) & ab(k_1-k_2+k_3-k_4)\\ a(k_1-k_2-k_3+k_4) & ab(k_1-k_2+k_3-k_4) & a^2(k_1+k_2+k_3+k_4)\end{bmatrix}\begin{Bmatrix}w\\\theta_x\\\theta_y\end{Bmatrix}=\begin{Bmatrix}P\\-Pb\\G\end{Bmatrix}$$

What can you conclude when all the springs are of the same stiffness?

1.5 Work and energy

Suppose that a particle, moving from A to B along some path in space, is subjected to a force **F** as shown in Fig. 1.5.1.†

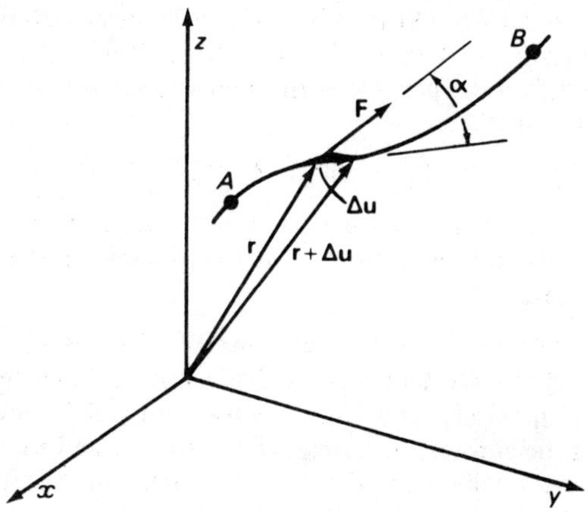

Fig. 1.5.1

The work which **F** performs in an infinitesimal displacement $\Delta\mathbf{u} = \Delta u\mathbf{i} + \Delta v\mathbf{j} + \Delta w\mathbf{k}$ is defined as

$$\Delta W = \mathbf{F} \cdot \Delta\mathbf{u} \tag{1.5.1}$$

or

$$\Delta W = F_x \Delta u + F_y \Delta v + F_z \Delta w\ . \tag{1.5.2}$$

That is, work is the product of a force and the displacement component in the direction of the force. The total work done by **F** in the displacement from A to B is then

$$W = \int_A^B \mathbf{F} \cdot \mathrm{d}\mathbf{u} = \int_A^B (F_x \mathrm{d}u + F_y \mathrm{d}v + F_z \mathrm{d}w) \tag{1.5.3}$$

By definition, work is a scalar quantity and will be positive or negative according as the senses of the displacement and force are the same or opposite to each other.

† To overcome the intuitive feeling that force and displacement should be in the same direction, recall that a stone thrown *horizontally* feels the *downward* force of gravity as it traverses its path. It is the acceleration of the particle which is in the direction of the force.

Recalling that a couple consists of a pair of equal, opposite parallel forces, we can extend definition (1.5.1) to give the work done by a couple **M** in an infinitesimal rotation $\Delta\boldsymbol{\theta}$ to be

$$\Delta W = \mathbf{M} \cdot \Delta\theta \ . \qquad (1.5.4)$$

In a finite rotation from A to B

$$W = \int_A^B \mathbf{M} \cdot \mathrm{d}\boldsymbol{\theta} = \int_A^B (M_x \mathrm{d}\theta_x + M_y \mathrm{d}\theta_y + M_z \mathrm{d}\theta_z) \ . \qquad (1.5.5)$$

When a particle is a constituent member of a mechanical system, it will generally be subjected to internal and external forces so that it will be convenient to regard the total work done on the particle as the sum of the work of the internal and external forces, that is,

$$W = W_{\text{int}} + W_{\text{ext}} \ . \qquad (1.5.6)$$

Any work done on a constitutent particle is work done on the system: if work is done on all constituent particles, the total work done on the system is the sum of all such works.

In the light of definition (1.5.1) we can conceive of forces which do no work. Thus, if an isolated particle is pressed against a smooth fixed surface, the force of reaction at the surface (or the force of constraint which prevents the particle from penetrating the surface), is normal to the surface and so does no work if the particle moves tangentially along the surface. Further, if the surface is moved with the particle held in contact with it, the forces of interaction, one on each body, are equal and opposite and so the positive work of one exactly cancels the negative work of the other. As a final example, the *relative* positions of two constituent particles of a rigid body do not change as the body is displaced so that their forces of interaction do no net work. The common factor in all these examples is the existence of *rigid* constraints between the particles and it is the work of the forces required to maintain such constraints which is zero. In a *deformable* body (whether elastic or not), the internal constraints are not rigid and we can imagine the particles displaced relative to each other when the body is disturbed. Here, the work of the internal forces is not zero: it is the existence of such non zero internal work which allows the use of energy methods for stress analysis.

To generalise the above results, suppose that a system of N particles experience infinitesimal displacments $\Delta\mathbf{u}_s = \Delta u_s\mathbf{i} + \Delta v_s\mathbf{j} + \Delta w_s\mathbf{k}, s = 1, 2 \ldots\ldots N$. If the system is unconstrained all the $\Delta\mathbf{u}_s$ can be arbitrary and independent and the total work done is

$$\Delta W = \sum_{s=1}^{N} \mathbf{F}_s \cdot \Delta\mathbf{u}_s = \sum_{s=1}^{N} (F_{x_s}\Delta u_s + F_{y_s}\Delta v_s + F_{z_s}\Delta w_s) \ . \qquad (1.5.7)$$

If the system is constrained to give it only $n < 3N$ degrees of freedom, the $\Delta\mathbf{u}_s$ etc. can no longer be given independent increments. However, if generalised coordinates q_i are introduced, the displacement can be expressed in terms of the

Δq_i which *are* independent. Equation (1.3.8) gives the relations implicit in the constraints between the Δu_s etc. and the Δq_i; the work equation (1.5.7) can then be written as

$$\Delta W = \sum_{s=1}^{N} \left[F_{x_s}\left(\sum_{j=1}^{n} \frac{\partial f_s}{\partial q_j} \Delta q_j \right) + F_{y_s}\left(\sum_{j=1}^{n} \frac{\partial g_s}{\partial q_j} \Delta q_j \right) + F_{z_s}\left(\sum_{j=1}^{n} \frac{\partial h_s}{\partial q_j} \Delta q_j \right) \right] . \tag{1.5.7b}$$

Now it does not matter whether the sum on j is carried out before the sum on s or vice versus, the result is the same, so that summing on s first, we can write

$$\Delta W = Q_1 \Delta q_1 + Q_2 \Delta q_2 + \ldots\ldots = \sum_{j=1}^{n} Q_j \Delta q_j \ . \tag{1.5.8}$$

Here, the Q_j are functions of the forces and current values of the q_j given by

$$Q_j = \sum_{s=1}^{n} \left(F_{x_s} \frac{\partial f_s}{\partial q_j} + F_{y_s} \frac{\partial g_s}{\partial q_j} + F_{z_s} \frac{\partial h_s}{\partial q_j} \right) \ . \tag{1.5.9}$$

Thus, for constrained systems, it is the quantities Q_j which are relevant for work calculations and not the original F's. The Q_j are called the components of the **generalised force** since if each is multiplied by the corresponding generalised coordinate increment the result is work. It is important to observe that by viewing the displacement of a constrained system as changes in the generalised coordinates, the geometric constraints are not violated: the forces of reaction thus do no work when the constraints are rigid. It is because such forces do not figure in any work calculations that energy methods have such a great practical advantage over methods in the mechanics of complex systems.†

We saw in Section 1.3 that the choice of the generalised coordinate system is not unique so that the above developments could be repeated for q_i', q_i'', etc. Clearly, the work ΔW must be the same each time since we are merely using different means for describing the same displacement. We can thus define the components of the generalised force as the coefficients of an invariant first order differential form. It usually happens in analytic problems that ΔW is the total differential of a function of the coordinates called the **work function**. Then

† Completely deformable bodies have an infinite number of degrees of freedom and their mechanics is described by partial differential equations (see Ch. 3). The assumptions which are incorporated in establishing simplified theories, as for beams or plates, effectively introduce certain internal rigid constraints (negligible shear effects for instance). The forces corresponding to such constraints do no work in displacements and so the attractions of energy methods for establishing approximate theories (Ch. 6) can be seen to stem from this consideration.

$$\Delta W = d\phi = \frac{\partial \phi}{\partial q_1} dq_1 + \frac{\partial \phi}{\partial q_2} dq_2 + \ldots\ldots = \sum_{i=1}^{n} \frac{\partial \phi}{\partial q_i} dq_i \tag{1.5.10}$$

and

$$W = \int_A^B d\phi = (\phi_B - \phi_A) \; . \tag{1.5.11}$$

The work done thus depends only on the end points of the path and not on the path itself: in particular, no net work is done in a closed path. Comparing equations (1.5.10) and (1.5.8) we have the important result that under these circumstances,

$$Q_i = \frac{\partial \phi \; (q_1 \ldots\ldots q_n)}{\partial q_i} \; . \tag{1.5.12}$$

It is usual in engineering literature to work in terms of the negative of the work function.

Then

$$V(q_1, q_2, \ldots\ldots q_n) = -W(q_1, q_2 \ldots\ldots q_n) \tag{1.5.13}$$

and

$$Q_i = -\frac{\partial V}{\partial q_i} \; . \tag{1.5.14}$$

The existence of forces derivable from a single scalar function of the coordinates is of considerable significance. Such forces are found to satisfy the law of conservation of energy and consequently they are said to be conservative: $V(q_1, q_2, \ldots\ldots q_n)$ is the **potential function** and its value is the **potential energy** of the forces.

Elastic bodies are those which recover their original shape completely when disturbing forces are removed. In a complete loading-unloading cycle, the internal forces produce zero net work since the configuration thereby follows a closed path: such forces are thus conservative and have a **potential energy of strain** or **strain energy.** For convenience we shall write the total potential for a fully conservative system as

$$V = U + \Omega \; . \tag{1.5.15}$$

Here, U is the strain energy stored in the system and Ω is the potential of the external loads and we can conceive of internal and external generalised force from

$$\frac{\partial V}{\partial q_i} = \frac{\partial U}{\partial q_i} + \frac{\partial \Omega}{\partial q_i} = -Q_{\text{int}} - Q_{\text{ext}} = -Q \; . \tag{1.5.16}$$

We have previously observed that we usually take the datum configuration of a mechanical system for granted so that any calculations relate to *changes* in configuration or displacements: we are thus concerned only with *changes* in the potential of forces and not with absolute values. For calculation purposes, the datum value can be chosen arbitrarily and usually we choose it to be zero.

The calculation of the work of the internal forces in a general deformable solid must await the review of the ideas of stress and strain in a continuum given in Chapter 3. However, for a simple one dimensional connexion between two particles, the necessary expression can be found by very elementary considerations. Thus, suppose that two particles attracted by internal forces are displaced by infinitesimal amounts as shown in Fig. 1.5.2.

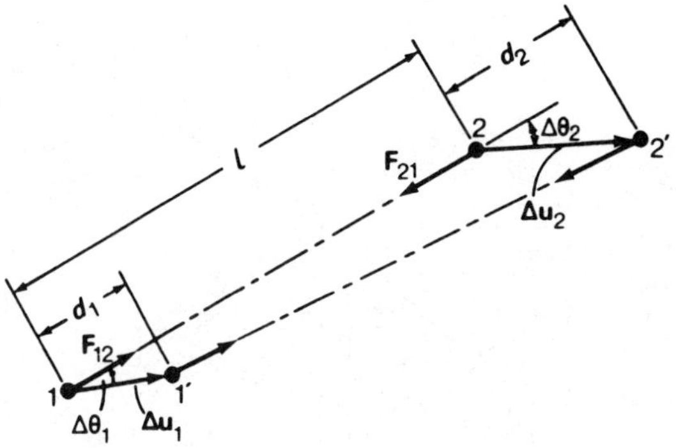

Fig. 1.5.2

$$\Delta W_{int} = \mathbf{F}_{12} \cdot \Delta \mathbf{u}_1 + \mathbf{F}_{21} \cdot \Delta \mathbf{u}_2$$
$$= F_{12} \Delta u_1 \cos \Delta\theta_1 - F_{21} \Delta u_2 \cos \Delta\theta_2 \ ,$$

but $$F_{12} = F_{21} = F$$

then $$\Delta W_{int} = F(\mathrm{d}_1 - \mathrm{d}_2)$$

or $$\Delta W_{int} = -F\Delta e \ , \qquad (1.5.17)$$

where Δe is the *increase* in distance between the particles or the *extension* of the link between them. It is important to notice the negative sign in Equation (1.5.17).

If the link is elastic, $\Delta W_{int} = -\Delta U$ where $U = U(e)$, the strain energy, which is some function of the extension e.

Then $$-\Delta W_{int} = \Delta U = \frac{\mathrm{d}U}{\mathrm{d}e}\Delta e \ . \qquad (1.5.18)$$

So that comparing equations (1.5.17) and (1.5.18),

$$\frac{\mathrm{d}U(e)}{\mathrm{d}e} = F \ , \qquad (1.5.19)$$

that is, the form of $U(e)$ defines the force-deformation characteristic for the link. When the extension is increased from zero to some current value e_c, the total strain energy stored will be

$$U = \int_0^{e_c} F\mathrm{d}e \ . \qquad (1.5.20)$$

In this expression we have arbitrarily taken the strain energy at the datum position to be zero which implies $F = 0$ when $e = 0$. We have already pointed out that since we are only concerned with *changes* in energy in statics calculations, there is no disadvantage in doing this. In general, a plot of F against e will be non-linear as shown in Fig. 1.5.3, then U is seen to be the area under the curve

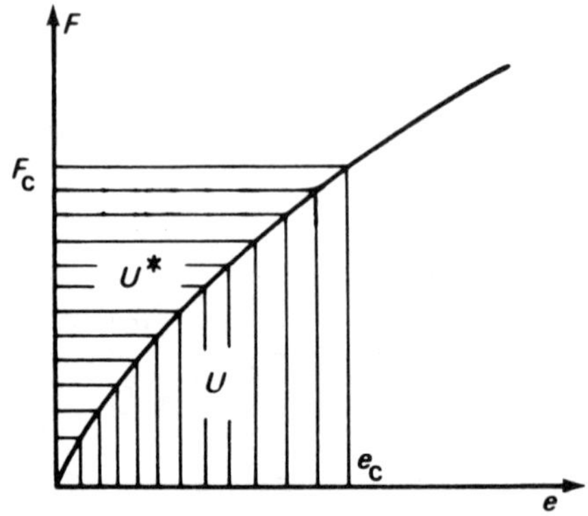

Fig. 1.5.3

If the link behaves according to Hooke's law, $F = ke$ where k is some constant, then equation (1.5.20) gives

$$U = \tfrac{1}{2}ke^2 \quad . \tag{1.5.21}$$

We can drop the subscript c if we understand that e is the current value of extension.

EXAMPLE 1.5.1

The rigid link of Fig. 1.5.4 is in equilibrium as shown. Determine expressions for the generalised force

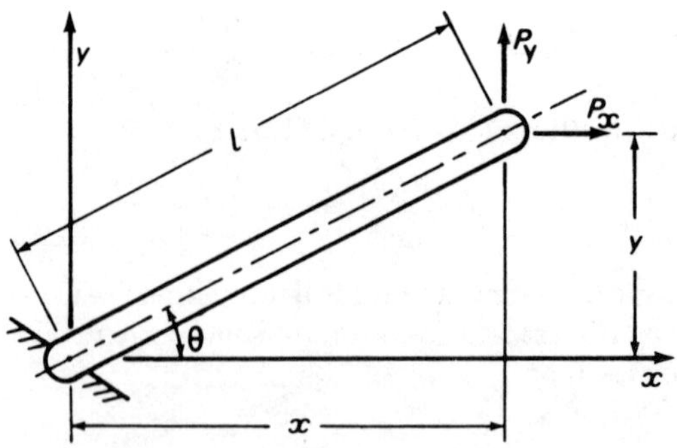

Fig. 1.5.4

The system has one degree of freedom and either θ, x or y will serve as a generalised coordinate.

If θ receives an increment $\Delta\theta$,

$$\Delta W = P_y \Delta y + P_x \Delta x \tag{a}$$

where

$$\Delta y = l \cos\theta \, \Delta\theta, \qquad \Delta x = -l \sin\theta \, \Delta\theta \ . \tag{b}$$

Using θ as the generalised coordinate

$$W = (P_y l \cos\theta - P_x l \sin\theta)\Delta\theta = \Theta \, \Delta\theta \ . \tag{c}$$

Then the generalised force corresponding to θ is

$$\Theta = P_y l \cos\theta - P_x l \sin\theta \ . \tag{d}$$

Using y as a generalised coordinate

$$\Delta W = (P_y - P_x \tan\theta)\Delta y = Y \Delta y \tag{e}$$

where the generalised force is now

$$Y = P_y - P_x \tan\theta \ . \tag{f}$$

Finally, if x is the generalised coordinate, the generalised force is

$$X = (P_x - P_y \cot\theta) \ . \tag{g}$$

We see that the expression for generalised force component depends on the choice of the generalised coordinate even though the actual forces are fixed.

EXAMPLE 1.5.2

Fig. 1.5.5 shows a rigid link in equilibrium. The datum position with no load is $\theta = 0$ and the displacement to the equilibrium position is small. Determine the generalised force component corresponding to θ.

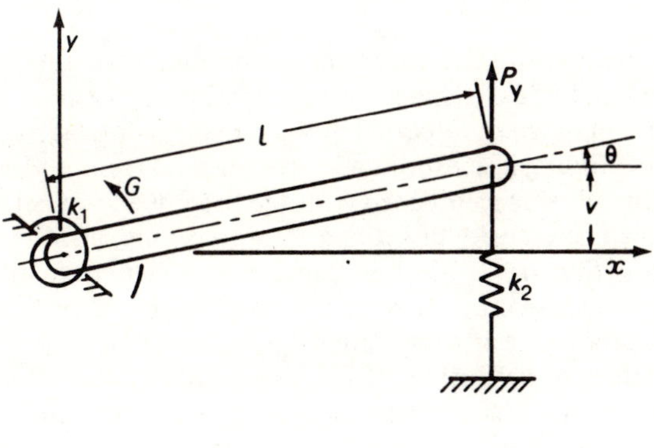

Fig. 1.5.5

The internal couple M_1 and force T_2 in springs k_1 and k_2 respectively do work in the infinitesimal displacement $\Delta\theta$ so that

$$\Delta W = \Delta W_{ext} + \Delta W_{int} = P_y \Delta v + G\Delta\theta - M_1 \Delta\theta - T_2 \Delta e \ . \quad \text{(a)}$$

But $M_1 = k_1\theta, \quad T_2 = k_2 e$

so $$\Delta W = (P_y l + G - k_1\theta - k_2 l^2\theta)\Delta\theta \quad \text{(b)}$$

or $$\Delta W = \Theta\Delta\theta \ . \quad \text{(c)}$$

The generalised force Θ is made up of an internal and external part

$$\Theta_{int} = -k_1\theta - k_2 l^2\theta \ , \quad \text{(d)}$$

$$\Theta_{ext} = P_y l + G \ , \quad \text{(e)}$$

$$\Theta = \Theta_{int} + \Theta_{ext} \quad \text{(f)}$$

and for compatibility $e = v = l\theta$.

Problems 1.5

1.5.1 If the system of Problem 1.3.1 supports a force F_3 corresponding to u_3 and a couple F_4 corresponding to u_4, determine the generalised external forces corresponding to the sets of generalised coordinates of Problem 1.3.1.

Ans: $$\{Q_a\}^t = \left[\left(\frac{1}{2}F_3 + \frac{1}{r_1}F_4\right) \quad \frac{1}{2}F_3\right] \ .$$

1.5.2 Forces F_1 and F_4 corresponding to u_1 and u_4 are applied to the frame of Problem 1.3.2. What are the components of generalised external force corresponding to the chosen generalised coordinates?

Ans: $$\{Q\}^t = [(F_1 + F_4 \tan\theta) \quad 0 \quad 0] \ .$$

1.6 Principles of virtual work and stationary total potential energy

The use of vector methods to determine the equilibrium configuration of a mechanical system under load requires us to pay attention to an individual particle or suitable sub-system of particles. We must take into account not only the active (applied) forces, but also the *reactive* ones called into play to ensure that the system displaces in accordance with the prescribed geometric constraints. For simple systems this is a very direct and easy procedure, but the example of the Hooke's joint in Section 1.4 showed that, as the system becomes more complex, this approach becomes increasingly cumbersome. Now from purely geometrical considerations a given system can take up any number of displaced configurations within the limitations of the constraints: however, only one of these will be the true one which corresponds to equilibrium between the actual applied loads and the induced reactions. This observation leads to an alternative procedure for identifying the true position. Thus, suppose that with the actual loads acting at their *fixed prescribed* values, work calculations are carried out as the system is *imagined* to be displaced through a series of geometrically possible positions. The constraints are not violated in these imagined changes of configuration, so that if they are rigid, the forces corresponding to them do no work. Then, if a criterion for identifying the true position is based on work, the forces at rigid constraints will not figure in any calculations.

The test range of geometrically possible, or *admissible,* configurations may be restricted to the *neighbourhood* of the true one so that they are arrived at by infinitesimal displacements from or *variations* to the actual equilibrium configuration. It is important to observe that during such variations, all the forces remain fixed at their actual equilibrium values (after all, we are merely comparing the relative merits of various configurations which *might* correspond to the actual forces). Analytically, variations which do not violate constraints are brought about by arbitrary independent infinitesimal changes, or variations δq_i, in the generalised coordinates q_i.† When a system experiences these variations in configuration, it is said to execute **virtual displacements** from its position of equilibrium. These displacements are called virtual because they are *imagined* to occur purely for comparison purposes. They need not bear any relation to any actual displacement Δq_i which might occur as a result of an actual change of load; they are merely required to satisfy the geometric constraints.

The work done by the actual forces through a virtual displacement is called **virtual work**, its value is given by equation (1.5.8) when the Δq_i are replaced by δq_i so that

$$\delta W = \sum_{j=1}^{n} Q_j \delta q_j \quad . \tag{1.6.1}$$

As we have pointed out, the forces at rigid constraints do not figure in equation (1.6.1). However, if there are any deformable members in the system, their internal forces will make a contribution.

The **principle of virtual work** is the means by which the true configuration is identified. It states that a mechanical system is in its equilibrium configuration if the virtual work of all the forces acting on it is zero in a virtual displacement. That is, for equilibrium

$$\delta W = \sum_{j=1}^{n} Q_j \delta q_j = 0 \quad . \tag{1.6.2}$$

Now, the δq_j are arbitrary, independent variations in the q_j so that equation (1.6.2) implies that

$$Q_j = 0 \quad j = 1, 2, \ldots\ldots n \quad . \tag{1.6.3}$$

Solution of these n simultaneous equations of equilibrium yields the values of the q's corresponding to the true equilibrium configuration.

† The symbol δq_i is used to denote a variation as opposed to an actual increment Δq_i in q_i. See Chapter 4.

When the system contains deformable members, equation (1.6.2) becomes

$$\delta W_{\text{int}} + \delta W_{\text{ext}} = 0 \qquad (1.6.4)$$

and equations (1.6.3) becomes

$$(Q_j)_{\text{int}} + (Q_j)_{\text{ext}} = 0 \qquad (1.6.5)$$

as the conditions for equilibrium.

It is important to observe that the principle of virtual work holds whether the forces are conservative or not. In particular, the internal forces in a deformable solid need not be conservative so that the principle is applicable to inelastic as well as elastic bodies; this has considerable significance for stress analysis. However, if any deformable bodies *are* elastic, linear or not, so that Hooke's law or some more complicated load-deformation law is appropriate, the internal forces will have a potential and the virtual work equation (1.6.4) takes the form

$$\delta W_{\text{ext}} = \delta U \; ; \qquad (1.6.6)$$

this is the form in which the principle is frequently quoted in books on structural analysis. If the external forces are also conservative, $\delta W_{\text{ext}} = -\delta\Omega$ and the principle takes the form

$$\delta V = \delta(U + \Omega) = 0 \; . \qquad (1.6.7)$$

Now V is a function of the generalised coordinates so that equation (1.6.7) is an analytic statement of the **principle of stationary total potential energy** which asserts: *of all the geometrically possible configurations which a mechanical system can take up, the true one, corresponding to equilibrium between the applied loads and the induced reactions, is that for which the total potential energy is stationary.*

Since $V = V(q_1, q_2, \ldots\ldots q_n)$,

$$\delta V = \frac{\partial V}{\partial q_1}\delta q_1 + \ldots\ldots = \sum_{j=1}^{n}\frac{\partial V}{\partial q_j}\delta q_j \; .\dagger \qquad (1.6.8)$$

Here, the δq_j are arbitrary and independent so that if $\delta V = 0$, we have

$$\frac{\partial V}{\partial q_j} = 0\,, \quad j = 1, 2, \ldots\ldots m \; . \qquad (1.6.9)$$

† Since $V = V(q_1, q_2, \ldots\ldots q_n)$, variations on the q's yield, by Taylor's series, a total variation in V of

$$\delta V^t = \sum_{i=1}^{n}\frac{\partial V}{\partial q_i}\delta q_i + \frac{1}{2}\sum_{i=1}^{n}\sum_{j=1}^{n}\frac{\partial^2 V}{\partial q_i \partial q_j}\delta q_i \delta q_j + 0(\delta^3)$$

or

$$\delta V^t = \delta V + \tfrac{1}{2}\delta^2 V + 0(\delta^3) \; .$$

To first order,

$$\delta V^t = \delta V = \sum_{i=1}^{n}\frac{\partial V}{\partial q_i}\delta q_i \; .$$

Thus the principle provides us with n equations of equilibrium expressed in terms of the applied loads and the generalised coordinates q_j from which the q_j, defining the equilibrium configuration, may be found.

It is important to observe that we made no restriction as to the magnitudes of the displacements involved when the system moved from the datum to the equilibrium configuration in these considerations: the *small* quantities involved were the virtual displacements which were merely devices used in a comparison process. When the displacements from datum to equilibrium configurations are 'large', certain geometric non-linearities arise so that even if Hookean materials are used, the overall problem is still non-linear†: however, the principle of virtual work still holds.

It is worth summarising, then, that the principle of virtual work holds whether elastic or inelastic deformations occur and whether displacements are large or small. If deformations are elastic, with a linear or non-linear stress-strain law, and the external forces are conservative, the principle of virtual work becomes the principle of stationary total potential energy. We thus have an extremely powerful tool of analysis. The principle is exact but we shall see in due course that very powerful approximate methods can be built upon it. In this way, numerical solutions can be obtained for engineering design problems which would otherwise be intractable.

EXAMPLE 1.6.1

Treat Example 1.4.1 by the method of virtual work.

We will use θ and ψ as our generalised coordinates. We will not determine the generalised forces by a formal application of equations (1.5.9) but calculate the virtual work directly.

Let θ and ψ receive variations $\delta\theta$, $\delta\psi$. Then in a virtual displacement

$$\begin{aligned}\delta W &= F_{1y}\delta(l_1 \sin\theta) + F_{1x}\delta(l_1\cos\theta)\\ &\quad + F_{2y}\delta(l_1\sin\theta + l_2\sin\psi)\\ &\quad + F_{2x}\delta(l_1\cos\theta + l_2\cos\psi) \qquad \text{(a)}\end{aligned}$$

i.e.

$$\begin{aligned}\delta W &= [(F_{1y} + F_{2y})l_1\cos\theta - (F_{1x} + F_{2x})l_1\sin\theta]\,\delta\theta\\ &\quad + [F_{2y}l_2\cos\psi - F_{2x}l_2\sin\psi]\,\delta\psi \qquad \text{(b)}\end{aligned}$$

or

$$\delta W = \Theta\delta\theta + \Psi\delta\psi\ . \qquad \text{(c)}$$

But $\delta W = 0$ for equilibrium and since $\delta\theta$ and $\delta\psi$ are arbitrary, we have

$$\tan\psi = F_{2y}/F_{2x}\ , \qquad \text{(d)}$$

$$\tan\theta = (F_{1y} + F_{2y})/(F_{1x} + F_{2x})\ . \qquad \text{(e)}$$

Thus, we have obtained the equilibrium configuration in a much more direct way than in Example 1.4.1.

† In such problems, questions of stability of equilibrium arise. We are not very much concerned with these in this book, but see Section 1.9.

EXAMPLE 1.6.2

Treat the Hooke's joint of Example 1.4.2 by the principle of virtual work.

The system has one degree of freedom. If the rotation of shafts 1 and 2 about their respective axes are ψ_1 and ψ_2, then in the neighbourhood of the *given* configuration, Fig. 1.4.6,

$$\delta\psi_2 = \delta\psi_1 \cos\theta \tag{a}$$

for compatibility.

In a virtual displacement $\delta\psi_1, \delta\psi_2$,

$$\delta W = 0 = M_1\delta\psi_1 - M_2\delta\psi_2 \tag{b}$$

or

$$0 = (M_1 - M_2\cos\theta)\delta\psi_1 \ , \tag{c}$$

but $\delta\psi_1$ is arbitrary so that

$$M = M_2 \cos\theta \ . \tag{d}$$

This result has been obtained with dramatically less effort than would have been required in the approach of Example 1.4.2.

EXAMPLE 1.6.3

Treat the system of Example 1.4.3 by the virtual work principle.

Since the equilibrium configuration is defined by the displacements u and v, they constitute generalised coordinates. Then, in a virtual displacement δu, δv,

$$\delta W = P_x\delta u + P_y\delta v - T_1\delta e_1 - T_2\delta e_2 - T_3\delta e_3 - T_4\delta e_4 \ . \tag{a}$$

But for compatibility (satisfaction of the geometric constraints)

$$\delta e_1 = -\delta e_3 = \delta u \tag{b}$$

$$-\delta e_2 = \delta e_4 = \delta v \ . \tag{c}$$

Then, since $T_1 = k_1 e_1$, etc., and $e_1 = u$ etc

$$\delta W = [P_x - (k_1 + k_3)u]\delta u + [P_y - (k_2 + k_4)v]\delta v \tag{d}$$

or

$$\delta W = X\delta u + Y\delta v \ . \tag{e}$$

But δu and δv are arbitrary and independent and since $\delta W = 0$ for equilibrium, $X = 0$, $Y = 0$ yield.

$$u = P_x/(k_1 + k_3) \tag{f}$$

$$v = P_y/(k_2 + k_4) \ . \tag{g}$$

The spring forces are then readily obtained.

We observe here that since the system was so sinple, there was no real computational advantage in using virtual work instead of the vector method.

EXAMPLE 1.6.4

Treat Example 1.4.4 by the principle of stationary total potential energy.

The system has two degrees of freedom u, θ. Then observing that for compatibility, $e_1 = (u - a\theta)$ etc.

$$V = \tfrac{1}{2}k_1(u - a\theta)^2 + \tfrac{1}{2}k_2u^2 + \tfrac{1}{2}k_3(u + b\theta)^2$$
$$- P_1\left(u - \frac{a\theta}{2}\right) - P_2u - P_3\left(u + \frac{b\theta}{2}\right) - M\theta \qquad \text{(a)}$$

i.e. $$V = V(u, \theta) = U(u, \theta) + \Omega(u, \theta) \quad . \qquad \text{(b)}$$

For equilibrium, $\delta V = 0$ so that

$$\frac{\partial V}{\partial u} = k_1(u - a\theta) + k_2u + k_3(u + b\theta) - P_1 - P_2 - P_3 = 0 \qquad \text{(c)}$$

$$\frac{\partial V}{\partial \theta} = -k_1a(u - a\theta) + k_3b(u + b\theta) + \frac{P_1a}{2} - \frac{P_3b}{2} - M = 0 \qquad \text{(d)}$$

that is $(k_1 + k_2 + k_3)u - (ak_1 - bk_3)\theta = P_1 + P_2 + P_3$ (e)

$$-(ak_1 - bk_3)u + (a^2k_1 + b^2k_2)\theta = M + P_3b/2 - P_1a/2 \qquad \text{(f)}$$

which are precisely equations (e) and (f) of Example 1.4.4.

If equations (e) and (f) are written in matrix form

$$[K]\{q\} = \{Q\} \qquad \text{(g)}$$

where $K_{11} = (k_1 + k_2 + k_3)$, $\{q\}^t = [u, \theta]$ etc. We notice:

(i) $U(u, \theta)$ is a homogeneous quadratic form in the generalised coordinates $u = q_1$, $\theta = q_2$.

That is $$U = \tfrac{1}{2}K_{11}q_1^2 + K_{12}q_1q_2 + \tfrac{1}{2}K_{22}q_2^2$$

or, referring to the appendix,

$$U = \tfrac{1}{2}\{q\}^t[K]\{q\} \quad . \qquad \text{(h)}$$

(ii) The *stiffness coefficient* K_{ij} is given by

$$K_{ij} = \frac{\partial^2 U}{\partial q_i \partial q_j} \quad . \qquad \text{(j)}$$

It has the physical significance of being the force required at q_i to maintain unit value for q_j, when all other q's are zero.

(iii) Equation (g) can be expressed in the form

$$\{Q\}_{\text{int}} + \{Q\}_{\text{ext}} = 0 \qquad \text{(k)}$$

which are clearly equations of equilibrium at, and in the direction of, each of the generalised coordinates.

The previous discussions and examples have shown that a significant feature of the principle of virtual work (and its derived principle of stationary total potential energy) is that the equilibrium configuration can be determined without explicitly taking into account the reactions at rigid constraints: this was ensured by employing *virtual* displacements; that is, those infinitesimal displacements which do not violate the constraints. Should we also need these reactions, supplementary calculations are necessary. Now if we recall that when a system is in equilibrium, each and every particle is in equilibrium under the action of *all* forces (active and reactive) acting on it, then the virtual work of all the forces on the particle is zero for *any* arbitrary infinitesimal displacement of the particle. Here, then, is a means of determining a selected reaction force if we can conceive of a virtual displacement which satisfies all the constraints *except* the one corresponding to the selected reaction: the virtual work of this constraint force must now figure in the calculations.

EXAMPLE 1.6.5

Use the virtual work principle to find the reactions at the supports of the beam of Fig. 1.6.1(a)

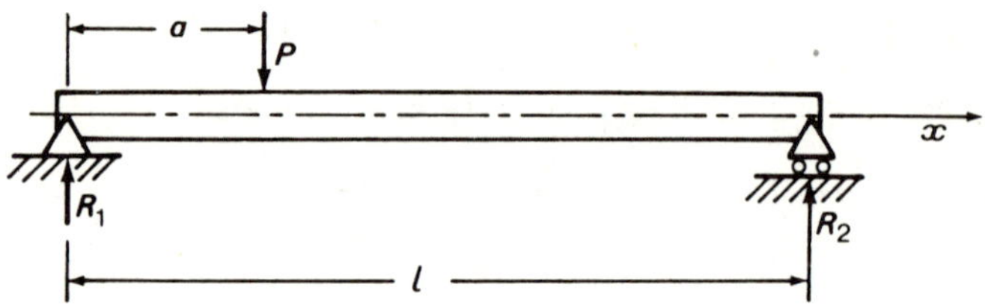

Fig. 1.6.1(a)

A rigid body virtual displacement violating the condition of zero displacement at R_1 is shown in Fig. 1.6.1(b).

Fig. 1.6.1(b)

Then $\quad \delta W = R_1 l \delta\theta - P(l - a)\delta\theta = 0 \,.$ (a)

But $\delta\theta$ is arbitrary $\quad \therefore R_1 = P(l - a)/l \,.$ (b)

From Fig. 1.6.1(c)

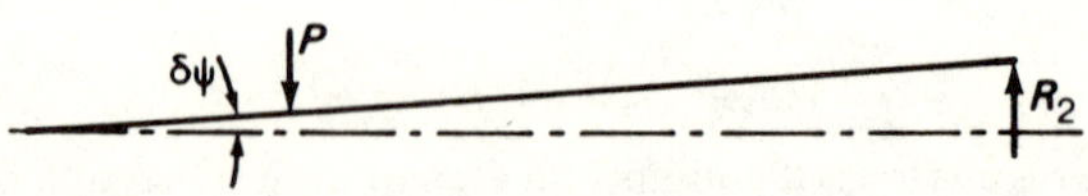

Fig. 1.6.1(c)

$$\delta W = R_2 l \delta\psi - Pa\delta\psi = 0 \ . \qquad \text{(c)}$$

But $\delta\psi$ is arbitrary $\quad \therefore R_2 = Pa/l \ . \qquad$ (d)

The technique demonstrated in this example is but a physical interpretation of an elegant method due to Lagrange which is more fully discussed in Chapter 4 and used in advanced problems later. In Example 1.6.4 our first task was to use the compatibility conditions to identify the generalised coordinates so that we could write $V = V(q_1, q_2, \ldots\ldots q_n)$. Determining $\delta V = 0$ was then an ordinary stationary value problem for a function of several independent variables. However, had we written $V = V(u_1, u_2, \ldots\ldots)$ where the u's are arbitrary coordinates, the task would have been to find the stationary value of this function, *subject* to some constraints $F_i(u_1, u_2, \ldots\ldots) = 0$, $i = 1, 2 \ldots\ldots m$. In effect, this was the procedure adopted in examples 1.6.1, 2, 3, but the method of Lagrange is more systematic.

The technique is to form a modified potential energy expression

$$\bar{V} = V(u_1, u_2, \ldots\ldots) + \sum_{i=1}^{m} \lambda_i F_i(u_1, u_2, \ldots\ldots) \qquad 1.6.10$$

where the λ_i are the Lagrange multipliers which are, as yet unknown. Then, $\delta\bar{V}$ is formed by variations (virtual changes) in the $u_1, u_2, \ldots\ldots \lambda_1, \lambda_2 \ldots\ldots$ as if they were free of constraints. The condition $\delta\bar{V} = 0$ then yields to the solution to the original constrained problem.

EXAMPLE 1.6.6

Treat Example 1.6.4 by the method of Lagrange multipliers.
The constraint equations are

$$e_1 - u + a\theta = 0 \qquad \text{(a)}$$

$$e_2 - u = 0 \qquad \text{(b)}$$

$$e_3 - u - b\theta = 0 \qquad \text{(c)}$$

Then

$$\bar{V} = \tfrac{1}{2}k_1 e_1^2 + \tfrac{1}{2}k_2 e_2^2 + \tfrac{1}{2}k_3 e_3^2 - P_1\left(u - \frac{a\theta}{2}\right) - P_2 u - P_3\left(u + \frac{b\theta}{2}\right)$$
$$- M\theta + \lambda_1(e_1 - u + a\theta) + \lambda_2(e_2 - u) + \lambda_3(e_3 - u - b\theta) \qquad \text{(d)}$$

$$\delta\bar{V} = k_1 e_1 \delta e_1 + k_2 e_2 \delta e_2 + k_3 e_3 \delta e_3 - P_1\left(\delta u - \frac{a\delta\theta}{2}\right) - P_2 \delta u$$
$$- P_3\left(\delta u + \frac{b\delta\theta}{2}\right) - M\delta\theta + \lambda_1(\delta e_1 - \delta u + a\delta\theta)$$
$$+ \lambda_2(\delta e_2 - \delta u) + \lambda_3(\delta e_3 - \delta u - b\delta\theta) \ . \qquad \text{(e)}$$

We notice that the coefficients of $\delta\lambda_1$ etc. are zero in view of (a), (b), (c). Rearranging,

$$\delta\bar{V} = (k_1e_1 + \lambda_1)\delta e_1 + (k_2e_2 + \lambda_2)\delta e_2 + (k_3e_3 + \lambda_3)\delta e_3 - (P_1 + P_2 + P_3 + \lambda_1 + \lambda_2 + \lambda_3)\delta u + \left(\frac{P_1a}{2} - \frac{P_3b}{2} + \lambda_1a - \lambda_3b - M\right)\delta\theta\ . \tag{f}$$

Then, for equilibrium, $\delta\bar{V} = 0$

and $$k_1e_1 + \lambda_1 = 0\,,\quad k_2e_2 + \lambda_2 = 0,\quad k_3e_3 + \lambda_3 = 0\ ; \tag{g}$$

$$P_1 + P_2 + P_3 + \lambda_1 + \lambda_2 + \lambda_3 = 0\ ; \tag{h}$$

$$\left(\frac{P_1}{2} + \lambda_1\right)a - \left(\frac{P_3}{2} + \lambda_3\right)b - M = 0\ . \tag{j}$$

Thus, equations (g), (h), (j) in conjunction with (a), (b), (c), determine the equilibrium configuration and the spring forces. The λ's are intermediaries which have a physical interpretation; in fact they are the spring forces with minus sign attached.

Problems 1.6

1.6.1 Solve Problems 1.4.3 and 1.4.4 by means of the principle of virtual work.

1.6.2 Formulate Problem 1.6.1 in terms of the potential energy principle.

1.6.3 Determine the stiffness equilibrium equations for the system of Fig. 1.6.2.

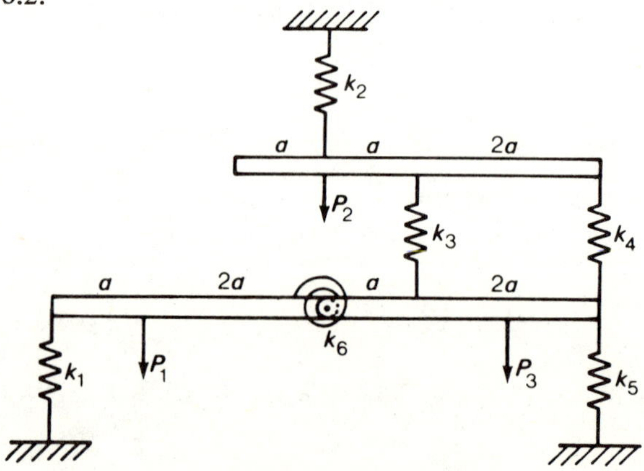

Fig. 1.6.2

1.6.4 Deduce the non-linear equation which defines the equilibrium configuration of the system shown in Fig. 1.6.3. Springs 1 and 2 have linear characteristics $F_1 = k_1e_1$, $M_2 = k_2\theta_2$ and the arrangement is such that they are free of strain with $\theta = 0$.

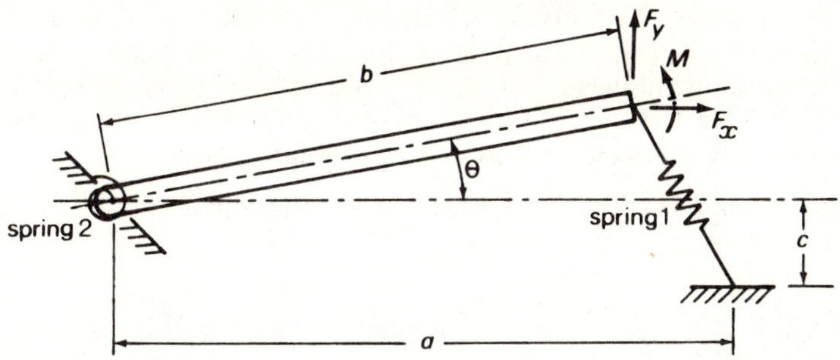

Fig. 1.6.3.

Ans: $0 = k_2\theta - M - F_y b\cos\theta + F_x b\sin\theta + k_1\{1 - [(a-b)^2 + c^2]^{1/2}[(a - b\cos\theta)^2 + (c + b\sin\theta)^2]^{-1/2}\}\{ab\sin\theta + bc\sin\theta\}$.

1.7 Complementary work and complementary energy

Fig. 1.5.3 showed that when a force-displacement relation was plotted with force as ordinate and displacement as abscissa, the area under the curve corresponded to the work done. Since work and energy provided such powerful tools for analysis, it seems reasonable to ask if the area above the curve is also of some use: we shall see in due course that it provides the basis of another powerful set of analytic tools complementary to those of work.

Referring to the particle displaced through the space curve of Fig. 1.5.1, the *complementary work* for a force increment is defined to be

$$\Delta W^* = \mathbf{u}\cdot\Delta\mathbf{F} \tag{1.7.1}$$

or

$$\Delta W^* = u\,\Delta F_x + v\,\Delta F_y + w\,\Delta F_z \tag{1.7.2}$$

where **u** is the displacement from the datum to some intermediate position **r**. In a total movement from A to B along the curve, the total complementary work will be

$$W^* = \int_A^B \mathbf{u}\cdot\mathrm{d}\mathbf{F} = \int_A^B (u\mathrm{d}F_x + v\mathrm{d}F_y + w\mathrm{d}F_z)\ . \tag{1.7.3}$$

It follows that if the 'force' is a couple,

$$\Delta W^* = \theta\cdot\Delta\mathbf{M}$$

and

$$W^* = \int_A^B \theta\cdot\mathrm{d}\mathbf{M} = \int_A^B (\theta_x\mathrm{d}M_x + \theta_y\mathrm{d}M_y + \theta_z\mathrm{d}M_z)\ . \tag{1.7.4}$$

For work calculations, the geometric constraints (compatibility conditions) are taken into account explicitly by the introduction of generalised (independently alterable) coordinates: the equilibrium equations are then consequences of the virtual work principle. In contrast, complementary work calculations are concerned with sets of forces in which the equilibrium conditions are explicitly taken into account and the idea of statically independent forces must be introduced: then, as we shall see in Section 1.8, the conditions of compatibility are consequences of the principle of virtual complementary work.

The plane system of Fig. 1.7.1 is in equilibrium under the set of forces F_{ix} etc. acting at points x_i, y_i,

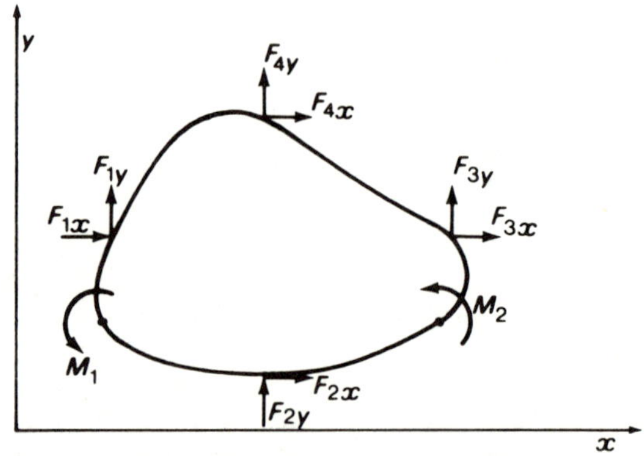

Fig. 1.7.1

For equilibrium

$$\Sigma F_x = \sum_{i=1}^{4} F_{ix} = 0$$

$$\Sigma F_y = \sum_{i=1}^{4} F_{iy} = 0$$

$$\Sigma M_0 = \sum_{i=1}^{4} (F_{iy}x_i - F_{ix}y_i) + M_1 + M_2 = 0 \ . \qquad (1.7.5)$$

With three independent equations relating ten quantities, three of the quantities are expressible in terms of the remaining seven: equilibrium is thus maintainable when seven of the forces are altered independently provided the remaining three are adjusted in accordance with equations (1.7.5). It is immaterial how the choice is made, but if some of the forces represent prescribed loads, they will naturally figure amongst the statically independent ones. If the number of reactive forces is greater than three, then some of these are also statically independent. We have previously described such reactions as redundant.

The examples of Section 1.4 showed that the analysis of complex systems requires the isolation of various sub-systems and a consideration of their equilibrium. The general three dimensional situation, then, is one in which we may be concerned with $3m$ forces, F_{ix}, F_{iy}, F_{iz} (some being the prescribed loads, some external reactions and some internal reactions) related by n independent equations of equilibrium. Of these forces, $r = (3m - n)$ will be statically independent and all the others will be expressible in terms of them by means of the n equations of equilibrium. If we denote the independent forces by $P_1, P_2, \ldots\ldots P_r$, then, we can write

$$\begin{aligned} F_{ix} &= F_i(P_1, P_2, \ldots\ldots P_r) \\ F_{iy} &= G_i(P_1, P_2, \ldots\ldots P_r) \qquad i = 1, 2, \ldots\ldots m \\ F_{iz} &= H_i(P_1, P_2, \ldots\ldots P_r) \;. \end{aligned} \tag{1.7.6}$$

When the P_j receive increments ΔP_j,

$$\Delta F_{ix} = \sum_{j=1}^{r} \frac{\partial F_i}{\partial P_j} \Delta P_j \qquad \text{etc.} \tag{1.7.7}$$

and the complementary work of all the force increments is

$$\Delta W^* = \sum_{i=1}^{m} (u_i \Delta F_{ix} + v_i \Delta F_{iy} + w_i \Delta F_{iz}) \tag{1.7.8}$$

or

$$\begin{aligned} \Delta W^* = \sum_{i=1}^{m} \Bigg[u_i \left(\sum_{j=1}^{r} \frac{\partial F_i}{\partial P_j} \Delta P_j \right) + v_i \left(\sum_{j=1}^{r} \frac{\partial G_i}{\partial P_j} \Delta P_j \right) \\ + w_i \left(\sum_{j=1}^{r} \frac{\partial H_i}{\partial P_j} \Delta P_j \right) \Bigg] \;. \end{aligned} \tag{1.7.9}$$

The summations can be carried out in any order so that

$$\Delta W^* = p_1 \Delta P_1 + p_2 \Delta P_2 + \ldots\ldots p_r \Delta P_r = \sum_{j=1}^{r} p_j \Delta P_j \tag{1.7.10}$$

where

$$p_j = \sum_{i=1}^{m} \left(u_i \frac{\partial F_i}{\partial P_j} + v_i \frac{\partial G_i}{\partial P_j} + w_i \frac{\partial H_i}{\partial P_j} \right) \;. \tag{1.7.11}$$

That is, the p_j are functions of the displacements and the rate of change of the F's with respect to the statically independent forces.

It may happen that ΔW^* is the exact differential of a function of the P_j which may be called the **complementary work function**. Integrals such as (1.7.3) are then independent of the load path and are functions of the end positions A and B alone: in particular, an integral over a closed path is zero.

Under these circumstances we can write

$$\Delta W^* = \mathrm{d}\phi^* = \sum_{j=1}^{r} \frac{\partial \phi^*}{\partial P_j} \Delta P_j \quad . \tag{1.7.12}$$

Comparing equation (1.7.10) and (1.7.12)

$$p_j = \frac{\partial \phi^*}{\partial P_j} \tag{1.7.13}$$

which is a quantity having the dimension of length or angle.

Now, we may work with the negative of ϕ^* and call it the complementary energy V^* by analogy with work function and potential energy.

Then
$$V^*(P_1, P_2, \ldots\ldots P_r) = -\phi^*(P_1, P_2, \ldots\ldots) \tag{1.7.14}$$

and
$$p_j = -\frac{\partial V^*}{\partial P_j} \quad . \tag{1.7.15}$$

When the system contains deformable members (elastic or not), the internal forces will perform complementary work and

$$W^* = W^*_{\text{ext}} + W^*_{\text{int}} \quad . \tag{1.7.16}$$

When both external and internal forces possess a complementary energy function, then we can write

$$V^* = U^* + \Omega^* \quad . \tag{1.7.17}$$

The form of U^*, called the complementary strain energy (or stress energy), determines the force-deformation characteristics of a deformable body and is given by experimental data.

Determining the complementary work of the internal forces in a general deformable solid must await the ideas of stress and strain in a continuum. However, for a simple one dimensional connexion between two particles, the necessary expression can be derived in an elementary manner analogous to that used for work in Section 1.5. Thus, referring to Fig. 1.5.2, if intermediate values of the displacements are $\mathbf{u}_1$ and $\mathbf{u}_2$,

$$\begin{aligned} \Delta W^*_{\text{int}} &= \mathbf{u}_1 \cdot \Delta \mathbf{F}_{12} + \mathbf{u}_2 \cdot \Delta \mathbf{F}_{21} \\ &= -(\mathbf{u}_2 - \mathbf{u}_1) \cdot \Delta \mathbf{F} \end{aligned}$$

or
$$\Delta W^*_{\text{int}} = -e\Delta F \tag{1.7.18}$$

where e is the intermediate *extension* of the link between the two particles. Expression (1.7.18) holds whether the link is elastic or not: however, if it does behave elastically, $\Delta W^*_{\text{int}} = -\Delta U^*$ where $U^* = U^*(F)$. Then,

$$-\Delta W^*_{\text{int}} = \Delta U^* = \frac{\mathrm{d}U^*}{\mathrm{d}F} \Delta F = e\Delta F \tag{1.7.19}$$

so that

$$\frac{dU^*}{dF} = e \ . \qquad (1.7.20)$$

The form of $U^*(F)$ thus determines the force-deformation characteristic of the link. When the extension increases from zero to some current value e_c corresponding to a current force F_c, the total complementary strain energy stored is

$$U^* = \int_0^{F_c} e\mathrm{d}F \ . \qquad (1.7.21)$$

By taking a zero e to correspond to a zero F, we can arbitrarily take our datum for U^* to be zero at zero F.† Fig. 1.5.3 is a plot of $e \sim F$ and we see that U^* is the area above the curve. If the link obeys Hooke's law, $F = ke$ and equation (1.7.21) yields

$$U^* = \frac{F^2}{2k} \qquad (1.7.22)$$

where we have dropped the subscript c on the understanding that F is merely some current value of force.

We observe that for the linear elastic case, U is *numerically* equal to U^*. Nevertheless, it is important to distinguish between the two quantities; U is properly a function of the elongation whilst U^* is properly a function of the internal force.

EXAMPLE 1.7.1

A rigid link, pinned to a rigid support at one end is in equilibrium in the configuration of Fig. 1.7.2. Find an expression for ΔW^* when increments ΔF_x, ΔF_y and ΔM are applied.

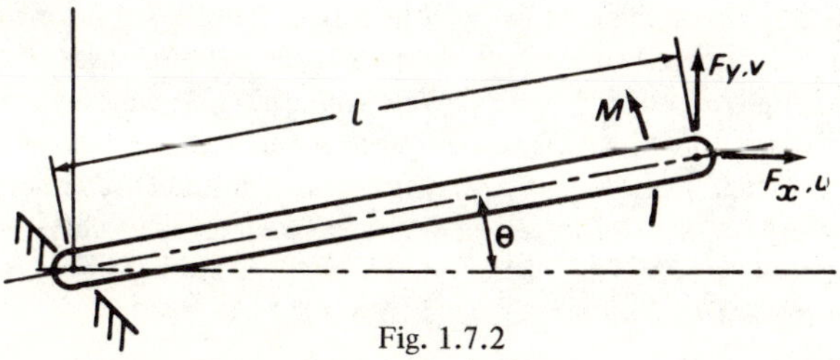

Fig. 1.7.2

For equilibrium in the displaced configuration, the datum position being $\theta = 0$,

$$M + F_y(l + u) = F_x v \ . \qquad \text{(a)}$$

Any increments must also be in equilibrium,

$$\therefore \Delta M + \Delta F_y(l + u) = \Delta F_x v \ . \qquad \text{(b)}$$

† Thus, we do not take into account any residual stresses locked in the system during manufacture.

Now
$$\Delta W^* = u\Delta F_x + v\Delta F_y + \theta\Delta M \ . \tag{c}$$

Regarding F_x and F_y as the statically independent quantities, equation (b) in (c) yields
$$\Delta W^* = (u + \theta v)\Delta F_x + [v - (l + u)\theta]\Delta F_y \tag{d}$$

1.8 Principles of virtual complementary work and stationary complementary energy

In Section 1.7 we saw that from the point of view of satisfying the requirements of equilibrium alone it is possible to regard some of the forces acting on a system (or sub-system) as independent. Any one of these selected forces can be altered without reference to the others but in such an event, the remaining (dependent) forces must of course change appropriately in order that the full set shall represent an equilibrium situation. Furthermore, if such a load change actually occurred, we would expect the configuration to alter in some manner compatible with the constraints. Up to now we have regarded the compatibility conditions to be derivable purely from geometric considerations: however, they may be obtained in an indirect manner by an approach which has especial significance for statically indeterminate deformable systems. In such systems, some of the forces which can be regarded as statically independent are not prescribed but are the forces induced at redundant reactions or in redundant members. The procedure is to carry out a mathematical experiment, analogous to that visualised for virtual work, which yields compatibility equations in terms of redundant forces.

A mechanical system is supposed to have a prescribed configuration imposed upon it (some displacement from its as-manufactured state). Then, several sets of forces, each consisting of applied loads and induced reactions and which are acceptable for equilibrium alone, are examined in turn and the true one, which produces the desired configuration, is singled out according to some criterion. The range of *admissible* force patterns is restricted to the neighbourhood of the true one so that they are arrived at by infinitesimal changes from, or *variations* to the actual forces: our datum for test changes is thus the *true* set of forces corresponding to the *true* configuration. We emphasise that such variations must constitute equilibriant infinitesimal force systems, but that otherwise they are quite arbitrary and need not bear any relation to any actual force change which may be contemplated: the configuration stays fixed at the true state whilst these force variations occur. Analytically, such changes are brought about by variations δP_j in the *statically independent* forces P_j.

The imaginary arbitrary infinitesimal changes of forces δP_j which do not violate equilibrium and which need not correspond to any actual changes ΔP_j called for by any actual change of configuration are called **virtual forces.**

From the definition of complementary work it follows that, with p_j of equation (1.7.10) corresponding to the actual displacements and the ΔP_j replaced by δP_j, we have

$$\delta W^* = p_1\delta P_1 + p_2\delta P_2 + \ldots\ldots = \sum_{j=1}^{r} p_j\delta P_j \tag{1.8.1}$$

as the *virtual* complementary work of the *virtual* forces through the *actual* displacements.

We have previously observed that the reactions at rigid constraints have no complementary work so that they do not figure in equation (1.8.1): however, there is a contribution from the virtual internal forces induced in deformable members.

When carrying out the imaginary comparison process above, the true set of forces is identified by the **principle of virtual complementary work** which states that *a mechanical system is in a compatible state of deformation if and only if the complementary virtual work of any virtual forces through the displacements is zero.*

That is, for compatibility

$$\delta W^* = \sum_{j=1}^{r} p_j \delta P_j = 0 \; . \tag{1.8.2}$$

But since the δP_j are arbitrary, equation 1.8.2 implies

$$p_j = 0 \text{ for compatibility} \; . \tag{1.8.3}$$

These conditions are true for any mechanical system, but when the members are elastic (linear or not), $\delta W^*_{\mathbf{int}} = -\delta U^*$ and equation (1.8.2) becomes

$$\delta W^*_{\mathbf{ext}} = \delta U^* \; . \tag{1.8.4}$$

If the external forces have a complementary potential Ω^*, $\delta W^*_{\mathbf{ext}} = -\delta\Omega^*$ and the principle of complementary virtual work becomes

$$\delta V^* = 0 \; . \tag{1.8.5}$$

This is the principle of stationary total complementary energy which asserts that V^* is stationary when the displacements corresponding to an equilibrium force system are compatible.

With $V^* = V^*(P_1, P_2 \ldots\ldots P_r)$, equation (1.8.5) is equivalent to

$$\frac{\partial V^*}{\partial P_j} = 0 \, , \quad j = 1, 2, \ldots\ldots r \tag{1.8.6}$$

which represents r equations of compatibility.

In practical applications some of the statically independent forces may actually be *prescribed* loads. If $P_k \ldots\ldots P_r$ are such forces, then $P_1 \ldots\ldots P_{k-1}$ must be redundant reactions, either internal to or on boundary points of the system. Since equilibrium requirements must be satisfied by all trial sets of virtual forces, $\delta P_k \ldots\ldots \delta P_r$ can only be non-zero if actual increments $\Delta P_k \ldots\ldots \Delta P_r$ occur: if actual increments are not contemplated, $\delta P_k = O, \ldots\ldots \delta P_r = O$ and the virtual complementary work principle provides a means of determining the redundant reactions. Suppose further that $P_1, P_2 \ldots\ldots P_g$, $g < k-1$ are external redundants, that is, they are the as yet unkown external forces called for at boundary points where displacements are prescribed. Then for variations in such forces, the expression (1.7.9) for complementary work degenerates to

$$\delta W^*_{e_u} = -\delta\Omega^*_u = \left(\sum_{i=1}^{g} u_i \delta P_i\right)_u . \tag{1.8.7}$$

Here, u_i is a prescribed displacement corresponding to δP_i and the subscript u emphasises that it is the displacement and not the force which is prescribed.

Equation (1.8.4) now becomes

$$\delta U^* - \left(\sum_{i=1}^{g} u_i \delta P_i\right)_u = 0 \tag{1.8.8}$$

and since the u_i remain fixed as the δP_i are imposed, this can be written as

$$\delta\left[U^* - \left(\sum_{j=1}^{g} u_i P_i\right)_u\right] = 0 = \delta V^* . \tag{1.8.9}$$

An important special case is that in which the boundary points are rigidly supported; the $(u_i)_u$ are now zero and equation (1.8.9) reduces to

$$\delta U^* = 0 . \tag{1.8.10}$$

For a linear elastic system U^* is numerically equal to U and (1.8.10) is a special form of **Castigliano's** famous **theorem of least work** to which we shall return in due course.

Once the redundant forces have been determined, all the forces at all the reactions and in all members are expressible in terms of the prescribed loads and they will have the true values corresponding to compatible displacements. In particular U^* will now be a function of these prescribed loads and the principle of complementary virtual work can be used to calculate the actual displacement at a selected station.

Thus with $U^* = U^*(P_k, \ldots\ldots P_r)$ alone,

$$\Delta U^* = \sum_{j=k}^{r} \frac{\partial U^*}{\partial P_j} \Delta P_j , \tag{1.8.11}$$

if actual increments $\Delta P_k \ldots\ldots \Delta P_r$ are imposed.

From the principle of virtual complementary work with $\delta P_k = \Delta P_k$ etc., we have, when the supports are rigid,

$$u_k \Delta P_k + \ldots\ldots u_r \Delta P_r = \frac{\partial U^*}{\partial P_k} \Delta P_k + \ldots\ldots \frac{\partial U^*}{\partial P_r} \Delta P_r$$

or

$$\sum_{j=k}^{r} \left(u_j - \frac{\partial U^*}{\partial P_j}\right) \Delta P_j = 0 . \tag{1.8.12}$$

But the ΔP_j are arbitrary and independent so that

$$u_j = \frac{\partial U^*}{\partial P_j}, \quad j = k, \ldots\ldots r \,. \tag{1.8.13}$$

That is, since all these calculations are based on complementary *work*, u_j is the displacement at P_j in the direction of P_j. For linear elastic systems $U^* = U$ and equation (1.8.13) is **Castigliano's theorem on deflexions.**

It is important to observe that in using equation (1.8.13):

(i) All redundant reactions (internal and external) have already been determined in terms of the prescribed loads;

(ii) There are no displacements at the supports.

The principle of virtual complementary work is applicable to elastic or inelastic systems, but it is restricted to situations in which the displacements from the datum to the equilibrium configuration are small. In this respect, it is not as powerful a tool as the virtual work principle which has been used very successfully in problems of large deflexions in structures where questions of stability of equilibrium arise. However, we may use virtual work arguments to show that the *virtual complementary work principle is a valid one for small displacements.*

First recall the illustration of Example 1.4.3 that for small displacements, the equilibrium relations for the deformed configuration are the same as for the datum position, that is say, there is no geometric non-linearity. Now, any set of virtual forces constitutes an equilibrium set so that if they are regarded as moving through the actual small displacements of the system, the principle of virtual work applies. But this choice of forces and displacements also defines complementary work, and the principle of virtual complementary work follows from the principle of virtual work.

EXAMPLE 1.8.1

A rigid link, pinned to a rigid support is in equilibrium as shown in Fig. 1.8.1. What information is given by $\delta W^* = 0$?

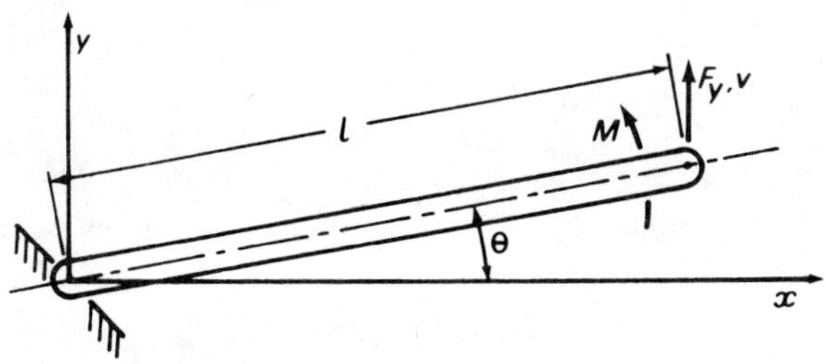

Fig. 1.8.1

$$\delta W^* = \theta\delta M + v\delta F_y \,. \tag{a}$$

But for equilibrium,

$$M + F_y l \cos\theta = 0\,, \tag{b}$$

and

$$\delta M + \delta F_y l \cos\theta = 0\ . \tag{c}$$

That is, we have one equation of equilibrium and so one dependent quantity. Substituting from (c) in (a), regarding F_y as the statically independent quantity,

$$\delta W^* = (-\theta l \cos\theta + v)\delta F_y = 0\ . \tag{d}$$

But δF_y is arbitrary so that the term in parentheses must be zero. The geometry of Fig. 1.8.1 shows that this is not generally true. However, if θ is small, $\cos\theta \doteqdot 1$ and $v = l\theta$. Thus, the principle of virtual complementary work has lead to a valid equation of geometry when displacements are small.

EXAMPLE 1.8.2

The ends of a rigid bell crank displace as shown in Fig. 1.8.2 under the action of the forces F_{1x} . . . . . and a torsion spring. The crank is pinned at the support and the spring provides a restraining moment $M = k\theta^n$. What does $\delta W^* = 0$ imply?

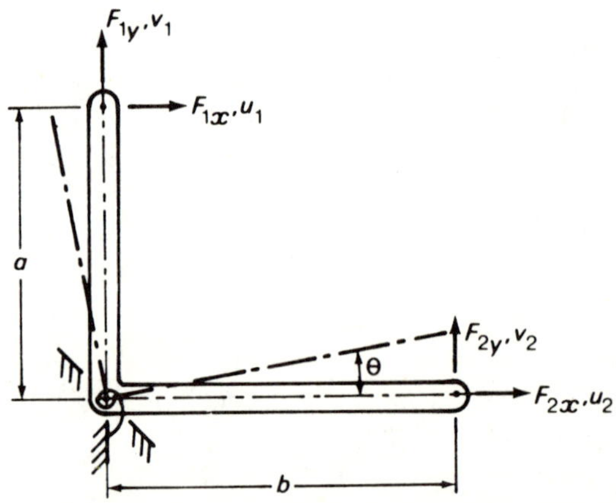

Fig. 1.8.2

$$\delta W^* = u_1\delta F_{1x} + v_1\delta F_{1y} + u_2\delta F_{2x} + v_2\delta F_{2y} - \theta\delta M\ . \tag{a}$$

For equilibrium in the displaced configuration, there is one equation relating δF_{1x} . . . . . δM,†

$$\delta F_{1y}u_1 - \delta F_{1x}(a + v_1) + \delta F_{2y}(b + u_2) - \delta F_{2x}v_2 - \delta M = 0 \tag{b}$$

† The reader should also consider allowing translation at the pin. The pin reactions would then enter into the discussion and equilibrium of forces would need to be taken into account in addition to equation (b).

Treating δM as the dependent quantity, (this choice is quite arbitrary) and substituting from (b) in (a)

$$\delta W^* = \delta F_{1x}[u_1 + \theta(a + v_1)] + \delta F_{1y}[v_1 - \theta u_1] + \delta F_{2x}[u_2 + \theta v_2] + \delta F_{2y}[v_2 - \theta(b + u_2)] = 0 \quad \text{(c)}$$

or $$\delta W^* = p_1\delta P_1 + p_2\delta P_2 + p_2\delta P_3 + p_4\delta P_4 = 0 \quad \text{(d)}$$

where $\delta P_1 = \delta F_{1x}$ etc. Then, since $\delta P_1 \ldots\ldots$ are arbitrary, $p_1 = 0 \ldots\ldots$ and again provided displacements are small, so that terms such as θv_1 are of second order, equation (c) yields

$$\begin{aligned} u_1 &= -a\theta\,, & v_1 &= 0 \\ u_2 &= 0 & v_2 &= b\theta \end{aligned} \quad \text{to first order} \quad \text{(e)}$$

which are valid equations of geometry.

We observe:

(i) The system is statically determinate and had $F_{1x} \ldots\ldots F_{2y}$ been *prescribed* instead of $u_1 \ldots\ldots v_2$, then $\delta M = 0$ from equation (b). This corresponds to the fact that the equilibrium force distribution in a statically determinate system is unique.

(ii) The non-linear characteristic of the spring did not pose any special difficulty.

EXAMPLE 1.8.3

A light rigid bar supported by three linear springs has small displacements u, θ prescribed. The force P and couple M of Fig. 1.8.3 are the means for obtaining u and θ. What information does $\delta W^* = 0$ provide?

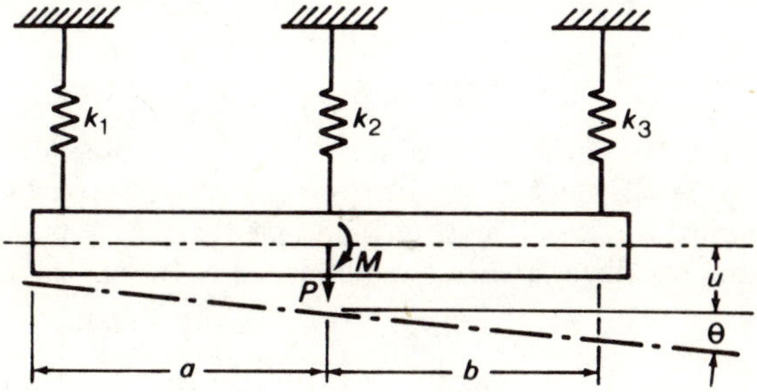

Fig. 1.8.3

Let the spring forces and extensions be F_1, F_2, F_3 and e_1, e_2, e_3, then

$$\delta W^* = -e_1\delta F_1 - e_2\delta F_2 - e_3\delta F_3 + u\delta P + \theta\delta M \quad \text{(a)}$$

But for equilibrium of the bar

$$F_1 + F_2 + F_3 - P = 0 \tag{b}$$

$$F_1 a - F_3 b + M = 0 \tag{c}$$

Two equations imply two statically dependent quantities. We will arbitrarily choose M, P and F_3 tb be statically independent. Then, equation (a) becomes

$$\delta W^* = 0 = \delta P(u - e_2) + \delta M\left(\theta - \frac{e_2 - e_1}{a}\right)$$

$$+ \delta F_3\left[e_3 - e_2 - \frac{b}{a}\left(e_2 - e_1\right)\right] . \tag{d}$$

Since δP, δM, δF, are arbitrary $\delta W^* = 0$ yields valid equations of geometry as before.

That is

$$e_1 = u - a\theta$$
$$e_2 = u$$
$$e_3 = u + b\theta . \tag{e}$$

Using Hooke's law together with equations (e), we find the induced spring forces. Using these forces in equations (b) and (c) we find the M and P called for to maintain the prescribed configuration.

$$F_1 = k_1(u - a\theta), \quad F_2 = k_2 u, \quad F_3 = k_3(u + b\theta)$$

$$P = (k_1 + k_2 + k_3)u + (bk_3 - ak_1)\theta ,$$

$$M = (bk_3 - ak_1)u + (a^2k_1 + b^2k_3)\theta .$$

EXAMPLE 1.8.4

Interpret Example 1.8.3 differently by supposing that M and P are *prescribed* so that u and θ are as yet unknown.

Again for equilibrium

$$F_1 + F_2 + F_3 = P \tag{a}$$

$$F_3 b - F_1 a = M . \tag{b}$$

Clearly P and M must be chosen as two of the statically independent quantities. We will take F_3 to be the third. If we form a set of virtual forces by taking variations in all the actual forces, we must now observe that $\delta P = 0$, $\delta M = 0$ since P and M are prescribed. Then, for equilibrium of the virtual forces,

$$\delta F_1 + \delta F_2 + \delta F_3 = 0 \tag{c}$$

$$b\delta F_3 - a\delta F_1 = 0 . \tag{d}$$

That is, the virtual forces constitute a set of *self equilibrating* forces independent of the applied loads. Then, choosing δF_3 as the independent force,

$$\delta W^* = \left[\frac{b(e_2 - e_1) - a(e_3 - e_2)}{a}\right]\delta F_3 . \tag{e}$$

But if $\delta W^* = 0$ and δF_3 arbitrary,

$$b(e_2 - e_1) = a(e_3 - e_2) \tag{f}$$

which is again a valid equation of geometry. However, if we use Hooke's law to express equation (f) in terms of the spring forces.

$$b\left(\frac{F_2}{k_2} - \frac{F_1}{k_1}\right) = a\left(\frac{F_3}{k_2} - \frac{F_2}{k_2}\right) \tag{g}$$

but these forces are constrained by the equilibrium equations (a) and (b) so with F_1 chosen as the redundant, we find

$$F_1\left[\frac{1}{k_1} + \frac{1}{k_2}\left(\frac{a+b}{b}\right)^2 + \frac{1}{k_3}\left(\frac{a}{b}\right)^2\right] = \frac{a+b}{bk_2}\left(P - \frac{M}{b}\right) - \frac{M_a}{k_3 b^2} \, . \tag{h}$$

In the special case where $b = a,\ k_1 = k_2 = k_3 = k$,

$$F_1 = P/3 - M/2a \, . \tag{j}$$

Substituting in (a) and (b) gives

$$F_2 = P/3 \, , \quad F_3 = P/3 + M/2a \, . \tag{k}$$

Thus, expressing the equations of compatibility in terms of the redundant force and prescribed loads yields the former. Subsequent use of the equilibrium equations determines the remaining spring forces and employing Hooke's law yields the displacements if they are required. The displacements corresponding to P and M may also be obtained by equation (1.8.13).

Now
$$U^* = \sum_{i=1}^{3} \frac{F_i^2}{2k_i} \tag{l}$$

and using results (j) and (k)

$$U^* = \frac{1}{2k}\left[\frac{P^2}{3} + \frac{M^2}{2a^2}\right] = U^*(P, M) \tag{m}$$

$$u = \frac{\partial U^*}{\partial P} = \frac{P}{3k} \, , \quad \theta = \frac{\partial U^*}{\partial M} = \frac{M}{2ka^2} \, . \tag{n}$$

It should be observed that the absence of a coupling term in MP in equation (m) is due solely to the particular characteristics of the system. In general, such a term will be present as in the example below.

EXAMPLE 1.8.5

The light rigid bar of Fig. 1.8.4 is pinned to a rigid support at the left hand end. Additional support is provided by springs k_1 and k_2 and the torsion spring k_3. Determine equations for the redundants.

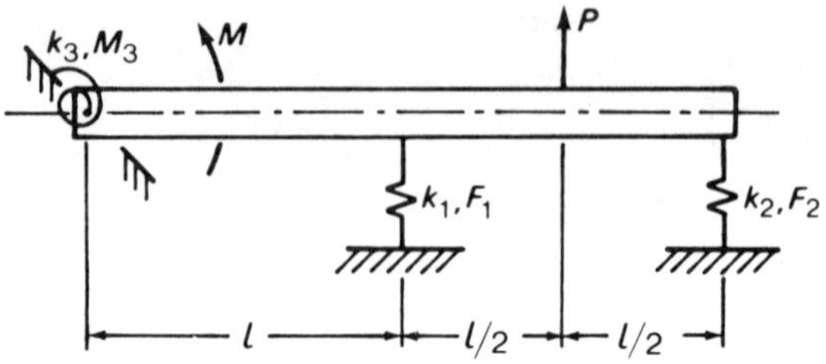

Fig. 1.8.4

All prescribed displacements are zero; the $(\Sigma \ldots\ldots)_u$ term in equation (1.8.9) is thus zero and

$$\delta V^* = \delta U^* = 0 , \tag{a}$$

Now
$$U^* = \frac{F_1^2}{2k_1} + \frac{F_2^2}{2k_2} + \frac{M_3^2}{2k_3} . \tag{b}$$

But, for equilibrium, F_1, F_2, M_3 are related by

$$M + \frac{3Pl}{2} - M_3 - F_1 l - 2F_2 l = 0 . \tag{c}$$

That is, one equation so that there must be one statically dependent qunatity. We (arbitrarily) choose M_3. Substituting for M_3 from (c) in (b) yields

$$U^* = U^*(F_1, F_2, P, M) . \tag{d}$$

Equation (a) for compatibility thus requires

$$\frac{\partial U^*}{\partial F_1} = 0 , \quad \frac{\partial U^*}{\partial F_2} = 0 . \tag{e}$$

Rather than determining expression (d) explicitly and then differentiating we may save a certain manipulative effort if we proceed thus:

$$\frac{\partial U^*}{\partial F_1} = \frac{F_1}{k_1} + \frac{M_3}{k_3}\frac{\partial M_3}{\partial F_1} = 0 \tag{f}$$

$$\frac{\partial U^*}{\partial F_2} = \frac{F_2}{k_2} + \frac{M_3}{k_3}\frac{\partial M_3}{\partial F_2} = 0 . \tag{g}$$

Using equations (c) in (f) and (g)

$$\left(\frac{1}{k_1} + \frac{l^2}{k_3}\right)F_1 + \frac{2l^2}{k_3}F_2 = \frac{l}{k_3}\left(M + \frac{3Pl}{2}\right) \tag{h}$$

$$\frac{2l^2}{k_3}F_1 + \left(\frac{1}{k_2} + \frac{4l^2}{k_3}\right)F_2 = \frac{2l}{k_3}\left(M + \frac{3Pl}{2}\right) . \tag{j}$$

These can be written in matrix notation as

$$[H]\{F\} = \{u\} \tag{k}$$

This is a matrix compatibility equation in terms of the redundants F_1 and F_2. $[H]$ is called the **flexibility matrix** of the system and, as demonstrated here, it is symmetric. The elements H_{ij} are the **flexibility influence coefficients** and represent the displacement at i due to a unit load at j.

Problems 1.8

1.8.1 For the system of Problem 1.4.3, choose force F_1 in spring 1 as the redundant quantity. Use the complementary energy principle to deduce an expression for F_1 and hence find F_2, F_3, F_4. Check the calculations by establishing whether the spring elongations due to these forces are all compatible.

1.8.2 Use the complementary energy principle to determine the spring forces in the system of Problem 1.4.4. Check the calculations using the condition of compatibility of the spring elongations.

1.8.3 Use the principle of virtual complementary work to establish valid equations of geometric fit for the system of Problem 1.6.4 when displacements are small.

1.8.4 Displacement components u, v, w in Problem 1.4.3 are supposed prescribed. Use the complementary energy principle to deduce the required forces P_x, P_y, P_z. Compare the results, when $k_1 = k_2 = k_3 = k_4$. With those obtained in Problem 1.4.3.

1.9 Stability of equilibrium

In the previous sections we examined mechanical systems occupying their equilibrium configuration but did not consider what may be termed the quality of their equilibrium. Now it is a matter of experience that when some systems are slightly disturbed from their equilibrium position, they tend to return to it when the disturbance is removed whilst others tend to depart even further from it. Clearly, if a loaded structure is in the latter condition, it is potentially dangerous so that a designer must guard against it. Problems relating to the stability of structures are of considerable practical significance but a full account of the subject is beyond the scope of this book. However, this first chapter would not be complete without some reference to the question of the stability of equilibrium.

A mechanical system is said to be in a **stable state of equilibrium** if, after the removal of some slight disturbance, it tends to return to its equilibrium configuration. If the slight disturbance results in the system departing even further from the equilibrium configuration, then it is **unstable**. One can conceive of an intermediate situation in which the slightly disturbed configuration is maintained when the disturbance is removed and we then have a state of **neutral** equilibrium. Evidently, the slight displacements contemplated here must be in accord with the geometric constraints so that they correspond to slight changes in the generalised coordinates of the system: a discussion of the stability of equilibrium can thus be based on virtual displacements.

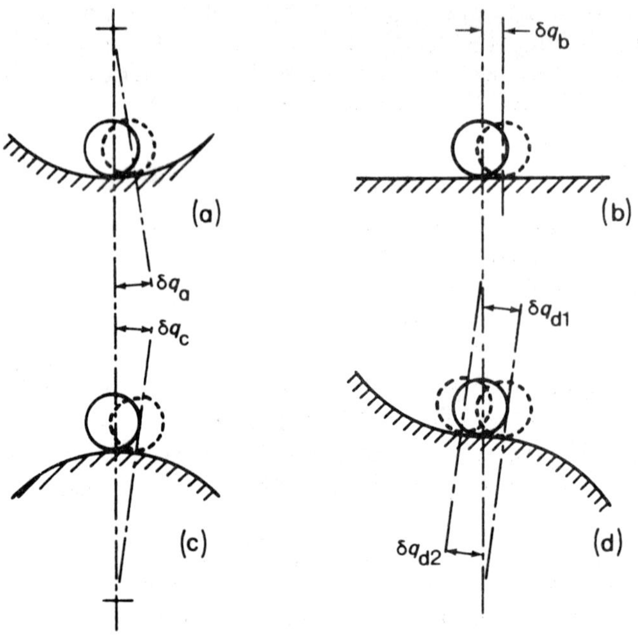

Fig. 1.9.1

To help fix ideas, the rigid sphere is in equilibrium on the rigid surfaces in each of the situations of Fig. 1.9.1(a) . . . (d). We readily see that if the ball bearing in the saucer of Fig. 1.9.1(a) is slightly disturbed through δq_a, it will tend to roll back down to the bottom of the saucer. In Fig. 1.9.1(b), the ball will stay in its new position whilst in Fig. 1.9.1(c) it will roll right off the inverted saucer once it is slightly disturbed. The situation in Fig. 1.9.1(d) is also unstable since a disturbance δq_{d1} will cause a runaway situation even though the system would tend to return to the equilibrium configuration after a disturbance δq_{d2}. The last example guides us to strengthen our statement on stability: a system is in a stable state of equilibrium if it tends to return to it from *any* geometrically admissible small disturbance.

Fig. 1.9.1 also illustrates how energy considerations describe the state of equilibrium. Thus the potential energy of the ball is increased as it climbs up the side of the saucer through δq_a: the ball being in contact with the surface of the saucer throughout makes the disturbance geometrically admissible. The principle of virtual work shows that the potential energy is stationary at equilibrium and we now see that it is a *minimum* when the equilibrium is stable. Conversely the potential energy is a maximum in the unstable configuration of Fig. 1.9.1(c), whilst in Fig. 1.9.1(d) it is neither a maximum nor a minimum and in Fig. 1.9.2 it is constant. We see that, because the potential energy is stationary at equilibrium, a discussion of stability involves a discussion of the higher order terms appearing in the increment of potential energy consequent upon a virtual displacement from the equilibrium configuration.

EXAMPLE 1.9.1

Discuss the equilibrium of the heavy rigid link of Fig. 1.9.2

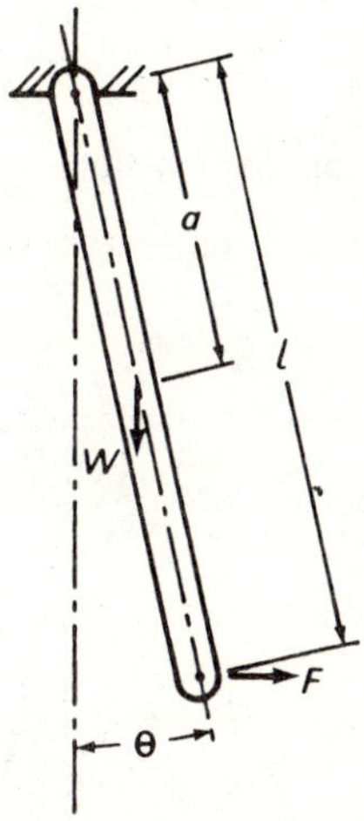

Fig. 1.9.2

The system has one degree of freedom and we choose θ as the generalized coordinate. The potential energy is

$$V = Wa(1 - \cos\theta) - Fl\sin\theta = V(\theta)\ . \tag{a}$$

To determine the equilibrium configuration and to ascertain its stability, let the system receive a virtual displacement $\delta\theta$, then the increment (total variation) in V is given by the Taylor expression

$$\Delta V = \frac{\mathrm{d}V}{\mathrm{d}\theta}\delta\theta + \frac{1}{2!}\left[\frac{\mathrm{d}^2 V}{\mathrm{d}\theta^2}(\delta\theta)^2\right] + \frac{1}{3!}\left[\frac{\mathrm{d}^3 V}{\mathrm{d}\theta^3}(\delta\theta)^3\right] + \ldots \tag{b}$$

or

$$\Delta V = \delta V + \frac{1}{2!}\delta^2 V + \frac{1}{3!}\delta^3 V + \ldots \tag{c}$$

where δV, $\delta^2 V$, $\delta^3 V$ etc, are the first, second and third etc. variations in V.

Now, the equilibrium configuration is defined by $\delta V = 0$, then, neglecting third order quantities,

$$\Delta V = \tfrac{1}{2}\delta^2 V\ . \tag{d}$$

Since $(\delta\theta)^2$ is essentially positive, the sign of ΔV is determined by the sign of $\mathrm{d}^2 V/\mathrm{d}\theta^2$; if it is positive ΔV is positive and the equilibrium is stable, otherwise it is unstable. Here

$$\frac{\mathrm{d}V}{\mathrm{d}\theta} = Wa\sin\theta - Fl\cos\theta \tag{e}$$

$$\frac{d^2 V}{d\theta^2} = Wa \cos\theta + Fl \sin\theta \tag{f}$$

so that for the equilibrium

$$\tan\theta = Fl/Wa \tag{g}$$

or

$$\theta = \theta_0 \text{ or } (\theta_0 + \pi) \tag{h}$$

with θ_0 as the principal value $\tan^{-1}(Fl/Wa)$.

The two possible equilibrium configurations are shown in Fig. 1.9.3(a) and (b)

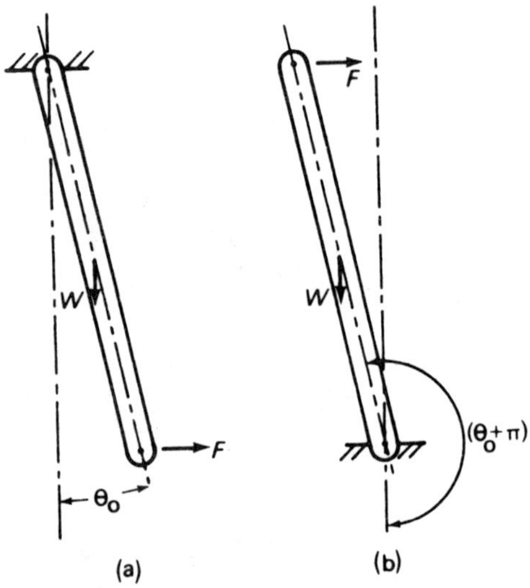

Fig. 1.9.3

For $\theta = \theta_0$,

$$\frac{d^2 V}{d\theta^2} = \sqrt{F^2 l^2 + W^2 a^2} \tag{j}$$

This is essentially positive so that the position is stable.

For $\theta = (\theta_0 + \pi)$,

$$\frac{d^2 V}{d\theta^2} = -\sqrt{F^2 l^2 + W^2 a^2}\ . \tag{k}$$

This is essentially negative so that the position is unstable.

The reader should examine Fig. 1.9.3 and verify that a slight disturbance of $\pm\delta\theta$ from θ_0 or $(\theta_0 + \pi)$ leads to an unbalanced restoring or upsetting moment respectively.

In the above example we had the more usual situation in which the second derivative of V was non-zero. When this is not the case, the higher order derivatives must be examined: if the lowest order remaining derivative is of even order, then according as it is positive or negative equilibrium is stable or unstable; if it is of odd order, a situation characterised by Fig. 1.9.1(d) obtains, and we again have instability.

For a system having many degrees of freedom, the potential energy is given by

$$V = V(q_1, q_2, \ldots \ldots q_n) \ . \tag{1.9.1}$$

Taylor's expansion gives the increment in V corresponding to a virtual displacement $\delta q_1, \delta q_2, \ldots\ldots$ as

$$\Delta V = \sum_{i=1}^{n} \frac{\partial V}{\partial q_i}\delta q_i + \frac{1}{2!}\sum_{i}^{n}\sum_{j}^{n}\frac{\partial^2 V}{\partial q_i \partial q_j}\delta q_i \delta q_j + \ldots\ldots \tag{1.9.2}$$

or

$$\Delta V = \delta V + \frac{1}{2!}\delta^2 V + R \ . \tag{1.9.3}$$

R is a third order small quantity. Now $\delta V = 0$ at equilibrium so that the sign of ΔV is governed by the sign of $\delta^2 V$. If we denote

$$a_{ij} = \frac{\partial^2 V}{\partial q_i \partial q_j} \tag{1.9.4}$$

$$\delta^2 V = \sum_i \sum_j a_{ij}\delta q_i \delta q_j \tag{1.9.5}$$

and introducing the matrix $[a]$ of the coefficients a_{ij} equation (1.9.5) can be written as

$$\delta^2 V = \{\delta q\}^t [a] \{\delta q\} \ . \tag{1.9.6}$$

For stability $\delta^2 V$ must be positive and this requires that $[a]$ be a positive definite matrix†. The coefficients a_{ij} are functions of the applied loads and the properties of the system so that positive definiteness of $[a]$ imposes the conditions which the loads must satisfy in order that the configuration be stable.

EXAMPLE 1.9.2

Two uniform rigid links, each of length l and weight W are pinned and supported in an upright position by the springs as shown in Fig. 1.9.4. When the links are vertical, there is just no force in any spring. Under what conditions is the vertical configuration stable?

The system has two degrees of freedom, θ_1 and θ_2 serve as generalised coordinates. We will assume θ_1 and θ_2 are sufficiently small in the test position so that the springs remain essentially horizontal. Then,

$$V = k_1(l \sin\theta_1)^2 + k_2(l\sin\theta_1 + l\sin\theta_2)^2 + \frac{Wl}{2}\cos\theta_1 + Wl\cos\theta_1 + \frac{l}{2}\cos\theta_2 \ . \tag{a}$$

† $[a]$ is positive definite if the determinant of $[a]$ and all its principal minors are positive. See Appendix.

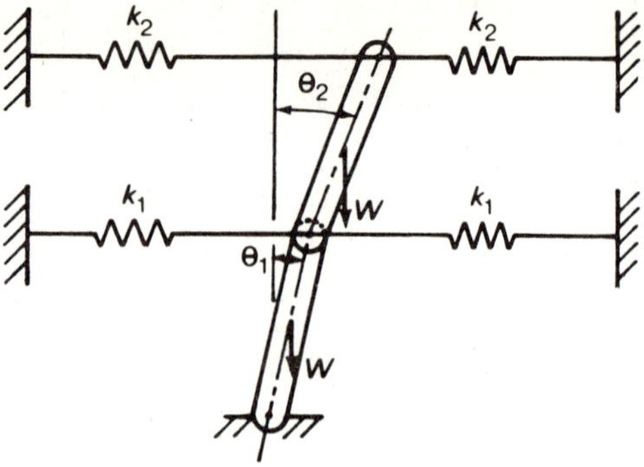

Fig. 1.9.4

The required derivatives are

$$\frac{\partial V}{\partial \theta_1} = (k_1 + k_2)l^2 \sin 2\theta_1 + 2k_2 l^2 \sin \theta_2 \cos \theta_1 - \frac{3Wl}{2} \sin \theta_1 \quad \text{(b)}$$

$$\frac{\partial^2 V}{\partial {\theta_1}^2} = 2(k_1 + k_2)l^2 \cos 2\theta_1 - 2k_2 l^2 \sin \theta_1 \sin \theta_2 - \frac{3Wl}{2} \cos \theta_1 \quad \text{(c)}$$

$$\frac{\partial V}{\partial \theta_2} = 2k_2 l^2 \sin \theta_1 \cos \theta_2 + k_2 l^2 \sin 2\theta_2 - \frac{Wl}{2} \sin \theta_2 \quad \text{(d)}$$

$$\frac{\partial^2 V}{\partial {\theta_2}^2} = -2k_2 l^2 \sin \theta_1 \sin \theta_2 + 2k_2 l^2 \cos 2\theta_2 - \frac{Wl}{2} \cos \theta_2 \quad \text{(e)}$$

$$\frac{\partial^2 V}{\partial \theta_1 \partial \theta_2} = 2k_2 l^2 \cos \theta_1 \cos \theta_2 \ . \quad \text{(f)}$$

For equilibrium, $\partial V/\partial\theta_1 = \partial V/\partial\theta_2 = 0$ and equations (b) and (d) show that $\theta_1 = \theta_2 = 0$ satisfy these requirements. To test the stability of this configuration we must generate matrix $[a]$ when $\theta_1 = \theta_2 = 0$. Then,

$$[a] = \begin{bmatrix} 2(k_1 + k_2)l^2 - \dfrac{3Wl}{2} & 2k_2 l^2 \\ 2k_2 l^2 & 2k_2 l^2 - \dfrac{Wl}{2} \end{bmatrix} \quad \text{(g)}$$

To simplify subsequent manipulation we will suppose $k_1 = k_2 = k/2$ and adopt the notation $r = kl/W$. Then one condition $|a| > 0$ for stability becomes

$$f(r) = 4r^3 - 10r + 3 > 0 \ . \quad \text{(h)}$$

If we plot $f(r) \sim r$ we have Fig. 1.9.5

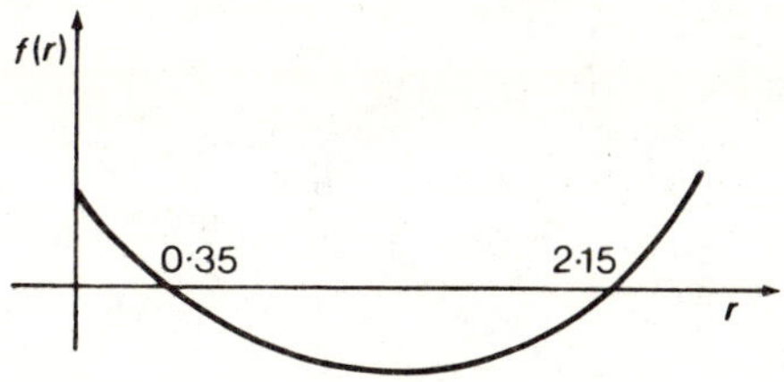

Fig. 1.9.5

At first sight it appears that the system is stable provided $r < 0.35$ or $r > 2.15$. However, examination of the signs of the principal minors of $[a]$ shows that they are negative if $r < 0.35$ so that we only have a stable situation for $r > 2.15$.

Problem 1.9

1.9.1 Fig. 1.9.6 shows a simple laboratory device for demonstrating instability. Show that if equilibrium is to be stable when the rod is vertical the limiting value of h is $2ka^2/mg$. The masses of the rod and base are negligible and the springs are precompressed equally. Each spring has stiffness k.

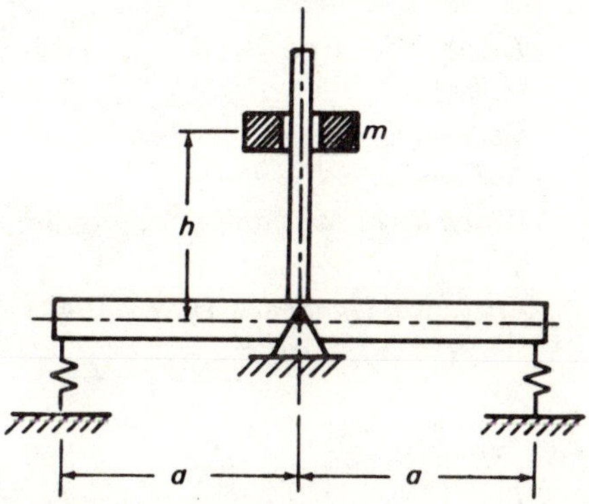

Fig. 1.9.6

1.9.2 Buckling of elastic columns is simulated by the system of Fig. 1.9.7. Each bar, of length l, is light and rigid and the two torsion springs, which are unstrained when the bars are horizontal, have rotational stiffness k Nm/radian. Discuss stability of equilibrium.

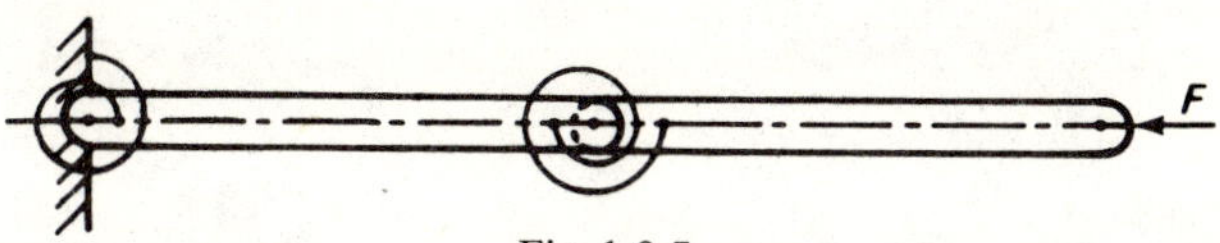

Fig. 1.9.7

Ans: $$[a] = \begin{bmatrix} (2k - Fl) & -k \\ -k & (k - Fl) \end{bmatrix} .$$

1.9.3 Solve Problem 1.9.1 if the springs have non-linear characteristics $F = k \sin e$ and there is no pre-compression nor back-lash in the system. e is the spring extension and k is a constant.

1.10 Further Reading

Charlton, T.M. — *Energy Principles in Applied Statics,* Blackie (1959)

Energy Principles in Theory of Structures, Oxford University Press, 1973

Gregory, M.S. — *An Introduction to Extremum Principles,* Butterworth (1969)

Lamb, H. — *Higher Mechanics,* Cambridge University Press (1920)

Lanczos, C. — *The Variational Principles of Mechanics,* 4th Edition, University of Toronto Press, (1970)

Langhaar, H.L. — *Energy Methods in Applied Mechanics,* John Wiley and Sons (1962)

Timoshenko, S. — *The History of Strength of Materials,* McGraw-Hill (1953)

Timoshenko, S. *and* Gere, J.M. — *Theory of Elastic Stability,* 2nd Edition, McGraw-Hill (1961)

Sommerfield, A. — Mechanics: *Lectures on Theoretical Physics,* Vol. 1, Academic Press (1964)

Synge, L. *and* Griffiths, B.A. — *Principles of Mechanics,* 3rd Edition, McGraw-Hill (1959)

Whittaker, E.T. — *Analytical Dynamics.* Dover (1944) and C.U.P. (1904)

Chapter 2

Bar Type Systems

2.1 Introduction

In describing the behaviour of real engineering systems in analytic terms, it is necessary to form a conceptual model of the real situation; an appropriately developed theory then relates to this model. Evidently, the degree of correlation between the actual behaviour of a system and the predictions made in the design calculations will depend crucially on the quality of the modelling: the greater the sophistication of the chosen model, the greater the precision that we can reasonably expect but, unfortunately, this almost inevitably means greater complexity in calculation. For many purposes, the idealisations into the lumped spring and rigid body systems of Chapter 1 will be adequate; however, all real mechanical systems consist of material distributed in a continuous manner so that the mechanical properties are correspondingly distributed: in strength and deformation calculations it is necessary to take this into account. The formal body of analytic techniques for treating such problems is continuum mechanics but, as we shall see in Chapter 5, an attempt to treat all systems from this generalised stand point leads to insuperable difficulties. Experience has shown that there are many engineering systems which can be treated, with satisfactory accuracy, as assemblages of deformable members whose individual mechanical behaviour can be described relatively easily by means of 'mechanics of materials.' From an analysis point of view, the significant feature of such an approach is that the system is deemed to have a finite number of degrees of freedom and not the infinite number which the distributed properties would imply. Many of the ideas developed in Chapter 1 are thus relevant. Important engineering systems amenable to such a treatment are the so-called skeletal structures which may consist of a single bar on an assemblage of bars experiencing axial load, bending, transverse shearing, torsion or some combination of these actions.

To treat such systems at this stage is valuable for a number of reasons:

(i) They are very widely used and so are important in their own right.

(ii) Their treatment provides a natural extension of the principles of Chapter 1 to realistic deformable systems.

(iii) Since they have a finite number of degrees of freedom, energy principles may be applied without the need for the calculus of variations. This contrasts with the treatment of the continuum problems discussed later.

(iv) In due course we shall seek approximate solutions to continuum problems by constraining them to have a finite number of degrees of freedom. The ideas developed now will also be relevant under those circumstances.

In this chapter we apply the principles of virtual work and virtual complementary work to determine the mechanical characteristics of individual deformable members and simple frames. We will see that the virtual work principle leads to equations of equilibrium expressed in terms of displacements whilst the virtual complementary work principle leads to equations of compatibility in terms of statically independent (redundant) forces. We wish to emphasise principles and limits of applicability so that the systems used for illustration will not be very complicated. The more formal discussion of the finite element method in Chapter 8 will show some of the systematic procedures which can be built upon these principles and which are essential in the treatment of complex situations.

2.2 The straight bar under axial load

This is the simplest load transmitting element. If the bar has a uniform cross-section along its entire length and the material has uniform properties, then referring to Fig. 2.2.1 it is reasonable to suppose that plane cross-sections remain plane after the loads are applied so that changes in configuration are characterised by the axial movement of such sections.

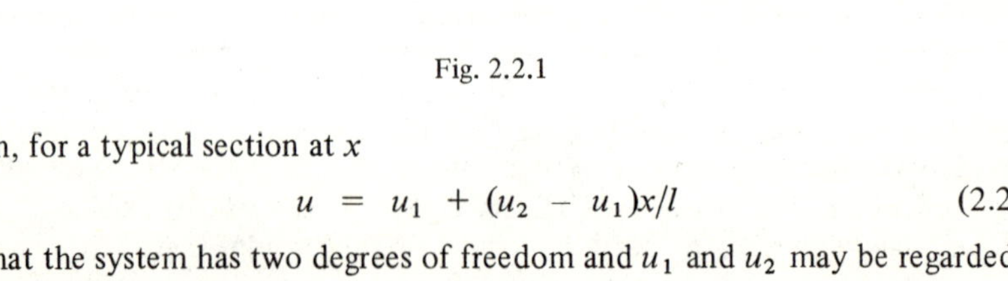

Fig. 2.2.1

Then, for a typical section at x

$$u = u_1 + (u_2 - u_1)x/l \tag{2.2.1}$$

so that the system has two degrees of freedom and u_1 and u_2 may be regarded as generalised coordinates.

If the bar is in equilibrium, then the work of P_1, P_2 and the internal forces through virtual displacements is zero. We have pointed out in Chapter 1 that the choice of virtual displacements is arbitrary, and we need only ensure that they are compatible. These requirements are clearly met if we allow virtual displacements $\delta u_1 = \delta u_2 = \delta u$ say, which correspond to an axial rigid body motion. Then, only the external forces P_1 and P_2 do work, so that we have

$$\delta W_e = P_1\delta u_1 + P_2\delta u_2 = (P_1 + P_2)\delta u = 0 \tag{2.2.2}$$

But δu is arbitrary so that the principle of virtual work gives the familiar result $P_1 = -P_2 = -P$ say.

Suppose, now, virtual displacements $\delta u_1 \neq \delta u_2$ are applied; a deformation of the bar is implied so that the internal forces also do work. Then

$$\delta W_e + \delta W_i = P_1 \delta u_1 + P_2 \delta u_2 + \delta W_i = P(\delta u_2 - \delta u_1) + \delta W_i = 0$$

or
$$\delta W_i = -P\delta e \tag{2.2.3}$$

where δe is the extension of the bar.

If the material of the bar is *elastic*, the work of the internal forces has a potential so that $\delta W_i = -\delta U$ and

$$\delta U = P\delta e \tag{2.2.4}$$

A tensile test on the bar could lead to a load elongation curve of the form shown in Fig. 2.2.2(a) and a stress strain curve as in Fig. 2.2.2(b).

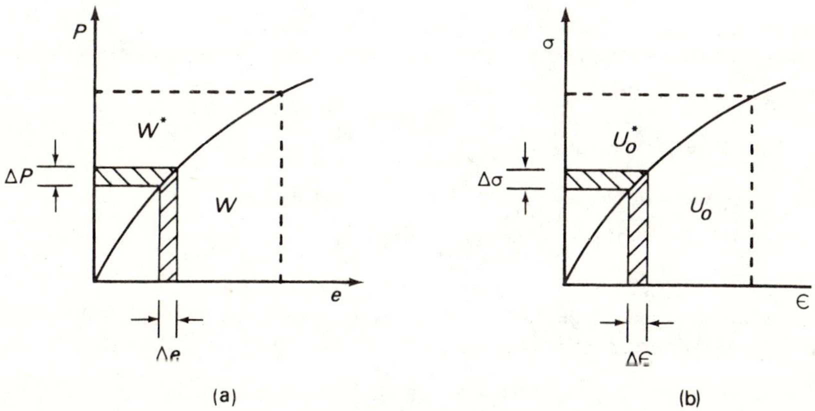

Fig. 2.2.2

Then equation 2.2.4 may be used to calculate the strain energy stored in the bar when it is stretched by the total amount e.

Since the choice of virtual displacements is arbitrary, we may allow δe to coincide with an *actual* increment Δe corresponding to an *actual* increment of force ΔP from some intermediate equilibrium value P_i so that

$$\Delta U = P_i \Delta e \ . \tag{2.2.5}$$

Then, to within an error which is of second order, ΔU is the area of the shaded vertical strip in Fig. 2.2.2(a). If P_i is *slowly* increased from zero to its final value P,† equilibrium is maintained throughout the loading process so that equation 2.2.5 holds throughout and

† So that dynamic effects are not involved.

$$U = \int_{e=0}^{e=e} P_i \mathrm{d}e \tag{2.2.6}$$

Recalling the elementary definitions of stress σ and strain ϵ for a bar in simple tension (or compression), $P = \sigma A$ and $e = \epsilon l$ where A and l are the cross sectional area and length respectively, then equation (2.2.6) becomes

$$U = \left(\int_{e=0}^{e=e} \sigma \mathrm{d}\epsilon\right) Al = \left(\int_{e=0}^{e=e} \sigma \mathrm{d}\epsilon\right) \text{vol.} \tag{2.2.7}$$

and denoting $U = U_0 \times \text{vol}$, we have the *strain energy per unit volume* or *strain energy density* defined to be

$$U_0 = \int_{\epsilon=0}^{\epsilon=\epsilon} \sigma \mathrm{d}\epsilon \ . \tag{2.2.8}$$

This is the area under the curve of Fig. 2.2.2(b).

From equation (2.2.4) we find that

$$\delta U_0 = \sigma \delta \epsilon \tag{2.2.9}$$

but $U_0 = U_0(\epsilon)$, that is, U_0 is a function of the state of strain (the deformed configuration) so that

$$\delta U_0 = \frac{\partial U_0}{\partial \epsilon} \delta \epsilon \ . \tag{2.2.10}$$

Comparing equations (2.2.9) and (2.2.10) we see that

$$\sigma = \frac{\partial U_0(\epsilon)}{\partial \epsilon} \tag{2.2.11}$$

so that the form of U_0 defines the stress-strain law for the material. When Hooke's law defines the material's behaviour, $\sigma = E\epsilon$ and equation (2.2.8) gives

$$U_0 = \frac{E\epsilon^2}{2} \tag{2.2.12}$$

and

$$U = \frac{E\epsilon^2}{2} \text{vol} = \frac{EA}{2l} e^2 \ . \tag{2.2.12a}$$

In more general situations U will not be constant but will be a function of position in the bar. To determine U under these circumstances, an integration through the volume in necessary. That is

$$U = \int U_0 \mathrm{d} \ \text{vol} \ . \tag{2.2.12b}$$

Example 2.2.1 illustrates this.

The complementary strain energy stored in the bar is calculated in a similar way. For equilibrium of the bar of Fig. 2.2.1, $P_1 = -P_2 = -P$ say, so that the complementary work of δP_1 and δP_2 through displacements u_1 and u_2 is

$$\delta W_e{}^* = u_1 \delta P_1 + u_2 \delta P_2 = \delta P(u_2 - u_1)$$

or

$$\delta W_e{}^* = e \delta P = \delta U^* \ . \tag{2.2.13}$$

Choosing δP to be an actual load increment ΔP,

$$\Delta U^* = e\Delta P \tag{2.2.13a}$$

and with the load increasing slowly from zero or its final value P

$$U^* = \int_{P=0}^{P=P} e\mathrm{d}P \,. \tag{2.2.14}$$

This is the area above the curve of Fig. 2.2.2(a).

Again, since $P = \sigma A$, $e = \epsilon l$,

$$U^* = \left(\int_{\sigma=0}^{\sigma=\sigma} \epsilon \mathrm{d}\sigma\right) \text{vol} \tag{2.2.15}$$

so that we can define

$$U_0^* = \int_{\sigma=0}^{\sigma=\sigma} \epsilon \mathrm{d}\sigma \tag{2.2.16}$$

to be the *complementary strain energy density*, which is the area above the curve of Fig. 2.2.2(b).

If U_0^* is expressible as a function of the state of stress, $U_0^* = U_0^*(\sigma)$ then,

$$\delta U_0^* = \frac{\partial U_0^*(\sigma)}{\partial \sigma}\delta\sigma \,. \tag{2.2.17}$$

But from equation 2.2.13,

$$\delta U_0^* = \epsilon\delta\sigma \tag{2.2.18}$$

so that

$$\epsilon = \frac{\partial U_0^*}{\partial \sigma} \,. \tag{2.2.19}$$

The form of U_0^* thus defines the constitutive equation for the material as an alternative to equation (2.2.11). If Hooke's law is obeyed, $\sigma = E\epsilon$ and equation (2.2.16) gives

$$U_0^* = \frac{\sigma^2}{2E} \tag{2.2.20}$$

and

$$U^* = \frac{\sigma^2}{2E}\text{vol} = \frac{P^2 l}{2EA} \,. \tag{2.2.20a}$$

In more general situations U_0^* will not be constant but a function of position in the bar. Then

$$U^* = \int U_0^* \,\mathrm{d}\,\text{vol}. \tag{2.2.21}$$

It is important to appreciate that in the various energy principles to be developed, the strain energy and strain energy density must properly be regarded as functions of the strains (or displacements) whilst the complementary strain energy density) must properly be regarded as functions of the stresses (or forces). When the stress-strain law for the material is non-linear or, as we shall see later, if initial stresses and strains exist, the distinction between strain energy and complementary strain energy is clear. However, when Hooke's law holds, U_0 and U_0^* are *numerically* equal: in fact

$$U_0 = \frac{E\epsilon^2}{2} = \frac{\sigma\epsilon}{2} = \frac{\sigma^2}{2E} = U_0^* \tag{2.2.22}$$

and

$$U = \frac{EA}{2l}e^2 = \frac{1}{2}Pe = \frac{P^2 l}{2EA} = U^* . \tag{2.2.23}$$

Whilst these results are useful for numerical purposes, they can be a source of confusion and the reader must, in theoretical developments, clearly observe when U_0 is used and when U_0^* is appropriate. Furthermore, equations (2.2.11) and (2.2.19) for the elastic stress-strain laws are valid whether Hooke's law obtains or not.

The effect of initial strains is most readily taken into account by rewriting the stress-strain law. Thus, suppose the bar experiences some initial strain (elongation) of ϵ_0 (e_0), which may be due to temperature rise or initial lack of fit in an assembly, then a plot of stress versus strain has the appearance of Fig. 2.2.3.

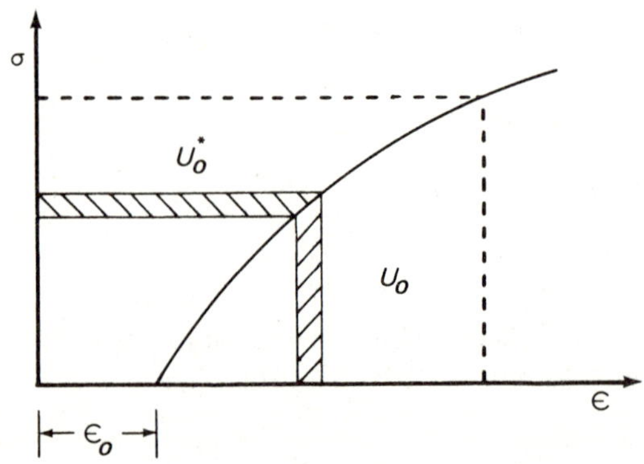

Fig. 2.2.3

Here

$$dU_0 = \sigma d\epsilon \tag{2.2.24}$$

and

$$U_0 = \int_0^\epsilon \sigma d\epsilon \tag{2.2.25}$$

which is the same area as before.

However,

$$dU_0^* = \epsilon d\sigma \tag{2.2.26}$$

and

$$U_0^* = \int_0^\sigma \epsilon d\sigma \tag{2.2.27}$$

so that this area is different from that implied by equation (2.2.16).

If Hooke's law obtains, we write it in the form

$$\sigma = E(\epsilon - \epsilon_0) \tag{2.2.28}$$

ϵ and ϵ_0 being the total and initial strains respectively.

Substituting in equations (2.2.25) and (2.2.27) yields

$$U_0 = \frac{E\epsilon^2}{2} - E\epsilon\epsilon_0 \tag{2.2.29}$$

and

$$U_0{}^* = \frac{\sigma^2}{2E} + \sigma\epsilon_0 \, . \tag{2.2.30}$$

Under these circumstances, the curve of Fig. 2.2.3 becomes a straight line and it is clear that even then, U_0 is not equal to $U_0{}^*$.

EXAMPLE 2.2.1

The tie bar of Fig. 2.2.4 is slightly tapered such that

$$A = A_0(1 + \alpha x/l)$$

where α is a small coefficient.

Calculate the total strain energy and complementary strain energy stored if Hooke's law is obeyed.

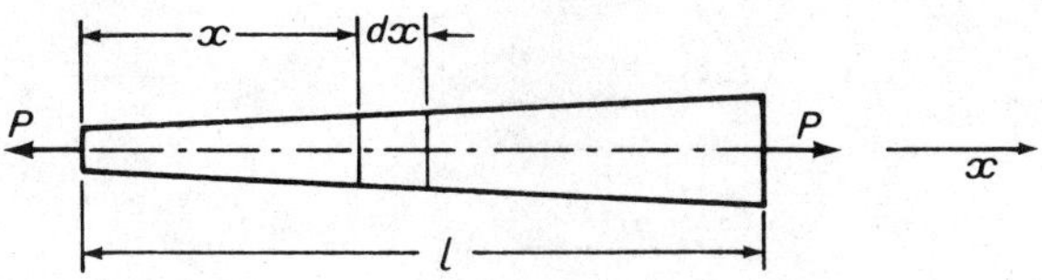

Fig. 2.2.4

Since the taper is slight, we can reasonably assume stress (and strain) to be uniform over a cross section even though they vary with length. Then at x

$$\sigma = \frac{P}{A_0(1 + \alpha x/l)} = \frac{Pl}{\alpha A_0(l/\alpha + x)}$$

so that

$$U^* = \int_0^l \frac{\sigma^2}{2E} A\mathrm{d}x = \int_0^l \frac{1}{2E} \frac{P^2 l^2 \alpha A_0(l/\alpha + x)}{\alpha^2 A_0{}^2 (l/\alpha + x)^2 l} \mathrm{d}x$$

$$= \frac{P^2 l}{2E\alpha A_0} \int_0^l \frac{1}{(l/\alpha + x)} \mathrm{d}x$$

$$= \frac{P^2 l}{2E\alpha A_0} \Big| \ln(l/\alpha + x) \Big|_0^l$$

or

$$U^* = \frac{P^2 l}{2EA_0} \cdot \frac{\ln(1 + \alpha)}{\alpha} \tag{a}$$

Since $U^* = U$ for a Hooke's law bar, an expression for U can be obtained directly from (a) but a supplementary calculation is necessary if it is to be expressed in terms of the deformation.

Now the total elongation e of the bar is given by:

$$e = \int_0^l \frac{\sigma}{E} \mathrm{d}x = \frac{Pl}{E\alpha A_0} \int_0^l \frac{\mathrm{d}x}{(l/\alpha + x)} = \frac{Pl}{E\alpha A_0} \ln(1 + \alpha) \ . \qquad \text{(b)}$$

Then substituting for P from (b) in (a) gives

$$U = \frac{Ee^2 A_0}{2l} \cdot \frac{\alpha}{\ln(1 + \alpha)} \qquad \text{(c)}$$

It will be instructive for the reader to show that by using the expansion for $\ln(1 + \alpha)$, in the limit when α tends to zero, equations (a) and (c) give the correct result for a uniform bar in tension.

EXAMPLE 2.2.2

Apply the principle of stationary total potential energy to the bar of Fig. 2.2.1 assuming the material to obey Hooke's law. Take the datum for the potential energy of P_1 and P_2 to be at $u_1 = 0$ and $u_2 = 0$ respectively.

$$V = U + \Omega \qquad \text{(a)}$$

now

$$U = \frac{EA}{2l} e^2 = \frac{EA}{2l} (u_2 - u_1)^2 \qquad \text{(b)}$$

or

$$U = \frac{EA}{2l} u_1^2 - \frac{EA}{l} u_1 u_2 + \frac{EA}{2l} u_2^2 \ . \qquad \text{(c)}$$

It is instructive to write this in the notation

$$U = \tfrac{1}{2} k_{11} u_1^2 + k_{12} u_1 u_2 + \tfrac{1}{2} k_{22} u_2^2 \qquad \text{(d)}$$

$$\Omega = -P_1 u_1 - P_2 u_2 \ . \qquad \text{(e)}$$

Thus we see that

$$V = V(u_1, u_2) \ .$$

Then for equilibrium $\delta V = 0$ requiring

$$\frac{\partial V}{\partial u_1} = k_{11} u_1 + k_{12} u_2 - P_1 = 0 \qquad \text{(f)}$$

$$\frac{\partial V}{\partial u_2} = k_{21} u_1 + k_{22} u_2 - P_2 = 0 \ . \qquad \text{(g)}$$

We have written k_{21} for k_{12} in equation (g) to provide a regularity of pattern. Using matrix notation we have

$$\frac{EA}{l} \begin{bmatrix} 1 & -1 \\ -1 & 1 \end{bmatrix} \begin{Bmatrix} u_1 \\ u_2 \end{Bmatrix} = \begin{Bmatrix} P_1 \\ P_2 \end{Bmatrix} \qquad \text{(h)}$$

or
$$[k]\{u\} = \{P\} . \tag{j}$$

Application of the P.S.T.P.E. has thus provided the load deformation characteristics of the bar.

$[k]$ is called the **stiffness matrix** (compare with Example 1.6.4) and it is clearly symmetric. $k_{11}, k_{22}, k_{12} = k_{21}$ are the **stiffness coefficients** where k_{ij} is the force required at i when a unit displacement is applied at j, all other displacements being zero.

Problems 2.2

2.2.1 A uniform bar, length l and cross sectional area A, is held at one end. It supports a distributed axial force *F/unit length* and a force P at the 'free' end. Show that the strain and complementary strain energies are given by

$$U = \frac{l}{2EA}(P^2 + PFl + F^2l^2/3)$$

when Hooke's law obtains.

2.2.2 The energy absorbing capacity of a structure is important for withstanding impact loading; a design feature which reduces this capacity is thus deleterious. Compare the strain energy stored in a stepped bar in tension with that stored in a plain bar: the former has a diameter D over a length L and d over l whilst the latter has diameter d over length $(L + l)$. Neglect stress concentration effects.

Ans: $$U_s/U_p = (1 + Ld^2/lD^2)/(1 + L/l) .$$

2.3 Torsion of a round bar and a simple shear stress state

When the uniform round bar of Fig. 2.3.1(a) sustains twisting moments only, symmetry arguments show that plane cross sections remain plane and rotate about the axis without distortion. Conditions do not vary along the bar so that its deformation is characterised by a rotation of a cross section

$$\theta = \theta_1 + (\theta_2 - \theta_1)x/l . \tag{2.3.1}$$

We see that the system has two degrees of freedom with θ_1 and θ_2 as generalised coordinates so that appropriate virtual displacements are $\delta\theta_1$ and $\delta\theta_2$.

The material feels a simple shear stress state as shown on the element isolated in Fig. 2.3.1(b).

If the bar receives a virtual displacement $\delta\theta_1 = \delta\theta_2$, it turns as a rigid body and the principle of virtual work gives the familiar result $T_1 = -T_2 = -T$, say, for equilibrium.

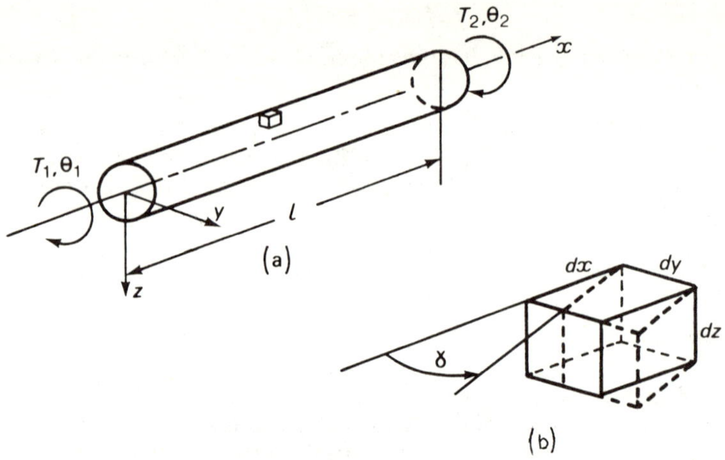

Fig. 2.3.1

Now let there be virtual displacements $\delta\theta_1 \neq \delta\theta_2$, then the bar is twisted and the internal forces also do work. Then for an *elastic* bar $\delta W_i = -\delta U$ and

$$\delta W_e = T_1\delta\theta_1 + T_2\delta\theta_2 = T(\delta\theta_2 - \delta\theta_1) = T\delta\theta = \delta U. \quad (2.3.2)$$

Choosing $\delta\theta$ to coincide with an actual increment of angle of twist $\Delta\theta$ corresponding to an actual increment of torque ΔT from an intermediate value T_i

$$\Delta U = T_i\Delta\theta \quad (2.3.3)$$

and as T increases *slowly* from zero to its final value

$$U = \int_{\theta=0}^{\theta=\theta} T_i d\theta \ . \quad (2.3.4)$$

This result holds whether the bar is solid or hollow. If it is a *thin walled tube*, thickness t, the shearing stress τ is very nearly constant, and equilibrium requirements alone are enough to give, for a diameter D

$$T = \frac{\pi^2 D^2 t}{2}\tau \ . \quad (2.3.5)$$

The geometry of deformation shows that

$$l\gamma = \frac{D\theta}{2} \quad (2.3.6)$$

so that equations (2.3.2), (2.3.5), (2.3.6) give

$$\delta U = \tau\delta\gamma \text{ vol} \quad (2.3.7)$$

and

$$U = \left(\int_{\gamma=0}^{\gamma=\gamma} \tau d\gamma\right) \text{vol} \ . \quad (2.3.8)$$

We may now define a shear strain energy density $U_0(\gamma)$ so that from (2.3.7)

$$\delta U_0 = \tau\delta\gamma = \frac{\partial U_0}{\partial\gamma}\delta\gamma \tag{2.3.9}$$

and

$$\tau = \frac{\partial U_0(\gamma)}{\partial\gamma} . \tag{2.3.10}$$

This is a constitutive relation for shearing of the material. When Hooke's law is appropriate, $\tau = G\gamma$ and equation (2.3.9) gives:

$$U_0 = \frac{G\gamma^2}{2} \tag{2.3.11}$$

where G is the shear modulus.

To find the total strain energy stored in a solid (or thick walled hollow) bar, we may either use equation (2.3.11) in

$$U = \int U_0 \,\mathrm{d\,vol} \tag{2.3.12}$$

or use the strength of materials result $T = GJ\theta/l$ in equation (2.3.4). Either way, we find, remembering that J is the polar second moment of area,

$$U = \frac{GJ\theta^2}{2l} . \tag{2.3.13}$$

It is important to notice that result (2.3.13) only holds for round bars of Hooke's law material. If the cross section of the bar is non-circular, the theory, which is summarised in Chapter 3, is much more complicated, but it can be shown that

$$U = \frac{GK\theta^2}{2l} \tag{2.3.14}$$

where K is a constant depending on the shape of the section

There is a clear parallel between the developments of this section and those of Section 2.2. We suggest that the reader should now complete the analogous arguments for complementary energy showing that with $U_0^* = U_0^*(\tau)$ as the complementary energy density,

$$\delta U_0^* = \gamma\delta\tau \tag{2.3.15}$$

$$\gamma = \frac{\partial U_0^*(\tau)}{\partial\tau} \tag{2.3.16}$$

$$U^* = \int U_0^* \,\mathrm{d\,vol} . \tag{2.3.17}$$

When Hooke's law holds,

$$U_0^* = \frac{\tau^2}{2G} \tag{2.3.18}$$

and for torsion of a circular section bar

$$U^* = \frac{T^2 l}{2GJ} . \tag{2.3.19}$$

For *numerical* work involving Hooke's law materials we have the alternative expressions

$$U_0 = \frac{G\gamma^2}{2} = \frac{1}{2}\tau\gamma = \frac{\tau^2}{2G} = U_0^* \qquad (2.3.20)$$

whilst for torsion of a round bar,

$$U = \frac{GJ\theta^2}{2l} = \frac{1}{2}T\theta = \frac{T^2 l}{2GJ} = U^* \qquad (2.3.21)$$

and torsion of a non-circular section bar

$$U = \frac{GK\theta^2}{2l} = \frac{1}{2}T\theta = \frac{T^2 l}{2GK} = U^* \; . \qquad (2.3.21)$$

If the round bar of Fig. 2.3.1 also supports an axial force *P*, the element of Fig. 2.3.1(b) is in the combined state of stress shown in Fig. 2.3.2.

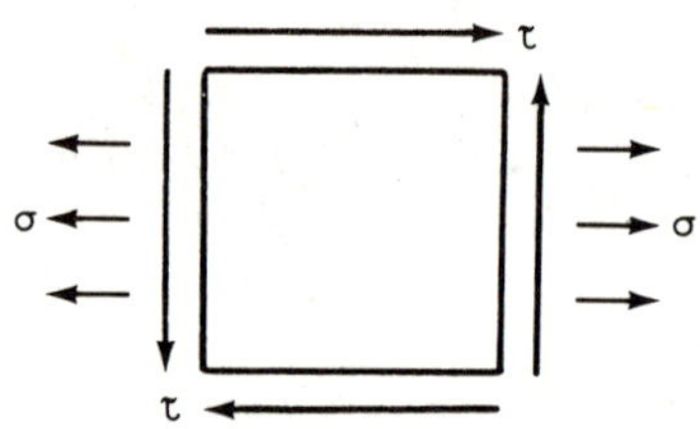

Fig. 2.3.2

Now the direct stress σ does not produce a shearing strain γ and the shearing stress τ does not produce a direct strain ϵ when the material is isotropic. Since there is no coupling between the two stress/strain components, we have for Hooke's law materials

$$U_0 = \frac{E\epsilon^2}{2} + \frac{G\gamma^2}{2} \qquad (2.3.22)$$

and

$$U_0^* = \frac{\sigma^2}{2E} + \frac{\tau^2}{2G} \; . \qquad (2.3.23)$$

2.4 The straight beam in bending

The bar of Fig. 2.4.1 is subjected to a pure bending action and the expressions for U and U^* may be obtained in a manner analogous to that used in Section 2.2. Then;

$$U = \int M d\theta \quad \text{and} \quad U^* = \int \Theta \, dM \qquad (2.4.1)$$

where $d\theta$ and Θ are rotations of one end of the bar relative to the other.

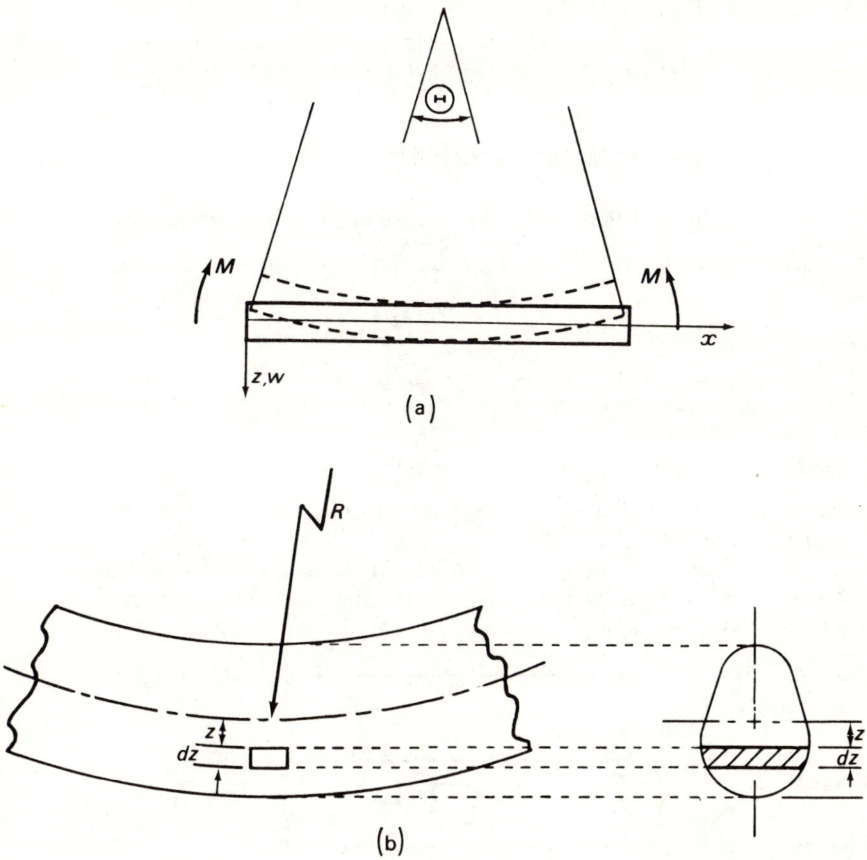

Fig. 2.4.1

Symmetry arguments show that plane cross sections of the bar remain plane during bending so that the axial direct strain varies linearly over the section. If, further, we make the usual mechanics of materials assumption that the only significant stress component is the axial one, we may use the expressions for strain energy density and complementary energy density developed in Section 2.2.†

In Fig. 2.4.1, bending takes place in the plane xOz,†† then the axial component of strain is

$$\epsilon = \epsilon_x = z/R \tag{2.4.2}$$

† See Section 6.3 also.

†† The usual beam theory shows that bending takes place in the same plane as that of the bending moment provided the moment acts about a principal axis of the section.

R is the radius of curvature of the neutral layer. Then,

$$U = \int U_0 \, \mathrm{d}\,\mathrm{vol} = \int_0^l \left[\int_A U_0(\epsilon) \mathrm{d}A \right] \mathrm{d}x$$

or

$$U = \int_0^l \left[\int_A U_0(z/R) \mathrm{d}A \right] \mathrm{d}x \; . \tag{2.4.3}$$

In this case of pure bending, the term in square brackets is independent of x. If the material obeys Hooke's law, U_0 is given by equation (2.2.12), so that

$$U = \int_0^l \left[\int_A \frac{E}{2} z^2/R^2 \, \mathrm{d}A \right] \mathrm{d}x \tag{2.4.4}$$

or

$$U = \frac{EIl}{2} \left(\frac{1}{R} \right)^2 \tag{2.4.5}$$

where I is the second moment of area defined by $I = \int_A z^2 \mathrm{d}A$.

When the bending moment varies along the beam $1/R$ is no longer constant, furthermore equilibrium requires the presence of a transverse shearing force. Evidently a transverse shearing stress τ will be induced on a vertical section so that a typical element in the beam experiences direct and shearing stress components as shown in Fig. 2.4.2.

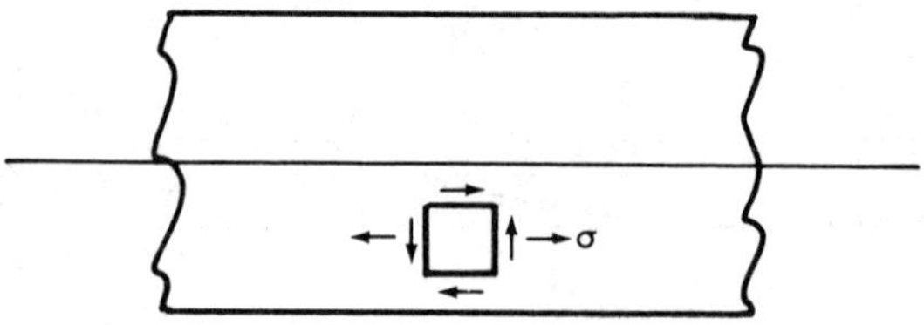

Fig. 2.4.2

As we pointed out in Section 2.3, there is no coupling effect between σ and τ when the material is isotropic so that

$$U_0 = \frac{E\epsilon^2}{2} + \frac{G\gamma^2}{2} \; . \tag{2.4.6}$$

Integrating through the volume of the bar† gives

$$U = \int U_0 \mathrm{d}\,\mathrm{vol} = U_{\mathrm{bending}} + U_{\mathrm{shearing}} \; . \tag{2.4.7}$$

If, additionally, the bar supports an axial resultant force P, we may write $\epsilon = \epsilon_b + \epsilon_d$ where ϵ_b is due to bending alone and ϵ_d is due to the direct load alone. Then

$$U_0 = \frac{E}{2}(\epsilon_b + \epsilon_d)^2 + \frac{G\gamma^2}{2} \; . \tag{2.4.8}$$

† To effect this integration, γ must be known as a function of position in the cross section. See Problem 2.4.1.

Integrating through the volume of the bar, we find that the contribution from the coupling term $\epsilon_d \epsilon_b$ cancels out and

$$U = U_{\text{bending}} + U_{\text{direct}} + U_{\text{shear}} \ . \tag{2.4.9}$$

It is important to observe that in general it is not possible merely to add strain energy terms in this way. The contribution of coupling terms caused by various combinations of applied loads only happen to cancel out in particular situations.

In the usual bar-type structure, the strain energy contributions due to shearing and axial load are small compared with that due to bending and are neglected. Then, we have

$$U = \int \frac{EI}{2}\left(\frac{1}{R}\right)^2 \mathrm{d}x \tag{2.4.10}$$

or

$$U = \int \frac{EI}{2}\left(\frac{d^2 w}{dx^2}\right)^2 \mathrm{d}x \ . \tag{2.4.11}$$

Here, $1/R$ is approximated by $(-\mathrm{d}^2 w/\mathrm{d}x^2)$ since the curvature is small and EI remains under the integral sign to allow for non-uniform sections. The minus sign is required when positive curvature is concave up and the axes are as in Fig. 2.4.1.†

The complementary strain energy is also readily calculated. Thus, for pure bending and a Hooke's law material

$$U^* = \int U_0^* \,\mathrm{d}\,\text{vol} = \int \frac{\sigma^2}{2E}\,\mathrm{d}\,\text{vol} = \int_0^l \left[\int_A \frac{M^2 z^2 \mathrm{d}A}{2EI^2}\right] \mathrm{d}x$$

or

$$U^* = \frac{M^2 l}{2EI} \ . \tag{2.4.12}$$

Again, when the bending moment varies along the beam and there is an axial force present, it is usually adequate to use

$$U^* = \int_0^l \frac{M^2}{2EI}\,\mathrm{d}x \ . \tag{2.4.13}$$

As in the cases considered previously, it may be convenient to use the fact that $U = U^*$ numerically when Hooke's law obtains.

EXAMPLE 2.4.1

Determine the strain energy stored in a uniform beam bent by end couples and forces alone as shown in Fig. 2.4.3. Assume that shear effects are negligible and that Hooke's law obtains.

† See also Section 6.3.

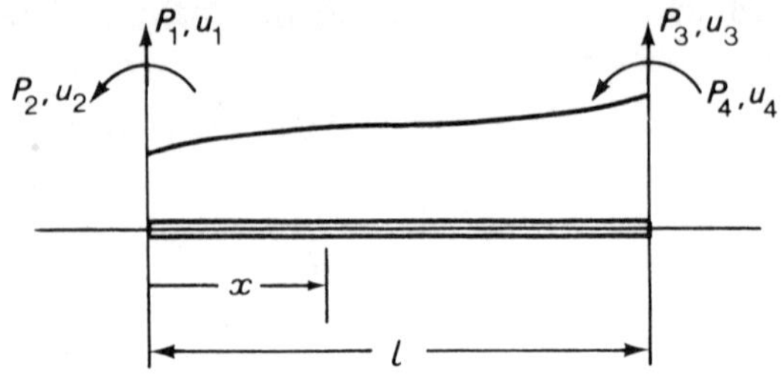

Fig. 2.4.3

$P_1 \ldots\ldots P_4$ are end forces and couples corresponding to end deflexions and slopes $u_1 \ldots\ldots u_4$.

Since there is no distributed transverse load, the governing differential equation for the deflexion curve, from equation (6.3.19) is

$$EI\frac{d^4 w}{dx^4} = 0 \tag{a}$$

Integrating equation (a) and adjusting the constants of integration to fit the conditions of the problem, we may readily show that

$$w = u_1\psi_1(x) + u_2\psi_2(x) + u_3\psi_3(x) + u_4\psi_4(x) \tag{b}$$

or

$$w = \sum_{i=1}^{4} u_i\psi_i(x) \tag{c}$$

where

$$\left.\begin{aligned} \psi_1(x) &= 1 - 3x^2/l^2 + 2x^3/l^3 \\ \psi_2(x) &= l(x/l - 2x^2/l^2 + x^3/l^3) \\ \psi_3(x) &= 3x^2/l^2 - 2x^3/l^3 \\ \psi_4(x) &= l(-x^2/l^2 + x^3/l^3)\ . \end{aligned}\right\} \tag{d}$$

Evidently the system has four degrees of freedom having $u_1 \ldots\ldots u_4$ as generalised coordinates and $\psi_1 \ldots\ldots \psi_4$ as coordinate functions (interpolation functions) which describe the deformed *shape*.

Now

$$U = \int_0^l \frac{EI}{2}\left(\frac{d^2w}{dx^2}\right)^2 dx \tag{e}$$

then substituting from (c) in (e), we have

$$U = \int_0^l \frac{EI}{2}\left(\sum_{i=1}^{4} u_i\psi_i''\right)^2 dx = \frac{1}{2}\sum_{i=1}^{4}\sum_{j=1}^{4}\left(\int_0^l EI\psi_i''\psi_j''u_iu_j\,dx\right).$$

If we denote

$$k_{ij} = \int_0^l EI\psi_i''\psi_j''\,dx = k_{ji}\ , \tag{f}$$

$$U = \frac{1}{2}\sum_{i=1}^{4}\sum_{j=1}^{4} k_{ij}u_iu_j \ . \tag{g}$$

Written in matrix terms $$U = \frac{1}{2}\{u\}^t[k]\{u\} \ . \tag{h}$$

This is the typical quadratic form for the strain energy stored in a linearly elastic system having a finite number of degrees of freedom as illustrated in Example 1.6.4. As in that example, the k_{ij}'s are the stiffness coefficients and if the integrals (f) are evaluated, we find that

$$[k] = \frac{EI}{l^3}\begin{bmatrix} 12 & 6l & -12 & 6l \\ & 4l^2 & -6l & 2l^2 \\ \text{symm} & & 12 & -6l \\ & & & 4l^2 \end{bmatrix} . \tag{j}$$

Problems 2.4

2.4.1 The bars of Problem 2.2.2 are subjected (i) to a given twisting moment and (ii) to a given bending moment. Compare the strain energies stored in each case.

Ans: $$U_s/U_p = (1 + Ld^4/lD^4)/(1 + L/l) \ .$$

2.4.2 A rotating shaft, diameter d and supported by several bearings, carries a heavy disk of moment of inertia I. If, when the shaft is rotating at ω rad/sec, a bearing distance l from the disk seizes, calculate the maximum shearing stress in the shaft assuming all the kinetic energy is stored as strain energy. Hooke's law material, shear modulus G.

Ans: $$\tau_{\max} = \frac{\omega}{2d}\sqrt{\frac{32GI}{\pi l}} \ .$$

2.4.3 The diameter of a shaft, of Hooke's law material, varies linearly from d_0 to $d_0/2$ over its length l. Show that the complementary strain energy stored when a torque T is being transmitted is

$$U^* = \frac{14}{3}\frac{T^2 l}{2GJ_0} \ .$$

2.4.4 A cantilever, length l has a rectangular section $b \times h$ (h = depth) and supports concentrated transverse and longitudinal forces F and T at its free end. Compare the bending, tension and shearing strain energies when $F = T$. Assume the usual parabolic distribution of shearing stress for bent beams.

2.4.5 A beam, length $3l$, is clamped at both ends and provided with additional spring supports at the third points; each spring has stiffness k. Denoting the deflexions and slopes at $x = l$ and $x = 2l$ by q_1, q_2, q_3, q_4 respectively, show that the strain energy stored in the system due to concentrated actions at q_1 etc, is

$$U = \tfrac{1}{2}\{q\}^t[K]\{q\}$$

where

$$[K] = \frac{EI}{l^3}\begin{bmatrix} (24 + kl^3/EI) & 0 & -12 & 6l \\ & 8l^2 & 6l & 2l^2 \\ & & (24 + kl^3/EI) & 0 \\ \text{symm} & & & 8l^3 \end{bmatrix}$$

2.5 Stress resultants and displacements in single bar structures: Castigliano's theorems

If the bars of Figs. 2.2.1 or 2.3.1 or 2.4.1 are regarded as *structures*, their function is to support some applied load and so they themselves must be adequately supported. Within the limits of the usual 'strength of materials' theories, the stress analysis of bars can be regarded as virtually complete when the stress resultants (bending moment, twisting moment, shear force and axial force) have been determined. Under these circumstances, we are only interested in the resultant moment and force which may be induced as reactions at the supports. Depending on the details of the system, these resultant actions may be statically determinate or indeterminate and, once they are known, the stress resultants can be readily determined at any section in the member. We now show how such systems may be treated by energy principles and shall arrive at particular forms of Castingliano's celebrated theorems.

Consider first the simple arrangement of Fig. 2.5.1. Here the left hand end of the bar is rigidly supported and cannot displace so that the force P, required there for equilibrium, does not do work nor complementary work. The virtual

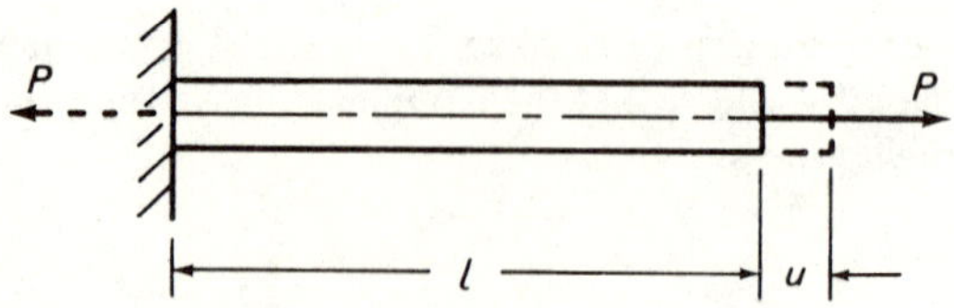

Fig. 2.5.1

work equation (2.2.5), in which actual force increments and displacements appear, thus becomes

$$\Delta U = P\Delta u \tag{2.5.1}$$

but since the elongation $e = u$, $U = U(u)$ so that

$$\Delta U = \frac{\partial U}{\partial u} \Delta u \quad . \tag{2.5.2}$$

Comparing (2.5.1) and (2.5.2)

$$P = \frac{\partial U}{\partial u} \quad . \tag{2.5.3}$$

This is a special case of **Castigliano's first theorem** and, although originally derived for linear elastic behaviour, this development shows that it is true whether Hooke's law obtains or not. In fact, it is an equation of equilibrium for the right hand end of the bar.

Similarly, from the complementary virtual work equation (2.2.13a)

$$\Delta U^* = u\Delta P \tag{2.5.4}$$

since the displacement at a rigid support is zero. But

$$U^* = U^*(P)$$

so that

$$\frac{\partial U^*}{\partial P} = u \quad , \tag{2.5.5}$$

which is true whether Hooke's law holds or not. Now if the behaviour *is* linear ealstic, we have seen that $U = U^*$ numerically so that if we express the strain energy as a function of the load, we have

$$u = \frac{\partial U}{\partial P} \quad . \tag{2.5.5a}$$

This is a special case of **Castigliano's second theorem**, his theorem on displacements.

We consider now situations in which the reactions are statically indeterminate so that some of them are redundant.

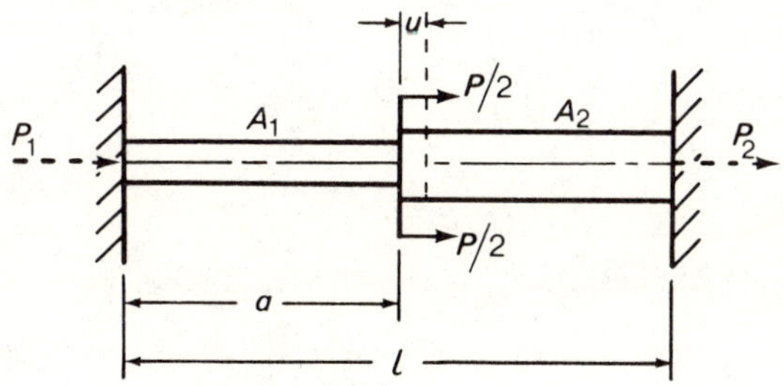

Fig. 2.5.2

The stepped bar of Fig. 2.5.2 has its ends rigidly clamped. Suppose the reactions at its ends (that is, the actions of the supports on the bars) are as shown, then for equilibrium

$$P_1 + P_2 + P = 0 \quad . \tag{2.5.6}$$

Since there is only one equation relating two unkown reactions, the system is statically indeterminate with one redundant reaction. There is only one degree of freedom with generalised coordinate u. If we are primarily interested in the displacement, the most direct approach is to use the principle of stationary total potential energy. For a Hooke's law material, we have, since P_1 and P_2 do no work,

$$V = \left[\frac{EA_1}{2a} + \frac{EA_2}{2(l - a)}\right]u^2 - Pu = V(u) \tag{2.5.7}$$

for $\delta V = 0$ $$\frac{\partial V}{\partial u} = 0 = \left[\frac{EA_1}{a} + \frac{EA_2}{(l - a)}\right]u - P$$

or $$u = P\left[\frac{EA_1}{a} + \frac{EA_2}{(l - a)}\right]^{-1} \tag{2.5.8}$$

defining the equilibrium configuration.

It is worth noting that, in this case, equation (2.5.3) would also have led to this result in a very direct way.

Knowing u, the strains and hence stresses in each portion of the bar are readily calculated and the reactions P_1 and P_2 obtained.

An alternative procedure, which may be more useful in more complex bar problems, is to use complementary energy. For overall equilibrium of the bar, equation (2.5.6) holds. For purposes of analysis, we may regard *either* of the unknown reactions as the statically independent (or redundant) one. We choose P_1 and write the equilibrium equation as

$$P_2 = -(P + P_1) . \tag{2.5.9}$$

Then, since the supports are rigid, we can write, see equation (1.8.9),

$$V^* = U^* = \frac{P_1^2 a}{2EA_1} + \frac{P_2^2(l - a)^2}{2EA_2} \tag{2.5.10}$$

in the form $$U^* = \frac{P_1^2 a}{2EA_1} + \frac{(P + P_1)^2(l - a)^2}{2EA_2} \tag{2.5.11}$$

or $$U^* = U^*(P, P_1) \tag{2.5.12}$$

Since P is prescribed, variations in U^* arise from variations in P_1 alone, that is

$$\delta U^* = \frac{\partial U^*}{\partial P_1}\delta P_1 . \tag{2.5.13}$$

With δP_1 arbitrary, we have for compatibility

$$\frac{\partial U^*}{\partial P_1} = 0 . \tag{2.5.14}$$

Substituting from equation (2.5.11) in (2.5.14)

$$P_1 = -P\left[1 + \frac{a}{l - a}\frac{A_2}{A_1}\right]^{-1}, \qquad (2.5.15)$$

P_2 is then given by equation (2.5.9).

Substituting from equation (2.5.15) in (2.5.11), we have U^* expressed in terms of applied load P alone so that equation (2.5.5) yields the displacement u directly. However, as we saw in Example 1.8.5, we can use an alternative approach which reduces the manipulative effort in more difficult situations. From equation (2.5.12) and the chain rule of differentiation,

$$\frac{\partial U^*(P)}{\partial P} = \frac{\partial U^*(P, P_1)}{\partial P} + \frac{\partial U^*(P, P_1)}{\partial P_1}\frac{\partial P_1}{\partial P}. \qquad (2.5.16)$$

Then, substituting from (2.5.11) and (2.5.15)

$$u = \frac{\partial U^*}{\partial P} = P\left[\frac{EA_1}{a} + \frac{EA_2}{(l - a)}\right]^{-1} \qquad (2.5.17)$$

which is precisely the result found by the potential energy method.

It is important to notice that, when Castigliano's second theorem is to be used to find displacements in a statically indeterminate system, the redundant reactions *must* be found in terms of the applied load *before* any attempt is made to form $\partial U/\partial P$. Failure to observe this leads to incorrect solutions and is a frequent cause of difficulty amongst students.

Having the results of Sections 2.3 and 2.4 at hand we can radily extend these procedures to a variety of bar problems. We shall proceed through a series of examples and, to this end, it will be useful to record that for a bar of Hooke's law material subjected to bending, torsion, shearing and axial load,

$$U^* = U = \int_0^l \frac{M_{yy}^{\ 2}}{2EI_{yy}}\,dx + \int_0^l \frac{M_{zz}^{\ 2}}{2EI_{zz}}\,dx + k^*\int_0^l \frac{T^2}{2GJ}\,dx + \int_0^l \frac{P^2}{2EA}\,dx$$

$$+ k_y\int_0^l \frac{F_y^{\ 2}}{2GA}\,dx + k_z\int_0^l \frac{F_z^{\ 2}}{2GA}\,dx + k^{**}\int_0^l \frac{F_yF_z}{2GA}\,dx. \qquad (2.5.18)$$

Here M_{yy} and M_{zz} are the bending moments about axes Oy and Oz respectively with I_{yy} and I_{zz} the appropriate second moments of area, T is the twisting moment, F_y and F_z are the shearing forces and P is the axial force at a section. A is the section area and J is the polar second moment area. E and G are Young's modulus and the shear modulus, k_y, k_z, k^* and k^{**} are constants which depend on the shape of the section. The interesting feature of this result is the uncoupling of all the energy terms which is only so when forces and moments are referred to the shear centre of the section. The coefficient k^{**} is zero if the section has one axis of symmetry, whilst the shear centre coincides with the centroid of the section when it has two axes of symmetry. The result follows from $U^* = \int U_0^*\,d\,vol$ where U_0^* involves the shearing and direct stresses expressed in terms of the stress resultants by means of the usual mechanics of materials theory for bars. Further interesting discussions on this topic appear in the book by Biezeno and Grammel.

We have previously pointed out that, in the usual bar structures, it is only bending and torsion energies which are significant: then we use the simpler expression

$$U^* = U = \int_0^l \frac{M_{yy}^2}{2EI_{yy}} dx + \int_0^l \frac{M_{zz}^2}{2EI_{zz}} dx + k^* \int_0^l \frac{T^2}{2GJ} dx \quad . (2.5.19)$$

EXAMPLE 2.5.1

The bar of Fig. 2.5.3 is cranked in the horizontal plane and supports a vertical point load and distributed load as shown. Determine the vertical deflexion at A and B.

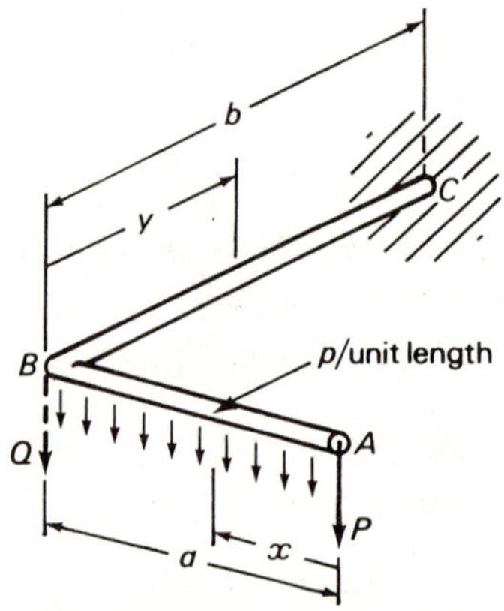

Fig. 2.5.3

The deflexion at A is found from a direct use of $\partial U/\partial P = \Delta_p$. There is no point load at B, then to determine the deflexion there, we introduce a fictitious point load Q at B in the direction of the required displacment's direction and form $(\partial U/\partial Q)_{Q=0}$. $\partial U/\partial Q$ must be determined *before* Q is set equal to zero of course.

To find Δ_P.

$$\Delta_P = \frac{\partial U}{\partial P} = \frac{\partial U_{AB}}{\partial P} + \frac{\partial U_{BC}}{\partial P} \quad . \tag{a}$$

Finding $\partial U_{AB}/\partial P$

$$M = -Px - \frac{px^2}{2} \tag{b}$$

$$\frac{\partial U_{AB}}{\partial P} = \int_0^a \frac{M}{EI}\frac{\partial M}{\partial P}dx = \int_0^a \frac{1}{EI}\left(-Px - \frac{px^2}{2}\right)(-x)\,dx$$

or

$$\frac{\partial U_{AB}}{\partial P} = \frac{1}{EI}\left(\frac{Pa^3}{3} + \frac{pa^4}{8}\right) . \tag{c}$$

Finding $\partial U_{BC}/\partial P$

The portion BC of the bar is in a state of combined bending and torsion with

$$M = -(P + pa)\,, \quad T = (Pa + pa^2/2) \tag{d}$$

Then

$$\frac{\partial U_{BC}}{\partial P} = \int_0^b \frac{M}{EI}\frac{\partial M}{\partial P}dy + \int_0^b \frac{T}{GJ}\frac{\partial T}{\partial P}dy . \tag{e}$$

Substituting from (d) in (e)

$$\frac{\partial U_{BC}}{\partial P} = \frac{(P + pa)b^3}{3EI} + \left(Pa + \frac{pa^2}{2}\right)\frac{ab}{GJ} . \tag{f}$$

From (c) and (f)

$$\Delta_P = \frac{1}{EI}\left[\frac{Pa^3}{3} + \frac{pa^4}{8} + \frac{(P + pa)b^3}{3}\right] + \left(\frac{Pa + pa^2}{2}\right)\frac{ab}{GJ} .$$

To find $(\Delta_Q)_{Q=0}$,

we observe that the expression for strain energy stored in portion AB, is independent of Q whereas in portion BC,

$$M = -(P + pa + Q)y\,, \quad T = \left(Pa + \frac{pa^2}{2}\right) \tag{g}$$

$$\frac{\partial M}{\partial Q} = -y\,, \qquad \frac{\partial T}{\partial Q} = 0 \tag{h}$$

so that

$$\frac{\partial U}{\partial Q} = \frac{\partial U_{BC}}{\partial Q} = \int_0^b \frac{(-1)(P + pa + Q)y(-y)dy}{EI} \tag{j}$$

and

$$\left(\frac{\partial U}{\partial Q}\right)_{Q=0} = \frac{(P + pa)b^3}{3EI} . \tag{k}$$

These results can of course be obtained readily by the usual direct methods of mechanics of materials.

EXAMPLE 2.5.2

Determine the horizontal and vertical components of displacement at the free end of the curved bar of Fig. 2.5.4. The radial distributed load is constant.

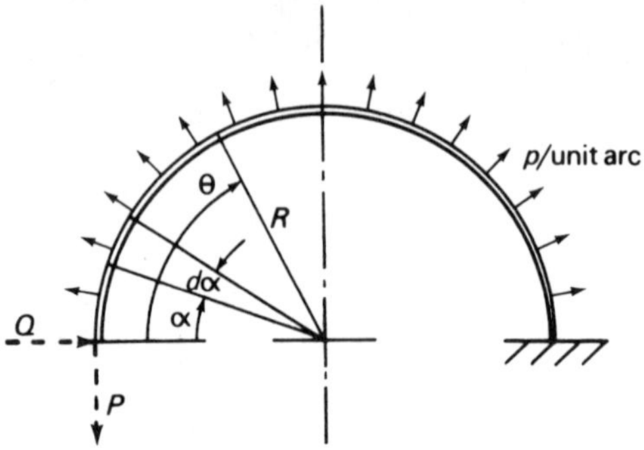

Fig. 2.5.4

Since there are no prescribed point loads at the free end, we introduce the fictitious ones, P, Q.

For *thin* curved bars, the bending curvature is proportional to the bending moment as for straight bars so that the problem can be treated by Castigliano's theorem using energy expression (2.5.19). Here, there is only one non-zero bending moment

$$M = \int_0^\theta R \sin(\theta - \alpha)\, pR\, d\alpha - QR \sin\theta - PR(1 - \cos\theta) \tag{a}$$

or
$$M = pR^2(1 - \cos\theta) - QR\sin\theta - PR(1 - \cos\theta) \tag{b}$$

$$\frac{\partial M}{\partial Q} = -R\sin\theta\,, \quad \frac{\partial M}{\partial P} = -R(1 - \cos\theta)\ . \tag{c}$$

Then
$$(\Delta_Q)_{\substack{P=0\\Q=0}} = \left(\int_0^\pi \frac{M}{EI}\frac{\partial M}{\partial Q} R\,d\theta\right)_{\substack{P=0\\Q=0}}$$

$$= \frac{1}{EI}\int_0^\pi pR^2(1 - \cos\theta)(-R\sin\theta)R\,d\theta$$

or
$$(\Delta_Q)_{\substack{P=0\\Q=0}} = -\frac{2pR^4}{EI} \tag{d}$$

$$(\Delta_P)_{\substack{P=0\\Q=0}} = \frac{1}{EI}\int_0^\pi pR^2(1 - \cos\theta)(-R)(1 - \cos\theta)R\,d\theta$$

or
$$(\Delta_P)_{\substack{P=0\\Q=0}} = \frac{-3\pi pR^4}{2EI}\ . \tag{e}$$

EXAMPLE 2.5.3

Deduce the flexibility compatibility equations for the clamped beam of Fig. 2.5.5.

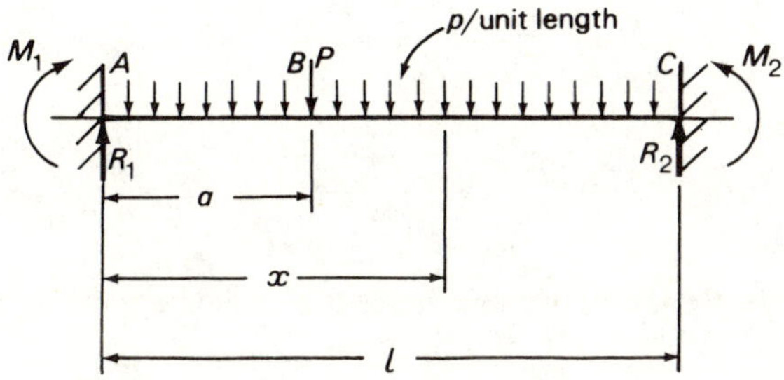

Fig. 2.5.5

For equilibrium

$$\Sigma F_y = R_1 + R_2 - P - pl = 0 \tag{a}$$

$$\Sigma M_c = M_1 + R_1 l - M_2 - P(l - a) - \frac{pl^2}{2} = 0\,. \tag{b}$$

Corresponding to two equations in four reactions, there are two redundancies. We choose these, quite arbitrarily, to be M_1 and R_1. Then

$$R_2 = P + pl - R_1 \tag{c}$$

$$M_2 = M_1 + R_1 l - P(l - a) - \frac{pl^2}{2}\,. \tag{d}$$

Since the prescribed displacements are zero, the variation in total complementary energy δV^* simplifies to δU^* and for a linearly elastic system $\delta U^* = \delta U$. [See equation (1.8.9).]

Now $$M = M_1 + R_1 x - \frac{px^2}{2} - P\langle x - a\rangle \tag{e}$$

where $\langle\,\rangle$ signifies that this term has no value for $x < a$.

From equation (2.5.19) and (e)

$$U^* = U = U(M_1, R_1, p, P)\,. \tag{f}$$

For compatibility $\delta U^* = 0$ and since M_1 and R_1 are the statically independent reactions, $\delta U^* = 0 = \delta U$ requires.

$$\frac{\partial U}{\partial M_1} = 0\,, \qquad \frac{\partial U}{\partial R_1} = 0\,. \tag{g}$$

Observing that $$\frac{\partial M}{\partial M_1} = 1\,, \qquad \frac{\partial M}{\partial R_1} = x$$

and $$\frac{\partial U}{\partial M_1} = \int \frac{M}{EI}\frac{\partial M}{\partial M_1}\,dx\,, \qquad \frac{\partial U}{\partial R_1} = \int \frac{M}{EI}\frac{\partial M}{\partial R_1}\,dx$$

carrying out the necessary integrations we have

$$\frac{M_1 l}{EI} + \frac{R_1 l^2}{2EI} = \frac{pl^3}{6EI} + \frac{P(l-a)^2}{2EI} \qquad \text{(h)}$$

$$\frac{M_1 l^2}{2EI} + \frac{R_1 l^3}{3EI} = \frac{pl^4}{8EI} + \frac{P}{6EI}(2l^3 - 3al^2 + a^3) \qquad \text{(j)}$$

or $$[H]\{F\} = \{u\} \qquad \text{(k)}$$

where $[H]$ is the *flexibility matrix* relative to the chosen redundants and is seen to be symmetric. $\{F\}^t = [M_1, R_1]$.

Equation (k) has the following physical significance: if the left hand end of the beam were released, p and P alone would cause a deflexion and slope to appear there. M_1 and R_1 alone would also cause an end deflexion and slope. Then equation (k) is a statement of the closure of the gap to effect satisfaction of compatibility at the end.

In the special case of $a = l/2$, we have the familiar results that when P acts alone,

$$M_1 = -\frac{Pl}{8}\,, \qquad R_1 = P/2\,,$$

when p acts alone

$$M_1 = -\frac{pl}{12}\,, \qquad R_1 = pl/2\,.$$

In both cases, equilibrium then yields the symmetric situation of $M_1 = M_2$, $R_1 = R_2$.

Problems 2.5

2.5.1 Determine the stresses and displacements in each element of the tube-rod assembly of Fig. 2.5.6. The same Hooke's law material is used throughout. Use, successively, the potential energy and complementary energy principles. Does one method have an advantage over the other in this case?

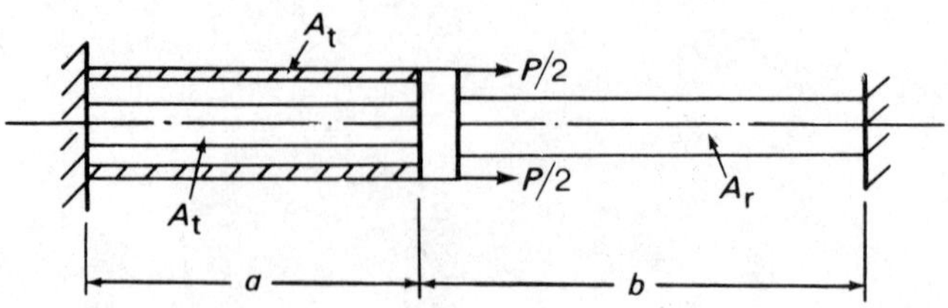

Fig. 2.5.6

2.5.2 A cantilever is given extra support by a rigid prop at its free end and carries a uniformly distributed load over its entire length. Use Castigliano's theorems to show that the prop reaction and mid span deflexion are given by

$$R_p = 3pl/8\,, \quad w_{ms} = pl^4/192EI$$

2.5.3 A particular spring suspension is sketched in Fig. 2.5.7. By considering only the bending strain energy in the semicircular portion to be significant, show that the displacement at O is always in the direction of the load and that the stiffness is independent of the direction of the load.

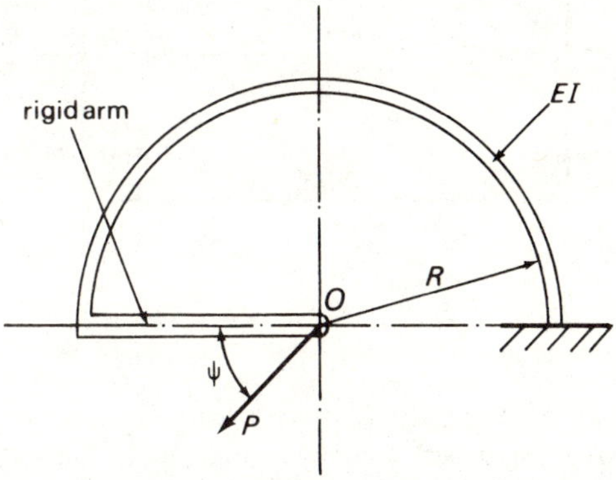

Fig. 2.5.7

2.5.4 The frame of Fig. 2.5.8 supports (i) a unit horizontal load at D, (ii) vertical loads P at B and C. Determine the horizontal displacement at D in each case and the horizontal thrust which would be induced at D if it were pinned to a rigid support instead of being on rollers.

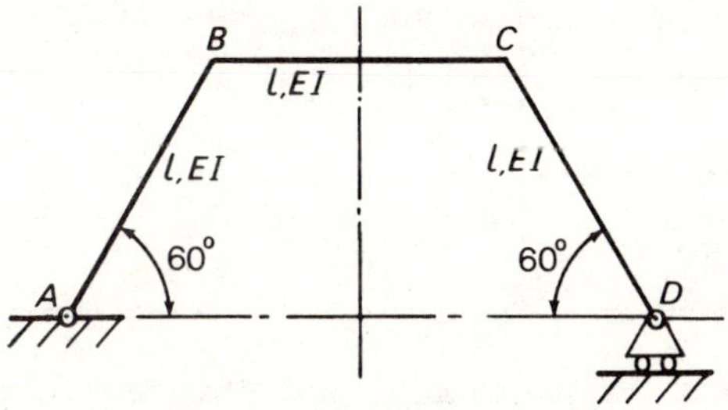

Fig. 2.5.8

Ans: $1.25l^3/EI$; $0.72Pl^3/EI$; $\sqrt{3}P/3$

2.5.5 A beam $ABCD$, length $3l$, is simply supported at its ends A, D and rests on springs, stiffness k, at each third span point B and C. Find all the reactions when a vertical force P is applied at B and $k = 6EI/l^3$.

Ans: $R_A = 0.208P$, $R_B = 0.542P$, $R_C = 0.292P$, $R_D = -0.042P$.

2.5.6 The clamped-clamped beam of Fig. 2.5.9 experiences differential settlement of its supports as shown. Show that the flexibility compatibility equation is

$$\frac{l}{EI}\begin{bmatrix} 6 & 3l \\ 3l & 2l^2 \end{bmatrix}\begin{Bmatrix} M_1 \\ R_1 \end{Bmatrix} = \begin{Bmatrix} 0 \\ -\Delta \end{Bmatrix}$$

and that $M_1 = 6EI\Delta/l^2$, $R_1 = -12EI\Delta/l^3$.

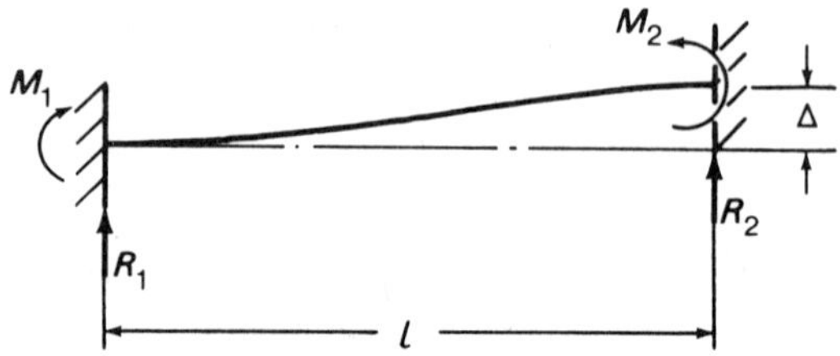

Fig. 2.5.9

2.6 Pin jointed frames

The simplest idealisation of a structural system into an assemblage of discrete members consists of straight bars connected together at their ends by frictionless pins and supporting loads applied at the pins. In such an arrangement, the bars experience purely axial forces with no bending, shearing or torsion. Fig. 2.6.1(a) illustrates a simple *plane frame* whilst Fig. 2.6.1(b) shows a *space frame*.

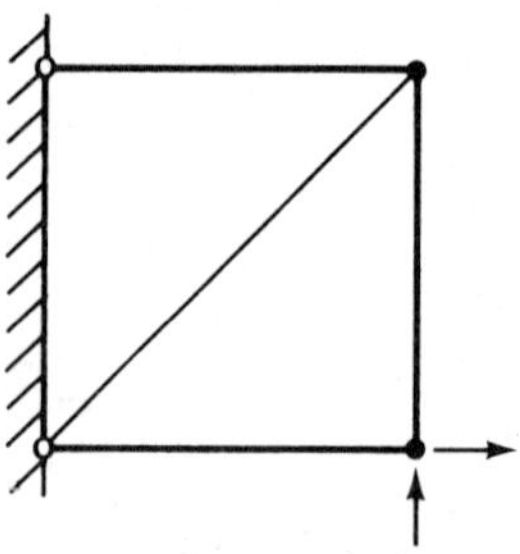

Fig. 2.6.1(a)

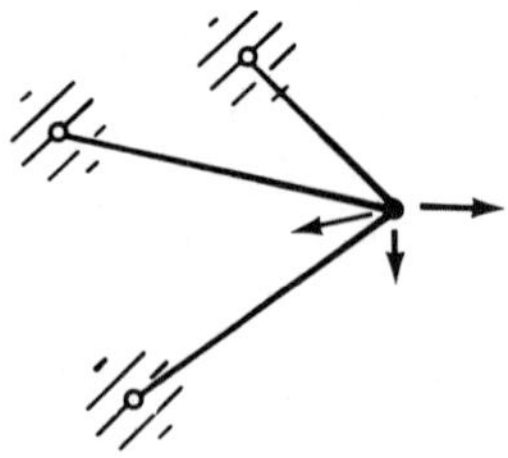

Fig. 2.6.1(b)

Generalising our experience with single bars, we see that the configuration of such systems is completely definable by the positions of the joints. Then, to describe the forces and deformations in the frame by means of the virtual work principle, it proves convenient to regard displacements as generalised coordinates. As in the case of single bars, any contemplated virtual displacements fall naturally into two classes:

(i) Joints are displaced such that their positions relative to each other do not change so that the frame moves as a rigid body. Such virtual displacements violate the geometric constraints at the supports and so require work terms corresponding to the reactions as well as the applied loads. If the number of independent virtual rigid body motions exactly equals the number of reactive elements, the support arrangement is statically determinate. If the number of motions is less than the number of reactions, the system is supported in a statically indeterminate manner whilst if the number is greater, the system is a mechanism and cannot support any loads.

(ii) The joints are displaced such that their relative positions change so that a deformation of the system occurs. The internal forces as well as the external ones do work in such virtual displacements and the principle of virtual work leads to the determination of the actual joint displacements and member forces due to some prescribed loading.

We can use these ideas to establish the statical determinancy or indeterminancy of a system. A plane frame having j joints has $2j$ degrees of freedom (one 'horizontal' and one 'vertical' displacement component at each joint); of these, three will correspond to rigid body motions (translations 'horizontally' and 'vertically' and rotation in the plane) and, by the principle of virtual work, give the three equations of overall equilibrium for the frame. There will then be $(2j - 3)$ further equations of equilibrium available. If there are s reactive elements at the supports, the system is supported in a statically indeterminate, or determinate manner or is a mechanism according as $s >$ or $=$ or < 3: if $s > 3$, then $r_c = s - 3$ is the number of external redundancies. Now the number of internal forces equals the number of bars, n say, so that the total forces to be determined is $(n + s)$ and the system thus appears to be statically determinate if $(n + s) = 2j$. However, we have already seen that there is external statical indeterminancy unless $s = 3$, so that a plane frame can only be statically determinate if $(n + 3) = 2j$. Furthermore, local regions of the frame must be examined since it is possible to have the correct value for n, but with too many bars in one region and too few in another. If $n + s > 2j$, the frame is statically indeterminate with internal or external redundant elements or both. Thus $n - (2j - s) = r$ is the total number of redundant elements (or degrees of indeterminacy) of which $r_e = s - 3$ is the number of redundant reactions and $r_i = (n + 3 - 2j)$ is the number of redundant bars. In the case of a space frame the condition for statical determinancy is $(n + s) = 3j$.

EXAMPLE 2.6.1

Examine the determinancy of the plane frame of Fig. 2.6.2.

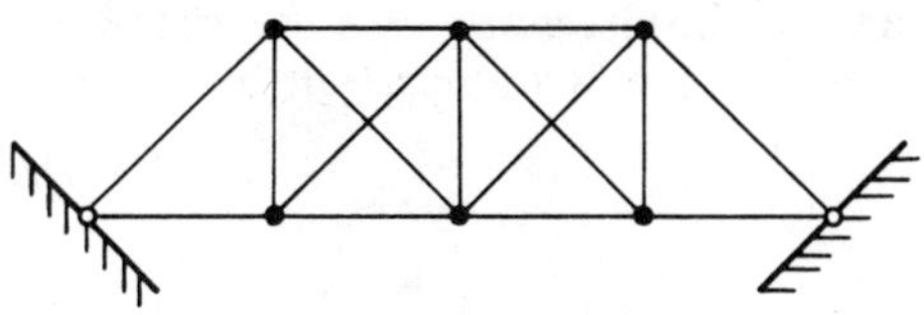

Fig. 2.6.2

Here, $s = 4$, $j = 8$, $n = 15$ so that $2j = 16$, $n + s = 19$ and $(n + s) - 2j = 3$. The frame is then statically indeterminate with three redundancies.

Number of external redundancies $r_e = 4 - 3 = 1$ so that the number of redundant bars is $3 - 1 = 2$.

Having used the virtual work principle to establish the determinancy of a frame, we now proceed to calculate forces and joint (or node) displacements. We first treat a statically determinate system. Two well known and very direct methods for calculating the member forces are the method of joints and the method of sections. It is valuable to see how these methods can be developed by virtual work arguments which themselves show the freedom with which virtual displacements may be chosen

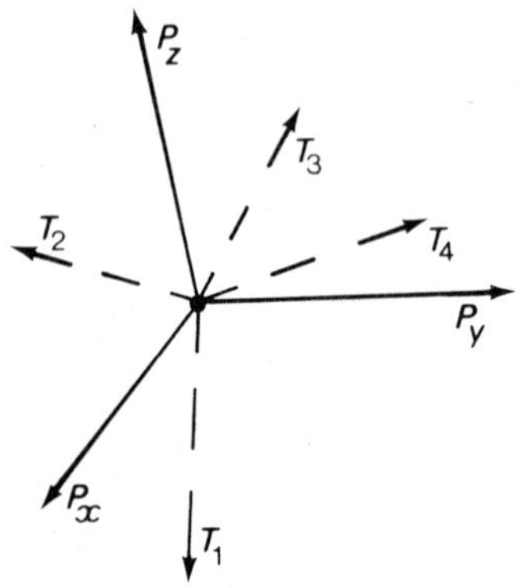

Fig. 2.6.3

Suppose some members in a space frame meet at the typical joint of Fig. 2.6.3 at which the externally applied load has Cartesian components P_x, P_y, P_z whilst the member forces (assumed for the present to be tensions) are T_1, T_2 . . . . . Let the frame receive a virtual displacement such that the selected joint moves δu, δv, δw in the coordinate directions whilst all other joints are held fixed. (Such a pattern obviously bears no relation to the actual displacements.) Then, for equilibrium of the frame

$$\delta W_e + \delta W_i = 0 \tag{2.6.1}$$

But for the selected virtual displacement, only the M_j members meeting at the joint contribute to δW_i. If the virtual elongations are δe_i, equation (2.6.1) becomes

$$P_x \delta u + P_y \delta v + P_z \delta w = \sum_{i=1}^{M_j} T_i \delta e_i \, . \tag{2.6.2}$$

Denoting the direction cosines defining the axis of member i by

$$\cos nx = l_i\,, \qquad \cos ny = m_i\,, \qquad \cos nz = n_i$$

$$\delta e_i = \delta u l_i + \delta v n_i + \delta w n_i\ . \tag{2.6.3}$$

Substituting from (2.6.3) in (2.6.2) and rearranging

$$0 = \left(P_x - \sum_{i=1}^{M_j} T_i l_i\right)\delta u + \left(P_y - \sum_{i=1}^{M_j} T_i m_i\right)\delta v + \left(P_z - \sum_{i=1}^{M_j} T_i n_i\right)\delta w\,. \tag{2.6.4}$$

Since the δu, δv, δw are arbitrary

$$P_x = \sum_{i=1}^{M_j} T_i l_i$$

$$P_y = \sum_{i=1}^{M_j} T_i m_i \tag{2.6.5}$$

$$P_z = \sum_{i=1}^{M_j} T_i n_i$$

which are the familiar equations of equilibrium for the joint. If similar virtual displacements are applied to each of the joints in turn, sets of equations of the form 2.6.5 are generated and all the bar forces determined.

Consider now the plane structure of Example (2.6.2) below in which only forces in members *BC, FC, FE* are required. This affords a further opportunity to emphasise the freedom of choice available for virtual displacements and how a judicious choice can save significant effort in calculation.

EXAMPLE 2.6.2

Use the principle of virtual work to obtain the reactions at the supports and the forces in members *BC, FC* and *FE* of the statically determinate structure of Fig. 2.6.4(a).

To find the reactions consider a rigid body virtual rotation $\delta\theta$ of the whole frame clockwise about D then from *P.V.W*,

$$R_A\, 3l\delta\theta - P 2l\delta\theta = 0$$

$$\therefore\ R_A = \tfrac{2}{3}P\,.$$

Consider now a rigid body virtual rotation $\delta\psi$ anticlockwise about A. Then from *P.V.W.*

$$R_D 3l\delta\psi - Pl\delta\psi = 0$$

$$R_D = \tfrac{1}{3}P\,.$$

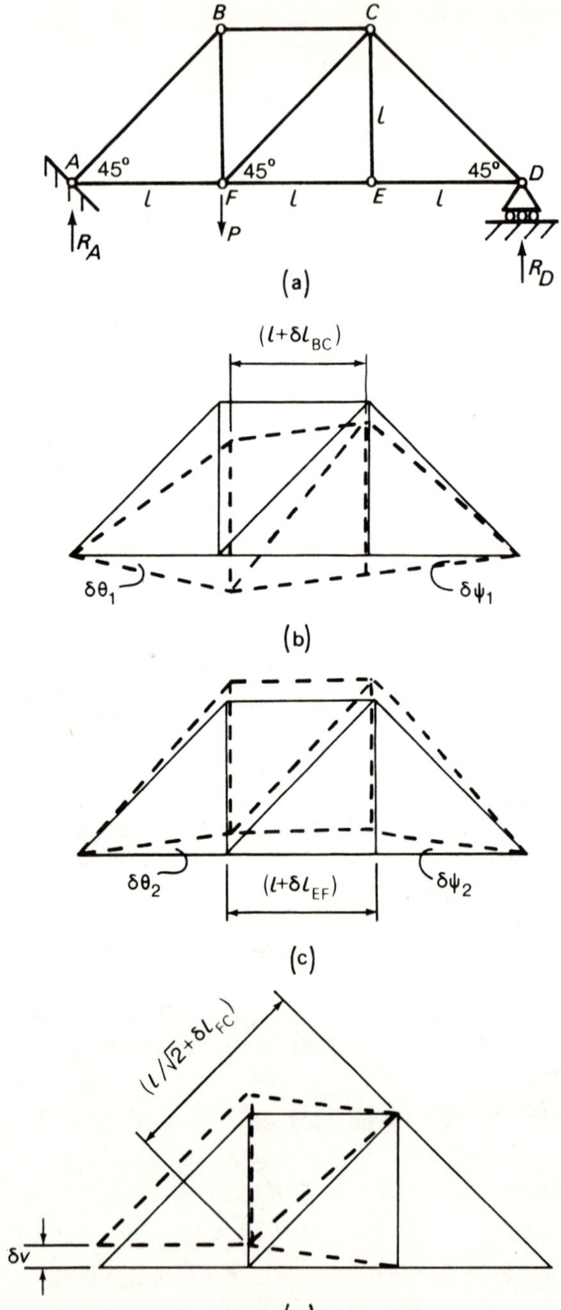
B
C
l
A
45°
45°
45°
D
l
F
l
E
l
P
R_A
R_D
(a)
(l+δl_BC)
δθ_1
δψ_1
(b)
δθ_2
δψ_2
(l+δl_EF)
(c)
(l/√2+δl_FC)
δv
(d)

Fig. 2.6.4

To determine the force in member *BC*, we assume the virtual displacement pattern of Fig. 2.6.4(b). Here, only member *BC* changes its length, the other parts of the frame move as rigid bodies and the virtual pattern clearly bears no relation to how the frame would actually deform.

From *P.V.W.*

$$Pl\delta\theta_1 = T_{BC}\delta l_{BC}$$

but from the geometry

$$\delta l_{BC} = -\tfrac{3}{2}l\delta\theta_1$$

$$\therefore T_{BC} = -\tfrac{2}{3}P.$$

In a similar manner, by using the virtual displacement pattern of Fig. 2.6.4(c) where only bar *FE* deforms,

$$T_{FE} = P/3 \ .$$

To obtain a virtual displacement involving a change in length of bar *FC* alone, we may use the pattern shown in Fig. 2.6.4(d). Then the *P,V.W.* gives,

$$T_{FE} = \sqrt{2}P/3 \ .$$

These are the results which would have been obtained directly by the method of sections on cutting members *BC, FE* and *FC.*

The forces in the members of a statically determinate structure are thus readily found, either by the usual rules of statics or by the virtual work principle as illustrated here, and it remains to determine the joint displacements. We could proceed as follows: having obtained the forces in the bars, their extensions can readily be calculated using the appropriate stress-strain law so that a geometric construction, such as the **Williot-Mohr** diagram, will give the deformed configuration of the whole frame. However, we can obtain joint displacements more accurately by a special application of the virtual work principle called the **unit dummy load** method.† Since for the time being we are only concerned with statically determinate structures, the form of the stress-strain law is immaterial but we are restricted to small displacements as will become clear below.

The procedure is as follows:

(i) Calculate the actual member forces T_{10}, T_{20} . . . . . T_{n0} due to the actual applied loads P_1, P_2 . . . . . P_m.

(ii) Calculate the actual member elongations e_{10}, e_{20}, ... e_{n0} corresponding to T_{10}, T_{20}, ... using the appropriate stress strain law for the material.

(iii) Remove the actual loads and apply a unit (dummy) load $P_1 = 1$ at the particular joint in the direction in which the actual displacement component due to the actual loads is required.

(iv) Calculate the member forces T_{11}, T_{21}. . . . . T_{n1} due to the unit dummy load.

† We shall meet a more general form of this method from the complementary virtual work principle later in this chapter.

The forces of (iv) above together with the unit dummy load and the necessary reactions at the rigid supports constitute a set of forces in equilibrium and so if they are multiplied by any system of corresponding bar virtual elongations and compatible joint virtual displacements, the corresponding virtual work must be zero. Now the *actual* bar elongations (already calculated) and the *actual* joint displacements (as yet unkown) are bound to be compatible so that, if *they* are used in conjunction with the unit dummy load and the bar forces corresponding to it, the principle of virtual work must apply so that

$$\delta W_e + \delta W_i = 0 . \tag{2.6.6}$$

But in this equation, the only external load doing work is the unit dummy one so that if the corresponding actual joint displacement component is u_0, equation (2.6.6) gives

$$1 u_0 - \left(\sum_{i=1}^{n} T_{i1} e_{io} \right) = 0$$

so that

$$u_0 = \sum_{i=1}^{n} T_{i1} e_{io} \tag{2.6.7}$$

where n is the number of bars in the frame.

To obtain the actual displacements at various joints, it is only necessary to apply a unit dummy load at each joint in turn and repeat the process.

It is important to realise that this special procedure is only applicable to *small* displacement systems. The reason for this is clear if we recall that the work calculations require the dummy bar forces to have the same line of action as the actual elongations etc. This would be the case for small deflexion theory since then the equilibrium conditions are always written down in terms of the undeformed configuration whereas, if gross distortions are contemplated, equilibrium conditions must be written down for the deformed position: under the latter circumstances, there is no assurance that the actual and dummy configurations would coincide. However, although *geometric* non-linearities are not allowed, the stress-strain law for the material need not be linear.

Frequently a structure supports just one load P_0 and the displacement corresponding to it is required. Now, if we use *actual* forces and *actual* bar elongations in equation 2.6.6, we have immediately

$$P_0 u_0 = \sum_{i=1}^{n} T_{io} e_{io}$$

which, if we multiply both sides of the equals sign by one half, is an equation of *actual* work for a single 'statically' applied load on a linear elastic system.

EXAMPLE 2.6.3

Use the dummy load method to find the horizontal displacement u for the frame of Fig. 2.6.5. The materials obey Hooke's law.

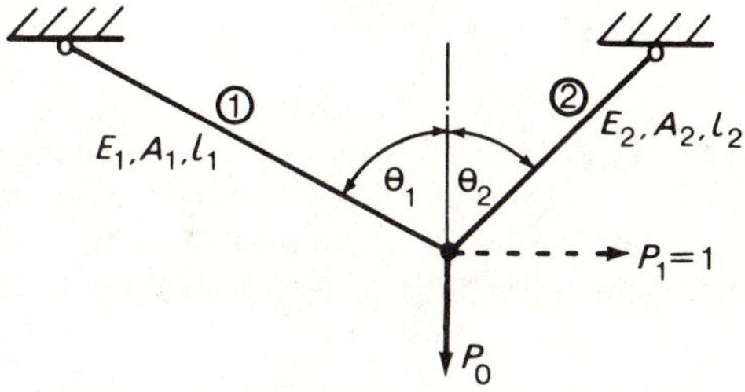

Fig. 2.6.5

From equilibrium, when P_0 is applied alone,

$$T_{10} = \frac{P_0}{\cos\theta_1 + \sin\theta_1 \cot\theta_2}, \qquad T_{20} = \frac{P_0}{\cos\theta_2 + \sin\theta_2 \cot\theta_1} \tag{a}$$

and from Hooke's law

$$e_{10} = \frac{T_{10} l_1}{E_1 A_1}, \qquad e_{20} = \frac{T_{20} l_1}{E_2 A_2}. \tag{b}$$

When P_1 is applied alone equilibrium gives

$$T_{11} = \frac{1}{\sin\theta_1 + \cos\theta_1 \tan\theta_2}, \qquad T_{21} = \frac{-1}{\sin\theta_2 + \cos\theta_2 \tan\theta_1}.$$

Then, from the *P.V.W.*

$$1u = T_{11} e_{10} + T_{21} e_{20}.$$

If $\theta_1 = \theta_2$, $l_1 - l_2$, $E_1 A_1 = E_2 A_2$, this gives $u = 0$ as we should expect from symmetry.

It is useful to observe that, in this approach, we have transferred the awkward geometric task of obtaining joint displacements from calculated bar elongations into the easier equilibrium task of finding bar forces due to a unit dummy load.

We have described the above special procedure for statically determinate systems since it provides valuable insight into the workings of the virtual work principle and into the freedom one has in a choice of virtual displacements. It can be readily extended to the analysis of statically indeterminate systems provided the principle of superposition holds: that is provided that in addition to the need for small displacements the material obeys Hooke's law. The interested reader should refer to Problem 2.6.3 and texts on structural theory† since we

† See the list for further reading at the end of this chapter.

shall not pursue this matter here. We proceed to develop methods having greater generality, including in due course what may be termed the 'true' unit dummy load method. At this stage our aim is to demonstrate which of these principles apply to situations involving:

(i) large or small displacements;

(ii) linear or non-linear stress-strain laws for the materials;

(iii) initial strains.

Accordingly to avoid the distractions of large amounts of manipulation, the physical situations described will be relatively simple. However it should be appreciated that the ideas themselves are immediately applicable to more complex systems: these we shall treat in a systematic manner in Chapter 8.

EXAMPLE 2.6.4

Discuss the forces and deformations in the symmetric frame of Fig. 2.6.6, by means of the principle of virtual work. Accept an arbitrary (but elastic) stress strain law, which is the same for the material of both inclined bars, and the possibility of large deflexions. The force P is independent of displacements.

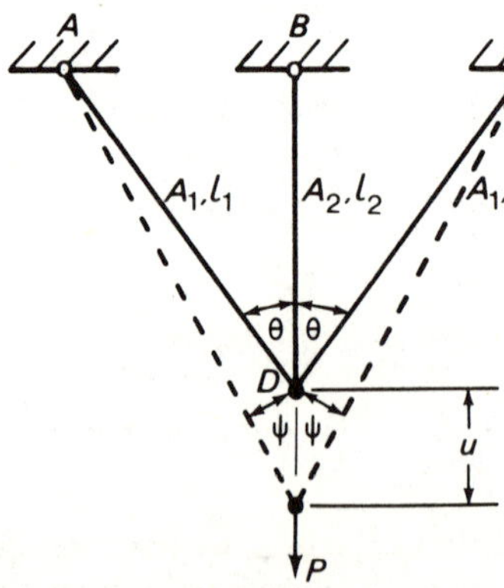

Fig. 2.6.6

The stress-strain law for the material is

$$\sigma = \frac{\partial U_0}{\partial \epsilon}.$$

The frame has one degree of freedom and either u or ψ may be used as a generalised coordinate. The extension of both inclined bars is e_1, that of the vertical bar is e_2.

Geometry of large deformation:

$$e_1 = l_1 \frac{\sin\theta}{\sin\psi} - 1 = \epsilon_1 l_1 \quad \text{(a)}$$

$$e_2 = u = l_1(\sin\theta \cot\psi - \cos\theta) = \epsilon_2 l_2 \;. \quad \text{(b)}$$

Since the frame is elastic and P independent of the displacements, we may use the *P.V.W.* in the special form of the *P.S.T.P.E.* Then

$$\delta V = \delta(U + \Omega) = 0 \quad \text{(c)}$$

is the condition for equilibrium.

Choosing ψ as our generalised coordinate, we have

$$V = 2U_1(\psi) + U_2(\psi) - Pl_1(\sin\theta \cot\psi - \cos\theta) \;. \quad \text{(d)}$$

Conditions are uniform within each bar so that

$$U = U_0 \,.\, \text{vol}$$

and $\quad V = 2U_{01}.A_1 l_1 + U_{02}.A_2 l_2 - Pl_1(\sin\theta \cot\psi - \cos\theta) \;. \quad$ (e)

For equilibrium,

$$\delta V = 0 = 2A_1 l_1 \frac{\partial U_{01}}{\partial\psi}\delta\psi + A_2 l_2 \frac{\partial U_{02}}{\partial\psi}\delta\psi + \frac{Pl_1 \sin\theta}{\sin^2\psi}\delta\psi \quad \text{(f)}$$

Now $\quad \dfrac{\partial U_{01}}{\partial\psi} = \dfrac{\partial U_{01}}{\partial\epsilon_1}\dfrac{\partial\epsilon_1}{\partial\psi} = \sigma_1 \dfrac{1}{l_1}\dfrac{\partial\epsilon_1}{\partial\psi} \quad$ (g)

and $\quad \dfrac{\partial U_{02}}{\partial\psi} = \dfrac{\partial U_{02}}{\partial e_2}\dfrac{\partial\epsilon_2}{\partial\psi} = \sigma_2 \dfrac{1}{l_2}\dfrac{\partial\epsilon_2}{\partial\psi} \quad$ (h)

and since $\delta\psi$ is arbitrary, equation (f) becomes

$$0 = 2T_1\left(-l_1 \frac{\sin\theta \cos\psi}{\sin^2\psi}\right) + T_2\left(-l_1 \frac{\sin\theta}{\sin^2\psi}\right) + \frac{Pl_1 \sin\theta}{\sin^2\psi} \quad \text{(j)}$$

whcrc $\quad T_1 = \sigma_1 A_1 \,, \quad T_2 = \sigma_2 A_2$

rearranging equation (j)

$$P = 2T_1 \cos\psi + T_2 \;. \quad \text{(k)}$$

Thus, the *P.V.W.* has given the equation of equilibrium at the joint in the deformed configuration (large deflexions) with an arbitrary stress-strain law.

To avoid unhelpful manipulative difficulties in the sequel, we shall now suppose that Hooke's law holds so that

$$T_1 = E_1 A_1 e_1 / l_1 \,, \quad T_2 = E_2 A_2 e_2 / l_2 \;. \quad \text{(l)}$$

Then, from equations (a), (b) and (l) in (k) we have

$$P = 2E_1 A_1(\sin\theta \cot\psi - \cos\psi) + E_2 A_2(\tan\theta \cot\psi - 1)$$

or $\quad C_1 \cot\psi + C_2 \cos\psi + C_3 = 0 \;. \quad$ (m)

This is a transendental equation in ψ, with C_1, C_2, C_3 constants, from which ψ and hence e_1, e_2, u, P_1, and P_2 may be found. It demonstrates that large deformation situations are non-linear even when the material obeys Hooke's law.

Let us suppose now that the displacements are, in fact, small, then $\psi = \theta + \Delta\theta$ where $\Delta\theta$ is small. Then, with the usual approximations for small angles, equation (k) becomes

$$P = 2T_1 \cos\theta + T_2 \ . \qquad \text{(n)}$$

That is, the equation of equilibrium takes precisely the form it would hold for the undeformed configuration which is a characteristic feature of small displacement theory. In a similar way, the compatibility equations (a) and (b) simplify to

$$e_1 = u\cos\theta \ , \qquad e_2 = u \ . \qquad \text{(o)}$$

Equations (n) and (o) together with Hooke's law and $E_1A_1 = E_2A_2 = EA$ give

$$u = Pl_2/EA(1 + 2\cos^3\theta) \qquad \text{(p)}$$

$$T_1 = P\cos^2\theta/(1 + 2\cos^3\theta) \ , \qquad T_2 = P/(1 + 2\cos^3\theta) \ . \qquad \text{(q)}$$

It may happen that some initial strains are built into a statically indeterminate structure. (Why not in a statically determinate one?) This effect may arise from initial lack of fit of the members (so that 'assembly stresses' are introduced) or from differential thermal expansions. The most direct way to take this into account is in the stress-strain law for the material as in the example below.

EXAMPLE 2.6.5

In the frame of Fig. 2.6.6, suppose the inclined members have an initial length $(l_1 + e_{10}) = (l/\cos\theta + e_{10})$ whilst the vertical member is $(l + e_{20})$. e_{10} and e_{20} may represent the extremes of manufacturing tolerance or may be due to temperature rise. Assuming Hooke's law and small displacements, discuss the stresses and deformation in the frame.

Here
$$\sigma = E(\epsilon - \epsilon_0) \qquad \text{(a)}$$

so that
$$U_0 = \int \sigma d\epsilon = \frac{E\epsilon^2}{2} - E\epsilon\epsilon_0 \ . \qquad \text{(b)}$$

Then
$$U = 2U_1 + U_2$$

or
$$U = 2A_1l_1\left(\frac{E\epsilon_1^{\ 2}}{2} - E\epsilon_1\epsilon_{10}\right) + A_2l_2\left(\frac{E\epsilon_2^{\ 2}}{2} - E\epsilon_2\epsilon_{20}\right) . \qquad \text{(c)}$$

To use the *P.V.W.* we must express U as a function of the generalised coordinate u.

Now $e_1 = u\cos\theta$, $e_2 = u$, then if we suppose that $E_1 = E_2$ and $A_1 = A_2$,

$$V = \frac{2A_1 l}{\cos\theta}\left[\frac{E}{2}\left(\frac{u\cos^2\theta}{l}\right)^2 - E\frac{u\cos^2\theta}{l}\epsilon_{10}\right]$$

$$+ Al\left[\frac{E}{2}\left(\frac{u}{l}\right)^2 - E\frac{u}{l}\epsilon_{20}\right] - Pu = V(u)\ . \qquad \text{(d)}$$

The equilibrium configuration is given by $\delta V = 0$. In forming δV, only u receives a variation; ϵ_{10} and ϵ_{20} are definite fixed quantities so that variations $\delta\epsilon_{10}$ and $\delta\epsilon_{20}$ do not occur. Solving, we have

$$u = \frac{[P + (EA/l)(e_{20} + 2e_{10}\cos^2\theta)]\,l}{EA(1 + \cos^3\theta)} \qquad \text{(e)}$$

from which

$$\sigma_1 = \frac{E\cos\theta}{l}(u\cos\theta - e_{10}) \qquad \text{(f)}$$

$$\sigma_2 = \frac{E}{l}(u - e_{20})\ . \qquad \text{(g)}$$

Here, we see that u can be non-zero even for a zero value of P. Then, in the event of $e_{10} = e_{20}\cos\theta$, $u = e_{20}$ from geometry and $\sigma_1 = \sigma_2 = 0$ as expected.

These examples illustrate that the range of applicability of the *P.V.W.* in stress analysis is very wide and so makes it a very powerful tool. Nevertheless, linear elastic frames are still a practically important class so that as a final application, we set up the load-deflexion equations for such a system having several degrees of freedom.

EXAMPLE 2.6.6

Determine the load-deflexion equations for the frame of Fig. 2.6.7(a) by means of the *P.V.W.* All members are of the same Hooke's law material and cross sectional area. The applied forces are independent of displacement.

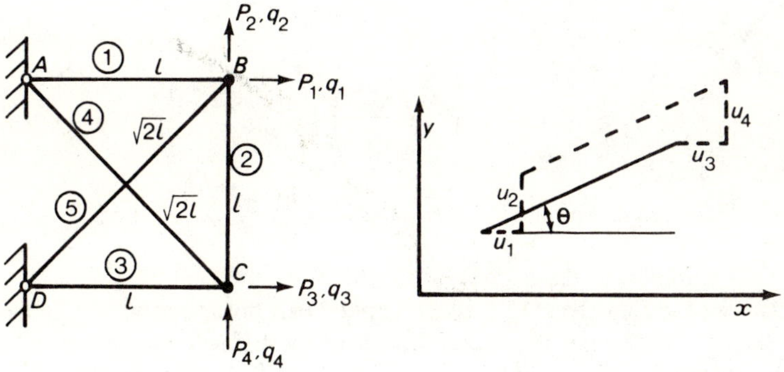

Fig. 2.6.7(a) Fig. 2.6.7(b)

The number of degrees of freedom for the frame is $2j - s = 8 - 4 = 4$. We choose $q_1 \ldots\ldots q_4$ as the generalised coordinates.

Geometry of deformation

For the typical inclined member of Fig. 2.6.7(b) the displacement-elongation relation is

$$e = (u_3 - u_1)\cos\theta + (u_4 - u_2)\sin\theta \ . \tag{a}$$

Since the bars are elastic and the loads independent of displacements, the *P.V.W.* becomes the *P.S.T.P.E.* Then

$$V = \sum_{i=1}^{5} U_i - \sum_{j=1}^{4} P_j q_j \ . \tag{b}$$

Now $U_i = EAe_i^2/2l_i$ so the first task is to use equation (a) to express the U_i as functions of $q_1 \ldots\ldots q_4$. Then

$$V = \frac{EA}{2l}q_1^2 + \frac{EA}{2l}(q_2 - q_4)^2 + \frac{EA}{2l}q_3^2 + \frac{EA}{2\sqrt{2}l}\left(\frac{q_1 + q_2}{\sqrt{2}}\right)^2 + \frac{EA}{2\sqrt{2}l}\left(\frac{q_3 - q_4}{\sqrt{2}}\right)^2 - \sum_{j=1}^{4} P_j q_j \tag{c}$$

i.e.

$$V = V(q_1, q_2, q_3, q_4) \ . \tag{d}$$

Then, the equilibrium configuration is identified by

$$\frac{\partial V}{\partial q_i} = 0 \ . \quad i = 1, 2, 3, 4 \ . \tag{e}$$

These constitute a set of simultaneous equilibrium equations for joints B and C.

Carrying out the differentiations, equating to zero and displaying the resulting equations in matrix form, we have:

$$\frac{EA}{l}\begin{bmatrix} 1 + \frac{\sqrt{2}}{4} & \frac{\sqrt{2}}{4} & 0 & 0 \\ \frac{\sqrt{2}}{4} & \left(1 + \frac{\sqrt{2}}{4}\right) & 0 & -1 \\ 0 & 0 & \left(1 + \frac{\sqrt{2}}{4}\right) & -\frac{\sqrt{2}}{4} \\ 0 & -1 & -\frac{\sqrt{2}}{4} & \left(1 + \frac{\sqrt{2}}{4}\right) \end{bmatrix} \begin{Bmatrix} q_1 \\ q_2 \\ q_3 \\ q_4 \end{Bmatrix} = \begin{Bmatrix} P_1 \\ P_2 \\ P_3 \\ P_4 \end{Bmatrix}$$

or

$$[K]\{q\} = \{P\} \ . \tag{g}$$

Once again we see that the *P.V.W.* when applied to a linear elastic system having a finite number of degrees of freedom has lead to a stiffness equilibrium relation between loads and displacements. Here $[K]$ is the stiffness matrix of the complete structure and $K_{ij} = K_{ji}$.

Having determined the q's from equations (g) the bar elongations may be found using equation (a). Use of Hooke's Law then provides the bar forces.

Example 2.2.2 showed that the characteristics of a single member can also be described by a stiffness relation

$$[k]\{u\} = \{F\} . \qquad \text{(h)}$$

In Chapter 8 on the finite element method, we shall see how the system stiffness matrix may be synthesised systematically from those of the individual elements.

The principle of virtual work provides a criterion for identifying the equilibrium configuration of a system so that the primary unkowns in the formulation of the problem are the generalised coordinates. As we have previously pointed out in Chapter 1, we usually take the datum configuration for granted so that the generalised coordinates are usually identified with the displacements of the system from its datum to its deformed position under load. The equations which give the required displacments are equilibrium equations and the procedure is applicable whether the material stress-strain law is linear or not, whether the displacements are small or not (that is linear or non-linear geometric effects) and whether initial strains are present or not. When the system is linear elastic, the required displacements are given by a set of linear simultaneous algebraic equations and the coefficients of the unknowns are identified as stiffness coefficients. Once the displacements have been obtained the stresses (or stress resultants) are then inferred from them using the appropriate strain-displacement and stress-strain relations. In later chapters, we shall use these ideas to obtain *approximate* solutions to more complex situations. However it is important to appreciate that this approximation does not come from the principle of virtual work which is itself *exact* but from assumptions, about the behaviour of the system, which facilitate solution.

It frequently happens in design practice that one is primarily interested in the stresses and not the displacements in a system. If the system is statically determinate, there is no special difficulty since the bar forces are obtainable directly; if it is statically indeterminate, however, our primary task reduces to finding the forces at redundant reactions or members since once they are known, determining the forces in the reamining members and reactions is again straightforward from equilibrium considerations. To treat such problems, the principle of virtual complementary work is appropriate since the primary unknowns in the formulation are the statically independent forces. In some situations, the number of redundant reactions and members may be significantly smaller than the number of degrees of freedom: use of the *P. V.C.W.* then has a computational advantage since there will be fewer unknowns to be found simultaneously compared with the use of the *P. V.W.* Furthermore if the displacement at only a few selected stations is required, a particular application of the *P. V.C.W.* (which is Castigliano's famous second theorem for the linear case) is especially convenient. A considerable attraction for bar systems is that the complementary energy is very easily expressed in terms of the applied loads and stress resultants in the members. As with the *P. V.W.*, our aim at this stage is to clarify the applicability of the principles involved and not to provide a catalogue of results for evermore complex systems. Accordingly the physical situations will again be relatively simple. It should be appreciated that precisely the same arguments will hold for more elaborate structures, but since the manipulations involved are considerable, they need to be carried out systematically.

We illustrate now that the principle of virtual complementary work leads to equations of compatibility expressed in terms of the wanted redundants, but that the displacements must be small. This contrasts with the principle of virtual work which yields equations of equilibrium in terms of wanted displacements even when displacements are large.

EXAMPLE 2.6.7

Discuss the behaviour of the system of Example 2.6.4 by means of the *P.V.C.W.*

Using the rules for statical determinancy we see that there is one redundancy. For equilibrium of joint D in the deformed configuration, we require

$$2T_1 \cos\psi + T_2 = P . \tag{a}$$

P is prescribed and we choose T_1 to be the statically independent bar force. Since the prescribed displacements at joints A, B, C are zero, the term $-(\Sigma T_i u_i)_u$ of equation (1.8.9) makes no contribution so that

$$V^* = U^* = 2U_1^* + U_2^* \tag{b}$$

then

$$V^* = 2U_{01}^* . A_1 l_1 + U_{02}^* . A_2 l_2 . \tag{c}$$

For compatibility, $\delta V^* = 0$ or

$$0 = 2(A_1 l_1)\delta U_{01}^* + (A_2 l_2)\delta U_{02} . \tag{d}$$

But, since the stress strain law for the material is $\epsilon = \partial U_0^* / \partial\sigma$,

$$\delta U_{01}^* = \frac{\partial U_{01}}{\partial T_1}\delta T_1 = \frac{1}{A_1}\frac{\partial U_{01}^*}{\partial\sigma_1}\delta T_1 = \frac{\epsilon_1}{A_1}\delta T_1 \tag{e}$$

and

$$\delta U_{02}^* = \frac{\partial U_{02}^*}{\partial T_2}\delta T_2 = \frac{1}{A_2}\frac{\partial U_{02}^*}{\partial\sigma_2}\delta T_2 = \frac{\epsilon_2}{A_2}\delta T_2 . \tag{f}$$

From equation (a), observing that $\delta P = 0$ since P is prescribed,

$$2\delta T_1 \cos\psi + \delta T_2 = 0 . \tag{g}$$

Using equations (e), (f), (g) in (d) gives, with $A_1 = A_2$,

$$0 = (e_1 - e_2 \cos\psi)2\delta T_1 \tag{h}$$

where e_1 and e_2 are the extensions of the bars.

Now δT_1 is arbitrary so that equation (h) requires $e_1 = e_2 \cos\psi$. If the displacements are large, this is *not* a valid equation of geometry however, if e_1 and e_2 are small, $\psi = (\theta + \Delta\theta)$ with $\Delta\theta$ small and we have

$$e_1 = e_2 \cos\theta . \tag{j}$$

This *is* a valid equation of compatibility for *small* displacements. Up to now, the form of the stress-strain law was arbitrary; its use in equation (h) would yield the required expression for the redundant member force. If for simplicity, we now assume Hooke's law to hold and that $E_1A_1 = E_2A_2 = EA$ say, equation (j) gives, when (a) is taken into account with $\psi = \theta$,

$$\frac{T_1 l_1}{EA} = \frac{(P - 2T_1 \cos\theta)}{EA} l_2 \cos\theta \tag{k}$$

with $l_2 = l_1 \cos\theta$, we have

$$T_1 = P\cos^2\theta/(1 + 2\cos^3\theta) \tag{l}$$

and substituting in (a)

$$T_2 = P/(1 + 2\cos^3\theta) . \tag{m}$$

The vertical displacement at joint D follows as

$$u = e_2 = Pl_2/EA(1 + 2\cos^3\theta) . \tag{n}$$

The results correspond exactly to those obtained in Example 2.6.4.

The simple frame of Examples 2.6.4 and 2.6.7 had just one degree of freedom and one redundant member. When we were only concerned with small displacements, neither the virtual work nor virtual complementary work principle had a special advantage for a complete solution. However, the frame of Example 2.6.6 had four degrees of freedom but only one redundant bar. It would appear that if we are only concerned with bar forces, there is an advantage in using the principle of virtual complementary work in that case.

EXAMPLE 2.6.8

Determine the bar forces in the frame of Example 2.6.6.

For equilibrium of joints B and C

$$\begin{aligned} P_1 - T_1 - \frac{1}{\sqrt{2}}T_5 &= 0 , \quad P_2 - T_2 - \frac{1}{\sqrt{2}}T_5 = 0 \\ P_3 - T_3 - \frac{1}{\sqrt{2}}T_4 &= 0 , \quad P_4 + T_2 + \frac{1}{\sqrt{2}}T_4 = 0 . \end{aligned} \tag{a}$$

Since the frame is rigidly supported, we have $V^* = U^*$, then selecting T_5 to be the statically independent (redundant) bar force, we have

$$\begin{aligned} V^* = U^* = \frac{1}{2EA}\Bigg\{&\left(P_1 - \frac{1}{\sqrt{2}}T_5\right)^2 l + \left(P_2 - \frac{1}{\sqrt{2}}T_5\right)^2 l \\ &+ [\sqrt{2}(P_2 + P_4) - T_5]^2\sqrt{2}l \\ &+ \left(P_3 - P_2 - P_4 - \frac{1}{\sqrt{2}}T_5\right)^2 l + T_5^2\sqrt{2}l\Bigg\} \end{aligned} \tag{b}$$

or

$$U^* = U^*(P_1, P_2, P_3, P_4, T_5) . \tag{c}$$

For compatibility $\delta V^* = \delta U^* = 0$ and since $P_1, \ldots\ldots P_4$ are prescribed, this implies

$$\frac{\partial U^*}{\partial T_5} = 0 \qquad \text{(d)}$$

which yields

$$0 = \frac{l}{EA}\left\{\left(P_1 - \frac{1}{\sqrt{2}}T_5\right)\left(-\frac{1}{\sqrt{2}}\right) + \left(P_2 - \frac{1}{\sqrt{2}}T_5\right)\left(-\frac{1}{\sqrt{2}}\right)\right.$$
$$+ [\sqrt{2}(P_2 + P_4) - T_5](-1)\sqrt{2}$$
$$\left. + \left(P_3 - P_2 - P_4 - \frac{1}{\sqrt{2}}T_5\right)\left(-\frac{1}{\sqrt{2}}\right) + T_5\sqrt{2}\right\} \qquad \text{(e)}$$

or $$T_5 = \frac{2}{3 + 4\sqrt{2}}[P_1 + 2\sqrt{2}P_2 + P_3 + (2\sqrt{2} - 1)P_4] \;. \qquad \text{(f)}$$

Substituting from (f) in (a) then gives all the remaining bar forces.

If now we seek the displacement at the point of application and in the direction of one of the applied loads $P_1 \ldots\ldots P_4$, we substitute from (f) in (b) to give $U^* = U^*(P_1, P_2, P_3, P_4)$ then the required displacement

$$u_i = \frac{\partial U^*}{\partial P_i} \;. \qquad \text{(g)}$$

It is instructive to write equation (d) in an alternative form. Thus

$$U^* = \sum_{i=1}^{5} U_i^* \qquad \text{(h)}$$

where $U_i^* = \frac{1}{2}(T_i^2 l_i / E_i A_i)$ is the complementary strain energy in a typical bar i. Then

$$\frac{\partial U^*}{\partial T_5} = \sum_{i=1}^{5} \frac{\partial U_i^*}{\partial T_i}\frac{\partial T_i}{\partial T_5} = \sum_{i=1}^{5} \frac{T_i l_i}{E_i A_i}\frac{\partial T_i}{\partial T_5} = 0 \qquad \text{(j)}$$

is the condition for compatibility. The displacement u_i is similarly obtained from

$$u_i = \sum_{i=1}^{5} \frac{T_i l_i}{E_i A_i}\frac{\partial T_i}{\partial P} \qquad \text{(k)}$$

after $T_1 \ldots\ldots T_5$ have been determined from (j) and (a).

Equation (k) represents the unit load method described previously and summarised in Equation (2.6.7).

Problems 2.6

2.6.1 Discuss the statical determinancy of the pin jointed frames of Fig. 2.6.8.

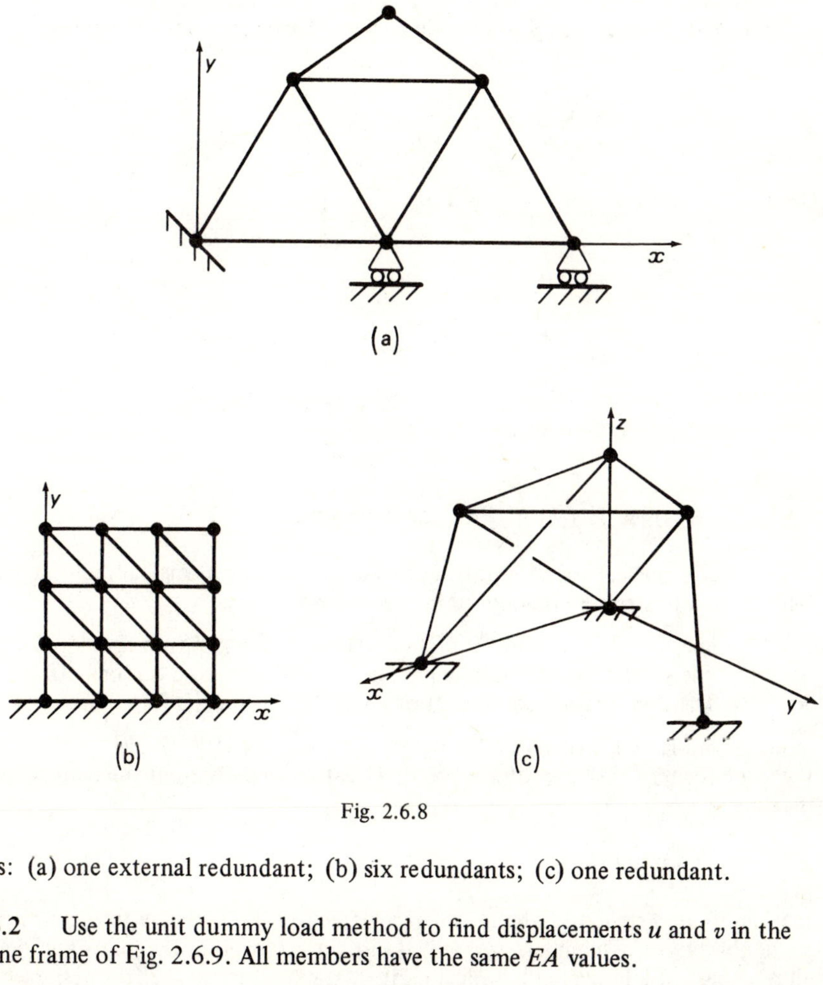

Fig. 2.6.8

Ans: (a) one external redundant; (b) six redundants; (c) one redundant.

2.6.2 Use the unit dummy load method to find displacements u and v in the plane frame of Fig. 2.6.9. All members have the same EA values.

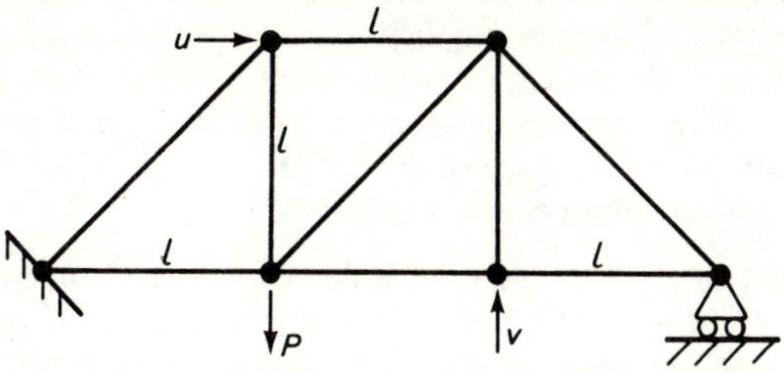

Fig. 2.6.9

Ans: $u = 8Pl/9EA\,, \quad v = -2Pl(3\sqrt{2} + 5)/9EA\ .$

2.6.3 The frame of Fig. 2.6.10 has two generalised coordinates q_1 and q_2. Establish the stiffness equilibrium equation and determine q_1 and q_2. Find the forces in the bars. Treat the same problem by the unit dummy load and complementary energy approaches to provide a check. Displacements are small and all bars have the same *EA* values.

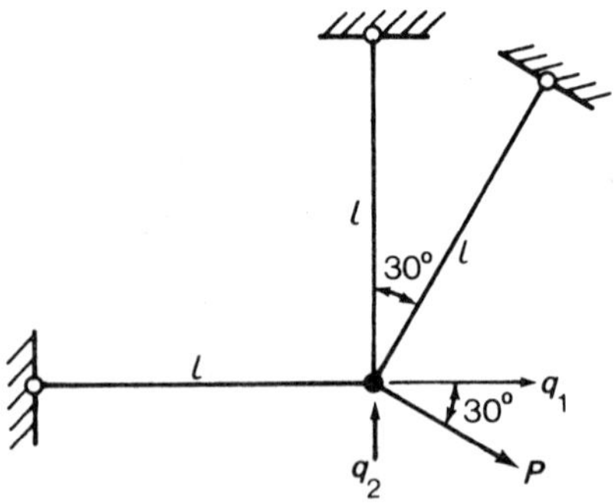

Fig. 2.6.10

Ans: $q_1 = 0.247Pl/EA\ ;\ q_2 = -0.286Pl/EA\ ;\quad T_{\text{vert}} = 0.286P$

2.6.4 The vertical bar of Problem 2.6.3 is too long by a small amount Δ. What assembly stresses are induced in the frame?

2.6.5 Bars 1, 2 . . . . . of Fig. 2.6.7(a) experience temperature rises T_1, T_2 . . . . Establish the stiffness equilibrium equations which take these thermal effects into account. The temperature coefficient of linear expansion is α for all bars.

(Hint: formulate an expression for the total potential energy V ; $\delta V = 0$ yields the same form of system stiffness matrix equation as before but the load vector is changed.)

2.7 Rigid jointed frames

In a pin jointed structure, the members are free to rotate relative to each other at the joints. If the members are welded or rivetted together the joints are 'rigid'; all members which meet at a typical joint rotate together so that the actions of the loads are transmitted by shearing, bending and torsion as well as by axial forces. The stress analysis of such frames is evidently more elaborate than for pin jointed structures. However, recalling the results of Problem 2.4.4, we see that in an analysis by work methods, it is usually only necessary to take into account the bending and torsion strain energy in the members.

It is necessary first to consider the statical determinancy of rigid jointed frames. A plane member transmits the actions of the applied load by means of three stress resultants, bending moment, shearing force and axial force. (In a space member, there are six stress resultants per member). At a given joint, there are three degrees of freedom in the plane (two translations and one rotation) and six degrees of freedom in space. Then proceeding in a manner similar to that of Section 2.6, a plane frame is found to be statically determinate provided

$$3m = 3j - 3 . \tag{2.7.1}$$

If $3m > 3j - 3$, the total number of statically independent stress resultants in the structure is

$$r_i = 3m - 3j + 3 . \tag{2.7.2}$$

Denoting the total number of reactive elements (forces and couples) at the supports by s, the structure is supported in a statically indeterminate manner if $s > 3$ and the number of external redundancies is

$$r_e = s - 3 . \tag{2.7.3}$$

Then, the total number of redundancies in an internally and externally statically indeterminate frame is

$$r = r_i + r_e = 3m - 3j + s . \tag{2.7.4}$$

We notice that s cannot be less than 3 or we have a mechanism and that s cannot be more than 3 at any given joint. Furthermore, it is necessary to check that no isolated sub-region of a frame constitutes a mechanism. In the case of a space frame, the degree of indeterminancy is given by

$$r = 6m - 6j + s . \tag{2.7.5}$$

It may happen that certain of the stress resultants are known to be zero at some sections: the bending moment at a pin joint is zero, for example. If the total number of such *releases* is k, equations 2.7.4 and 2.7.5 become

$$r = 3m - 3j + s - k \qquad 2.7.4(a)$$

$$r = 6m - 6j + s - k \qquad 2.7.5(a)$$

EXAMPLE 2.7.1

How many redundants in the frame of Fig. 2.7.1(a)?

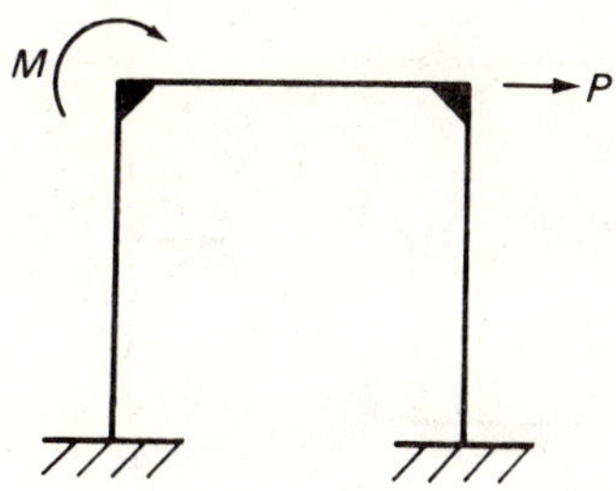

Fig. 2.7.1(a)

Here, $m = 3, j = 4, s = 6$ so that the number of redundancies is

$$r = 3 \times 3 - 3 \times 4 + 6 = 3 . \qquad \text{(a)}$$

The number of external redundancies

$$r_e = 6 - 3 = 3 . \qquad \text{(b)}$$

Thus, once the reactive elements have been found, the stress resultants can be determined in all the members since $r_i = 0$.

In principle, the choice of redundant quantities is arbitrary. Figs 2.7.1(b), (c) and (d) show some alternatives

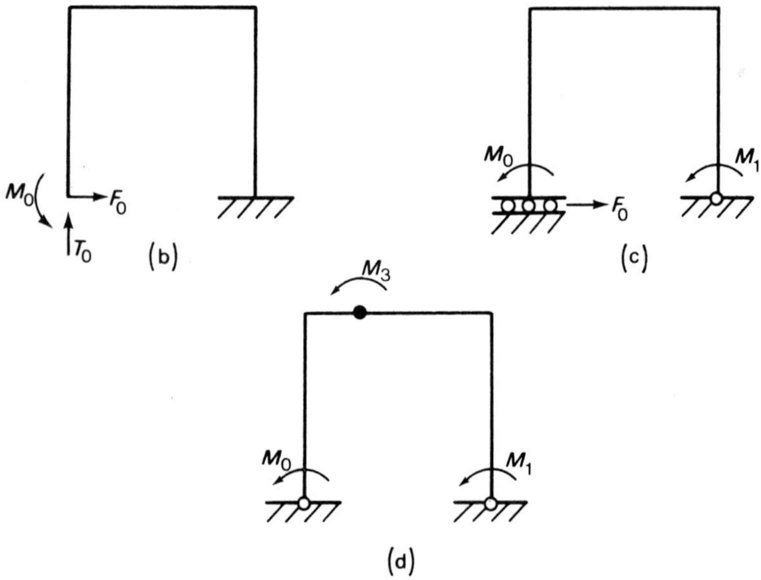

Fig. 2.7.1

EXAMPLE 2.7.2

Discuss the redundants in the frame of Fig. 2.7.2(a)

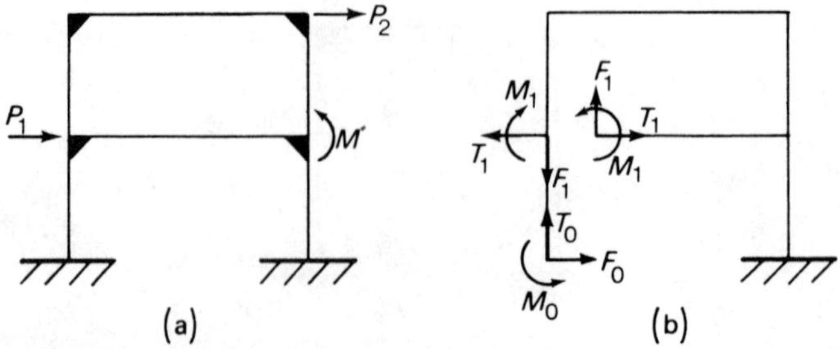

Fig. 2.7.2

Here $m = 6, j = 6, s = 6$

$$r = 6 \times 6 - 6 \times 6 + 6 = 6 \tag{a}$$

and

$$r_e = 6 - 3 = 3 \; . \tag{b}$$

Thus there are three external and three internal redundancies. A possible choice is shown in Fig. 2.7.2(b).

EXAMPLE 2.7.3

The circular ring of Fig. 2.7.3(a) is supported on flat surface. Discuss the redundancies.

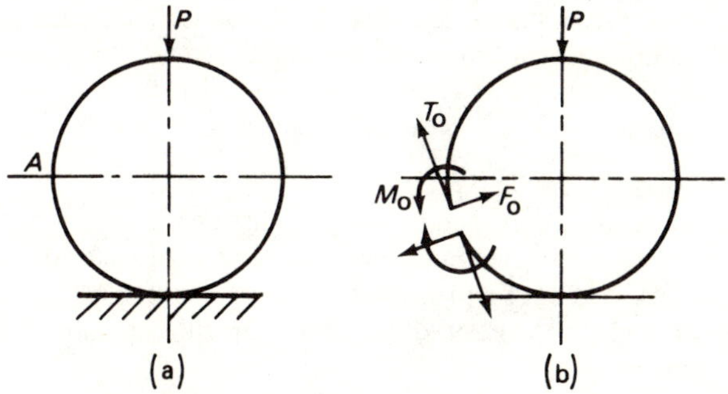

Fig. 2.7.3

Here the single normal reaction at the surface is adequate to maintain overall equilibrium so there are no external redundancies.

To apply formula (2.7.4), imagine the ring to be represented by a p-sided polygon with p rigid joints. Now imagine one side represents the rigid support to which two joints are attached, then $m = p - 1, j = p, s = 6$ so that

$$r = 3(p - 1) - 3p + s = 3 \; . \tag{a}$$

This is independent of p and the same as for the portal frame of Example 2.7.1. Now imagine p to increase indefinitely so that in the limit we have a circular ring and the two support joints coalesce giving $s = 3$. Then, there is no external redundancy and $r = r_i$. Thus, there are three internal redundancies in this 'doubly connected' system and they can be the stress resultants shown at the cut in Fig. 2.7.3(b). Notice that of the $s = 3$ support elements, only the vertical force needs to be non-zero for equilibrium.

These examples show that to determine the degree of redundancy in a structure, it is only necessary to make cuts (releases) at the supports and in the members in such a way that a singly connected chain or 'tree' of members is formed and the redundants are released as summarised in Fig. 2.7.4.

We have mentioned on several occasions that, in principle, the choice of redundants is arbitrary. On purely statical grounds this is true: however computational

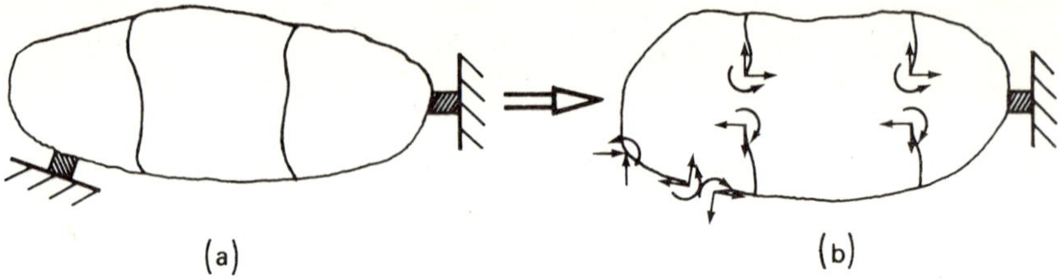

Fig. 2.7.4

aspects cannot be ignored for practical work. Thus, a suitable choice of redundants, perhaps taking account of symmetry or skew symmetry as discussed in Section 2.8, can reduce the computational effort significantly. On the other hand, a poor choice of redundants may lead to ill-conditioned equations with their attendant computational problems. A full discussion of this topic is beyond the scope of this book and the interested reader should consult texts on structural mechanics. The following observations will be helpful.

(a) Take advantage of symmetry.

(b) The reduced structure formed by making cuts and inserting joints should have stress-resultants, due to the prescribed loads, which are as close as possible to the true ones. In the case of a plane frame, where bending actions dominate, this is a matter of having the 'reduced' bending moment diagrams closely resembling the true ones.

(c) The effect of each released unkown redundant should be significant in as small a portion of the structure as possible.

We now illustrate the application of the principles of virtual work and virtual complementary work to rigid jointed frames.

EXAMPLE 2.7.4

Use the principle of virtual work to establish the load-displacement relations for the frame of Fig. 2.7.5(a). The material obeys Hooke's Law and EI is constant throughout.

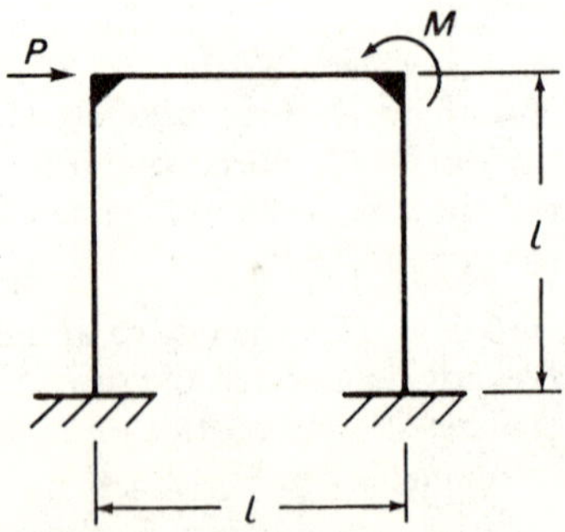

Fig. 2.7.5(a)

Regarding the frame as an assemblage of beams with local (member) coordinates $_1u_1$, $_1u_2$ . . . . . as shown in Fig. 2.7.5(b), we may use the expression for strain energy developed in Example 2.4.1. However, before we can properly say that the strain energy in the frame of Fig. 2.7.5(a) is the sum of the strain energies in the individual beams, we must take account of the constraints in the system.

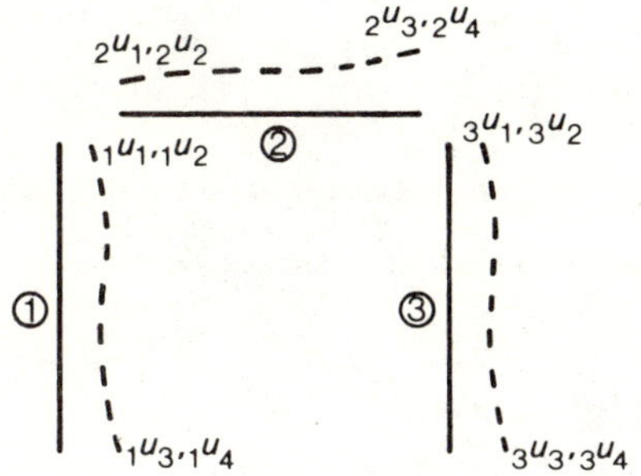

Fig. 2.7.5(b)

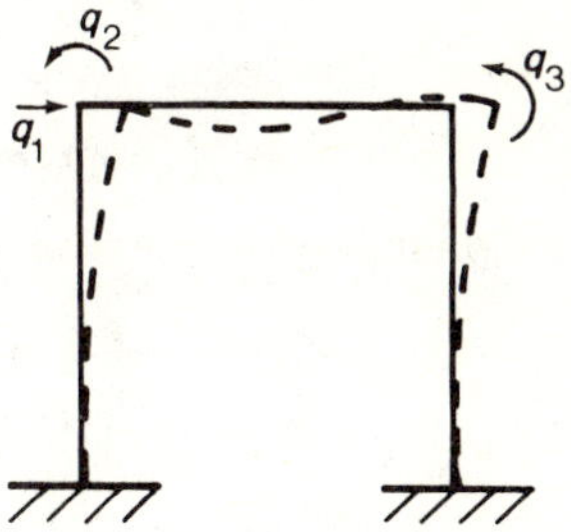

Fig. 2.7.5(c)

The external constraints on the system require that

$$_1u_3 = {}_1u_4 = {}_3u_3 = {}_3u_4 = 0 \qquad \text{(a)}$$

The neglect of the shear and direct strain energies compared with that due to bending corresponds to the introduction of internal constraints. Then, for compatibility, we have

$$_1u_1 = {}_3u_1 = q_1 \quad \text{(say)} \qquad \text{(b)}$$

$$_2u_1 = {}_2u_3 = 0 \qquad \text{(c)}$$

$$_1u_2 = {}_2u_2 = q_2 \quad \text{(say)} \qquad \text{(d)}$$

$$_2u_4 - {}_3u_2 = q_3 \quad \text{(say)} \; . \qquad \text{(e)}$$

The deformed shape of the frame then has the appearance of Fig. 2.7.5(c) and q_1, q_2, q_3 serve as generalised coordinates.

In Example 2.4.1, we saw that in a typical member (or *finite element*) of the frame, the strain energy is given by

$$U_m = \frac{1}{2}\left(\sum_{i=1}^{4}\sum_{j=1}^{4} k_{ij}u_iu_j\right)_m \qquad \text{(f)}$$

Then, incorporating the compatibility conditions (a) . . . . . (e), we find the total potential energy to be

$$V = U_1 + U_2 + U_3 - Pq_1 - Mq_3 \qquad \text{(g)}$$

or

$$\begin{aligned} V = {} & \tfrac{1}{2}[{}_1k_{11}q_1^2 + ({}_1k_{12} + {}_1k_{21})q_1q_2 + {}_1k_{22}q_2^2] \\ & + \tfrac{1}{2}[{}_2k_{22}q_2^2 + ({}_2k_{24} + {}_2k_{42})q_2q_3 + {}_2k_{44}q_3^2] \\ & + \tfrac{1}{2}[{}_3k_{11}q_1^2 + ({}_3k_{12} + {}_3k_{21})q_1q_3 + {}_3k_{22}q_3^2] \\ & - Pq_1 - Mq_3 \, . \end{aligned} \tag{h}$$

For equilibrium, $\delta V = 0$ and observing ${}_mk_{ij} = {}_mk_{ji}$ as seen in example 2.4.1,

$$\begin{aligned} \theta = {} & \delta q_1[({}_1k_{11} + {}_3k_{11})q_1 + {}_1k_{12}q_2 + {}_3k_{12}q_3 - P] \\ & + \delta q_2[{}_1k_{12}q_1 + ({}_1k_{22} + {}_2k_{22})q_2 + {}_2k_{24}q_3] \\ & + \delta q_3[{}_3k_{12}q_1 + {}_2k_{24}q_2 + ({}_2k_{44} + {}_3k_{22})q_3 - M] \, . \end{aligned} \tag{j}$$

But the virtual displacements δq_i are arbitrary so that equation (j) yields

$$\begin{aligned} ({}_1k_{11} + {}_3k_{11})q_1 + {}_1k_{12}q_2 + {}_3k_{12}q_3 &= P \\ {}_1k_{12}q_1 + ({}_1k_{22} + {}_2k_{22})q_2 + {}_2k_{24}q_3 &= 0 \\ {}_3k_{12}q_1 + {}_2k_{24}q_2 + ({}_2k_{44} + {}_3k_{22})q_3 &= M \end{aligned}$$

as the three stiffness equilibrium equations at the coordinates q_1, q_2, q_3. It is important to note that the subscripts 1, 2 etc. of k_{12} etc. refer to coordinates u_1 etc. *local* to each element. In matrix form

$$[K]\{q\} = \{Q\} \tag{l}$$

and it is clear that $K_{ij} = K_{ji}$.

Thus, although the behaviour of the individual member is more complex here than in Examples 1.6.4 and 2.6.6, their linear behaviour has lead to the same generalised form of stiffness equilibrium equation (l).

Using the explicit form of $[k]$ established in Example 2.4.1, we find here that the 'assembled' stiffness matrix is

$$[K] = \frac{EI}{l^3}\begin{bmatrix} 24 & 6l & 6l \\ 6l & 8l^2 & 2l^2 \\ 6l & 2l^2 & 8l^2 \end{bmatrix} . \tag{m}$$

Once the q's have been determined from equation (l), equations (a) to (e) yield the member (or element) displacements and $\{F\} = [k]\{u\}$ yields the member end moments and shears. On the basis of the usual beam theory, the stress analysis of the frame is then complete.

It is clear that as the system becomes increasingly complex so the need for a systematic procedure becomes more pressing. The examples so far have shown that matrix notation is eminently suited to a description of structure problems and this provides the analytic framework within which the systematic application of these theories may be carried out. In Chapter 8 we use this systematic form in connexion with finite element analysis of continuum problems. For further discussions of the so-called matrix theory of structures, the reader should consult one of the books cited at the end of this chapter.

EXAMPLE 2.7.5

Treat the frame of Example 2.7.4 by means of the principle of virtual complementary work.

There are three redundant reactions which we choose to be M_0, T_0 and F_0 in Fig. 2.7.6. Once they have been found, the bending moment in any member may be found from equilibrium conditions.

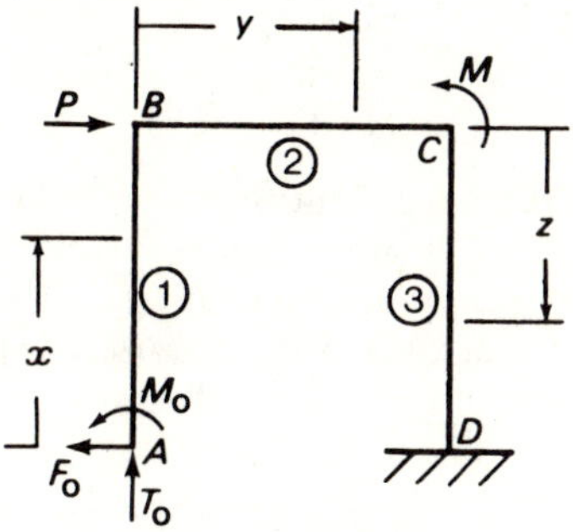

Fig. 2.7.6

Equilibrium conditions give the bending moments to be

Member AB $$M = F_0 x - M_0 \tag{a}$$

then $$\frac{\partial M}{\partial F_0} = x \, , \quad \frac{\partial M}{\partial M_0} = -1 \, , \quad \frac{\partial M}{\partial T_0} = 0 \tag{b}$$

Member BC $$M = F_0 l - M_0 + T_0 y \tag{c}$$

then $$\frac{\partial M}{\partial F_0} = l \, , \quad \frac{\partial M}{\partial M_0} = -1 \, , \quad \frac{\partial M}{\partial T_0} = y \tag{d}$$

Member CD $$M = F_0(l - z) - M_0 + T_0 l + Pz - M \tag{e}$$

$$\frac{\partial M}{\partial F_0} = (l - z) \, , \quad \frac{\partial M}{\partial M_0} = -1 \, , \quad \frac{\partial M}{\partial T_0} = l \, . \tag{f}$$

Since the supports are rigid, $V^* = U^*$ and

$$U^* = \sum_{\text{members}} \int \frac{M^2 ds}{2EI} \, . \tag{g}$$

Using equations (a), (c), (e) in (g) yields

$$U^* = U^*(F_0, M_0, T_0, P, M) \tag{h}$$

so that the compatibility condition $\delta U^* = 0$ gives

$$\frac{\partial U^*}{\partial F_0} = \frac{\partial U^*}{\partial M_0} = \frac{\partial U^*}{\partial T_0} = 0 \, . \tag{j}$$

Carrying out the necessary integrations and differentiations, we have

$$\begin{bmatrix} \dfrac{5l^3}{3EI} & \dfrac{l^3}{EI} & -\dfrac{2l^2}{EI} \\ \dfrac{l^3}{EI} & \dfrac{4l^3}{3EI} & -\dfrac{3l^2}{2EI} \\ -\dfrac{2l^2}{EI} & -\dfrac{3l^2}{2EI} & \dfrac{3l}{EI} \end{bmatrix} \begin{Bmatrix} F_0 \\ T_0 \\ M_0 \end{Bmatrix} = \begin{Bmatrix} \dfrac{Ml^2}{2EI} - \dfrac{Pl^3}{6EI} \\ \dfrac{Ml^2}{EI} - \dfrac{Pl^3}{2EI} \\ \dfrac{Ml}{EI} - \dfrac{Pl^2}{2EI} \end{Bmatrix} \tag{k}$$

or

$$[H]\{F\} = \{u\} . \tag{l}$$

Again, the principle of virtual complementary work has yielded a flexibility compatibility relation. $[H]$ is the flexibility matrix of the structure with respect to the chosen redundants' directions. Equation (l) represents the physical condition of closing, by the actions of F_0, M_0 and T_0 alone, the gaps which would open up at the cut at A due to P and M alone.

EXAMPLE 2.7.6

The frame of Example 2.7.5 is actually cut at A and loaded by a *known* $M_0 = 1$ alone. Use Castigliano's theorem to calculate the displacement corresponding to M_0 and confirm that it is equal to H_{33} in equation (l) of Example 2.7.5.

Again noting that $V^* = U^*$ in this case and that

$$U^* = \sum_{m=1}^{3} U_m^* = 3 \times \frac{M_0^2 l}{2EI} ,$$

we have

$$\left(\frac{\partial U^*}{\partial M_0}\right)_{M_0 = 1} = \frac{3l}{EI} .$$

Problems 2.7

2.7.1 The beam of Problem 2.4.5 carries a concentrated force P and a couple M at l and $2l$ from the left hand support respectively. Show that the stiffness equilibrium equation for the beam is

$$[K]\{q\} = \{Q\}$$

where $[K]$ is given in problem 2.4.5 and $\{Q\}^t = [P\,0\,0\,M]$.

2.7.2 For a beam supporting a distributed load $p(x)$, the deflected shape is not given by equations (b), (c), (d) of Example 2.4.1. However, for an *approximate* solution, we may suppose that the shape *is* given by these equations†. On this basis, deduce the generalised forces corresponding to the deflexion and slope at the free end of the cantilever (i) when $p(x) = p_0 = $ const, (ii) when

† The approximate solution of difficult stressing problems is the subject of Chapters 7 and 8.

$p(x) = p_0(1 - x/l)$. The length of the cantilever is l and x is measured from the clamped end. Write down the corresponding stiffness equilibrium equations.

Ans: $p_0 l/2\,,\; -p_0 l^2/12\,;\quad 3p_0 l/20\,,\; -p_0 l^2/30$.

2.7.3 Fig. 2.7.7 shows a chain link for which the rod diameter d is small compared with R. Show that the maximum bending moment occurs at the load point and is given by

$$M_{max} = PR(l + 2R)/2(l + \pi R)$$

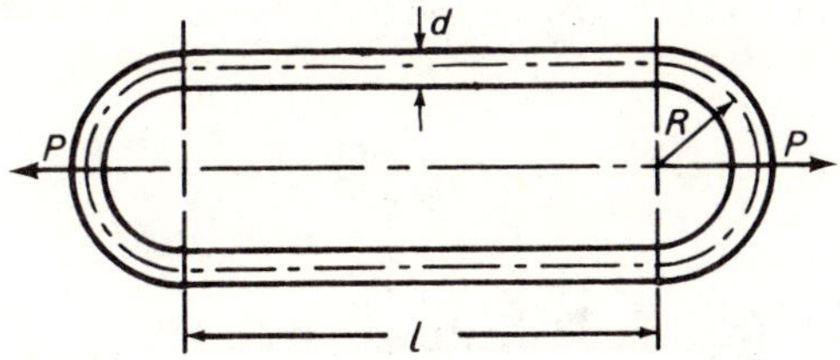

Fig. 2.7.7

2.7.4 A circular ring, diameter D and appropriate bending stiffness EI, is loaded by forces P at diametrically opposite points. Calculate the changes in diameter of the ring (i) in the direction of the loads (ii) at right angles to the load. If the ring is stiffened by a *rigid* diametral rod at right angles to the line of the forces, show that the change in diameter (i) is now

$$\Delta = 0.03PD^3/8EI\ .$$

2.7.5 A circular hoop of thin rectangular section is spun about the axis AA of Fig. 2.7.8 at ω rad/sec. Show that the horizontal diameter decreases by

$$\Delta = 2\rho\omega^2 R^5/Eh^2$$

where ρ is the density of the material and E is its Young's modulus.

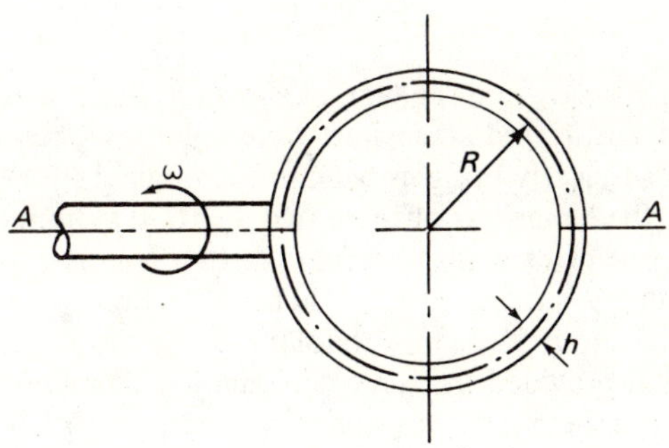

Fig. 2.7.8

2.8 Symmetric structures, symmetric and skew-symmetric load arrangements

When a structure enjoys symmetry of geometry and mechanical characteristics with respect to one, two or three axes, and the load arrangement is either symmetric or skew-symmetric with respect to the same axes, certain broad conclusions can be drawn regarding behaviour which can significantly reduce the computation involved in a stress or deformation analysis. Certain of the con clusions are true whether the stress-strain law for the material is elastic or not whilst others are valid only for linear elastic situations.

First consider the simply supported beam of Fig. 2.8.1(a). It is symmetric about axis *SS* and the load is distributed symmetrically with respect to the same axis.

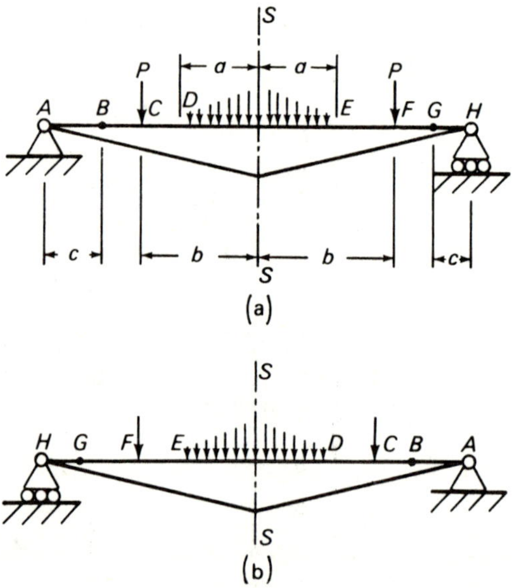

Fig. 2.8.1

When the system is viewed from the rear, as shown in Fig. 2.8.1(b) we may, for purposes of analysis, regard it as a new system; however reversing the labelling *HGFEDCBA* clearly bring us back to the original situation of Fig. 2.8.1(a). Evidently the behaviour of these two identical systems must be identical so that the displacements or stresses at two symmetrically placed stations such as *B* and *G* must be identical. The deflection curve of the beam must be symmetrical about *SS* so that the slope at the centre is zero. Similar reasoning shows that, provided the force-deformation characteristics are the same for members symmetrically positioned with respect to *BD* in the frame of Fig. 2.8.2, the displacements at *A* and *C* are the same whilst the forces in members *AD* and *CD* are the same etc.

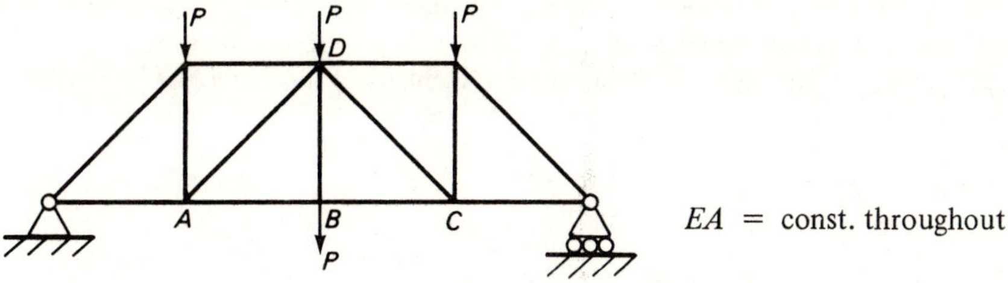

Fig. 2.8.2

In a stiffness analysis of these structures, it is only necessary to consider the sub-systems of Fig. 2.8.3(a) and (b). The symmetry arguments define adequate geometric constraints to be imposed at the axis of symmetry so that rigid body movements are precluded without restricting the natural displacement pattern of the sub-system. Thus in Fig. 2.8.3(a), the vertical displacement at A and the horizontal displacement and slope at B are zero: in Fig. 2.8.3(b), the vertical displacement at A and the horizontal displacements at B and C are zero. In the latter case, taking advantage of symmetry almost halves the number of displacement components to be found.

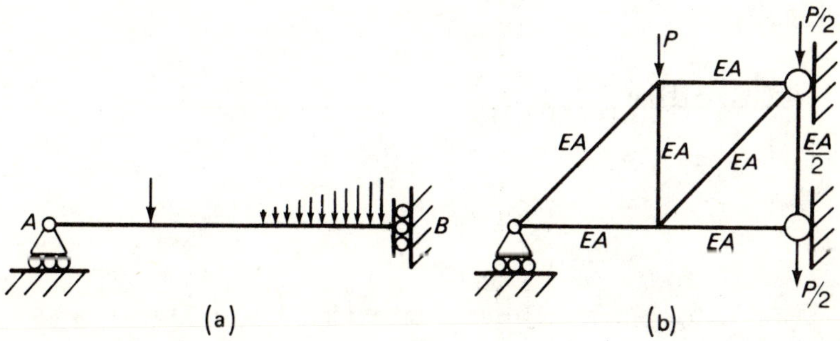

Fig. 2.8.3

The uniform beam of Fig. 2.8.4 has four supports; two of them are redundant and their determination appears to involve the solution of a pair of simultaneous equations. However, for the symmetrical loading shown, $R_1 = R_4$ and $R_2 = R_3$ so that effectively there is only one redundant to find.

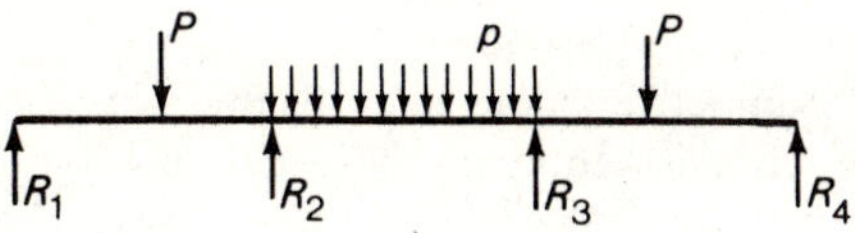

Fig. 2.8.4

It is important to observe that the above arguments proceed independently of the precise form of the force-deformation characteristics of the members: it is only necessary that symmetrically placed members in the frame (or elemental lengths in the beam) behave identically. When the force deformation characteristics of the members are skew-symmetric, as shown in Fig. 2.8.5, we can draw useful conclusions, complementary to the above, for symmetric structures supporting skew-symmetric loads.

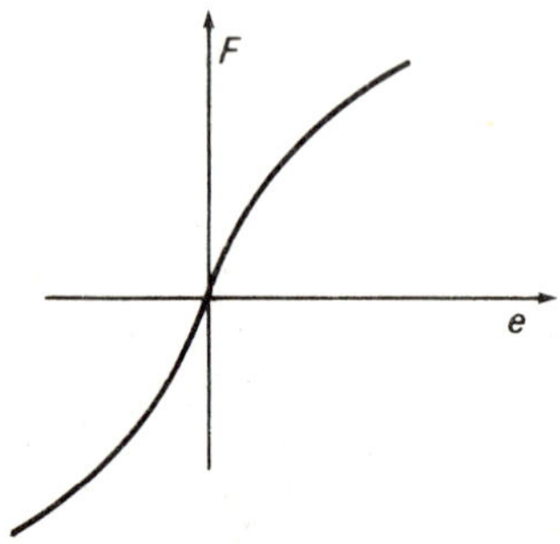

Fig. 2.8.5

The rules of statics indicate the presence of one redundant support for the beam of Fig. 2.8.6(a).

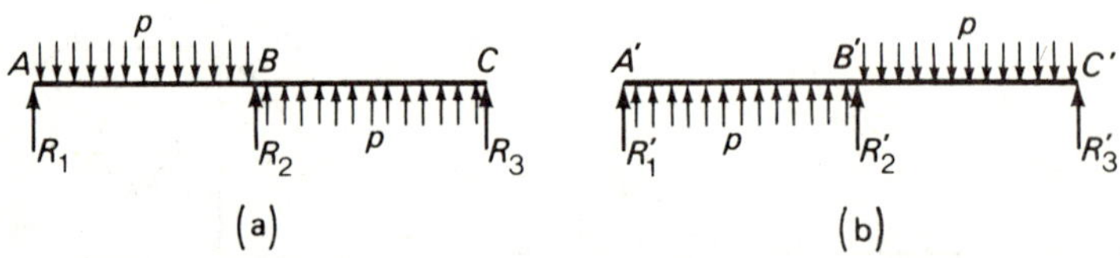

Fig. 2.8.6

It is clear that superposing the load system of Fig. 2.8.5(b) onto that of Fig. 2.8.5(a) results in a no-load arrangement and, if all elements of the beam behave elastically according to Fig. 2.8.5, all deflexions and reactions should be zero. However, the system of Fig. 2.8.6(b) is merely a view from the rear of the system of Fig. 2.8.6(a) so that $R_1 = R_3'$, $R_3 = R_1'$, $R_2 = R_2'$: superposing the two load patterns gives $R_1 + R_1' = 0$, $R_3 + R_3' = 0$, $R_2 + R_2' = 0$. This requires that $R_3 = -R_1$, and $R_2 = 0$ in the original system so that, in fact, it is statically determinate and the usual calculations for a redundant reaction are not necessary. These arguments also show that the deflexions at corresponding stations in *AB* and *BC* must be equal and opposite.

Thus we see that if a structure has an axis of symmetry, and is loaded either by a symmetric or skew symmetric arrangement of forces, the stress analysis is much simplified. The scope for applying these ideas is considerably widened when the structure is linearly elastic so that the principle of superposition holds. Now, any load arrangement can be decomposed into symmetric and skew-symmetric components, so that if the stress analysis corresponding to each is carried out separately, with the benefits described above, superposition gives the desired result for the actual arrangement. These arguments also apply when the deformation producing agency is other than a set of mechanical loads: initial strains due to symmetrical thermal or lack of fit effects can be accommodated in an analogous manner.

EXAMPLE 2.8.1

Discuss the redundancies in the 'bow girder' of Fig. 2.8.7.

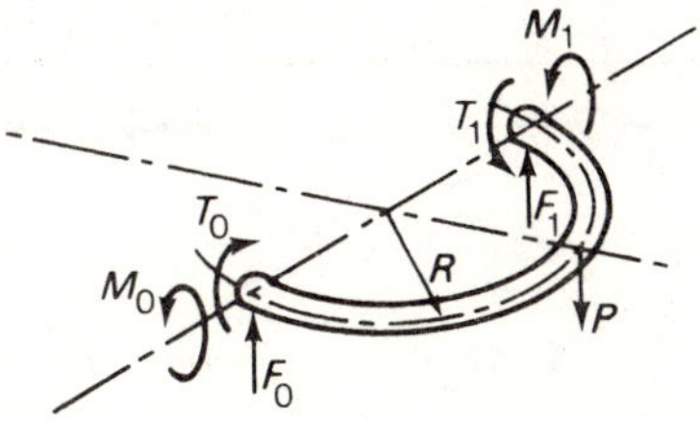

Fig. 2.8.7

A bow girder is a beam curved in plan and loaded out of its plane. The beam of Fig. 2.8.7 is clamped at both ends and loaded symmetrically.

The rules of statics indicate three redundant reactions from the fixing moments M_0, M_1, the twisting moments T_0, T_1 and the shear reactions F_0, F_1. However, from symmetry, $F_0 = F_1 = P/2, M_0 = M_1 = PR/2$ and $T_0 = T_1$. Effectively, there is only one redundant T_0 which can be found by means of the principle of virtual complementary work.

EXAMPLE 2.8.2

Represent the load on the beam of Fig. 2.8.8a by symmetric and skew symmetric components.

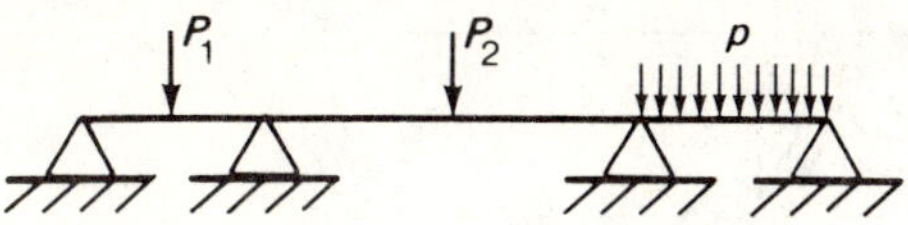

Fig. 2.8.8(a)

The simulation is achieved by a superposition of the systems of Fig. 2.8.8(b) . . . . (f).

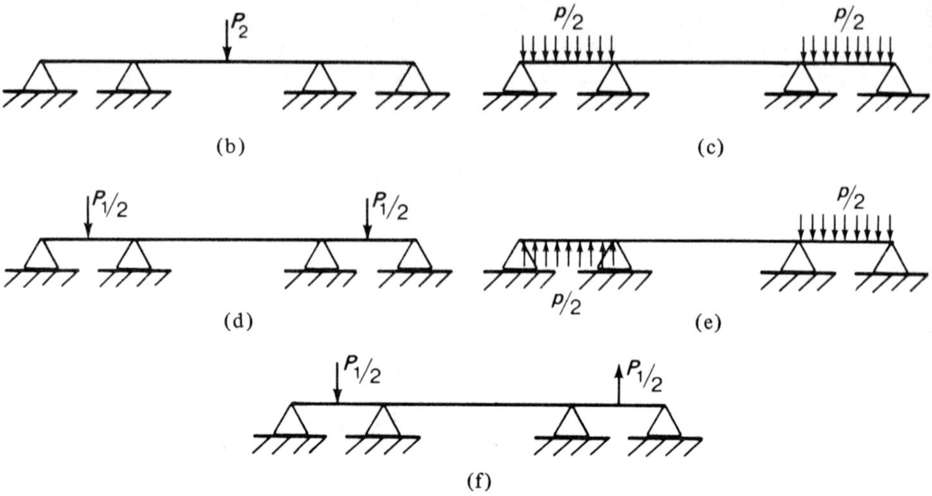

Fig. 2.8.8

Problems 2.8

2.8.1 Determine the twisting moment at the support in Example 2.8.1.

Ans: $$T_0 = -0.182PR\ .$$

2.8.2 Show that the bending moment distribution in the ring of Fig. 2.8.9 is given by

$$M = (PR/2\pi)[1 - (\cos\theta)/2 - \theta\sin\theta]\ .$$

Choose the bending moment and axial force at section C as the redundant quantities.

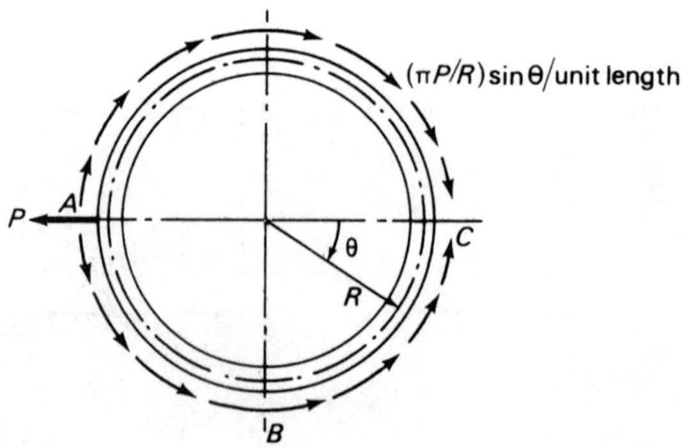

Fig. 2.8.9

2.9 Further reading

The literature on mechanics of materials and theory of structures is very extensive. The books and articles listed below comprise a very small sample.

Andrews, E.S. — *Elastic Stresses in Structures.* (This is a translation of a book by A. Castigliano). Scott, Greenwood (1919).

Argyris, J.H. *and* Kelsey, S. — *Energy Theorems and Structural Analysis,* Butterworth (1960).

Biezeno, C.B. *and* Grammel, R. — *Engineering Dynamics, Vols 1 and 2,* (translated by M.L. Meyer), Blackie (1955).

Brown, E.H. — *Structural Analysis,* Longmans (1967).

Case, J. *and* Chilver, A.H. — *Strength of Materials,* Edward Arnold (1959).

Charlton, T.M. — *Principles of Structural Analysis,* Longmans (1969).

Dym, C.L. *and* Shames, I.H. — *Solid Mechanics: a Variational Approach,* McGraw Hill (1973).

Hoff, N.J. — *The Analysis of Structures,* John Wiley (1956).

Langhaar, H.L. — *Energy Methods in Applied Mechanics,* John Wiley (1962).

Livesley, R.K. — *Matrix Methods of Structural Analysis,* Pergamon (1964).

Neal, B.G. — *Structural Theorems and their Applications,* Pergamon Press (1964).

Przemieniecki, J.S. — *Theory of Matrix Structural Analysis,* McGraw Hill (1968).

Shaw, F.S. — *Virtual Displacements and Analysis of Structures,* Prentice Hall (1972).

Westergaad, H.M. — 'On the Method of Complementary Energy', *Trans. A.S.C.E.,* Vol 107, p 765 (1942).

Williams, D. — 'The Relations between the Energy Theorems Applicable in Structural Theory.' *Phil. Mag., Series 7, Vol* **26**, *No 177 p 617,* (1938).

Chapter 3

Principles of Solid Continuum Mechanics

3.1 Introduction

Strength and deformation calculations based on the theories of 'mechanics of materials' are adequate for a wide range of significant engineering situations, the bars and frames of Chapter 2 being important examples. However, the successful design of nuclear reactors, jet engines, space craft and other advanced systems requires more refined theories of structural behaviour: indeed, with increasing demands for conservation of natural resources and economic pressures for more efficient performance, even the design of relatively simple systems requires greater precision than was previously acceptable. Continuum mechanics of solids is the body of analytic techniques which provides for this increased precision in the prediction of structural behaviour.

A continuum is a mathematical model (or idealisation) of real materials in which matter is assumed to be continuously distributed throughout the region of the solid and whose atomistic structure is ignored. Under these circumstances material particles in the body are identified with geometric points in the continuum and it is meaningful to speak of some characteristic property of the material at a *point;* this is important from an analytic point of view since one may then use the mathematical idea of limiting values. Thus, the mass density of a body *at a point* is the mass per unit volume at the point: that is, if a point P is surrounded by a portion of body, having mass Δm and volume ΔV, imagining the volume to decrease indefinitely, the density at P is defined to be

$$\rho = \lim_{\Delta V \to 0} \left(\frac{\Delta m}{\Delta V} \right) \tag{3.1.1}$$

Under these circumstances, any changes in density from point to point will be smooth so that ρ will be a continuous function of position in the solid. In a similar way, we can speak of the state of stress or strain or strain energy density to be functions of position.

In this chapter, we develop those concepts and principles necessary for the formulation of the general three dimensional problem in stress analysis. We shall use direct geometric and mechanical arguments in developing the various theories involved in contrast to the application of variational principles to formulation which appear in Chapter 6. Further we shall, on the basis of these general principles, summarise the treatment of a few special classes of problem of practical significance.

In developing these general theories, a substantial amount of mathematical manipulation is involved and because of this it is rather easy to lose track of the main thread of the argument. It has been found that the use of **indicial** (or **subscript** or **tensor**) **notation** is extremely advantageous in simplifying the presentation, reducing the amount of writing and revealing any regularity of pattern in the results. The reader not familiar with this notation is referred to the appendix wherein is given sufficient information for our purpose: the relatively small amount of effort required to gain the necessary skill will be amply repaid. We shall work mainly in terms of Cartesian coordinates so that we can proceed without the need for a formal knowledge of tensor calculus.

3.2 Configuration, state of strain and compatibility in a continuum

When a solid body changes its configuration as a result of some disturbance, we usually suppose that no splits or overlappings develop in the material. The new configuration is thus defined by displacements which are continuous functions of position in the solid. In general, passage to the new configuration may involve rigid body displacements and deformation of the body: in stress analysis, we are usually only interested in the deformations since it is they and not the rigid body motions which are related to the stresses. It is important to observe that the analysis of strain is pure geometry and makes no assumptions regarding the mechanical properties of the solid, other than it can be represented adequately by a continuum.

Suppose a body occupying some region V of space in its datum configuration is displaced to V', then a typical material point $P(x_1x_2x_3)$ will be displaced to some new position $P'(x_1x_2x_3)$ as shown in Fig. 3.2.1.

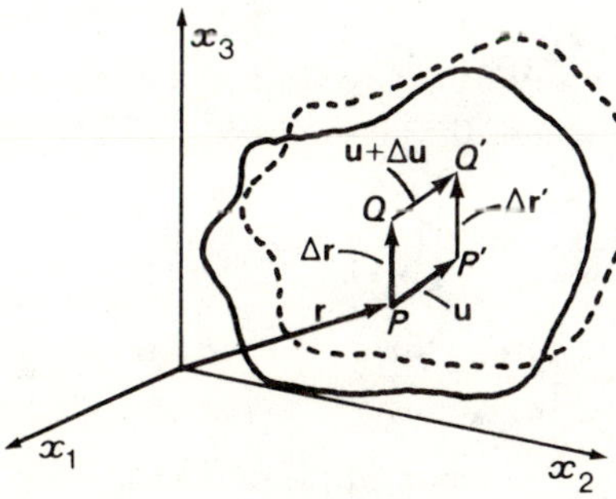

Fig. 3.2.1

The displacement vector $\mathbf{u}(x_1, x_2, x_3)$ has components $u_i(x_1, x_2, x_3)$, $i = 1, 2, 3$ and if these are known functions of position, the deformed shape of the solid may be constructed. A second point $Q(x_i + \Delta x_i)$, neighbouring on P, will be displaced through $\mathbf{u} + \Delta\mathbf{u}$ to Q': it is the *relative* displacements of a pair of neighbouring points which is significant for the analysis of strain.

The relative positions of P and Q and P' and Q' are given by the vectors $\Delta\mathbf{r}$ and $\Delta\mathbf{r}'$ respectively. The magnitudes of $\Delta\mathbf{r}$ and $\Delta\mathbf{r}'$ are given by

$$(\Delta r)^2 = \Delta x_i \Delta x_i \text{ and } (\Delta r')^2 = (\Delta x_i + \Delta u_i)(\Delta x_i + \Delta u_i) \qquad i = 1, 2, 3\,. \tag{3.2.1}$$

A measure of deformation is then

$$(\Delta r')^2 - (\Delta r)^2 = 2\Delta x_i \Delta u_i + \Delta u_i \Delta u_i \qquad i = 1, 2, 3 \tag{3.2.2}$$

where the repeated subscript implies a summation in accordance with the convention of the Appendix.

Using the linear part of Taylor's expansion, which is justified since the Δx_j are small,

$$\Delta u_i = \frac{\partial u_i}{\partial x_j}\Delta x_j \qquad i, j = 1, 2, 3\,. \tag{3.2.3}$$

Substituting from equation (3.2.3) in (3.2.2) yields

$$(\Delta r')^2 - (\Delta r)^2 = 2\frac{\partial u_i}{\partial x_j}\Delta x_i \Delta x_j + \frac{\partial u_i}{\partial x_j}\frac{\partial u_i}{\partial x_k}\Delta x_j \Delta x_k \,\dagger\,. \tag{3.2.4}$$

In the first term on the right hand side of equation (3.2.4), we can replace i by k as a dummy subscript, then noting that

$$\frac{\partial u_k}{\partial x_j}\Delta x_k \Delta x_j = \frac{1}{2}\left(\frac{\partial u_k}{\partial x_j} + \frac{\partial u_j}{\partial x_k}\right)\Delta x_j \Delta x_k\,,$$

we can write

$$(\Delta r')^2 - (\Delta r)^2 = 2E_{jk}\Delta x_j \Delta x_k\,, \qquad j, k = 1, 2, 3 \tag{3.2.5}$$

where E_{jk} is defined by

$$E_{jk} = \frac{1}{2}\left(\frac{\partial u_k}{\partial x_j} + \frac{\partial u_j}{\partial x_k}\right) + \frac{1}{2}\frac{\partial u_i}{\partial x_j}\frac{\partial u_i}{\partial x_k} = E_{kj}\,. \tag{3.2.6}$$

In unabridged notation, two of the nine equations (3.2.6) are

$$E_{xx} = \frac{\partial u}{\partial x} + \frac{1}{2}\left[\left(\frac{\partial u}{\partial x}\right)^2 + \left(\frac{\partial v}{\partial x}\right)^2 + \left(\frac{\partial w}{\partial x}\right)^2\right]. \tag{3.2.6a}$$

$$E_{xy} = \frac{1}{2}\left(\frac{\partial u}{\partial y} + \frac{\partial v}{\partial x}\right) + \frac{1}{2}\left[\frac{\partial u}{\partial x}\frac{\partial u}{\partial y} + \frac{\partial v}{\partial x}\frac{\partial v}{\partial y} + \frac{\partial w}{\partial x}\frac{\partial w}{\partial y}\right] \tag{3.2.6b}$$

where u, v, w are the x, y, z components of displacement.

Defining the quantity e by

$$e = \frac{\Delta r' - \Delta r}{\Delta r} \tag{3.2.7}$$

† k has been introduced as a dummy subscript here since expressions such as

$$\frac{\partial u_i}{\partial x_j}\frac{\partial u_i}{\partial x_j}\Delta x_j \Delta x_j$$

have no meaning.

it represents a unit elongation of the original infinitesimal line element $\Delta\mathbf{r}$.† A little manipulation shows that

$$e + \tfrac{1}{2}e^2 = \tfrac{1}{2}[(\Delta r')^2 - (\Delta r)^2]/\Delta r^2 \; . \qquad (3.2.8)$$

Now $\Delta\mathbf{r}$ is arbitrarily orientated in space, then denoting the direction cosines of $\Delta\mathbf{r}$ relative to the x_i axes by n_i, $i = 1, 2, 3$, $\Delta x_i = n_i \Delta r$. Equations (3.2.4), (3.2.7) and (3.2.8) give

$$e + \tfrac{1}{2}e^2 = E_{jk} n_j n_k \; . \qquad (3.2.9)$$

For a given deformation, e is a definite quantity independent of the choice of coordinates so that $E_{jk} n_j n_k$ is invariant with respect to rotation of axes. This tells us that the E_{jk} are the components of a second order tensor: it is called **Green's strain tensor.** When the strain components are small, the product terms in equations (3.2.5) are negligible compared with the others, and we have **Cauchy's infinitesimal strain tensor** having components

$$\epsilon_{kj} = \frac{1}{2}\left(\frac{\partial u_k}{\partial x_j} + \frac{\partial u_j}{\partial x_k}\right) = \frac{1}{2}(u_{k,j} + u_{j,k}) \; . \qquad (3.2.10)$$

Under these circumstances it is possible to attribute a simple geometric meaning to the nine components ϵ_{kj}. For small strains $e^2 \ll e$ and equation (3.2.9) becomes

$$e = \epsilon_{jk} n_j n_k \; . \qquad (3.2.11)$$

Now the elemental vector $\Delta\mathbf{r}$ has arbitrary orientation in these discussions, then suppose it is chosen parallel to, say, the x_1 direction, so that $n_1 = 1, n_2 = n_3 = 0$. Equations (3.2.10) and (3.2.11) yield

$$e_1 = \epsilon_{11} = \frac{\partial u_1}{\partial x_1} \; . \qquad (3.2.12)$$

In a similar way

$$\epsilon_{22} = \frac{\partial u_2}{\partial x_2} \quad \text{and} \quad \epsilon_{33} = \frac{\partial u_3}{\partial x_3} \qquad (3.2.13)$$

so that at the point of interest ϵ_{11}, ϵ_{22}, ϵ_{33} are unit elongations in the x_1, x_2, x_3 directions respectively. They are also called the direct or normal strain components in the x_1, x_2, x_3 directions.

To obtain a geometric meaning for ϵ_{12} etc. we consider a pair of vectors $\Delta\mathbf{r}_1$ and $\Delta\mathbf{r}_2$ with components Δx_{1j}, Δx_{2j} respectively, initially orthogonal and joining points P and Q and P and R respectively as shown in Fig. 3.2.2. After deformation, the vectors became $\Delta\mathbf{r}_1'$ and $\Delta\mathbf{r}_2'$ having components $(\Delta x_{1i} + \Delta u_{1i})$ and $(\Delta x_{2i} + \Delta u_{2i})$ respectively when point P is displaced through $\mathbf{u}$.

† It should be noted that the vector $(\Delta\mathbf{r}' - \Delta\mathbf{r})$ is not necessarily parallel to $\Delta\mathbf{r}$; e must not therefore be confused with the elementary unit elongation of a bar in tension. The latter is only a special case.

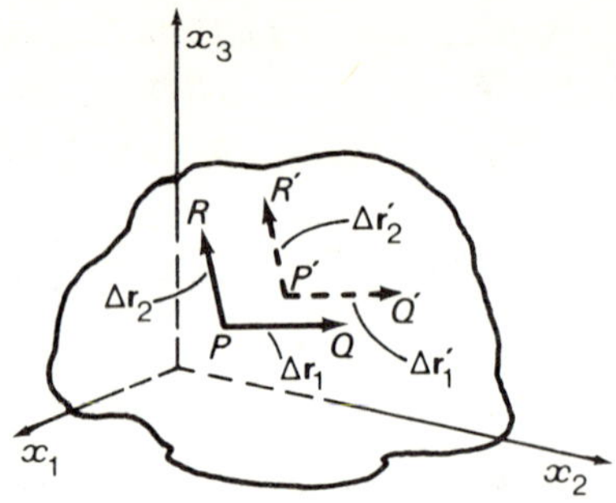

Fig. 3.2.2

From vector algebra (see Appendix), the inner product of $\Delta\mathbf{r}_1'$ and $\Delta\mathbf{r}_2'$ is

$$\Delta\mathbf{r}_1' \cdot \Delta\mathbf{r}_2' = \Delta r_1' \Delta r_2' \cos\theta = (\Delta x_{1i} + \Delta u_{1i})(\Delta x_{2i} + \Delta u_{2i}) . \tag{3.2.14}$$

Here, θ is the angle between $\Delta\mathbf{r}_1'$ and $\Delta\mathbf{r}_2'$, and there is a summation implied by the repeated subscript *i.*

$$\Delta r_1' \Delta r_2' \cos\theta = \Delta x_{1i}\Delta x_{2i} + \Delta x_{1i}\Delta u_{2i} + \Delta u_{1i}\Delta x_{2i} + \Delta u_{1i}\Delta u_{2i} . \tag{3.2.15}$$

When the strains are small, the term $\Delta u_{1i} \;\; \Delta u_{2i}$ is negligible compared with the others, then observing that from Taylor's expansion to a first order,

$$\Delta u_{1i} = \frac{\partial u_i}{\partial x_j}\Delta x_{1j} , \qquad \Delta u_{2i} = \frac{\partial u_i}{\partial x_j}\Delta x_{2j} \tag{3.2.16}$$

equation (3.2.15) becomes

$$\Delta r_1' \Delta r_2' \cos\theta = \Delta x_{1i}\Delta x_{2i} + \frac{\partial u_i}{\partial x_j}\Delta x_{1i}\Delta x_{2j} + \frac{\partial u_i}{\partial x_j}\Delta x_{2i}\Delta x_{1j} . \tag{3.2.17}$$

Interchanging the dummy subscripts *i* and *j* in the middle term on the right hand side and introducing the Kronecker δ†, we have

$$\Delta r_1' \Delta r_2' \cos\theta = \left(\delta_{ij} + \frac{\partial u_i}{\partial x_j} + \frac{\partial u_j}{\partial x_i}\right)\Delta x_{2i}\Delta x_{1j} . \tag{3.2.18}$$

Denoting by γ the *reduction* in the original right angle between $\Delta\mathbf{r}_1$ and $\Delta\mathbf{r}_2$ and observing that strains are small,

$$\cos\theta = \cos(\pi/2 - \gamma) = \sin\gamma \doteqdot \gamma \tag{3.2.19}$$

† The Kronecker delta has the property

$$\delta_{ij} = \begin{matrix} 1 & \text{if } i = j \\ 0 & \text{if } i \neq j \end{matrix}$$

See Appendix.

and using equation (3.2.18),

$$\gamma = \left(\delta_{ij} + \frac{\partial u_i}{\partial x_j} + \frac{\partial u_j}{\partial x_i}\right) \frac{\Delta x_{2i} \Delta x_{1j}}{\Delta r_1{}' \Delta r_2{}'} \quad . \tag{3.2.20}$$

In the preceeding discussion, the orientation of the orthogonal pair of vectors $\Delta \mathbf{r}_1$, $\Delta \mathbf{r}_2$ is arbitrary. Suppose now we choose them to be parallel to the x_1 and x_2 axes respectively; then their relative displacements will be as shown in Fig. 3.2.3.

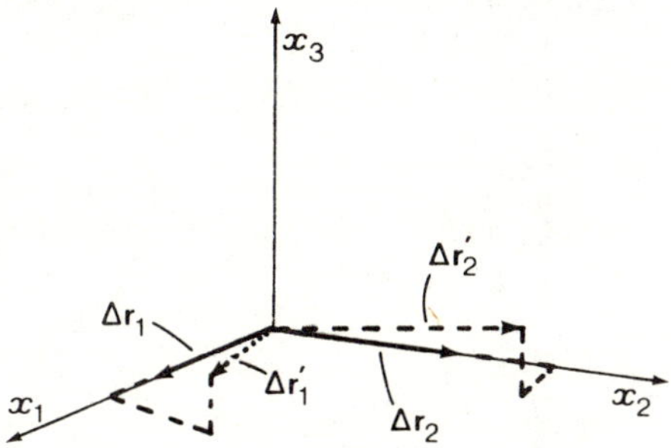

Fig. 3.2.3

In this situation the summations on i and j in equation (3.2.20) yield non-zero terms only when $i = 2$ and $j = 1$ since $\Delta \mathbf{r}_1$ and $\Delta \mathbf{r}_2$ only have non-zero components Δx_{11} and Δx_{22} respectively. Furthermore, since $\delta_{12} = 0$ by definition,

$$\gamma = \left(\frac{\partial u_2}{\partial x_1} + \frac{\partial u_1}{\partial x_2}\right) \frac{\Delta x_{22} \Delta x_{11}}{\Delta r_1{}' \Delta r_2{}'} \quad . \tag{3.2.21}$$

From Fig. 3.2.3

$$\Delta r_1{}' = \sqrt{(\Delta x_{11} + \Delta u_{11})^2 + \Delta u_{12}{}^2 + \Delta u_{13}{}^2} \tag{3.2.22}$$

$$\Delta r_2{}' = \sqrt{\Delta u_{21}{}^2 + (\Delta x_{22} + \Delta u_{22})^2 + \Delta u_{23}{}^2} \tag{3.2.23}$$

so that when the strains are small, terms such as $\Delta u_{11}{}^2$ can be neglected compared with the others yielding

$$\Delta r_1{}' = \Delta x_{11} \sqrt{(1 + 2\Delta u_{11}/\Delta x_{11})}\,, \quad \Delta r_2{}' = \Delta x_{22} \sqrt{(1 + 2\Delta u_{22}/\Delta x_{22})} \quad . \tag{3.2.24}$$

Again since strains are small, $\Delta u_{11}/\Delta x_{11}$ and $\Delta u_{22}/\Delta x_{22}$ are negligible compared with unity so that equation (3.2.21) becomes;

$$\gamma = \left(\frac{\partial u_1}{\partial x_2} + \frac{\partial u_2}{\partial x_1}\right) = 2\epsilon_{12} = 2\epsilon_{21} \quad . \tag{3.2.25}$$

Viewing along the x_3 axis towards the origin as in Fig. 3.2.4, we recognise γ as the engineering shearing strain with respect to the x_1 and x_2 directions; $\epsilon_{12} (= \epsilon_{21})$ is thus half this engineering shearing strain. In a similar way $\epsilon_{23} = \epsilon_{32}$ and $\epsilon_{31} = \epsilon_{13}$ are found to be half the engineering shearing strains with respect to the x_2x_3 and x_1x_3 directions respectively. The set of nine symmetric quantities ϵ_{ij} thus completely describes the state of infinitesimal strain at the point and may be displayed as a two dimensional array.

$$[\epsilon] = \begin{bmatrix} \epsilon_{11} & \epsilon_{12} & \epsilon_{13} \\ \epsilon_{21} & \epsilon_{22} & \epsilon_{23} \\ \epsilon_{31} & \epsilon_{32} & \epsilon_{33} \end{bmatrix} \qquad (3.2.26)$$

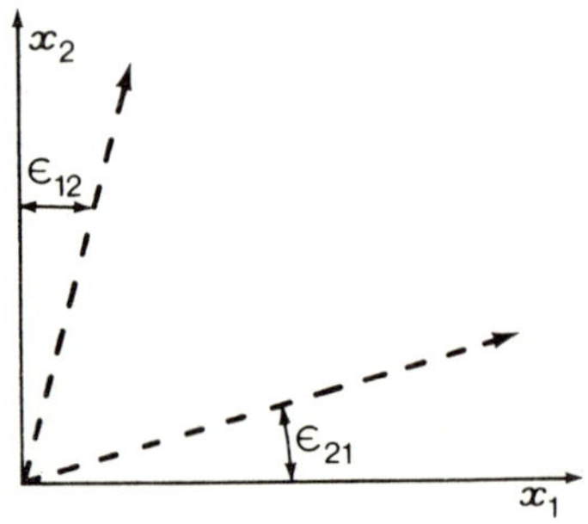

Fig. 3.2.4

In unabridged notation, the strain components are

$$\left.\begin{aligned} \epsilon_x &= \frac{\partial u}{\partial x}, \quad \epsilon_y = \frac{\partial v}{\partial y}, \quad \epsilon_z = \frac{\partial w}{\partial z} \\ \epsilon_{xy} &= \frac{1}{2}\gamma_{xy} = \frac{1}{2}\left(\frac{\partial u}{\partial y} + \frac{\partial v}{\partial x}\right) = \frac{1}{2}\gamma_{yx} = \epsilon_{yx} \\ \epsilon_{yz} &= \frac{1}{2}\gamma_{yz} = \frac{1}{2}\left(\frac{\partial v}{\partial z} + \frac{\partial w}{\partial y}\right) = \frac{1}{2}\gamma_{zy} = \epsilon_{zy} \\ \epsilon_{zx} &= \frac{1}{2}\gamma_{zx} = \frac{1}{2}\left(\frac{\partial w}{\partial x} + \frac{\partial u}{\partial z}\right) = \frac{1}{2}\gamma_{xz} = \epsilon_{xz} \end{aligned}\right\} \qquad (3.2.27)$$

where u, v, w are the x, y, z components of displacement.

The components ϵ_{ij} define the state of deformation with reference to the x_i axes. It is clear that since e of equation (3.2.11) is an invariant quantity, had we used a different set of axes x_p', $p = 1, 2, 3$ inclined to the x_i, a similar analysis to the above would lead to

$$e = \epsilon_{pq}' n_p' n_q' \qquad p,q = 1, 2, 3 \ . \qquad (3.2.28)$$

Here the n_p' are the direction cosines of $\Delta\mathbf{r}$ relative to these new axes and the ϵ_{pq}' are the strain components with respect to them. Equations (3.2.11) and (3.2.28) show that

$$\epsilon_{pq}' n_p' n_q' = \epsilon_{ij} n_i n_j \ . \qquad (3.2.29)$$

Denoting the cosine of the angle between the 'new' x_p' and the 'old' x_i by l_{pi}, coordinate geometry shows that

$$n_i = l_{pi} n_p' , \quad n_j = l_{qj} n_q' . \tag{3.2.30}$$

Then, equation (3.2.29) becomes

$$\epsilon_{pq}' = l_{pi} l_{qj} \epsilon_{ij} . \tag{3.2.31}$$

The fact that the ϵ_{ij} transform to ϵ_{pq}' according to the particular relation of equation (3.2.31), shows that they are components of a second order tensor (see appendix)†.

The previous analysis has shown that, in general, an arbitrarily oriented infinitesimal vector $\Delta \mathbf{r}$ embedded at a point in a solid changes its length and turns when the solid is deformed. We may ask, is there some orientation for $\Delta \mathbf{r}$ such that it experiences a change of length without rotation?

Now in general, $\Delta \mathbf{r}$ becomes $\Delta \mathbf{r}' = [\Delta \mathbf{r} + \Delta(\Delta \mathbf{r})]$ having a magnitude $[\Delta r + \Delta(\Delta r)]$. Then, from the inner product of the vectors

$$[\Delta r + \Delta(\Delta r)]^2 = (\Delta x_i + \Delta u_i)(\Delta x_i + \Delta u_i) .$$

Neglecting second order terms this reduces to

$$2 \Delta r \Delta(\Delta r) = 2 \Delta x_i \Delta u_i$$

so that

$$e = \frac{\Delta(\Delta r)}{\Delta r} = \frac{\Delta r' - \Delta r}{\Delta r} = \frac{\Delta x_i \Delta u_i}{(\Delta r)^2} . \tag{3.2.32}$$

But from equation (3.2.11)

$$e = \epsilon_{ij} n_i n_j = \epsilon_{ij} \frac{\Delta x_i \Delta x_j}{(\Delta r)^2} . \tag{3.2.11a}$$

Comparing equations (3.2.11a) and (3.2.32)

$$(\Delta u_i - \epsilon_{ij} \Delta x_j) \frac{\Delta x_i}{(\Delta r)^2} = 0$$

so that since $\Delta \mathbf{r}$ and hence Δx_i are arbitrary

$$\Delta u_i = \epsilon_{ij} \Delta x_j . \tag{3.2.33}$$

But if there is an orientation for $\Delta \mathbf{r}$ such that no rotation occurs, $\Delta u_i = e \Delta x_i = e \delta_{ij} \Delta x_j$ so that

$$(\epsilon_{ij} - e \delta_{ij}) \Delta x_j = 0$$

or dividing by Δr

$$(\epsilon_{ij} - e \delta_{ij}) n_j = 0 \quad i, j = 1, 2, 3 . \tag{3.2.34}$$

This is a set of three simultaneous equations in the three direction cosines for the special direction. Since all the right hand sides are zero, a non-trivial solution for the n_j exists only if the determinant of the coefficients is zero.

† Here we see one of the benefits of working in terms of the ϵ_{ij}. The set $\epsilon_x, \epsilon_y, \epsilon_z, \gamma_{xy} = \gamma_{yx}, \gamma_{yz} = \gamma_{zy}, \gamma_{zx} = \gamma_{xz}$ are not the components of a second order tensor. However, the transformation equation can be applied to $\epsilon_x, \epsilon_y, \epsilon_y, \frac{1}{2}\gamma_{xy}, \frac{1}{2}\gamma_{yz}, \frac{1}{2}\gamma_{zx}$.

That is if

$$\begin{vmatrix} (\epsilon_{11} - e) & \epsilon_{12} & \epsilon_{12} \\ \epsilon_{21} & (\epsilon_{22} - e) & \epsilon_{23} \\ \epsilon_{31} & \epsilon_{32} & (\epsilon_{33} - e) \end{vmatrix} = 0 \qquad (3.2.35)$$

Expanding this determinant yields a cubic in e

$$e^3 - E_1 e^2 + E_2 e - E_3 = 0 \qquad (3.2.36)$$

where $E_1 = \epsilon_{11} + \epsilon_{22} + \epsilon_{33}$ (3.2.37)

$$E_2 = \begin{vmatrix} \epsilon_{22} & \epsilon_{23} \\ \epsilon_{32} & \epsilon_{33} \end{vmatrix} + \begin{vmatrix} \epsilon_{11} & \epsilon_{13} \\ \epsilon_{31} & \epsilon_{33} \end{vmatrix} + \begin{vmatrix} \epsilon_{11} & \epsilon_{12} \\ \epsilon_{21} & \epsilon_{22} \end{vmatrix} \qquad (3.2.38)$$

$$E_3 = \begin{vmatrix} \epsilon_{11} & \epsilon_{12} & \epsilon_{13} \\ \epsilon_{21} & \epsilon_{22} & \epsilon_{23} \\ \epsilon_{31} & \epsilon_{32} & \epsilon_{33} \end{vmatrix} . \qquad (3.2.39)$$

For a given deformation, e is a definite quantity independent of the choice of coordinate so that E_1, E_2 and E_3 in equation (3.2.36) must also be so independent. Thus although the ϵ_{ij} are different for different choices of axes, their combinations in equations (3.2.37), (3.2.38) and (3.2.39) are not; E_1, E_2 and E_3 are called the **first, second** and **third invariants of the strain tensor.**

Since equation (3.2.36) is a cubic in e, there are three possible values of e (denoted by $\epsilon_1, \epsilon_2, \epsilon_3$) which satisfy the condition of no rotation of the vector $\Delta\mathbf{r}$. If each $\epsilon_1, \epsilon_2, \epsilon_3$ is substituted in equations (3.2.34) in turn and the orthogonality condition

$$n_i n_i = 1 \qquad (3.2.39)$$

is used, three sets of direction cosines defining three mutually perpendicular directions are obtained. These are called the principal directions of strain and the corresponding direct strains $\epsilon_1, \epsilon_2, \epsilon_3$ are the principal strain components at the point of interest. It may be shown that

(i) The roots of equation (3.2.36) are all real.

(ii) Of the three roots, one represents the greatest and another the least direct strain for any direction at the point of interest.

(iii) If any pair of principal strains are equal, then any direction in their plane is a principal direction. If all three principal strains are equal, then any direction at the point is a principal direction.

(iv) By the nature of the discussion, (no rotation of $\Delta\mathbf{r}$), the shearing strains relative to any pair of principal directions are zero.

The preceeding analysis has shown that the small deformation at a particular point in a continuum is characterised by the nine symmetric quantities ϵ_{ij} and that provided a displacement vector $\mathbf{u}$, defining the displaced configuration, is known as a continuous function of position in the solid, the ϵ_{ij} may be determined by means of equations (3.2.10). The reverse procedure is not so easy: we cannot choose six functions at random and say that they constitute a possible strain field. In mathematical terms, this is explained by equations (3.2.10) where the six strain components are seen to be expressible in terms of only three displacement component functions; clearly there must be some connexion between the ϵ_{ij}. We can also readily see that this must be so on physical grounds; thus imagine the solid to comprise a large number of blocks all fitting together; then if we attribute to each block in turn some arbitrary choice of its ϵ_{ij}, the blocks will not continue to fit each other after the imagined deformation, but gaps will open up and the blocks may tend to pierce into each other.

Thus, if a set of functions is to represent a proper set of strain components, they must be capable of integration [from equations (3.2.10)] to give a continuous distribution of displacement (that is, no splits or overlaps implied) in a solid: in fact, they must satisfy certain *compatibility conditions.* For a full discussion of this topic, the reader should consult one of the texts listed at the end of the chapter. For our purpose, it is sufficient to notice that with

$$\epsilon_{ij} = \tfrac{1}{2}(u_{i,j} + u_{j,i}) \tag{3.2.40}$$

differentiating partially with respect to x_k and x_l leads to

$$\epsilon_{ij,kl} = \tfrac{1}{2}(u_{i,jkl} + u_{j,ikl}) \,. \tag{3.2.41}$$

Interchanging subscripts,

$$\epsilon_{kl,ij} = \tfrac{1}{2}(u_{k,lij} + u_{l,kji}) \tag{3.2.42}$$

$$\epsilon_{jl,ik} = \tfrac{1}{2}(u_{j,lik} + u_{l,jik}) \tag{3.2.43}$$

$$\epsilon_{ik,jl} = \tfrac{1}{2}(u_{i,kjl} + u_{k,ijl}) \tag{3.2.44}$$

then we find the so-called St. Venant equations of compatibility to be

$$\epsilon_{ij,kl} + \epsilon_{kl,ij} - \epsilon_{jl,ik} - \epsilon_{ik,jl} = 0 \,. \tag{3.2.45}$$

At first sight, there appear to be 81, (3^4), equations here. However, closer examination shows that some are identities whilst, because of symmetry, others are repetitions. Then, only six are found to be independent and, written in unabriged notation, they are

$$2\frac{\partial^2 \epsilon_{xy}}{\partial x \partial y} = \frac{\partial^2 \epsilon_{xx}}{\partial y^2} + \frac{\partial^2 \epsilon_{yy}}{\partial x^2} \tag{3.2.46}$$

$$2\frac{\partial^2 \epsilon_{yz}}{\partial y \partial z} = \frac{\partial^2 \epsilon_{yy}}{\partial z^2} + \frac{\partial^2 \epsilon_{zz}}{\partial y^2} \tag{3.2.47}$$

$$2\frac{\partial^2 \epsilon_{zx}}{\partial y \partial x} = \frac{\partial^2 \epsilon_{zz}}{\partial x^2} + \frac{\partial^2 \epsilon_{xx}}{\partial z^2} \tag{3.2.48}$$

$$\frac{\partial^2 \epsilon_{xx}}{\partial y \partial z} = \frac{\partial}{\partial x}\left(-\frac{\partial \epsilon_{yz}}{\partial x} + \frac{\partial \epsilon_{zx}}{\partial y} + \frac{\partial \epsilon_{xy}}{\partial z}\right) \tag{3.2.49}$$

$$\frac{\partial^2 \epsilon_{yy}}{\partial z \partial x} = \frac{\partial}{\partial y}\left(-\frac{\partial \epsilon_{zx}}{\partial y} + \frac{\partial \epsilon_{xy}}{\partial z} + \frac{\partial \epsilon_{yz}}{\partial x}\right) \tag{3.2.50}$$

$$\frac{\partial^2 \epsilon_{zz}}{\partial x \partial y} = \frac{\partial}{\partial z}\left(-\frac{\partial \epsilon_{xy}}{\partial z} + \frac{\partial \epsilon_{yz}}{\partial x} + \frac{\partial \epsilon_{zx}}{\partial y}\right) \,. \tag{3.2.51}$$

For a solid body with no cavities (that is a 'singly connected region'), these conditions are sufficient to ensure that the ϵ_{ij} correspond to a continuous distribution of displacements. When the solid contains cavities, that is, it is 'multiply connected', further considerations are necessary.

To conclude this section, we reiterate that the analysis of strain is purely geometric and requires no assumptions about the material properties other than that it can adequately be represented as a continuum. The theory developed here thus describes deformations in steel, polymers, concrete or soils but with the restriction that such deformations be small. The theory for large strains is much more involved.

EXAMPLE 3.2.1

Deduce an expression for volume strain **(dilatation)** at a point and define **strain deviations.**

Imagine an infinitesimal rectangular parallelapiped of sides Δs_1, Δs_2, Δs_3 lined up with the directions of the principal strains ϵ_1, ϵ_2, ϵ_3. After straining, we have a new parallelapiped of sides $\Delta s_1(1 + \epsilon_1)$, $\Delta s_2(1 + \epsilon_2)$, $\Delta s_3(1 + \epsilon_3)$. The old and new volumes are

$$\Delta V = \Delta s_1 \Delta s_2 \Delta s_3\,, \quad \Delta V' = \Delta s_1 \Delta s_2 \Delta s_3 (1 + \epsilon_1)(1 + \epsilon_2)(1 + \epsilon_3)\,. \qquad \text{(a)}$$

Neglecting terms such as $\epsilon_1 \epsilon_2$ and higher order in (a), we have

$$\text{Vol. strain} = \frac{\Delta V' - \Delta V}{\Delta V} = \epsilon_1 + \epsilon_2 + \epsilon_3\,. \qquad \text{(b)}$$

The right had side of equation (b) is the sum of direct strains in three mutually perpendicular directions. But from equation (3.2.37), this quantity is the first invariant,

$$\therefore \text{Volume strain} = \epsilon_1 + \epsilon_2 + \epsilon_3 = \epsilon_{kk} = E_1\,.$$

For small strains, the governing equations are linear so that the principle of superposition of small effects holds. In modelling the behaviour of real materials, it proves convenient to regard the actual state of strain as a superposition of a **spherical** component and a **deviatoric** component.

Denoting the **mean** strain by

$$\epsilon_m = \tfrac{1}{3}\epsilon_{kk} \qquad \text{(d)}$$

the strain deviation component $\hat{\epsilon}_{ij}$ is defined as

$$\hat{\epsilon}_{ij} = \epsilon_{ij} - \delta_{ij}\epsilon_m\,. \qquad \text{(e)}$$

Writing equations (e) in matrix form, we have

$$\begin{bmatrix} \epsilon_{11} & \epsilon_{12} & \epsilon_{13} \\ \epsilon_{21} & \epsilon_{22} & \epsilon_{23} \\ \epsilon_{31} & \epsilon_{32} & \epsilon_{33} \end{bmatrix} = \begin{bmatrix} (\epsilon_{11} - \epsilon_m) & \epsilon_{12} & \epsilon_{13} \\ \epsilon_{21} & (\epsilon_{22} - \epsilon_m) & \epsilon_{23} \\ \epsilon_{31} & \epsilon_{32} & (\epsilon_{33} - \epsilon_m) \end{bmatrix} + \begin{bmatrix} \epsilon_m & 0 & 0 \\ 0 & \epsilon_m & 0 \\ 0 & 0 & \epsilon_m \end{bmatrix} , \qquad \text{(f)}$$

which can be re-expressed as

$$[\text{Actual State}] = [\text{Deviatoric State}] + [\text{Spherical State}] . \qquad \text{(g)}$$

The deviatoric state of strain is to do with distortion, or change of shape, without change of volume; the spherical state of strain is to do with a change of volume without distortion. To regard a state of strain as a superposition of these two proves convenient in the stress analysis of certain elastic-plastic or viscoelastic materials. (See Problem 3.2.2).

EXAMPLE 3.2.2

Discuss a state of plane strain.

If, say, the x_3 component of displacement u_3 is zero whilst u_1 and u_2 are functions of x_1 and x_2 alone, the body is in a state of plane-strain parallel to the x_1Ox_2 plane.

From the definition, one principal strain $\epsilon_3 = 0$ so that the other two principal strain directions lie in the x_1Ox_2 plane. Such a state of strain can be visualised to exist in a cylindrical body, loaded uniformly with respect to its axis, and having its end faces constrained not to move axially.

Examining compatibility equations (3.2.46) to (3.2.51), we see that under these circumstances, the only equation which is not satisfied identically is

$$2\frac{\partial^2 \epsilon_{12}}{\partial x_1 \partial x_2} = \frac{\partial^2 \epsilon_{11}}{\partial x_2{}^2} + \frac{\partial^2 \epsilon_{22}}{\partial x_1{}^2} .$$

EXAMPLE 3.2.3

The state of strain at a point of interest in a solid is defined by the matrix below. Determine the principal strains.

$$[\epsilon] = \begin{bmatrix} 2.74 & -2.12 & 4.07 \\ -2.12 & -3.69 & -3.02 \\ 4.07 & -3.02 & 5.05 \end{bmatrix} \times 100\mu \text{ strain}^{\dagger} .$$

† 1μ strain $= 10^{-6}$

The invariants of this strain tensor are found to be

$$E_1 = 11.48\,, \quad E_2 = 12.10\,, \quad E_3 = -5.64$$

where for convenience, the 100μ strain factor has been removed. The cubic yielding the factored principal strains is thus

$$\epsilon'^3 - 11.48\epsilon'^2 + 12.10\epsilon' + 5.64 = 0\,.$$

Solution gives the three principal strains as

$$\epsilon_1 = 10.21 \times 10^2\,\mu\text{ strain}, \quad \epsilon_2 = 1.61 \times 10^2\,\mu\text{ strain}$$

$$\epsilon_3 = -0.34 \times 10^2\,\mu\text{ strain}\,.$$

Using equations (3.2.34) together with the orthogonality condition $n_i n_i = 1$ for $\epsilon = \epsilon_1, \epsilon_2, \epsilon_3$ in turn, we find the direction cosines to be

$$\overset{1}{n}_1 = 0.51996\,, \quad \overset{1}{n}_2 = -0.49230\,, \quad \overset{1}{n}_3 = 0.69806$$

$$\overset{2}{n}_1 = 0.34257\,, \quad \overset{2}{n}_2 = 0.86870\,, \quad \overset{2}{n}_3 = 0.35753$$

$$\overset{3}{n}_1 = -0.78248\,, \quad \overset{3}{n}_2 = 0.05324\,, \quad \overset{3}{n}_3 = 0.62039\,.$$

It will be instructive for the reader to evaluate $\overset{1}{n}_i\overset{2}{n}_i, \overset{2}{n}_i\overset{3}{n}_i, \overset{3}{n}_i\overset{1}{n}_i$ and show that, to the accuracy used here, they are zero.

This result shows that the three principal directions are mutually perpendicular, a result which the reader is asked to prove more generally in Problem 3.2.1. Example 3.3.2, concerned with principal stresses, shows a completely parallel set of calculations.

Problems 3.2

3.2.1 When the three principal stresses at a point have distinct values ϵ_1, ϵ_2, ϵ_3 equation (3.2.34) yields

$$(\epsilon_{ij} - \epsilon_1\delta_{ij})\overset{1}{n}_j = 0 \tag{a}$$

$$(\epsilon_{ij} - \epsilon_2\delta_{ij})\overset{2}{n}_j = 0 \tag{b}$$

$$(\epsilon_{ij} - \epsilon_3\delta_{ij})\overset{3}{n}_j = 0\,. \tag{c}$$

Multiply (a) by $\overset{2}{n}_i$ and sum on i, multiply (b) by $\overset{1}{n}_i$ and sum on i, etc. show that

$$(\epsilon_1 - \epsilon_2)\overset{1}{n}_i\overset{2}{n}_i = 0 \text{ etc.} \tag{d}$$

This shows that the principal directions are mutually perpendicular. What can you say about the states $\epsilon_1 = \epsilon_2$ or $\epsilon_2 = \epsilon_3$ or $\epsilon_3 = \epsilon_1$ or $\epsilon_1 = \epsilon_2 = \epsilon_3$?

3.2.2 Working in terms of the principal strains, show that the invariants of the strain deviator are

$$\hat{E}_1 = 0 \tag{a}$$

$$\hat{E}_2 = -\tfrac{1}{6}[(\epsilon_1 - \epsilon_2)^2 + (\epsilon_2 - \epsilon_3)^2 + (\epsilon_3 - \epsilon_1)^2] \tag{b}$$

$$\hat{E}_3 = (\epsilon_1 - \epsilon_m)(\epsilon_2 - \epsilon_m)(\epsilon_3 - \epsilon_m) \ . \tag{c}$$

Equation (a) shows that volume strain is zero for the deviator state of strain.

3.2.3 The components of a strain tensor are

$$[\epsilon] = \begin{bmatrix} 3 & 5 & 8 \\ 5 & 1 & 0 \\ 8 & 0 & 2 \end{bmatrix} \times 10^3 \, \mu\,\text{strain} \ . \tag{a}$$

Determine the principal strains and their directions. Check your calculations by evaluating the invariants using (a) and the values of the principal strains and by confirming the orthogonality of the principal directions.

3.2.4 Electrical resistance strain gauges are devices for measuring the direct strains at the surface of a solid. In application, they are usually mounted onto a structure surface in the form of 'rosettes' such as that shown in Fig. 3.2.5. Using equation (3.2.31), show that

$$\epsilon_x = \epsilon_1, \quad \epsilon_y = \tfrac{1}{3}[2(\epsilon_2 - \epsilon_3) - \epsilon_1]$$

$$\gamma_{xy} = \frac{2\sqrt{3}}{3}(\epsilon_3 - \epsilon_2) \ .$$

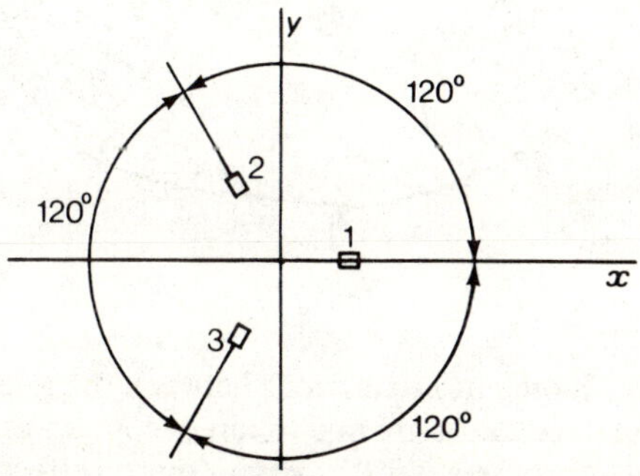

Fig. 3.2.5

3.2.5 Experiments indicated that a state of plane strain existed in the neighbourhood of a point of interest in a solid and that

$$\epsilon_x = 2 + c_1(x^2 + y^2) + (x^4 + y^4)$$

$$\epsilon_y = 3 + (x^2 + y^2) + (x^4 + y^4)$$

$$\gamma_{xy} = 4 + c_2 xy(x^2 + y^2 + 6) \ .$$

Determine the values of constants c_1 and c_2 in order that these equations should represent a compatible state of strain.

3.3 Forces, state of stress and equilibrium in a continuum

A continuum idealization of a solid body makes it inappropriate to regard the interactions between regions of a body, or several bodies, as forces between particles as described in Chapter 1; a new method of description is required. Interactions are of two kinds; forces at a distance such as gravity or magnetic forces, and forces due to contact between surfaces. The former are briefly termed **body forces** having an intensity of force per unit volume or per unit mass at a point; the latter are **surface forces** and the notion of stress as a measure of their intensity is central to the development of continuum mechanics. In the relatively simple states of stress assumed in the bars of Chapter 2, relatively simple reasoning is involved: to generalise these ideas to a three dimensional situation requires more careful consideration.

Suppose that a solid body, in equilibrium under the action of surface forces and body forces, occupies a region V of space as shown in Fig. 3.3.1.

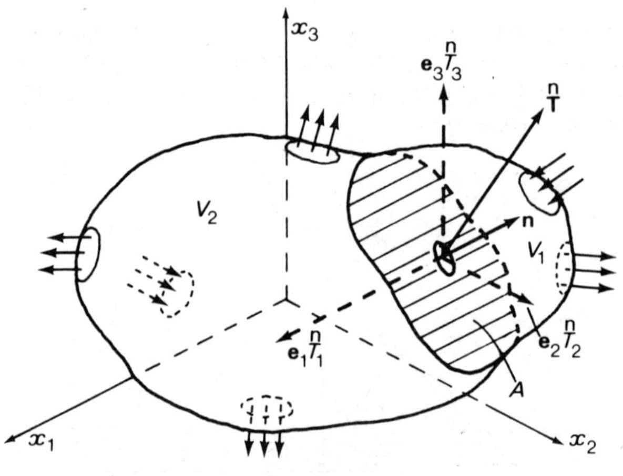

Fig. 3.3.1

The prescribed forces represent interactions between the given body and those external to it. The surface forces are due to direct contact with other bodies whilst the body forces may represent gravity or inertia effects. The interaction between the material in portions V_1 and V_2 separated by an imaginary surface A can also be viewed as an action at a distance and a surface action across A. For ordinary engineering structures, the former effect is negligible compared with that of the externally applied forces so that, in continuum mechanics, interest centres on the surface action across A as a measure of the transmission of force through the body.

The imaginary surface need not be restricted to a plane, but may even be imagined to form a closed surface enclosing a sub-region of the body. If we take an interest in a particular point P lying on this surface, an elemental area ΔA of A, surrounding P, will have some orientation defined by a unit normal vector $\mathbf{n}$†; the positive side of ΔA being defined as that to which $\mathbf{n}$ points. The action across ΔA will be equivalent to a resultant force $\Delta\mathbf{F}$ through P and a couple $\Delta\mathbf{G}$ about some axis through P. Then, the central assumption of continuum mechanics, the **stress principle of Cauchy and Euler**, asserts that in the limit as ΔA shrinks indefinitely, $\Delta\mathbf{F}/\Delta A$ tends to a limiting value whilst $\Delta\mathbf{G}/\Delta A$ tends to zero. Then

$$\overset{n}{\mathbf{T}} = \lim_{\Delta A \to 0} \left(\frac{\Delta F}{\Delta A}\right) \tag{3.3.1}$$

is called the **stress** or **traction vector** on A at P, being the force per unit surface of A at P. There are three points to notice about $\overset{n}{\mathbf{T}}$:

(i) It is not necessarily normal nor tangential to the surface A at P.

(ii) Its magnitude and orientation depend on the orientation of ΔA as well as the position of P in the solid; this fact is emphasised by the superscript n.

(iii) As drawn in Fig. 3.3.1, $\overset{n}{\mathbf{T}}$ represents the positive action at P of the part V_1 on V_2 across surface A. The action of V_2 on V_1 is $\overset{-n}{\mathbf{T}} = -\overset{n}{\mathbf{T}}$.

It follows from this discussion that, if we are to say that we know completely the state of stress at P, we must know $\overset{n}{\mathbf{T}}$ on *any* imaginary surface drawn through P.

$\overset{n}{\mathbf{T}}$ being a vector, it may be resolved into components. One possibility is a resolution into $\mathbf{N}$ and $\mathbf{S}$, components normal and tangential to the surface respectively; a graphical construction due to Mohr is then helpful to describe the relations between $\mathbf{N}$ and $\mathbf{S}$ as functions of the orientation of ΔA. An alternative procedure is to resolve $\overset{n}{\mathbf{T}}$ into its Cartesian components; with $\mathbf{e}_i$ being a unit vector in the x_i direction,

$$\overset{n}{\mathbf{T}} = \mathbf{e}_i \overset{n}{T}_i \tag{3.3.2}$$

as shown in Fig. 3.3.1. The $\overset{n}{T}_i$ are the magnitudes of the components which again are not necessarily normal nor tangential to ΔA at P. It is worth observing that the above discussion applies whether the surface A comprises a part of the actual surface of the solid or an imaginary slice through it.

We now come quite naturally to the double subscript notation for the stress components at a point. Suppose that a face A, through some point of interest, is so chosen that $\mathbf{n}$ is parallel to x_1 so that $\overset{n}{\mathbf{T}} = \overset{1}{\mathbf{T}}$ and resolution into Cartesian components gives one normal and two tangential components as shown in Fig. 3.3.2.

† When A is a closed surface, $\mathbf{n}$ is a unit *outward* normal vector. That is, it points *away from* and *not into* the enclosed volume.

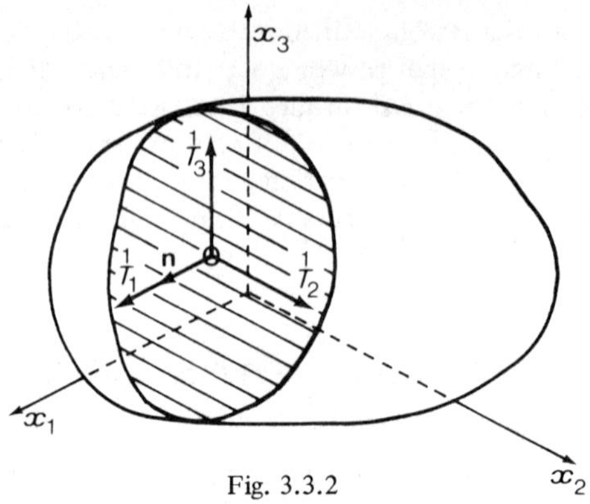

Fig. 3.3.2

It is usual to improve the notation by writing

$$\overset{1}{T}_1 = \sigma_{11}\,, \quad \overset{1}{T}_2 = \sigma_{12}\,, \quad \overset{1}{T}_3 = \sigma_{13} \tag{3.3.3}$$

In a similar way we arrive at $\sigma_{22}, \sigma_{23} \ldots\ldots \sigma_{31}$ on the faces of the parallelapiped isolated in Fig. 3.3.3.(a)

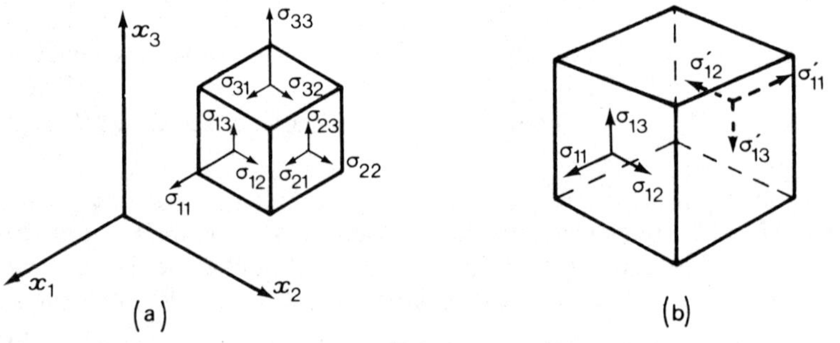

Fig. 3.3.3

The σ_{ij}'s are the magnitudes of the Cartesian components of the tractions, or stress vectors, on the Cartesian faces. $\sigma_{11}, \sigma_{22}, \sigma_{33}$ are the normal or direct stress components whilst σ_{12}, σ_{13} etc. are the tangential or shearing stress components. Almost invariably, the second subscript refers to the direction of the component whilst the first refers to the direction normal to the face on which it acts: we say *almost* invariably since some authors use the opposite notation and the reader should be aware of this. Positive stress components have the senses shown in Fig. 3.3.3(b). Thus σ_{11} is positive because it acts in a positive direction on a positive face (its normal is in the positive x_1 direction); σ_{11}' is positive because it acts in a negative direction on a negative face. All the components shown in Fig. 3.3.3(b) are positive; the remaining components will observe a similar pattern but, to avoid confusing the diagram, they have not been drawn in. It is worth emphasising that, by isolating the elemental parallelapiped as a free body as shown in Fig. 3.3.3, the σ_{ij} represent the action of the material outside the element on the material inside it.

There is a variety of notation used for the components of stress. Thus in the literature we may find σ_{11} variously written as σ_x, X_x, xx, σ_{xx} etc. We have already pointed out that, for general developments, the subscript notation is especially convenient. However when treating specific problems we will adopt a notation which is very widely used and equation (3.3.3) shows, in matrix form, its correspondence with the subscript notation.

$$\begin{bmatrix} \sigma_{11} & \sigma_{12} & \sigma_{13} \\ \sigma_{21} & \sigma_{22} & \sigma_{23} \\ \sigma_{31} & \sigma_{32} & \sigma_{33} \end{bmatrix} \equiv \begin{bmatrix} \sigma_x & \tau_{xy} & \tau_{xz} \\ \tau_{yx} & \sigma_y & \tau_{yz} \\ \tau_{zx} & \tau_{zy} & \sigma_z \end{bmatrix} . \tag{3.3.3}$$

To know the **state of stress** at a point, means that we know the $\overset{n}{\mathbf{T}}$ for *any* surface element through a point of interest: since there is an infinite number of differently orientated faces through a given point, this seems an impossible task. However, we now show that provided we know the stress components σ_{ij} associated with any three mutually perpendicular faces at a point, we can readily calculate the stress vector associated with any other face through the same point. The result is **Cauchy' formula** which is, perhaps, the only formula which needs to be remembered for the analysis of stress at a point.

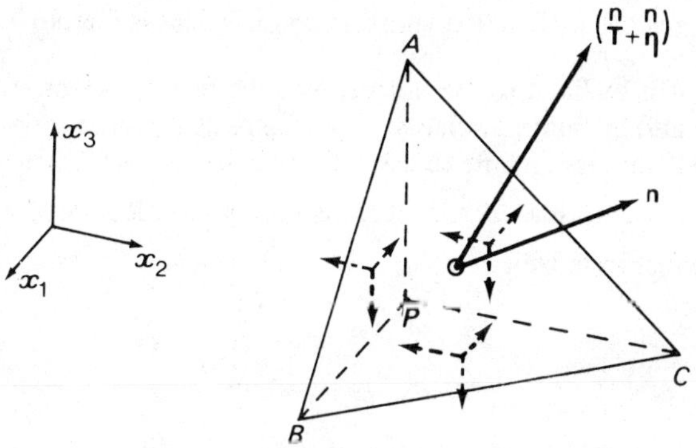

Fig. 3.3.4

Referring to Fig. 3.3.4, suppose we know the σ_{ij} associated with faces *APB, APC* and *BPC* parallel to the coordinate planes at point $P(x_i)$. Then, our objective is to determine the traction $\overset{n}{\mathbf{T}}$ on some other specified face through P in terms of the σ_{ij}. To this end, we draw face *ABC*, parallel to the specified face at some infinitesimal distance h from it so that we have defined a tetrahedron *PABC* of height h. On the areas *APB* etc, the average values of the stress components will be slightly different from the σ_{ij} on the faces *at* point P: call them $(\sigma_{ij} + \eta_{ij})$. The average traction on *ABC* will be slightly different from that on the specified face : call its components $(\overset{n}{T}_i + \overset{n}{\eta}_i)$. A body force may also be present, then if its intensity at $P(x_i)$ is $\overline{\mathbf{F}} = \overline{\mathbf{e}}_i F_i$ per unit volume, its average value within *PABC* will have components $(F_i + \eta_i{}^*)$. If we suppose that the σ_{ij} etc. vary continuously throughout the solid, the η_{ij} etc. are infinitesimals. Denoting the area of ABC by ΔA, the areas of APC, APB and BPC are respectively $\Delta A \cos(nx_1) = \Delta A n_1$, $\Delta A \cos(nx_2) = \Delta A n_2$, $\Delta A \cos(nx_3) = \Delta A n_3$ and the volume of $PABC$ is $\frac{1}{3}(\Delta A)h$. With $\mathbf{n}$ a unit vector, its components are the direction cosines n_i of the normal to ABC.

For equilibrium of the x_1 components of the forces on the element,

$$(\overset{n}{T}_1 + \overset{n}{\eta}_1)\Delta A + (F_1 + \eta_1{}^*)\tfrac{1}{3}h\Delta A - (\sigma_{11} + \eta_{11})n_1\Delta A$$

$$- (\sigma_{21} + \eta_{21})n_2\Delta A - (\sigma_{31} + \eta_{31})n_3\Delta A = 0 \ . \tag{3.3.4}$$

Dividing throughout by ΔA and taking the limit as $h \to 0$ so that face ABC comes into coincidence with the specified face, the body force contribution disappears and we have

$$\overset{n}{T}_1 = \sigma_{11}n_1 + \sigma_{21}n_2 + \sigma_{31}n_3 \ . \tag{3.3.5}$$

By considering equilibrium of forces in the x_2 and x_3 directions, we find that the traction on any face $\mathbf{n}$ through a point has components

$$\overset{n}{T}_i = \sigma_{ji}n_j \qquad i, j = 1, 2, 3 \tag{3.3.6}$$

where we recall that there is a summation on j. This is **Cauchy's formula.**

In general, $\overset{n}{\mathbf{T}}$ is inclined to the normal $\mathbf{n}$ of the face on which it acts and its magnitude and inclination relative to $\mathbf{n}$ change as the orientation of the face changes. It seems reasonable to ask if there is some orientation of face for which $\overset{n}{\mathbf{T}}$ happens to be normal to it. If there is such a face, it is clearly free of shear and with $\overset{n}{\mathbf{T}}$ parallel to $\mathbf{n}$, we can write

$$\overset{n}{\mathbf{T}} = \sigma\mathbf{n} \tag{3.3.7}$$

or

$$\overset{n}{T}_i = \sigma n_i \qquad i = 1, 2, 3 \tag{3.3.8}$$

where σ is some scalar. But equations (3.3.6) are still true for this special direction so that

$$\sigma_{ji}n_j = \sigma n_i = \sigma\delta_{ij}n_j$$

or

$$(\sigma_{ji} - \sigma\delta_{ij})n_j = 0 \tag{3.3.9}$$

where δ_{ij} is the Kronecker δ.

This is a set of homogeneous linear algebraic equations in the direction cosines: and a non-trivial solution, satisfying the orthogonality condition $n_i n_i = 1$, is obtained only if the determinant of the coefficients vanishes. That is if

$$\begin{vmatrix} (\sigma_{11} - \sigma) & \sigma_{12} & \sigma_{13} \\ \sigma_{21} & (\sigma_{22} - \sigma) & \sigma_{23} \\ \sigma_{31} & \sigma_{32} & (\sigma_{33} - \sigma) \end{vmatrix} = 0 \ . \tag{3.3.10}$$

Expanding the determinant yields a cubic in σ

$$\sigma^3 - I_1\sigma^2 + I_2\sigma - I_3 = 0 \tag{3.3.11}$$

where $$I_1 = \sigma_{11} + \sigma_{22} + \sigma_{33} \tag{3.3.12}$$

$$I_2 = \begin{vmatrix} \sigma_{22} & \sigma_{23} \\ \sigma_{32} & \sigma_{33} \end{vmatrix} + \begin{vmatrix} \sigma_{11} & \sigma_{13} \\ \sigma_{31} & \sigma_{33} \end{vmatrix} + \begin{vmatrix} \sigma_{11} & \sigma_{12} \\ \sigma_{21} & \sigma_{22} \end{vmatrix} \tag{3.3.13}$$

$$I_3 = \begin{vmatrix} \sigma_{11} & \sigma_{12} & \sigma_{13} \\ \sigma_{21} & \sigma_{22} & \sigma_{23} \\ \sigma_{31} & \sigma_{32} & \sigma_{33} \end{vmatrix} . \tag{3.3.14}$$

We shall prove below that $\sigma_{ij} = \sigma_{ji}$, corresponding to the familiar statement that 'cross shears are equal'. The matrix of coefficients in equations (3.3.9) is thus symmetric and it may be shown that all three roots $\sigma_1, \sigma_2, \sigma_3$ of equation (3.3.11) are real; they are called the principal stresses at the point. Of the three, one will be the greatest direct stress for any face through the point, whilst one will be the least. Substituting each value in turn in equations (3.3.9) and using $n_i n_i = 1$ yields three triads of direction cosines corresponding to three principal directions; it may be shown that they are mutually perpendicular. The principal stresses are definite physical quantities independent of the choice of co-ordinate axes so that I_1, I_2, I_3 in equation (3.3.11) must also be so independent: they are the **first, second and third invariants of the stress components.**

In the above analysis, we used a coordinate system x_i with which were associated stress components σ_{ij}. Evidently, the discussion is still valid for coordinates x_p' with stress components σ_{pq}': we now deduce a relation between σ_{ij} and σ_{pq}'. Suppose the 'new' coordinates x_p' are obtained from the 'old' coordinates x_i by rotation of axes, then from geometry

$$x_p' = l_{pi}x_i \quad p,i = 1, 2, 3 \tag{3.3.15}$$

where the $l_{pi} = \cos(x_p', x_i)$ are the direction cosines of the new axes relative to the old ones. Choose a unit vector **n** to coincide with x_p' (that is, the chosen face is normal to x_p'), and another unit vector **t** to coincide with x_q' (one of the other two 'new' axes lying *in* the chosen face). Then, the component of traction $\overset{n}{\mathbf{T}} = \overset{p}{\mathbf{T}}$ on face **n** in direction **t** is given by

$$\overset{p}{T}_q = \sigma_{pq}' = \overset{n}{\mathbf{T}} \cdot \mathbf{t} = \overset{n}{T}_i t_i = \sigma_{ji} n_j t_i \tag{3.3.16}$$

where Cauchy's formula has been used for the last term. But

$$n_j = \cos(n, x_j) = \cos(x_p', x_j) = l_{pj} \text{ and } t_i = \cos(t, x_i) = \cos(x_q' x_i) = l_{qi}$$

so that $$\sigma_{pq}' = l_{pj} l_{qi} \sigma_{ji} .$$

Here, i and j are dummy subscripts and may be interchanged, then accepting that $\sigma_{ij} = \sigma_{ji}$ from below, we may write

$$\sigma_{pq}' = l_{pi} l_{qj} \sigma_{ij} . \tag{3.3.17}$$

This law of transformation of the stress components shows that they are the components of a second order tensor. (See Appendix).

So far, this account of stresses has been concerned with conditions *at a point*: we now establish an analytic description of how stresses vary from point to point within the solid. To this end, we consider the equilibrium of a rectangular parallelapiped element, $dx_1\,dx_2\,dx_3$, with one corner at point $P(x_i)$ and generated by imaginary cuts in the solid parallel to the coordinate axes. Fig. 3.3.5 shows the element isolated as a free body with the actions on it of the material outside i We first consider equilibrium of forces in the x_1 direction; to avoid undue confusion on the diagram, only those stress components making a contribution to this relation are actually labelled.

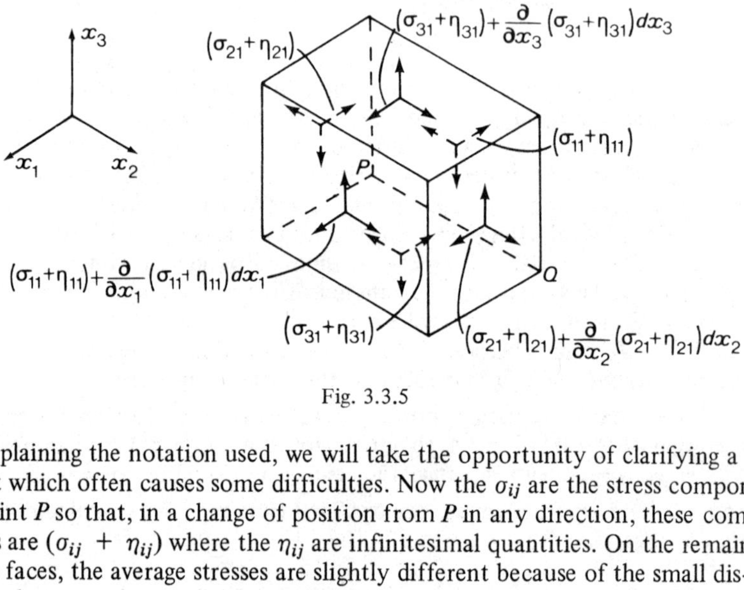

Fig. 3.3.5

In explaining the notation used, we will take the opportunity of clarifying a point which often causes some difficulties. Now the σ_{ij} are the stress components at point P so that, in a change of position from P in any direction, these components are $(\sigma_{ij} + \eta_{ij})$ where the η_{ij} are infinitesimal quantities. On the remaining three faces, the average stresses are slightly different because of the small distances between the parallel faces. As will be seen below, the η_{ij} contributions to equilibrium are of higher order and so may be neglected compared with the other terms so that the contributions of the η_{ij} are often ignored completely, but without explanation. Allowing for changes in the σ_{ij} from face to parallel face whilst apparently ignoring changes over a given face is a common source of confusion for the newcomer to the subject: it is hoped that these remarks will help to overcome these difficulties. With $(\mathbf{F} + \eta^*)$ as the average intensity of body force, equilibrium of forces in the x_1 direction requires

$$\left[(\sigma_{11} + \eta_{11}) + \frac{\partial}{\partial x_1}(\sigma_{11} + \eta_{11})dx_1\right]dx_2\,dx_3 - (\sigma_{11} + \eta_{11})dx_2\,dx_3$$

$$+\left[(\sigma_{21} + \eta_{21}) + \frac{\partial}{\partial x_2}(\sigma_{21} + \eta_{21})dx_2\right]dx_1\,dx_3 - (\sigma_{21} + \eta_{21})dx_1\,dx_3$$

$$+\left[(\sigma_{31} + \eta_{31}) + \frac{\partial}{\partial x_3}(\sigma_{31} + \eta_{31})dx_3\right]dx_1\,dx_2 - (\sigma_{31} + \eta_{31})dx_1\,dx_2$$

$$+ (F_1 + \eta_1{}^*)dx_1\,dx_2\,dx_3 = 0 \,.$$

Neglecting higher order terms then yields

$$\frac{\partial \sigma_{11}}{\partial x_1} + \frac{\partial \sigma_{21}}{\partial x_2} + \frac{\partial \sigma_{31}}{\partial x_3} + F_1 = 0 \, . \qquad (3.3.18)$$

Similar equations are obtained for the x_2 and x_3 directions. Summarising for all three directions,

$$\frac{\partial \sigma_{ji}}{\partial x_j} + F_i = 0$$

or

$$\sigma_{ji,j} + F_i = 0 \qquad i,j = 1, 2, 3 \, . \qquad (3.3.19)$$

In an analogous manner, equilibrium of moment about PQ in Fig. 3.3.5 requires

$$\sigma_{13} = \sigma_{31} \, .$$

In a similar way, we find $\sigma_{12} = \sigma_{21}$ and $\sigma_{23} = \sigma_{32}$, or

$$\sigma_{ij} = \sigma_{ji} \qquad i,j = 1, 2, 3 \, . \qquad (3.3.20)$$

Equations (3.3.19) and (3.3.20) show that six stress components are related by only three differential equations of equilibrium: we thus have on a continuum level, the characteristic difficulty of stress analysis – the situation is statically indeterminate. Equilibrium relations alone are thus not sufficient to yield a solution to a problem in stress analysis; the deformations must be taken into account and this requires some stress strain relation for the material of the solid.

It may happen that stress components σ_{33}, σ_{31} and σ_{32} are zero whilst the remaining components are independent of the x_3 coordinate. Such a situation, which is very nearly physically realisable in a thin plate loaded in its plane, is called a state of **plane stress.** Important simplifications to the theory occur under these conditions and they are briefly discussed in Section 3.7.

EXAMPLE 3.3.1

Deduce an expression for the so-called **octahedral shearing stress** and show its relation to a deviator state of stress.

At any point in a solid, three mutually perpendicular principal stresses exist. A plane equally inclined to the principal directions is called an **octahedral plane.** By letting the coordinate axes coincide with the principal directions, Cauchy's formula gives the component of the stress vector on such a plane as

$$\overset{n}{T}_i = \sigma_i n_i \qquad \text{no sum on } i \qquad \text{(a)}$$

and the magnitude of $\overset{n}{\mathbf{T}}$ is given by

$$(\overset{n}{T})^2 = (\sigma_1 n_1)^2 + (\sigma_2 n_2)^2 + (\sigma_3 n_3)^3 \, . \qquad \text{(b)}$$

But $n_1 = n_2 = n_3 = \pm 1/\sqrt{3}$ for an octahedral plane, so

$$(\overset{n}{T})^2 = \tfrac{1}{3}(\sigma_1 + \sigma_2 + \sigma_3)^2 \, . \qquad \text{(c)}$$

The magnitude of the normal component of $\overset{n}{\mathbf{T}}$ is given by

$$\sigma_n = \overset{n}{T}_i n_i = [\sigma_1 n_1^2 + \sigma_2 n_2^2 + \sigma_3 n_3^2] \tag{d}$$

or

$$\sigma_n = \tfrac{1}{3}(\sigma_1 + \sigma_2 + \sigma_3)\,; \tag{e}$$

then the shearing stress magnitude is

$$\tau_{\text{oct}}^2 = (\overset{n}{T})^2 - \sigma_n^2 \tag{e}$$

or

$$\tau_{\text{oct}}^2 = \tfrac{1}{3}(\sigma_1^2 + \sigma_2^2 + \sigma_3^2) - \tfrac{1}{9}(\sigma_1 + \sigma_2 + \sigma_3)^2 \,.$$

After some manipulation, we have

$$\tau_{\text{oct}} = \tfrac{1}{3}\sqrt{(\sigma_1 - \sigma_2)^2 + (\sigma_2 - \sigma_3)^2 + (\sigma_3 - \sigma_1)^2} \,. \tag{g}$$

The stress deviator tensor is defined as

$$\hat{\sigma}_{ij} = \sigma_{ij} - \sigma_m \delta_{ij} \tag{h}$$

where $\sigma_m = \tfrac{1}{3}(\sigma_1 + \sigma_2 + \sigma_3)$ is called the **mean stress.** (j)

Using the procedure previously described, the principal deviator components are given by roots of

$$\hat{\sigma}^3 - J_2\hat{\sigma} - J_3 = 0 \,. \tag{k}$$

The coefficient J_1 of $\hat{\sigma}^2$ is zero from definition (h) and the second invariant J_2 is

$$J_2 = \tfrac{1}{6}[(\sigma_1 - \sigma_2)^2 + (\sigma_2 - \sigma_3)^2 + (\sigma_3 - \sigma_1)^2] \,. \tag{l}$$

Comparing (g) and (l) we have

$$\tau_{\text{oct}}^2 = \tfrac{2}{3}J_2 \,. \tag{m}$$

EXAMPLE 3.3.2

The state of stress at a point of interest in a solid is defined by

$$[\sigma] = \begin{bmatrix} 3 & 5 & 8 \\ 5 & 1 & 0 \\ 8 & 0 & 2 \end{bmatrix} \times 10 \text{ N/mm}^2 \,. \tag{a}$$

Determine the principal stresses and their directions.

Temporarily leaving out the common factor 10, we have

$$I_1 = 6\,, \quad I_2 = -78\,, \quad I_3 = -108 \tag{b}$$

so that the principal stresses are given by

$$\sigma'^3 - 6\sigma'^2 - 78\sigma' + 108 = 0 \,. \tag{c}$$

The roots of this cubic are found to be

$$\sigma_1' = 11.825\,, \quad \sigma_2' = 1.285\,, \quad \sigma_3' = -7.105 \tag{d}$$

so that the principal stresses proper are

$$\sigma_1 = 118.25 \text{ N/mm}^2 , \quad \sigma_2 = 12.85 \text{ N/mm}^2 , \quad \sigma_3 = -71.05 \text{ N/mm}^2 .$$

For the first principal stress, equations (3.3.9) become

$$\begin{aligned} -8.83 \overset{1}{n}_1 + 5.00 \overset{1}{n}_2 + 8.00 \overset{1}{n}_3 &= 0 \\ 5.00 \overset{1}{n}_1 + 10.83 \overset{1}{n}_2 + 0 &= 0 \\ 8.00 \overset{1}{n}_1 + 0 - 9.83 \overset{1}{n}_3 &= 0 . \end{aligned}$$

From these equations

$$\begin{aligned} \overset{1}{n}_3 &= 0.814 \overset{1}{n}_1 \\ \overset{1}{n}_2 &= 0.462 \overset{1}{n}_1 . \end{aligned} \qquad \text{(f)}$$

Then, since $\overset{1}{n}_i \overset{1}{n}_i = 1$

$$\overset{1}{n}_1 = 0.730 , \quad \overset{1}{n}_2 = 0.337 , \quad \overset{1}{n}_3 = 0.594 . \qquad \text{(g)}$$

Similarly $\overset{2}{n}_1 = 0.049 , \quad \overset{2}{n}_2 = 0.845 , \quad \overset{2}{n}_3 = -0.552$

$$\overset{3}{n}_1 = 0.682 , \quad \overset{3}{n}_2 = -0.421 , \quad \overset{3}{n}_3 = -0.599 . \qquad \text{(h)}$$

It may be shown that $\overset{\alpha}{n}_i \overset{\beta}{n}_j = \delta_{ij}$, to the accuracy employed here, confirming that the principal directions are mutually orthogonal.

Problems 3.3

3.3.1 The state of stress at a point is given

$$[\sigma] = \begin{bmatrix} \frac{1}{4} & -\sqrt{6} & -\sqrt{\frac{3}{4}} \\ -\sqrt{6} & 2 & -\sqrt{2} \\ -\sqrt{\frac{3}{4}} & -\sqrt{2} & \frac{3}{4} \end{bmatrix} \times 10 \text{ N/mm}^2 .$$

Determine the following quantities

(i) The principal stresses.

(ii) The octahedral stress components (see Example 3.3.1).

(iii) The spherical and deviator stress tensors (see Example 3.3.1).

(iv) The maximum shearing stress component (which may be shown to be half the difference between the greatest and least principal stresses).

3.3.2 The state of stress at a point may be described as the superposition of the spherical and deviator tensors S_0 and S (see Example 3.3.1). Show that the invariants of the latter two tensors are

$$I_{0_1} = 3\sigma_m, \quad I_{0_2} = 3\sigma_m^2, \quad I_{0_3} = \sigma_m^3$$
$$\hat{I}_1 = 0, \quad \hat{I}_2 = -\tfrac{1}{6}\{(\sigma_1 - \sigma_2)^2 + (\sigma_2 - \sigma_3)^2 + (\sigma_3 - \sigma_1)^2\}$$
$$\hat{I}_3 = \frac{1}{27}\{(2\sigma_1 - \sigma_2 - \sigma_3)(2\sigma_2 - \sigma_3 - \sigma_1)(2\sigma_3 - \sigma_1 - \sigma_2)\}$$

where σ_m is the mean stress and σ_1, σ_2, σ_3 are the principal stresses. Notice that $\hat{I}_2 = -J_2$; the quantity which was introduced in Example 3.3.1.

3.3.3 The results of an experimental study of a machine part suggest that the state of stress in a region of interest is given by

$$[\sigma] = A\begin{bmatrix} \tfrac{3}{2}x_1^2x_2^2 & -x_1x_2^3 & 0 \\ -x_1x_2^3 & \tfrac{1}{4}x_2^4 & 0 \\ 0 & 0 & 0 \end{bmatrix} \text{N/mm}^2$$

where A is a parameter depending on the load. Show that the equations of equilibrium are satisfied.

3.4 Constitutive relations for materials

A constitutive equation is one which describes some property of a material. It may relate to electrical resistance or heat transfer characteristics; mechanical behaviour is characterised by the stress-strain relations for the material. The developments of Chapter 1 show that when discrete systems are statically indeterminate, establishing the equilibrium configuration requires a consideration of the force-deformation characteristics of the individual members. To describe the detailed behaviour within individual members is the purpose of continuum mechanics and a stress-strain law for the material is necessary for this purpose.

The stress-strain characteristics of real engineering materials are extremely complex and there is no single constitutive law available which is applicable to all materials under all conditions. To allow calculations to proceed, therefore, it is necessary to develop models of material behaviour; the more simplified the model, the more limited is the range of problems for which calculation yields results having adequate accuracy for the purpose at hand.

There are three broad classes of material behaviour significant for stress analysis.

(i) The material is **elastic**. Deformations due to stress disappear when the stress is removed. Elastic stress-strain laws may or may not be linear, but fortunately, strain is proportional to stress for a practically useful range of stress levels in a wide selection of engineering materials. The behaviour of such materials is described by Hooke's law. For static analyses, time does not enter into the discussion explicitly.

(ii) The material behaves **plastically**. Here, deformations do not disappear when the stresses are removed so that permanent or residual strains remain. Usually, relatively high stress levels are required to cause plastic deformation, and below these levels, Hooke's law often obtains. Time does not enter into the discussion explicitly.

(iii) The material is **viscoelastic or viscoplastic**. Now, deformations are functions of time even when the loads are steady so that creep occurs. Depending on the material, deformations may or may not disappear when loads are removed.

We shall only be concerned with linear elastic behaviour: then in the light of the notions of states of stress and strain, a natural generalisation of the simple Hooke's law relation Force = stiffness x deformation is

$$\sigma_{ij} = c_{ijkl}\epsilon_{kl} \qquad i,j = 1, 2, 3 \;. \tag{3.4.1}$$

The c_{ijkl} are elasticity coefficients and each component of stress is proportional to each component of strain. Since the stress and strain tensors are symmetric there are, at most, thirty six independent c_{ijkl}'s. Furthermore, energy considerations show that they are symmetric so that, in the most general form of Hooke's law, there are twenty-one elastic coefficients. In general we would expect the coefficients to depend on position in the solid and the choice of coordinate axes relative to the body. However, when the solid is **homogeneous**, the c_{ijkl} are constants and not functions of position. When the material has some regularity of internal structure, the c_{ijkl} depend on the choice of axes. However although many real materials are crystalline in character, the crystals are so small and randomly orientated that the gross behaviour of a representative region of material is **isotropic**, that is, the elastic properties are the same in all directions. We thus find that the idealisation of material into a homogeneous isotropic continuum provides a good representation of a wide range of important engineering materials.† By carrying out various coordinate transformations and demanding invariance of the c_{ijkl}'s, it may be shown that only two independent elastic coefficients are required for isotropy. Then, Hooke's law for isotropic, homogeneous materials becomes

$$\sigma_{ij} = \lambda\epsilon_{kk}\delta_{ij} + 2G\epsilon_{ij} \qquad i,j,k = 1, 2, 3 \tag{3.4.2}$$

where λ and G are Lamé's constants. G corresponds to the shear modulus of the engineering literature; it and λ can be expressed in terms of Young's modulus E and Poisson's ratio ν by

$$\lambda = \frac{\nu E}{(1 + \nu)(1 - 2\nu)}, \qquad G = \frac{E}{2(1 + \nu)} \;. \tag{3.4.3}$$

Equations (3.4.2) may be solved for the ϵ_{ij} in terms of the σ_{ij} to give

$$\epsilon_{ij} = \frac{1 + \nu}{E}\sigma_{ij} - \frac{\nu}{E}\sigma_{kk}\delta_{ij} \;. \tag{3.4.4}$$

† Composite materials are formulated to provide desirable directional properties on a macroscopic level. Fibre reinforced materials belong to this class and a continuum idealisation must provide for non-homogeneity and non-isotropic behaviour.

Written in unabridged notation, we have

$$\epsilon_x = \frac{1}{E}[\sigma_x - \nu(\sigma_y + \sigma_z)]$$

$$\epsilon_y = \frac{1}{E}[\sigma_y - \nu(\sigma_z + \sigma_x)]$$

$$\epsilon_z = \frac{1}{E}[\sigma_z - \nu(\sigma_x + \sigma_y)] \tag{3.4.5}$$

$$\gamma_{xy} = \tau_{xy}/G, \quad \gamma_{yz} = \tau_{yz}/G, \quad \gamma_{zx} = \tau_{zx}/G .$$

By allowing the coordinate axes to coincide with the principal directions of strain, equations (3.4.2) or (3.4.4) show that the principal directions of stress coincide with the principal directions of strain in an isotropic solid.

Hooke's law may be expressed in terms of the spherical and deviator stress tensors introduced in Examples 3.2.1 and 3.3.1. Beginning with equations (3.4.2) the reader may show that

$$\sigma_{kk} = 3K\epsilon_{kk}$$

$$\hat{\sigma}_{ij} = 2G\hat{\epsilon}_{ij} \tag{3.4.6}$$

where $K = E/3(1 - 2\nu)$ is the so-called Bulk modulus.

This mode of presentation is valuable for a more complete description of material behaviour. Thus, the Von-Mises model for an elastic-perfectly plastic material postulates that the first of equations (3.4.6) holds at all stress levels whilst the second holds only for as long as the quantity $\sigma_e = \sqrt{\tfrac{1}{2}[(\sigma_1 - \sigma_2)^2 + (\sigma_2 - \sigma_3)^2 + (\sigma_3 - \sigma_1)^2]}$ is less than some critical value σ_y. If $\sigma_e = \sigma_y$, plastic flow occurs whilst $\sigma_e > \sigma_y$ is not a physically realisable state; σ_y is the yield stress in a simple tension test. Ordinary mild steel is well represented by this model provided the strains are not too large; work hardening effects become significant when large deformations occur. Characterisation of viscoelastic behaviour is also facilitated by the use of the spherical and deviator tensors.

EXAMPLE 3.4.1

Deduce the matrix of elastic constants for an orthotropic material.

It is convenient to introduce the notation $\sigma_1 = \sigma_{11}$, $\sigma_2 = \sigma_{22}$, $\sigma_3 = \sigma_{33}$, $\sigma_4 = \sigma_{23}$, $\sigma_5 = \sigma_{31}$, $\sigma_6 = \sigma_{12}$, $\epsilon_1 = \epsilon_{11}$, $\epsilon_2 = \epsilon_{22}$ . . . . . $\epsilon_6 = \epsilon_{12}$. Now, Hooke's law can be

$$\sigma_i = c_{ij}\epsilon_j \tag{a}$$

or

$$\{\sigma\} = [c]\{\epsilon\} . \tag{b}$$

Suppose the internal structure of the material is such that there is elastic symmetry about the chosen coordinate axes $x_1x_2x_3$. Such a material is said to be orthotropic. If we impose the coordinate transformation $x_1 = x_1', x_2 = x_2'$, $x_3 = -x_3'$ or

$$\begin{Bmatrix} x_1' \\ x_2' \\ x_3' \end{Bmatrix} = \begin{bmatrix} 1 & 0 & 0 \\ 0 & 1 & 0 \\ 0 & 0 & -1 \end{bmatrix} \begin{Bmatrix} x_1 \\ x_2 \\ x_3 \end{Bmatrix} \tag{c}$$

from equations (a), we can write

$$\sigma_i' = c_{ij}\epsilon_j' . \tag{d}$$

Here, the c_{ij}'s are the same as those in equation (a) since we require elastic symmetry. But if we apply this transformation to equations (3.2.31) and (3.3.17) we find

$$\begin{aligned} \sigma_r' &= \sigma_r , \quad \epsilon_r' = \epsilon_r , \quad r = 1, 2, 3, 6 \\ \sigma_4' &= -\sigma_4 , \quad \sigma_5' = -\sigma_5 , \\ \epsilon_4' &= -\epsilon_4 , \quad \epsilon_5' = -\epsilon_5 . \end{aligned} \tag{e}$$

Furthermore, if we impose the transformation

$$\begin{Bmatrix} x_1'' \\ x_2'' \\ x_3'' \end{Bmatrix} = \begin{bmatrix} -1 & 0 & 0 \\ 0 & 1 & 0 \\ 0 & 0 & 1 \end{bmatrix} \begin{Bmatrix} x_1 \\ x_2 \\ x_3 \end{Bmatrix}, \tag{f}$$

Equations (3.2.31) and (3.3.17) show that the stress and strain components are unchanged except for

$$\sigma_5' = -\sigma_5 , \quad \sigma_6' = -\sigma_6 , \quad \epsilon_5' = -\epsilon_5 , \quad \epsilon_6' = -\epsilon_6 . \tag{g}$$

Substituting from (e) and (g) in (b), we find that for an orthotropic material

$$[c] = \begin{bmatrix} c_{11} & c_{12} & c_{13} & 0 & 0 & 0 \\ c_{21} & c_{22} & c_{23} & 0 & 0 & 0 \\ c_{31} & c_{32} & c_{33} & 0 & 0 & 0 \\ 0 & 0 & 0 & c_{44} & 0 & 0 \\ 0 & 0 & 0 & 0 & c_{55} & 0 \\ 0 & 0 & 0 & 0 & 0 & c_{66} \end{bmatrix} . \tag{h}$$

It is interesting to observe that symmetry with respect to transformations (c) and (f) imply symmetry with respect to

$$\begin{Bmatrix} x_1''' \\ x_2''' \\ x_3''' \end{Bmatrix} = \begin{bmatrix} 1 & 0 & 0 \\ 0 & -1 & 0 \\ 0 & 0 & 1 \end{bmatrix} \begin{Bmatrix} x_1 \\ x_2 \\ x_3 \end{Bmatrix} \tag{j}$$

The form of $[c]$ shows that there is no coupling of effects between the direct and shearing stresses and strains associated with the planes of elastic symmetry in an orthotropic material.

Problems 3.4

3.4.1 Show that the elastic constants are related as follows:

$$\lambda = K - \frac{2}{3}G = \frac{2G\nu}{1 - 2\nu}$$

$$G = \frac{3KE}{9K - E} = \frac{\lambda(1 - 2\nu)}{2\nu} .$$

3.4.2 Show that for an incompressible, isotropic linear elastic material $K = \infty, \nu = 1/2$ and $G = E/3$.

3.5 Formulation of problems in elastic stress analysis

To describe the deformed configuration of a continuum, we required three functions of position $u_i(x_1, x_2, x_3)$ which define the components of displacement of the material points from some datum position. Equations (3.2.40) then yield the strain components and subsequent application of Hooke's law gives the stress components. However, whilst any continuous functions $u_i(x_1, x_2, x_3)$ must imply a compatible strain field, they need not necessarily imply a stress distribution in equilibrium with the given loads: for such equilibrium to be assured, further restrictions must be imposed on the u_i. Thus, the formulation of any stressing problem requires the manipulation of three essential ingredients:

(i) Equilibrium requirements must be met in the interior and on the surface of the solid.

(ii) The strain field must be compatible so that displacements are continuous throughout the interior of the solid and satisfy any geometric constraints at the surface.

(iii) Some stress-strain law is required to describe the mechanical behaviour of the material.

These ingredients correspond exactly to those required to treat the statically indeterminate situations of Chapters 1 and 2.

Three broad classes of physical situations present themselves and their description in analytic terms leads to three classes of boundary value problems.

Class 1 Determine the distribution of stresses and displacements in a solid in equilibrium under the action of body forces specified throughout the volume V of the solid and displacements prescribed over its surface S.

Class 2 Determine the distribution of stresses and displacements when the body forces are prescribed throughout V and tractions are specified over the entire surface S.

Class 3 Determine the distribution of stresses and displacements when body forces are prescribed throughout V, the tractions are specified over part S_T of the surface and displacements are specified over the remaining part S_u of the surface

The nature of Class 1 problems suggest that some advantage may accrue from formulating the problem entirely in terms of displacements so that the u_i's are taken as the basic unkowns. Then, using Hooke's law (3.4.2) and the strain-displacement relations (3.2.40) the equilibrium equations may be expressed in terms of the u_i's as

$$G\nabla^2 u_i + (\lambda + G)u_{j,ji} + F_i = 0 \quad i,j = 1, 2, 3 \ . \qquad (3.5.1)$$

Solutions $u_i(x_1, x_2, x_3)$ to these equations, due to Navier, must satisfay the prescribed boundary displacements

$$u_i{}^* = f_i(x_1, x_2, x_3) \quad i = 1, 2, 3 \ , \qquad (3.5.2)$$

on S.† The reader should observe that, since the problem has been posed entirely in terms of displacements, the compatibility equations are satisfied automatically and so need not be adjoined as explicit conditions on the solution. Equations (3.5.1) and (3.5.2) thus define the boundary value problem.

Class 2 problems are conveniently formulated entirely in terms of the stress components or mathematical devices called stress functions introduced below. If we take the stress components as the basic unknowns they must , of course, satisfy the equations of equilibrium

$$\sigma_{ji,j} + F_i = 0 \ . \qquad (3.5.3)$$

Using Hooke's law yields the strains, but since equilibrium stresses can be conceived which do not imply continuous displacements, the equations of compatibility *must* be introduced as explicit conditions in this formulation. Expressing equations (3.2.45) or (3.2.46) to (3.2.51) in terms of the stresses, some manipulation leads to the Beltrami-Michell equations of compatibility

$$\nabla^2 \sigma_{ij} + \frac{1}{1+\nu}\sigma_{kk,ij} + \frac{\nu}{1-\nu}\delta_{ij}F_{k,k} + F_{i,j} + F_{j,i} = 0 \ . \qquad (3.5.4)$$

A stress field which is a solution to equations (3.5.3) and (3.5.4) in the interior of the solid must also satisfy

$$\overset{n}{T_i}{}^* = \sigma_{ji} n_j \qquad (3.5.5)$$

at the surface.††

All three sets of equations are required to define the boundary value problem; however, we must recall that for overall equilibrium of the solid, the resultant body force and resultant surface loading are related and cannot be selected without reference to each other.

Class 3 type problems may also be expressed in terms of displacements. Now, however, the u_i will be prescribed over part S_u of S whilst on S_T $\overset{n}{T_i}{}^* = \sigma_{ji}n_j$ prescribed implies

$$[\lambda\delta_{ij}u_{k,k} + G(u_{i,j} + u_{j,i})]n_j = \text{prescribed} \ . \qquad (3.5.6)$$

† The asterisk indicates that the u_i are prescribed.

†† The asterisk indicates that the surface tractions are prescribed.

It is clear from this brief summary that to seek an analytic solution to these boundary value problems in their most general form presents formidable mathematical difficulties. Indeed, whether a solution exists at all is a matter requiring investigation. Furthermore, if the solid contains cavities or holes, so that it is a multiply connected region, the solutions to the differential equations are subject to further conditions. For our purpose, however, it is sufficient to accept that in the small displacement theory of elasticity, a solution does indeed exist and it is unique. Even so, to make progress in calculations of practical significance, we must resort to approximation methods of some kind. Possibilities open to us are as follows

(i) We may attempt to develop, *ab initio,* an approximate theory appropriate to a particular problem, or class of problems, in the manner of the mechanics of materials. For design purposes we then idealise the real situation (sometimes grossly) to fit this special theory. For many purposes, this is adequate.

(ii) We may approach the problem by the so-called inverse or semi-inverse methods. In the former treatment, some solution to the differential equations is accepted and the corresponding problem situation is examined to see if it has any practical significance. Since the equations are linear, a superposition of several solutions may be examined. In the semi-inverse method, as used by St. Venant for his theory of bars, some assumptions are made about the distribution of displacements or stresses or both, whilst leaving sufficient freedom to satisfy the equations of elasticity fully. This requires a considerable degree of insight, but success leads to much simpler boundary value problems and a practically useful range of solutions. However, even here, the boundary conditions may be such that formulae for stresses or displacements cannot be obtained and solutions can only be obtained numerically.

(iii) On the basis of simplifying assumptions an approximate theory embodied in a consistently specified boundary value problem may be obtained. Here, the use of variational principles are especially advantageous as in the classic applications to the theory of plates and shells. This approach is at a higher level of idealisation than in (i) above but again the equations may be such that direct solution to yield formulae for stresses and displacements may not be possible (see Chapter 6). This brings us to

(iv) The properly formulated boundary value problem is solved approximately. In this book we are concerned with methods based on variational principles, but finite difference, collocation and other techniques are available.

Whichever approach is adopted, its range of applicability is widened by taking advantage of a fundamentally important principle due to St. Venant.† This can be expressed in more than one way, but essentially it state that:

> If some distribution of forces acting over a limited portion of body (perhaps a small part of its surface) is replaced by a different but equipollent distribution over the same portion, then in parts of the body sufficiently far

† The reader will find an illuminating discussion of St. Venant's principle in 'Foundations of Solid Mechanics' by Y.C. Fung cited in the reading list. Reference to interesting work by Goodier, Hoff and others will be found.

removed from it, the effects of the two distributions of forces is essentially the same.

To gain further insight into these matters, the remainder of this chapter is devoted to a discussion of some concrete examples.

EXAMPLE 3.5.1

An examination of experimental data suggests that the state of stress in a region of a machine part is defined by

$$[\sigma] = c\begin{bmatrix} x_2^2 + \nu(x_1^2 - x_2^2) & -2\nu x_1 x_2 & 0 \\ -2\nu x_1 x_2 & x_1^2 + \nu(x_2^2 - x_1^2) & 0 \\ 0 & 0 & \nu(x_1^2 + x_2^2) \end{bmatrix}, \tag{a}$$

c being a constant and body forces being negligible. Does this represent a true state of stress?

Substituting from (a) in the equilbrium equations $\sigma_{ji,j} = 0$ shows that they are satisfied.

Substituting from (a) in the compatibility equations $\nabla^2\sigma_{ij} + [1/(1+\nu)]\,\sigma_{kk,ij} = 0$ shows that they are violated. Equation (a) does not, therefore, represent a true state of stress even though equilibrium is satisfied in the interior.

Problems 3.5

3.5.1 In an isotropic solid a temperature rise T causes normal strains $\epsilon_{11}^{\circ} = \epsilon_{22}^{\circ} = \epsilon_{33}^{\circ} = \alpha T$ and no shearing strains; α is the linear thermal coefficient. Denoting *total* strains by ϵ_{ij}, Hooke's law then becomes

$$\epsilon_{ij} = \frac{1+\nu}{E}\sigma_{ij} - \frac{\nu}{E}\sigma_{kk}\delta_{ij} + \alpha T\delta_{ij}\,.$$

Show that $\quad \sigma_{ij} = \lambda\epsilon_{kk}\delta_{ij} + 2G\epsilon_{ij} - (3\lambda + 2G)\alpha T\delta_{ij}\,.$

3.6 The torsion of prismatic bars by end tractions

Suppose a uniform, circular section bar is loaded by twisting moments applied to its end faces as shown in Fig. 3.6.1. Symmetry arguments show that the bar twists about its axis with cross sections remaining plane and not distorting in their plane so that they rotate like rigid disks. Conditions are uniform with respect to the bar's length so that the rate of change of twist with respect to length, or the angle of twist per unit length, is a constant α. Then, viewing a cross-section as in Fig. 3.6.2 a typical point P is displaced to P' so that†

$$u = -\alpha zy, \quad v = \alpha zx, \quad w = 0 \tag{3.6.1}$$

when the right hand end of the bar is held fixed.

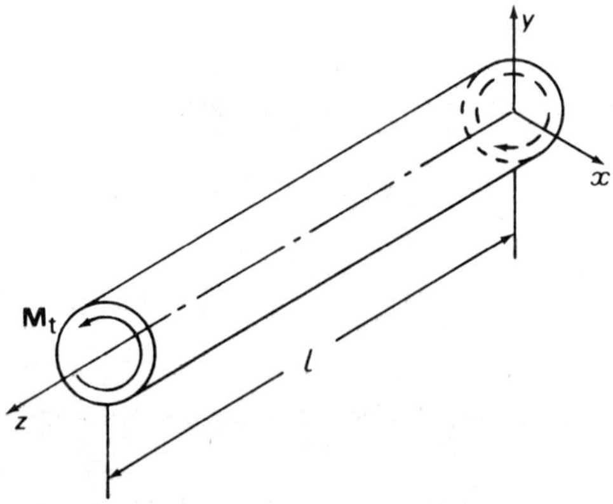

Fig. 3.6.1

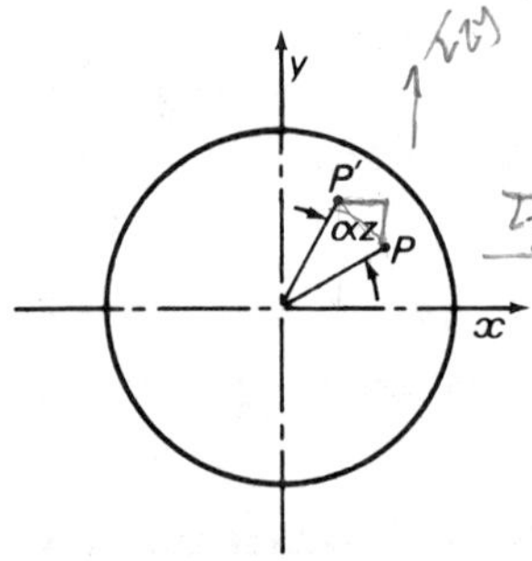

Fig. 3.6.2

The displacements (3.6.1) imply a stress distribution

$$\tau_{zx} = -G\alpha y, \quad \tau_{zy} = G\alpha x$$
$$\sigma_x = \sigma_y = \sigma_z = \tau_{xy} = 0 \tag{3.6.2}$$

when the strain displacement equations (3.2.10) and Hooke's law are used. Substitution from equations (3.6.2) in (3.5.3) readily shows that the stress distribution satisfies the equations of equilibrium in the interior of the bar when body forces F_i are neglifible. The specification of the problem requires that stresses (3.6.2) should also imply zero surface traction over the entire lateral surface of the bar. Substitution in equations (3.5.5) for this surface shows that $\overset{n}{T}_i{}^*$ is indeed zero. Again from the specification, the resultant action of the tractions on the end face should reduce to a twisting moment $\mathbf{M}_t$ about the axis of the bar. Substituting from equations (3.6.2) confirms that

$$\iint \sigma_z \mathrm{d}x\mathrm{d}y = \iint \sigma_z x \mathrm{d}x\mathrm{d}y = \iint \sigma_z y \mathrm{d}x\mathrm{d}y = 0$$

$$\iint \tau_{zx} \mathrm{d}x\mathrm{d}y = \iint \tau_{zy} \mathrm{d}x\mathrm{d}y = 0$$

$$\iint (\tau_{zy} x - \tau_{zx} y)\mathrm{d}x\mathrm{d}y = G\alpha J = M_t \neq 0 \ . \tag{3.6.3}$$

† When discussing specific problems, there is no special advantage in using the indicial notation.

Thus, all the equations of elasticity are correctly satisfied by solution (3.6.1) and denoting $\tau^2 = \tau_{zx}^2 + \tau_{zy}^2$ we have the familiar result

$$\frac{M_t}{J} = \frac{\tau}{r} = G\alpha \tag{3.6.4}$$

where J is the polar second moment of area of the section and $r^2 = x^2 + y^2$.

When the cross section of the bar is not circular, the displacement pattern (3.6.1) does not yield $\overset{n}{T}_i{}^* = 0$ for the lateral surface so that results (3.6.2) no longer hold. Observation shows that the cross section of a non-circular section does not remain plane on twisting but warps out of its plane so that the w of equations (3.6.1) is not zero. If we assume that the warping is the same for all cross sections and proportional to the angle of twist, and that u and v are still given by the first two of equations (3.6.1), we have

$$u = -\alpha zy\,, \quad v = \alpha zx\,, \quad w = \alpha\psi(x, y)\;. \tag{3.6.5}$$

Here, $\psi(x, y)$ is some function to be determined such that the governing differential equations and boundary conditions are satisfied. That is, it is the part of the guess which allows sufficient freedom in the selection to satisfy the equations of elasticity.

Equations (3.6.5) imply a stress distribution

$$\tau_{zx} = G\alpha\left(\frac{\partial\psi}{\partial x} - y\right), \quad \tau_{zy} = G\alpha\left(\frac{\partial\psi}{\partial y} + x\right)$$

$$\sigma_x = \sigma_y = \sigma_z = \tau_{xy} = 0\;. \tag{3.6.6}$$

Substitution in equilibrium equations (3.5.3) with $F_i = 0$ shows that the first two are satisfied identically whereas the third requires

$$\frac{\partial\tau_{xz}}{\partial x} + \frac{\partial\tau_{yz}}{\partial y} = 0 \tag{3.6.7}$$

or using equations (3.6.6)

$$\frac{\partial^2\psi}{\partial x^2} + \frac{\partial^2\psi}{\partial y^2} = \nabla^2\psi = 0\;. \tag{3.6.8}$$

Solutions to equation (3.6.8) must yield stresses which correspond to zero traction on the lateral surfaces, that is, $\sigma_{ji}n_j = \overset{n}{T}_i{}^* = 0, i, j = 1, 2, 3$ on this surface. The first two equations are satisfied identically whereas the third requires

$$\left(\frac{\partial\psi}{\partial x} - y\right)\cos(nx) + \left(\frac{\partial\psi}{\partial y} + x\right)\cos(ny) = 0\;. \tag{3.6.9}$$

The end view of the bar in Fig. 3.6.3 shows that

$$\cos(nx) = \frac{dy}{ds} = \frac{dx}{dn}\,,$$

$$\cos(ny) = -\frac{dx}{ds} = \frac{dy}{dn}\,; \tag{3.6.10}$$

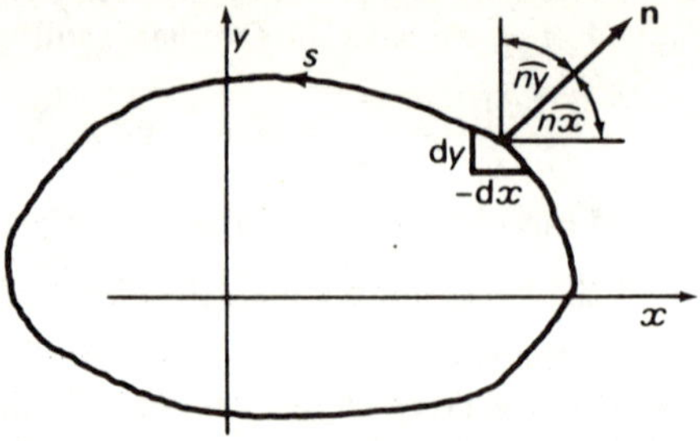

Fig. 3.6.3

substituting from equations (3.6.10) into (3.6.9) and using the chain rule of differentiation, the condition to be satisfied by ψ along the boundary of the cross section becomes

$$\frac{\partial \psi}{\partial n} = y \cos(nx) - x \cos(ny) . \tag{3.6.11}$$

It may readily be shown that stresses (3.6.6) yield a resultant action on the end faces which reduces to a pure couple about the axis.

$$M_t = G\alpha \iint \left[x^2 + y^2 + x\frac{\partial \psi}{\partial y} - y\frac{\partial \psi}{\partial x} \right] \mathrm{d}x\mathrm{d}y \tag{3.6.12}$$

or

$$M_t = G\alpha K \tag{3.6.13}$$

K is called the torsion constant for the bar and is a function of the cross section geometry alone.

The reader should observe that, since the formulation has been in terms of displacements, the equations of compatibility do not appear explicitly.

Prandtl proposed an alternative formulation to the problem which proceeds as follows. Assuming that the stress distribution is of the form (3.6.6), the only equation of internal equilibrium not satisfied identically is equation (3.6.7). The form of this equation suggests that a function $\phi(x, y)$ exists which is related to the stress components by

$$\tau_{zx} = \frac{\partial \phi}{\partial y}, \quad \tau_{zy} = -\frac{\partial \phi}{\partial x} \tag{3.6.14}$$

$\phi(x, y)$ is called a **stress function** and stresses calculated from any well behaved function in accordance with equations (3.6.14) are bound to satisfy internal equilibrium. However, since the strains must be compatible and equilibrium must also be satisfied on the surface of bar, further restrictions must be imposed on ϕ. Now from equations (3.6.6) and (3.6.14)

$$\frac{\partial \phi}{\partial y} = G\alpha\left(\frac{\partial \psi}{\partial x} - y\right), \quad -\frac{\partial \phi}{\partial x} = G\alpha\left(\frac{\partial \psi}{\partial y} + x\right). \tag{3.6.15}$$

Differentiating the first of these equations with respect to y and the second with respect to x, observing that for continuous $\psi(x, y)$ and hence compatibility of strains, $\partial^2 \psi/\partial x\partial y = \partial^2 \psi/\partial y\partial x$,

$$\frac{\partial^2 \phi}{\partial x^2} + \frac{\partial^2 \phi}{\partial y^2} + 2G\alpha = 0 \tag{3.6.16}$$

or

$$\nabla^2 \phi + 2G\alpha = 0 \tag{3.6.17}$$

as the equation defining ϕ in the interior of the bar. On the lateral surface of the bar, we require $\overset{n}{T_i}{}^* = \sigma_{ji}n_j$, $i,j = 1, 2, 3$ as before. The first two of these conditions are satisfied automatically whilst the third requires

$$\frac{\partial\phi}{\partial y}\frac{\mathrm{d}y}{\mathrm{d}s} + \left(-\frac{\partial\phi}{\partial x}\right)\left(-\frac{\mathrm{d}x}{\mathrm{d}s}\right) = 0$$

or

$$\frac{\mathrm{d}\phi}{\mathrm{d}s} = 0$$

or

$$\phi = \text{constant}$$

on the boundary of the section. For solid bars, there is only one boundary and we can choose

$$\phi = 0 \tag{3.6.18}$$

on the boundary without affecting the stresses. For a multiply connected section, further conditions must be imposed on ϕ. For the end faces, the tractions must result in a pure couple about the axis of the bar. Proceeding as for equations (3.6.3) we find that this is indeed the case and

$$M_t = 2\iint \phi \,\mathrm{d}x\mathrm{d}y\ . \tag{3.6.19}$$

Thus, by making certain assumptions about the displacement or stress field in the bar, we have found that a truly three dimensional problem has been described in terms of only one unkown function. This in turn is a function of position in the cross section alone and independent of the axial position in the bar. For the above solutions to hold true throughout the bar, it is necessary that the tractions applied to the end faces be distributed in precisely the manner called for by ϕ or ψ in the solution to the boundary value problem. If the end tractions are not applied precisely in this way but do at least reduce to a pure couple, the stress distribution near the ends of the bar is much more complicated than above. In regions away from the ends, however, the above theory describes the situation very well; as one would expect from St. Venant's principle.†

The mathematics of the torsion problem is dramatically less complicated than for the general three dimensional situation. Even so, the determination of stresses in and torsion constants for the bars of awkward shaped sections

† When the ends of the bar are so constrained that warping is restricted or prevented altogether, axial direct stresses are induced. The applicability of St. Venant's principle to such situations has been discussed by Hoff. An outline of this work is given by Y.C. Fung in the text cited at the end of this chapter.

found in practice is still difficult. More often than not, approximate solutions have to be found and when the formulation is in terms of the Prandtl stress function, this process is facilitated by invoking the so-called **soap film** or **membrane analogy** as an aid to visualising the form of the solution.

The basis of the analogy is the fact that the differential equation describing the small deflexion of a membrane supporting transverse pressure is of exactly the same form as equation (3.6.17) Thus, suppose we have an initially flat thin membrane stretched in such a way that it is tensioned uniformly in all directions in its plane.† When pressure is applied to one side of the membrane, it will distend so that some additional tension is induced in it. However, if we suppose that the initial tension is 'large' and the pressure is 'small', this additional tension will be negligible and the resulting theory is simplified. Taking the initially flat membrane to lie in the xOy plane of Fig. 3.6.4, its distended shape will be described by some function $w = w(x, y)$ where w is the small deflexion in the z direction.

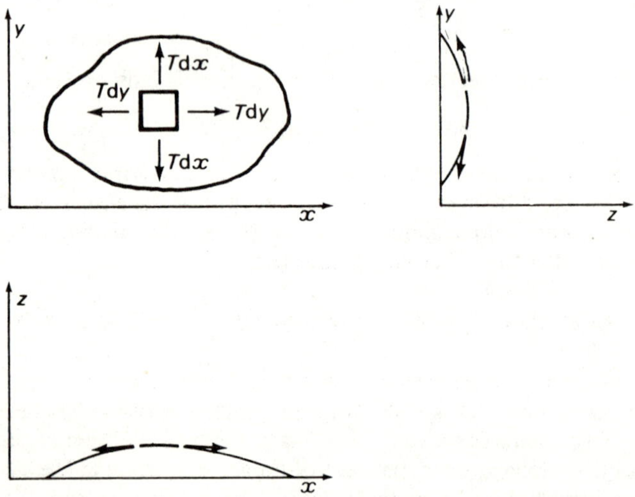

Fig. 3.6.4

Let the initial tension be T per unit length of an imaginary cut in the membrane and the uniform pressure be p per unit area. Then, observing that deflexions and slopes are small and that T is sensibly constant, equilibrium of forces in the z direction on element dxdy of Fig. 3.6.4 requires

$$p\,\mathrm{d}x\mathrm{d}y + T\left(\frac{\partial w}{\partial x} + \frac{\partial^2 w}{\partial x^2}\mathrm{d}x\right)\mathrm{d}y + T\left(\frac{\partial w}{\partial y} + \frac{\partial^2 w}{\partial y^2}\mathrm{d}y\right)\mathrm{d}x$$

$$- T\frac{\partial w}{\partial x}\mathrm{d}y - T\frac{\partial w}{\partial y}\mathrm{d}x = 0$$

† Such a situation can be achieved by drawing a soap film over an opening.

so that
$$\frac{\partial^2 w}{\partial x^2} + \frac{\partial^2 w}{\partial y^2} + \frac{p}{T} = 0 \tag{3.6.20}$$

or
$$\nabla^2 w + \frac{p}{T} = 0 \ . \tag{3.6.21}$$

This equation has exactly the same form as equation (3.6.17). Furthermore, with the membrane drawn across a plane opening, $w = 0$ around its edges, so that the boundary value problem for the small deflexion of a pressurised membrane is the same as that for the elastic torsion of bars formulated in terms of the Prandtl stress function.† This analogy was exploited by Taylor and Griffiths in a celebrated experimental determination of the torsion properties of various sections; the slopes of the experimental membrane gave the stress distribution whilst twice the volume under it gave the torsion constant. The availability of modern computing methods has rendered such an experimental approach redundant, but the analogy is still a powerful aid to visualising the solution to the problem as we shall see in Example 3.6.3.

EXAMPLE 3.6.1

The boundary of a cross section of a bar is defined by

$$f(x, y) = 0 \tag{a}$$

where $f(x, y)$ is a solution to

$$\nabla^2 f = \beta = \text{constant} \tag{b}$$

Find a suitable stress function.

From equations (b) and (3.6.16)

$$2\frac{G\alpha}{\beta}\nabla^2 f = 2G\alpha = -\nabla^2\phi \tag{c}$$

$$\phi = -2\frac{G\alpha}{\beta} f(x, y) \tag{d}$$

is a suitable stress function. It clearly satisfies both equations (3.6.17) and (3.6.18).

EXAMPLE 3.6.2

Test the procedure of Example 3.6.1 in the case of an elliptic section having semi-major and minor axes a and b respectively.

This section is defined by

$$f(x, y) = x^2/a^2 + y^2/b^2 - 1 = 0 \tag{a}$$

now $\nabla^2 f = 2/a^2 + 2/b^2 = \beta$ say. (b)

† Referring to the books by Timoshenko and Goodier or Hetenyi cited at the end of this chapter, it will be seen how the analogy may be extended to multiply connected regions.

A suitable stress function is thus

$$\phi = -\frac{G\alpha a^2 b^2}{a^2 + b^2}(x^2/a^2 + y^2/b^2 - 1) . \qquad \text{(c)}$$

Substituting in equation (3.6.19) gives

$$M_t = \frac{\pi a^3 b^3 G\alpha}{(a^2 + b^2)} . \qquad \text{(d)}$$

The reader should verify that the maximum shearing stress occurs at the end of the minor axis and is given by

$$\tau_{\max} = \frac{2G\alpha a^2 b}{a^2 + b^2} \qquad \text{(e)}$$

EXAMPLE 3.6.3

Exploit the membrane analogy to describe the torsion of a blade having a 'thin' rectangular section.

In view of the 'thinness' of the section, the contours of the analogous membrane will be virtually parallel to the 'long' edges except for localised zones near the 'short' edges as shown in Fig. 3.6.5.

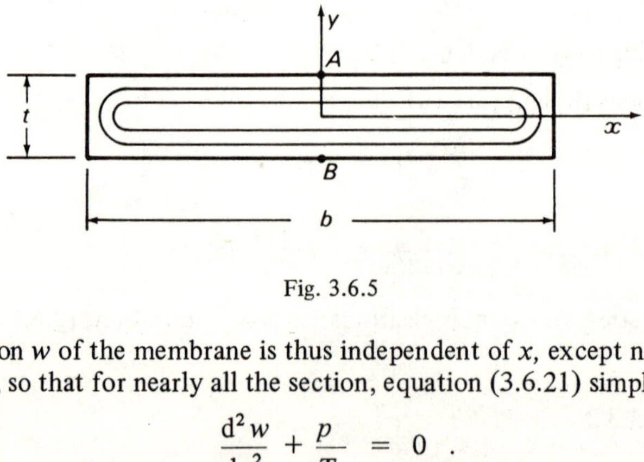

Fig. 3.6.5

The deflexion w of the membrane is thus independent of x, except near the short edges, so that for nearly all the section, equation (3.6.21) simplifies to

$$\frac{d^2 w}{dy^2} + \frac{p}{T} = 0 . \qquad \text{(a)}$$

Integrating and observing that $dw/dx = 0$ at $y = 0$ (from symmetry) and that $w = 0$ at $y = \pm t/2$, we have

$$w = \frac{p}{2T}\left(\frac{t^2}{4} - y^2\right) \qquad \text{(b)}$$

$$\frac{dw}{dy} = -\frac{p}{T}y . \qquad \text{(c)}$$

Replacing w by ϕ and p/T by $2G\alpha$, equations (b) and (c) evidently describe the distribution of the stress function (except near the short edges). The greatest slope of the membrane is dw/dy measured at the mid point of one of the long sides. Numerically, this is the same for points A and B so that the greatest shearing stress is

$$|\tau_{zx}|_{y=\pm t/2} = \left|\frac{d\phi}{dy}\right|_{y=\pm t/2} = G\alpha t \ . \tag{d}$$

The torsion constant is given by

$$K = \frac{M_t}{G\alpha} = \frac{2}{G\alpha}\iint \phi \, dx dy = \frac{2b}{G\alpha}\int_{-t/2}^{t/2} \phi dy$$

or

$$K = \frac{bt^3}{b} \ . \tag{e}$$

In evaluating the volume under the membrane, we have ignored the fact that its deflexion varies significantly with x as well as y in the small zones near the short edges. By visualising the membrane, we can see that the error implied here becomes significant as we move from 'thin' sections to 'square' ones. In the latter case, this simplified approach would not be adequate. (See Timoshenko and Goodier's book).

Problems 3.6

3.6.1 Show that

$$\phi = -G\alpha\left[\frac{x^2 + y^2}{2} - \frac{1}{2a}(x^3 - 3xy^2) - \frac{2}{27}a^2\right]$$

is suitable as a Prandtl stress function and that it describes the problem of torsion of a bar having the equilateral triangle section of Fig. 3.6.6.

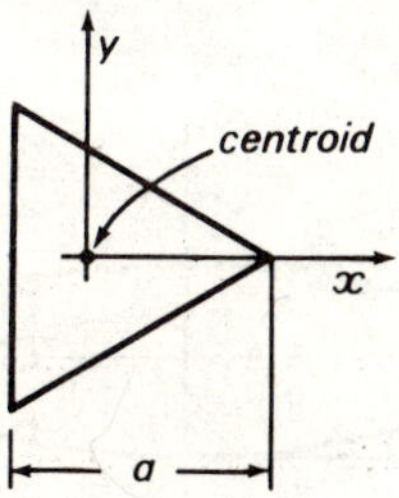

Fig. 3.6.6.

Show that the maximum shearing stress and torsion constant are

$$\tau_{max} = \frac{G\alpha a}{2}, \quad K = \frac{a^4}{15\sqrt{3}} \ .$$

3.6.2 If the warping function ψ is transformed to $\psi_1 = xy - \psi$, show that

$$\nabla^2 \psi_1 = 0 \tag{a}$$

and that the boundary conditions for torsion of a rectangular section $2a \times 2b$ become

$$\frac{\partial \psi_1}{\partial x} = 0 \text{ at } x = \pm a\,, \qquad \frac{\partial \psi_1}{\partial y} = 2x \text{ on } y = \pm b\,. \tag{b}$$

By separation of the variables, show that the solution to (a) becomes

$$\psi_1 = \sum_n A_n \sin k_n x \sinh k_n y\ . \tag{c}$$

A_n and k_n are constants and the origin of coordinates is taken at the centre of the section.

3.6.3 A thin blade has a symmetrical section bounded by parabolae giving the thickness at distance x from the centre by

$$t = t_0[1 - (2x/b)^2] \tag{a}$$

as shown in Fig. 3.6.7. Assuming that at any level x = constant, the analogous membrane for torsion has the form

$$w = \frac{p}{2T}\left[\frac{t^2}{4} - y^2\right], \tag{b}$$

show that an approximation to the torsion constant is

$$K = \frac{16}{105} b t_0^3\ . \tag{c}$$

10/4 NOTE $\frac{\partial^2 \omega}{\partial x^2} + \frac{\partial^2 \omega}{\partial y^2} + \frac{p}{2T}$ is not satisfied here!! why?? How??

22/4

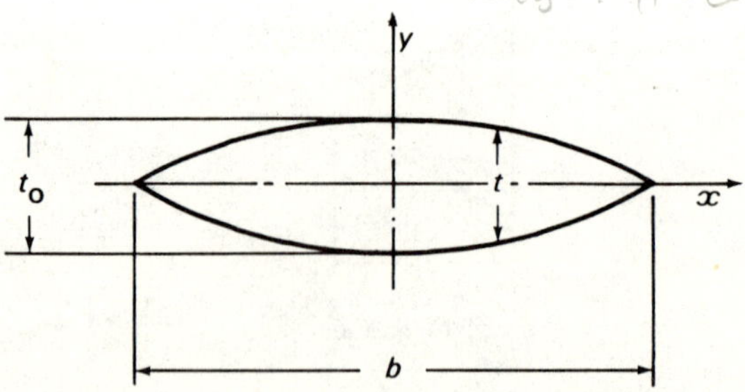

Fig. 3.6.7

3.6.4 For the torsion of a rectangular section $x = \pm a$, $y = \pm b$, choosing the Prandtl stress function in the form

$$\phi = \sum_{n=1}^{\infty} a_n Y_n \cos\frac{n\pi x}{2a}, \qquad n = \text{odd} \tag{a}$$

the boundary condition $\phi = 0$ is satisfied along $x = \pm a$; Y_n is a function of y alone. By substituting from (a) in equation 3.6.17, show that Y_n is given by

$$Y_n = A_n \sinh\frac{n\pi y}{a} + B_n \cosh\frac{n\pi y}{2a} + \frac{32G\alpha a^2}{n^3\pi^3 a_n}(-1)^{n-1} . \tag{b}$$

Determine A_n and B_n from the requirement that Y_n must be symmetric in y and $\phi = 0$ along $y = \pm b$. Show that the maximum stress for a square section bar is $\tau_{max} = 1.35G\alpha a$. Accept that $2G\alpha$ may be represented by

$$2G\alpha = \sum_{n=1}^{\infty} \frac{8G\alpha}{n\pi}(-1)^{\frac{n-1}{2}} \cos\frac{n\pi x}{a}, \qquad n = \text{odd} \tag{c}$$

3.7 Plane strain and plane stress situations

The torsion of bars discussed in Section 3.6 is an important example of a three dimensional system in which the stresses and displacements are described by a two dimensional theory. We now describe two other situations of practical importance where a plane theory obtains: physically, the situations are very different, but we treat them together since it turns out that the mathematics is the same for both.

A state of plane strain is said to exist with respect to the xOy plane in a body if it is deformed in such a way that

$$w = 0, \qquad u = u(x, y), \qquad v = v(x, y) . \tag{3.7.1}$$

This displacement pattern implies that

$$\epsilon_z = \gamma_{zx} = \gamma_{zy} = 0 \tag{3.7.2}$$

whilst ϵ_x, ϵ_y and γ_{xy} are functions of x and y alone. Invoking Hooke's law for the material then yields

$$\sigma_z = \nu(\sigma_x + \sigma_y), \qquad \tau_{zx} = \tau_{zy} = 0 \tag{3.7.3}$$

whilst σ_x, σ_y and τ_{xy} are functions of x and y alone. This situation obtains in a prismatic solid for which the loading does not vary with axial position and the end sections are rigidly constrained in the axial direction as shown in Fig. 3.7.1(a).

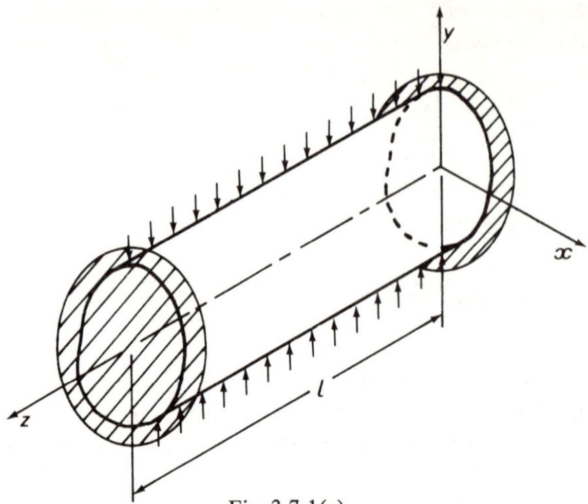

Fig. 3.7.1(a)

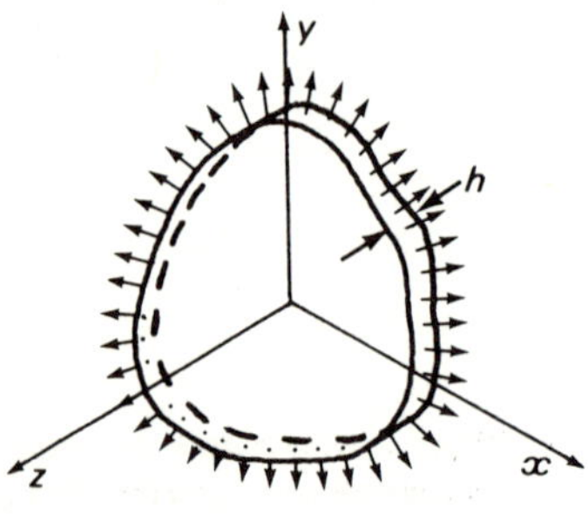

Fig. 3.7.1(b)

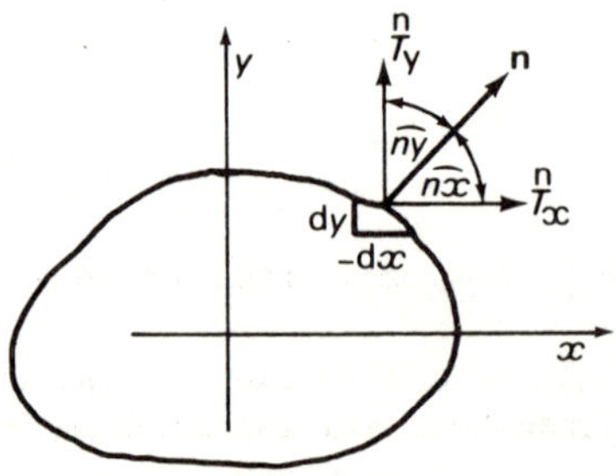

Fig. 3.7.1(c)

It may be readily verified that under these circumstances, only one of the compatibility equations (3.2.46) to (3.2.51) and only two of the equilibrium equations (3.3.19) are not satisfied automatically and we are left with the requirements

$$\frac{\partial^2 \epsilon_x}{\partial y^2} + \frac{\partial^2 \epsilon_y}{\partial x^2} = \frac{\partial^2 \gamma_{xy}}{\partial x \partial y} \tag{3.7.4}$$

$$\frac{\partial \sigma_x}{\partial x} + \frac{\partial \tau_{xy}}{\partial y} + F_x = 0 \tag{3.7.5}$$

$$\frac{\partial \tau_{yx}}{\partial x} + \frac{\partial \sigma_y}{\partial y} + F_y = 0 \tag{3.7.6}$$

to be met in the interior of the solid. These equations must be solved to satisfy boundary conditions which may be either

$$u = u^*(x, y)\,, \qquad v = v^*(x, y) \tag{3.7.7}$$

$$\text{or} \quad \overset{n}{T}_x = \overset{n}{T}{}_x^* = \sigma_x l + \tau_{xy} m, \quad \overset{n}{T}_y = \overset{n}{T}{}^*_y = \tau_{xy} l + \sigma_y m \tag{3.7.8}$$

along the entire edge of the cross section of the solid. In some problems, conditions (3.7.7) may apply along part of the edge and conditions (3.7.8) on the remainder. The asterisk indicates that the quantity is prescribed whilst in Fig. 3.7.1(c) $l = \cos(nx)$ and $m = \cos(ny)$.

A state of plane stress with respect to the xOy plane is said to exist if the situation is such that

$$\sigma_z = \tau_{zx} = \tau_{zy} = 0 \tag{3.7.9}$$

and σ_x, σ_y and τ_{xy} are functions of x and y alone. Suppose we have a thin flat plate stretched (not bent) by tractions applied to its edges alone and distributed smoothly and symmetrically with respect to the middle surface of the plate as shown in Fig. 3.7.1(b). Conditions (3.7.9) are clearly satisfied on the faces $z = \pm h/2$ and if h is small, it is reasonable to suppose that σ_z, τ_{zx}, τ_{xy} do not build up to any significant values within the plate either. A detailed examination shows that equations (3.7.9) and the conditions of equilibrium and compatibility can all be satisfied if σ_x, σ_y and τ_{xy} vary parabolically through the thickness, that is, if their distribution has the form $a(x, y) + z^2 b(x, y)$. Since z can never be greater than $h/2$, the second term can be made very small compared with the first by having a very thin plate and, for practical purposes, it will be adequate to know the averages of the stresses over the thickness. Provided we interpret $\overset{n}{T}_x \ldots\ldots \sigma_x \ldots\ldots \tau_{xy} \ldots\ldots u, v$ as such averages, we find that the problem is again described by equations (3.7.4) to (3.7.8).

When boundary tractions are specified, it is an advantage to work in terms of the stresses alone and the introduction of a stress function due to G.B. Airy provides a very elegant and practical device for the solution of the problem.

First, let us suppose that the body forces are conservative having a potential $\Omega(x, y)$ so that their x and y components are given by

$$F_x = -\frac{\partial \Omega}{\partial x}, \quad F_y = -\frac{\partial \Omega}{\partial y}. \tag{3.7.10}$$

The equations of equilibrium can then be written as

$$\frac{\partial}{\partial x}(\sigma_x - \Omega) + \frac{\partial \tau_{xy}}{\partial y} = 0$$

$$\frac{\partial \tau_{xy}}{\partial x} + \frac{\partial}{\partial y}(\sigma_y - \Omega) = 0. \tag{3.7.11}$$

We can readily verify by substitution that if the stresses are derived from a function $\chi(x, y)$ by means of

$$\sigma_x = \Omega + \frac{\partial^2 \chi}{\partial y^2}, \quad \sigma_y = \Omega + \frac{\partial^2 \chi}{\partial x^2}, \quad \tau_{xy} = -\frac{\partial^2 \chi}{\partial x \partial y} \tag{3.7.12}$$

equations (3.7.11) are satisfied automatically. The question is how to determine $\chi(x, y)$. By definitions (3.7.12), any well behaved $\chi(x, y)$ will yield a stress field satisfying equilibrium in the interior of the solid. However, the strains implied must be compatible if $\chi(x, y)$ is to represent any *true* stressing problem and must satisfy the traction boundary conditions if it is to describe a *particular* problem.

For plane strain, Hooke's law takes the form

$$\begin{aligned} E\epsilon_x &= (1 - \nu^2)\sigma_x - \nu(1 + \nu)\sigma_y , \\ E\epsilon_y &= (1 - \nu^2)\sigma_y - \nu(1 + \nu)\sigma_x , \\ G\gamma_{xy} &= \tau_{xy} \end{aligned} \tag{3.7.13}$$

and, the compatibility condition (3.7.4) becomes

$$\nabla^2(\sigma_x + \sigma_y) = -\frac{1}{1 - \nu}\left(\frac{\partial F_x}{\partial x} + \frac{\partial F_y}{\partial y}\right) . \tag{3.7.14}$$

Expressed in terms of the Airy function, we have

$$\nabla^4\chi = \frac{\partial^4\chi}{\partial x^4} + 2\frac{\partial^4\chi}{\partial x^2\partial y^2} + \frac{\partial^4\chi}{\partial y^4} = -\left(\frac{1 - 2\nu}{1 - \nu}\right)\left(\frac{\partial^2\Omega}{\partial x^2} + \frac{\partial^2\Omega}{\partial y^2}\right) . \tag{3.7.15}$$

For plane stress, Hooke's law takes the form

$$E\epsilon_x = \sigma_x - \nu\sigma_y , \quad E\epsilon_y = \sigma_y - \nu\sigma_x , \quad G\gamma_{xy} = \tau_{xy} \tag{3.7.16}$$

and the compatibility condition becomes

$$\nabla^2(\sigma_x + \sigma_y) = -(1 + \nu)\left(\frac{\partial F_x}{\partial x} + \frac{\partial F_y}{\partial y}\right) \tag{3.7.17}$$

or

$$\nabla^4\chi = \nabla^4\chi = -(1 - \nu)\left(\frac{\partial^2\Omega}{\partial x^2} + \frac{\partial^2\Omega}{\partial y^2}\right) . \tag{3.7.18}$$

If the body forces are constant or vanish, we have the interesting result that for both plane stress and plane strain situations

$$\nabla^2(\sigma_x + \sigma_y) = 0 \tag{3.7.19}$$

and

$$\nabla^4\chi = 0 \tag{3.7.20}$$

independent of the elastic constants of the material.

Equation (3.7.20) is the **biharmonic equation.** Any biharmonic function is thus the true solution to *some* elasticity problem in as much as it implies stresses in equilibrium and strains which are compatible. All that remains is to express the traction boundary conditions in terms of $\chi(x, y)$.

For the remainder of this development, we will suppose body forces to be absent, then equations (3.7.8) can be written.†

$$\overset{n}{T}_x{}^* = \frac{\partial^2 \chi}{\partial y^2} l - \frac{\partial^2 \chi}{\partial x \partial y} m = \frac{\partial^2 \chi}{\partial y^2}\frac{dy}{ds} + \frac{\partial^2 \chi}{\partial x \partial y}\frac{dx}{ds} = \frac{d}{ds}\left(\frac{\partial \chi}{\partial y}\right) \quad (3.7.21)$$

$$\overset{n}{T}_y{}^* = -\frac{\partial^2 \chi}{\partial x \partial y} l + \frac{\partial^2 \chi}{\partial x^2} m = -\frac{\partial^2 \chi}{\partial x \partial y}\frac{dy}{ds} - \frac{\partial^2 \chi}{\partial x^2}\frac{dx}{ds} = -\frac{d}{ds}\left(\frac{\partial \chi}{\partial x}\right). \quad (3.7.22)$$

Then, along the edge of the region,

$$\frac{\partial \chi}{\partial x} = -\int \overset{n}{T}_y{}^* \, ds + \alpha = A + \alpha \quad (3.7.23)$$

$$\frac{\partial \chi}{\partial y} = \int \overset{n}{T}_x{}^* \, ds + \beta = B + \beta . \quad (3.7.24)$$

Here, A and B are known boundary loading functions whilst α and β are constants of integration. Since from the calculus,

$$\frac{d\chi}{ds} = \frac{\partial \chi}{\partial x}\frac{dx}{ds} + \frac{\partial \chi}{\partial y}\frac{dy}{ds} \quad (3.7.25)$$

substituting from equations (3.7.23) and (3.7.24) and integrating with respect to s, we have

$$\chi = \int (Bm - Al)\, ds + \alpha x + \beta y + \gamma \quad (3.7.26)$$

with γ a constant of integration.

Observing that
$$\frac{\partial \chi}{\partial n} = \frac{\partial \chi}{\partial x}\frac{dx}{dn} + \frac{\partial \chi}{\partial y}\frac{dy}{dn} \quad (3.7.27)$$

equations (3.7.23) and (3.7.24) may be combined to give

$$\frac{\partial \chi}{\partial n} = (A + \alpha)l + (B + \beta)m . \quad (3.7.28)$$

To summarise, both the problems of plane stress and plane strain with edge tractions prescribed are defined by the boundary value problem

$$\nabla^4 \chi = 0 \text{ in the region}$$

$$\chi \text{ and } \frac{\partial \chi}{\partial n} \text{ specified on the boundary} . \quad (3.7.29)$$

When the region is multiply-connected, further conditions must be imposed on χ for each internal boundary. (See Sokolnikoff or Fung for example).

† Fig. 3.7.1(c) shows that

$$\cos(nx) = l = \frac{dy}{ds} = \frac{dx}{dn}, \quad \cos(ny) = m = -\frac{dx}{ds} = \frac{dy}{dn}$$

The use of curvilinear coordinates can simplify the stressing problem for certain boundary shapes. The general development of the appropriate theory is beyond the scope of this book and the reader should consult one of those cited at the end of the chapter. However, stressed objects having boundaries comprised of circular arcs and radial straight lines are of such practical importance that we now briefly summarise the relevant equations for plane situations in terms of polar coordinates.

Denoting the radial and tangential components of displacement, strain and stress by u, ϵ_r, σ_r, v, ϵ_θ, σ_θ, $\gamma_{r\theta}$, $\tau_{r\theta}$, it may be shown that

$$\epsilon_r = \frac{\partial u}{\partial r}, \qquad \epsilon_\theta = \frac{1}{r}\frac{\partial v}{\partial \theta} + \frac{u}{r}, \qquad \gamma_{r\theta} = \frac{1}{r}\frac{\partial u}{\partial \theta} + \frac{\partial v}{\partial r} - \frac{v}{r}. \tag{3.7.30}$$

Hooke's law takes the form (3.7.13) and (3.7.16) with σ_x replaced by σ_r etc., and the equations of equilibrium become

$$\frac{\partial \sigma_r}{\partial r} + \frac{\sigma_r - \sigma_\theta}{r} + \frac{1}{r}\frac{\partial \tau_{r\theta}}{\partial \theta} + F_r = 0$$

$$\frac{1}{r}\frac{\partial \sigma_\theta}{\partial \theta} + \frac{2}{r}\tau_{r\theta} + \frac{1}{r}\frac{\partial \tau_{r\theta}}{\partial r} + F_\theta = 0\,. \tag{3.7.31}$$

When body forces are absent equations (3.7.31) are satisfied identically when the stress components are derived from a function $\chi(r, \theta)$ according to

$$\sigma_r = \frac{1}{r}\frac{\partial \chi}{\partial r} + \frac{1}{r^2}\frac{\partial^2 \chi}{\partial \theta^2}, \qquad \sigma_\theta = \frac{\partial^2 \chi}{\partial r^2}, \qquad \tau_{r\theta} = -\frac{\partial}{\partial r}\left(\frac{1}{r}\frac{\partial \chi}{\partial \theta}\right). \tag{3.7.32}$$

Observing that $(\sigma_x + \sigma_y) = (\sigma_r + \sigma_\theta)$ from the first invariant and using the polar coordinate form of the Laplace operator, ∇^2, the equation of compatibility (3.7.19) becomes

$$\left(\frac{\partial^2}{\partial r^2} + \frac{1}{r}\frac{\partial}{\partial r} + \frac{1}{r^2}\frac{\partial^2}{\partial \theta^2}\right)(\sigma_r + \sigma_\theta) = 0\,. \tag{3.7.33}$$

Substituting from (3.7.32) in (3.7.33), we have the biharmonic equation in polar coordinates as

$$\left(\frac{\partial^2}{\partial r^2} + \frac{1}{r}\frac{\partial}{\partial r} + \frac{1}{r^2}\frac{\partial^2}{\partial \theta^2}\right)\left(\frac{\partial^2 \chi}{\partial r^2} + \frac{1}{r}\frac{\partial \chi}{\partial r} + \frac{1}{r^2}\frac{\partial^2 \chi}{\partial \theta^2}\right) = 0\,. \tag{3.7.34}$$

In 1899, Michell obtained a general solution to this equation in the form

$$\begin{aligned}\chi = {} & a_0 \ln r + b_0 r^2 + c_0 r^2 \ln r + d_0 r^2 \theta + e_0 \theta + f_0 r\theta \sin\theta + g_0 r\theta \cos\theta \\ & + (a_1 r + b_1 r^3 + c_1 r^{-1} + d_1 r \ln r)\sin\theta \\ & + (a_1' r + b_1' r^3 + c_1' r^{-1} + d_1' r \ln r)\cos\theta \\ & + \sum_{n=2}^{\infty} (a_n r^n + b_n r^{n+2} + c_n r^{-n} + d_n r^{-n+2})\sin n\theta \\ & + \sum_{r=2}^{\infty} (a_n' r^n + b_n' r^{n+2} + c_n' r^{-n} + d_n' r^{-n+2})\cos n\theta\,.\end{aligned} \tag{3.7.35}$$

By adjusting the values of the coefficients a_0, b_0 . . . . . , a number of solutions to practically important problems may be found.

EXAMPLE 3.7.1 Cf Sechler Prob 8.3

Accepting that

$$\chi = \frac{p}{8c^3}\left[x^2(y^3 - 3c^2y - 2c^3) - \frac{1}{5}y^3(y^2 - 2c^2)\right] \qquad \text{(a)}$$

is appropriate as a stress function, determine the stress field it generates. Show how St. Venant's principle may be invoked in order that equation (a) represents the stress distribution in a cantilever of unit thickness occupying $x > 0$, $-c \geqslant y \geqslant c$ and supporting a uniformly distributed pressure on its top edge $y = c$.

Using equations (3.7.12) with $\Omega = 0$, equation (a) generates

$$\sigma_x = \frac{p}{4c^3}\left[\left(3x^2 + \frac{6}{5}c^2\right)y - 2y^3\right] \qquad \text{(b)}$$

$$\sigma_y = \frac{p}{4c^3}\left[3x^2y - \frac{1}{5}(10y^3 - 6c^2y)\right] \qquad \text{(c)}$$

$$\tau_{xy} = \frac{3px}{4c^3}(y^2 - c^2)\ . \qquad \text{(d)}$$

Along the top edge, $y = c$, we find $\sigma_y = -p$, $\tau_{xy} = 0$. Along the lower edge $y = -c$, $\sigma_y = 0$, $\tau_{xy} = 0$. Then using equations (3.7.8) the applied traction along the top edge is $\overset{n}{T}_y = -p$, $\overset{n}{T}_x = 0$ and along the lower edge $\overset{n}{T}_y = 0$, $\overset{n}{T}_x = 0$. Along vertical edge $x = 0$, we find

$$\overset{n}{T}_x = -\frac{p}{4c^3}\left(\frac{6}{5}c^2y - 2y^3\right) \neq 0 \qquad \text{(e)}$$

$$\overset{n}{T}_y = 0\ . \qquad \text{(f)}$$

Thus, the stress function (a) implies a non-zero x component of traction along edge $x = 0$, however, from

$$\left|\int \sigma_x\,\mathrm{d}y\right|_{x=0} \quad \text{and} \quad \left|\int \sigma_x y\,\mathrm{d}y\right|_{x=0} \qquad \text{(g)}$$

we find that (e) implies zero resultant force and moment.

Rewriting equation (b) as

$$\sigma_x = \frac{3px^2y}{4c^3}\left[1 + \frac{\frac{2}{5}c^2y - \frac{2}{3}y^3}{x^2}\right]. \qquad \text{(h)}$$

Since $y \not> c$, the second term in square brackets becomes small compared with the first as x increases. Since $4c^3/3 = I$, the second moment of area of the section, expression (h) thus tends to

Is $I \neq \frac{1}{12}h^3 = \frac{2}{3}c^3$??

$$\text{for } x \gg y. \qquad \sigma_x = \frac{px^2y}{I} \qquad \text{(j)}$$

7/4/82

CHECK $\sigma_x = \frac{1}{2}\frac{px^2y}{I}$ AS $M = \frac{px^2}{2}$.

Equations (d) and (j) are the usual mechanics of materials result for the shearing and bending stresses in a cantilever supporting a uniformly distributed load p and having its free edge at $x = 0$. Thus, provided we invoke St. Venant's principle, equation (a) describes this situation away from the neighbourhood of $x = 0$.

EXAMPLE 3.7.2

A thin washer is shrunk onto a solid disk of the same thickness and material as shown in Fig. 3.7.2. If, before assembly, the radius of the hole in the washer is Δ smaller than the radius of the disk, adapt Michell's solution to describe the stresses in the assembly. The interference Δ is small compared with a.

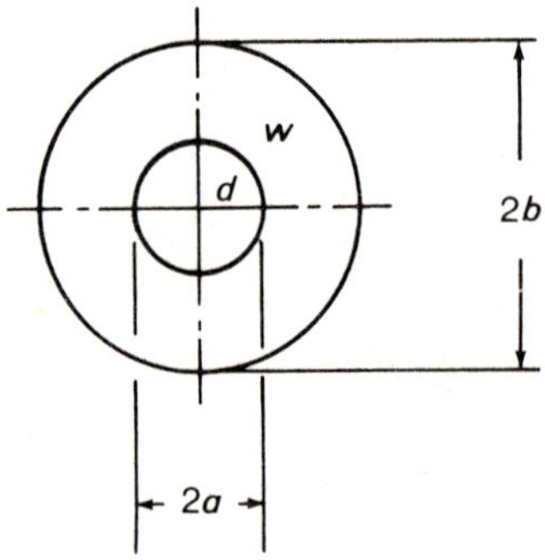

Fig. 3.7.2

The situation is axisymmetric so that it is appropriate to use the terms independent of θ in equation (3.7.35). That is we use

$$\chi(r) = \tfrac{1}{2}Ar^2 + B\ln r + Cr^2\ln r\ . \tag{a}$$

The stress field generated is then

$$\sigma_r = \frac{1}{r}\frac{d\chi}{dr} = A + B/r^2 + C(2\ln r + 1) \tag{b}$$

$$\sigma_\theta = \frac{d^2\chi}{dr^2} = A - B/r^2 + C(3 + 2\ln r) \tag{c}$$

$$\tau_{r\theta} = 0\ . \tag{d}$$

The shrink fit causes some interface pressure p at the interface. Then, isolating the disk and observing that $\sigma_r = -p$ at $r = a$ and that σ_r and σ_θ must be finite at $r = 0$, equations (b) and (c) yield

$$A_d = -p, \quad B_d = 0, \quad C_d = 0 \tag{e}$$

so that

$$(\sigma_r)_d = (\sigma_\theta)_d = -p\ . \tag{f}$$

For the washer, the traction boundary conditions $\sigma_r = -p$ at $r = a$ and $\sigma_r = 0$ at $r = b$ are not sufficient to determine A_w, B_w and C_w. However, consideration of displacements shows that C_w must be zero since otherwise

$$\epsilon_r = \frac{\mathrm{d}u}{\mathrm{d}r} = \frac{1}{E}[\sigma_r - \nu\sigma_\theta] \text{ and } \epsilon_\theta = \frac{u}{r} = \frac{1}{E}[\sigma_\theta - \nu\sigma_r]$$

are not consistent. Then, it turns out that

$$(\sigma_r)_w = -p\left[\frac{(b/r)^2 - 1}{(b/a)^2 - 1}\right] \tag{g}$$

$$(\sigma_\theta)_w = p\left[\frac{(b/r)^2 + 1}{(b/a)^2 - 1}\right]. \tag{h}$$

As yet the interface pressure p is unknown. However, for compatibility between the disk and washer at $r = a$,

$$(u_w - u_\mathrm{d})_{r=a} = \Delta . \tag{j}$$

Solution yields

$$p = \frac{E\Delta}{2a[1 - (a/b)^2]} \tag{k}$$

and hence σ_r, σ_θ and u in disk and washer.

Problems 3.7

3.7.1 Determine the constants A, B, C, D and E in

$$\chi = x^2(Ay^3 + By + C) + Dy^3 + Ey^5$$

in order that it shall represent the stress distribution in the beam of Fig. 3.7.3.

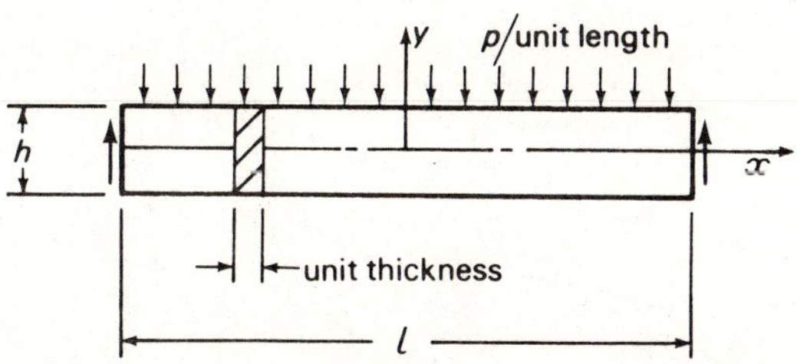

Fig. 3.7.3

3.7.2 The partial ring of Fig. 3.7.4 is of unit thickness. Accepting that

$$\chi = (Ar^2 + B/r + Cr + Dr\ln r)\sin\theta$$

is suitable as an Airy stress function, determine the values of A, B, C and D and show that

$$(\sigma_r)_{\theta=\pi/2} = \frac{P}{r^3}\frac{(a^2 + b^2)r^2 - a^2b^2 - r^4}{(a^2 - b^2) + (a^2 + b^2)\ln(b/a)}.$$

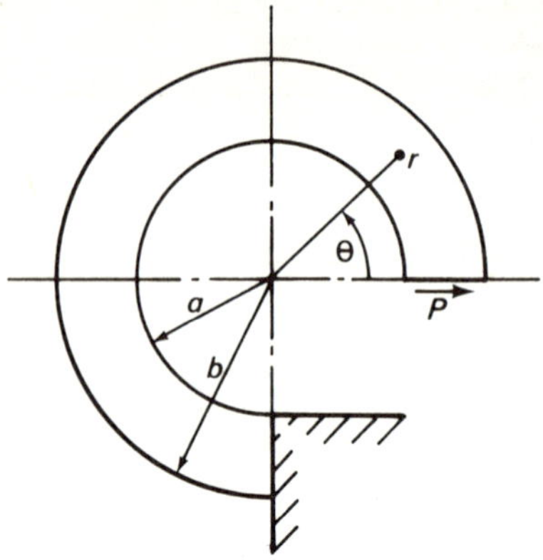

Fig. 3.7.4

3.7.3 Given that

$$\chi = r\theta(A \sin\theta + B\cos\theta) \qquad \text{(a)}$$

is suitable as a stress function, determine the nature of the stress field it generates, A and B are constants.

A parting-off tool for a lathe is of unit thickness and is shown schematically in Fig. 3.7.5. Show that equation (a) satisfies the boundary conditions along RS and ST and by considering the equilibrium of RST, determine A and B.

Contrast the expressions for the normal and shearing stresses at point O, on section mn, given by this approach and, those according to mechanics of materials.

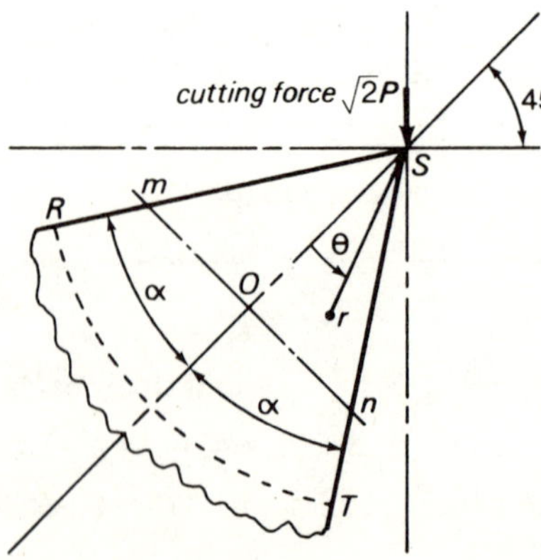

Fig. 3.7.5

3.8 Further Reading

Biezeno, C.B. *and* Grammell, R. — *Engineering Dynamics,* Blackie (1955)

Bland, D.R. — *The Theory of linear viscoelasticity,* Pergamon Press (1960).

Flugge, W. — *Viscoelasticity,* Blaisdell (1967).

Ford, H. *and* Alexander, J.A. — *Advanced Mechanics of Materials,* Longmans (1963)

Fung, Y.C. — *Foundations of Solid Mechanics,* Prentice Hall (1965).

Fung, Y.C. — *A First Course in Contunuum Mechanics,* Prentice Hall (1969).

Hetenyi, M. — *Handbook of Experimental Stress Analysis,* John Wiley and Sons (1950).

Hunter, S.C — *Mechanics of Continuous Media,* Ellis Horwood, (1976)

Johnson, W. *and* Mellor, P.R. — *Plasticity for Mechanical Engineers,* Van Nostrand (1962).

Love, A.E.H. — *A Treatise on the Mathematical Theory of Elasticity,* 4th Edition. Cambridge University Press (1927).

Mendelson, A. — *Plasticity: Theory and Application.* Macmillan (1962).

Sokolnikoff, I.S. — *The Mathematical Theory of Elasticity,* 2nd Edition. McGraw Hill (1956).

Timoshenko, S. *and* Goodier, J.N. — *Theory of Elasticity,* 3rd Edition, McGraw Hill (1970).

Chapter 4

Variational Calculus - Principles

4.1 Introduction

The variational calculus is mainly concerned with the stationary or 'extreme' values of certain definite integrals called functionals. These integrals involve functions which are unknown at the beginning of the analysis, and the purpose of the variational calculus is to determine what conditions are imposed on them when the integrals attain stationary values. Then, hopefully, satisfaction of these conditions will actually yield the functions themselves. We have seen that problems of static equilibrium of discrete systems can be characterised by the stationary value of a function of one or more variables; this is the domain of the ordinary calculus. When these stationary (extremum) principles are extended to more general stress analysis situations and other problems in continuum mechanics, then the calculus of variations is the appropriate mathematical tool. The direct formulation of problems in continuum mechanics entails consideration of equilibrium and continuity in conjunction with some constitutive equation describing the mechanical behaviour of the continuum. This seems a straightforward procedure, so that one may well ask 'What advantage is there in formulating continuum problems in accordance with some extremal principle?' An illuminating discussion of this topic will be found in the book by Lanczos†, but for our purposes, the following observations will suffice:

(i) Our earlier experience with discrete systems encourages us to expect some advantage.

(ii) We will see in due course that extremum principles can lead to the governing differential equations *together with* the appropriate boundary conditions. This is especially valuable in formulating an approximate theory as was shown by Kirchhoff when he resolved an anomaly in the theory of plates (see Chapter 6).

(iii) The so-called 'direct methods' of the variational calculus provide many powerful techniques for obtaining numerical solutions to very complex problems having significance in engineering practice.

In this chapter, we give only an outline of the variational calculus. For a more detailed understanding, the further reading list should be consulted.

† See the list of further reading for Chapter 1 (page 68).

4.2 Stationary value of functions of one variable

Certain of the concepts of the differential calculus are applicable to the variational calculus. It will therefore prove useful if we first review some of the basic ideas relating to the maxima and minima (or extreme values) of functions.

If a function of one variable, $f(x)$, is continuous and differentiable in a closed interval $a \leqslant x \leqslant b$, except perhaps at a finite number of points, then it attains an absolute maximum (or minimum value) at either

(a) A point where $f'(x) = 0$† (a stationary point), or

(b) At an end point of the interval, or

(c) At one of the points (if any exist) at which $f(x)$ is not differentiable.

In the context of stress analysis we could imagine $f(x)$ to define the deflexion curve of a beam in the region $0 \leqslant x \leqslant l$. As an example, suppose, we are discussing a cantilever, clamped at $x = 0$, supporting upward and downward loads, then depending on the precise nature of the loads, the maximum deflexion [absolute extreme value of $f(x)$] could either occur somewhere along the span where $f'(x) = 0$, a stationary point, or at the free end (an end point of the interval).

We see therefore that to determine the absolute extreme value of a function, we have to compare its values at the stationary point (or points) with the end point values. It follows from this immediately that if $f'(x)$ does not attain zero in the range, then the absolute extreme value is at one of the end points: this would be the case for a cantilever carrying a uniformly distributed downward load.

The condition $f'(x) = 0$ is *necessary* for a stationary value of $f(x)$ but not *sufficient* for us to determine whether a local maximum or minimum has been located; to this end a further test is required. Then, if $f''(x) < 0$, a maximum occurs whereas if $f''(x) > 0$, a minimum occurs. If $f''(x) = 0$ and the first non-zero derivative at the point of interest is of even order, then if its value is positive (negative) a minimum (maximum) is defined whereas if the first non zero derivative is of odd order, there is neither a maximum nor a minimum at the point. This latter situation corresponds to a point of inflexion on a beam.

Problems 4.2

4.2.1 The intensity of heat due to a point source varies directly as its strength and inversely as the square of the distance from it. Two sources A and B have strengths S_A and S_B. Show that the heat intensity at a point P between A and B is a minimum when $AP/PB = {}^3\sqrt{(S_A/S_B)}$.

4.2.2 The deflexion curve for a simply supported beam carrying a point load P at $x = a$ from one end is given by

$$w = \frac{P}{6EI}x\left(1 - \frac{a}{l}\right)(2al - a^2 - x^2) + \frac{P}{2EI}\langle x - a\rangle^3 .$$

† f' denotes df/dx, f'' denotes d^2f/dx^2 etc.

l is the length of the beam, EI the bending stiffness. The brackets <> signify that the term is to be ignored if $x < a$.

Determine the position of maximum deflexion (i) when $a = l/4$ and (ii) when $a = 3l/4$.

4.3 Stationary value of functions of more than one variable

In practical stress analysis work, we are frequently concerned with two or three dimensional problems so that we must know something about the stationary values of functions of several variables.

Suppose $w = f(x_1, x_2 \ldots\ldots x_n)$ is continuous and differentiable in a given closed region, then again we can say that w attains a greatest or least value either in the interior or on the boundary of the region. If $w = f(x_1, x_2)$, a function of two variables, it could, perhaps, represent the deflexion of the middle surface of a bent plate: the region of definition of w is then the extent of the plate and the maximum deflexion could be either at the edge or in the interior. The necessary condition for w to have a stationary value within the region is that

$$\frac{\partial f}{\partial x_1} = \frac{\partial f}{\partial x_2} = \ldots\ldots \frac{\partial f}{\partial x_n} = 0\ . \tag{4.3.1}$$

Alternatively, since the total differential of f is

$$\mathrm{d}f = \frac{\partial f}{\partial x_1}\mathrm{d}x_1 + \frac{\partial f}{\partial x_2}\mathrm{d}x_2 + \ldots\ldots \frac{\partial f}{\partial x_n}\mathrm{d}x_n\ ,$$

a necessary condition is that

$$\mathrm{d}f = 0\ . \tag{4.3.2}$$

The *sufficiency* conditions are more complicated than for the one-dimensional case: for a function of two variables, $f(x, y)$, in addition to $\mathrm{d}f = 0$, we need

$$\frac{\partial^2 f}{\partial x^2} < 0 \text{ and } \frac{\partial^2 f}{\partial x^2} \cdot \frac{\partial^2 f}{\partial y^2} - \left(\frac{\partial^2 f}{\partial x \partial y}\right)^2 > 0 \tag{4.3.3}$$

for a maximum, or

$$\frac{\partial^2 f}{\partial x^2} > 0 \text{ and } \frac{\partial^2 f}{\partial x^2} \cdot \frac{\partial^2 f}{\partial y^2} - \left(\frac{\partial^2 f}{\partial x \partial y}\right)^2 > 0 \tag{4.3.4}$$

for a minimum.

If

$$\frac{\partial^2 f}{\partial x^2} \cdot \frac{\partial^2 f}{\partial y^2} - \left(\frac{\partial^2 f}{\partial x \partial y}\right)^2 < 0\ ,$$

the stationary value is neither a maximum nor a minimum, whilst if

$$\frac{\partial^2 f}{\partial x^2} \cdot \frac{\partial^2 f}{\partial y^2} - \left(\frac{\partial^2 f}{\partial x \partial y}\right)^2 = 0\ ,$$

a further examination is necessary.

For functions of two variables, we can visualise these conditions geometrically. Thus $w = f(x, y)$ defines a surface above the xy plane: if this surface has smooth hills and valleys, the crests of the hills and the bottoms of the valleys represent the stationary values of w. We can also see that at these points, a tangent plane is horizontal and does not cut the surface in the neighbourhood of each point.

If the surface has a saddle point, then there is no local maximum or minimum and the tangent plane would cut into the surface in the neighbourhood of the point.

This discussion shows that when we speak of stationery values for $f(x, y)$ we are temporarily concerned with the local shape of $f(x, y)$ in the neighbourhood of a stationary point, (crests or troughs); because of this, it is more precise to refer to a **relative** maximum or minimum for $f(x, y)$.

Problems 4.3

4.3.1 Locate and establish the character of the stationary points for the functions:

(i) $f_1(x, y) = x^2 + 2cxy + y^2$. (ii) $f_2(x, y) = x^2y^2 - x^2 - y^2$.

(iii) $f_3(x, y) = x^2 - xy + y^2 - 2x + y$.

(iv) $f_4(x, y) = x^4 - 2x^2y^2 + y^4$.

Ans:

(i) $c^2 < 1$, min at $(0, 0)$; if $c^2 > 1$, saddle point a $(0, 0)$.
(ii) Max at $(0, 0)$; saddle points at $(1, 1)$, $(1, -1)$, $(-1, 1)$, $(-1, -1)$.
(iii) Min at $(1, 0)$.
(iv) Min along $y \pm x = 0$.

4.4 Stationary value problems with constraints

It frequently happens in practical applications that we wish to find the extreme value of a function subject to some subsidiary condition or constraint. As an example, we may wish to find the dimensions of a rectangular region of specified perimeter subject to the requirement that the area shall be a maximum. To tackle this problem, we could proceed as follows:

Let the region have sides x, y than area $A = xy$ and perimeter $P = 2x + 2y$. With P fixed, we can express y in terms of x by $y = P/2 - x$ giving

$$A = x(P/2 - x) .$$

We are thus left with the task of finding an ordinary stationary value of a function of one variable; the subsidiary condition having been taken into account.

To identify the stationary value,

$$\frac{\mathrm{d}A}{\mathrm{d}x} = \frac{P}{2} - 2x = 0,$$

so that $x = P/4$ and $y = P/2 - x = P/4$. $d^2A/\mathrm{d}x^2$ is readily seen to be negative so that this pair of values for x and y actually maximises A.

Whilst the above approach is quite correct, we seem to give undue emphasis to x. Furthermore, in other problems, to solve for one variable in terms of the other may well be inconvenient if not impossible. We are able to avoid such difficulties by using a very elegant approach due to Lagrange. To fix ideas, we will again consider functions of two variables. Then typically, our task will be to find the stationary values of $w = F(x, y)$ subject to the constraint $G(x, y) = 0$; thus x and y are no longer independent. Lagrange's method is to form a new function $H(x, y)$ such that

$$H(x, y) = F(x, y) + \lambda G(x, y), \tag{4.4.1}$$

where λ is, as yet, an unkown constant parameter. Then the conditions for a stationary value of this new function $H(x, y)$, treating it as if x and y *were* independent, together with the subsidiary condition $G(x, y) = 0$, will allow us to find the stationary points of the original function of interest, $w = F(x, y)$, as well as the value of λ. That is, the relations

$$\left.\begin{aligned} \frac{\partial H}{\partial x} &= 0 = \frac{\partial F}{\partial x} + \lambda\frac{\partial G}{\partial x} \\ \frac{\partial H}{\partial y} &= 0 = \frac{\partial F}{\partial y} + \lambda\frac{\partial G}{\partial y} \\ G(x, y) &= 0\,. \end{aligned}\right\} \tag{4.4.2}$$

This is Lagrange's method of undetermined multipliers and λ is called Lagrange's multiplier. For the method to succeed we require that at least one of $\partial G/\partial x$ and $\partial G/\partial y$ should not be zero at the points where $F(x, y)$ is stationary: that is stationary points of $F(x, y)$ must not coincide with a singular point of $G(x, y) = 0$.

We will illustrate the use of the method by re-solving the problem of the rectangular region.

To maximise $A = xy$ subject to the restriction that $P - 2x - 2y = 0$, we form

$$H(x, y) = xy + \lambda(P - 2x - 2y).$$

Then

$$\frac{\partial H}{\partial x} = y - 2\lambda = 0$$

$$\frac{\partial H}{\partial y} = x - 2\lambda = 0$$

$$P - 2x - 2y = 0$$

Evidently $x = y = 2\lambda$ so that $\lambda = P/8$ and $x = y = P/4$.

Here we see that no undue emphasis was placed on either x or y, but to achieve this symmetric treatment of the problem, it proved necessary to introduce an extra parameter, λ. However, in many practical applications, this is a small price to pay.

Problems 4.4

4.4.1 A long piece of sheet metal, 30 cm wide, is to be formed into a trough of trapezoidal section. Determine the length to be bent up at each end and the angle of the sides for maximum capacity.
Ans: 10 cm, 60°.

4.4.2 Minimise $f(x, y, z) = x^2 + 4y^2 + 16z^2$ subject to the constraints (i) $xyz = 1$, (ii) $xy = 1$, (iii) $x = 1$.

Ans:
(i) at $x^2 = 4, \quad y^2 = 1, \quad z^2 = \frac{1}{4}, \quad f = 12.$

(ii) at $x^2 = 2, \quad y^2 = \frac{1}{2}, \quad x = 0, \quad f = 4.$

(iii) at $x = 1, \quad y = 0, \quad z = 0, \quad f = 1.$

4.5 Generalisation of Lagrange's method

The method of Lagrange multipliers can be readily extended to determine the stationary values of a function of several variables subject to several conditions of constraint.

Suppose we wish to extremise the function $w = F(x_1, x_2, \ldots\ldots x_n)$ subject to the constraints $G_1(x_1, x_2 \ldots\ldots x_n) = 0, G_2(x_1, x_2, \ldots\ldots x_n) = 0, \ldots\ldots$ $G_k(x_1, x_2, \ldots\ldots x_n) = 0$. It is worth noting that there must be fewer equations of constraint than there are variables x_i, that is, we must have $k < n$. Then, Lagrange's rule becomes:

Form a new function

$$H(x_1, x_2 \ldots\ldots x_n) = F(x_1, x_2 \ldots\ldots x_n) + \sum_{i=1}^{k} \lambda_i G_i(x_1, x_2, \ldots\ldots x_n). \tag{4.5.1}$$

Then, the conditions for a stationary value of this new function $H(x_1 \ldots\ldots x_n)$, treating it as if the $x_1, x_2 \ldots\ldots x_n$ were independent, together with all the subsidiary conditions $G_1(x_1, x_2, \ldots\ldots x_n) = 0$ and $G_k(x_1, x_2 \ldots\ldots x_n) = 0$, will allow us to find the stationary points of the original function of interest, $w = F(x_1, x_2 \ldots\ldots x_n)$, as well as the values of all the $\lambda_1, \lambda_2, \ldots\ldots \lambda_k$. That is, we use the relations

$$\left.\begin{aligned} \frac{\partial H}{\partial x_1} &= 0 = \frac{\partial F}{\partial x_1} + \sum_{i=1}^{k} \lambda_i \frac{\partial G_i}{\partial x_1} \\ &\vdots \\ \frac{\partial H}{\partial x_n} &= 0 = \frac{\partial F}{\partial x_n} + \sum_{i=1}^{k} \lambda_i \frac{\partial G_i}{\partial x_n} \end{aligned}\right\} \tag{4.5.2}$$

$$\left.\begin{aligned} G_1(x_1, \ldots\ldots x_n) &= 0 \\ &\vdots \\ G_k(x_1, \ldots\ldots x_n) &= 0 \end{aligned}\right\} \tag{4.5.3}$$

For the method to succeed, we require that the extreme values of $F(x_1, \ldots\ldots x_n)$ do not occur at singular points of any of the $G_i(x_1 \ldots\ldots x_n) = 0$.

EXAMPLE 4.5.1

Find the greatest and least values of z on the ellipse formed by the intersection of the plane $x + y + z = 1$ and the ellipsoid $16x^2 + 4y^2 + z^2 = 16$.

Expressed in the terms used above, we are required to extremise $w = F(x, y, z) = z$ subject to the constraints $G_1(x, y, z) = x + y + z - 1 = 0$ and $G_2(x, y, z) = 16x^2 + 4y^2 + z^2 - 16 = 0$.

Now $H(x, y, z) = z + \lambda_1(x + y + z - 1) + \lambda_2(16x^2 + 4y^2 + z^2 - 16)$.

$$\frac{\partial H}{\partial x} = \lambda_1 + 32\lambda_2 x = 0 \,. \tag{a}$$

$$\frac{\partial H}{\partial y} = \lambda_1 + 8\lambda_2 y = 0 \,. \tag{b}$$

$$\frac{\partial H}{\partial z} = 1 + \lambda_1 + 2\lambda_2 z = 0 \,. \tag{c}$$

$$\frac{\partial H}{\partial \lambda_1} = x + y + z - 1 = 0 \,. \tag{d}$$

$$\frac{\partial H}{\partial \lambda_2} = 16x^2 + 4y^2 + z^2 - 16 = 0 \,. \tag{e}$$

We could now solve for all the variables in the usual way. However, we can save some work in this particular problem if we notice that since λ_1 and λ_2 cannot be zero, then equations (a) and (b) are consistent if $y = 4x$: equation (d) then gives $x = (1 - z)/5$. Equation (e) can now be written as a function of z alone, namely

$$21z^2 - 32z - 64 = 0,$$

giving $z = 8/3$ and $-8/7$ as the maximum and minimum values.

Problems 4.5

4.5.1 The plane $Ax + By + Cz = 0$ intersects the ellipsoid $(x/a)^2 + (y/b)^2 + (z/c)^2 = 1$. Show that the square of the least distance r from the origin to the curve of intersection is given by

$$\frac{A^2a^2}{a^2 - r^2} + \frac{B^2b^2}{b^2 - r^2} + \frac{C^2c^2}{c^2 - r^2} = 0 .$$

4.6 The concept of a functional and its variation

We have seen in the earlier sections of this chapter that an important practical application of the differential calculus is the determination of the stationary values of functions of a finite number of variables. In physical terms, we can imagine that we are discussing systems with a finite number of degrees of freedom. The calculus of variations, on the other hand, is concerned with the stationary values of **functionals.** A functional is a quantity the value of which depends on the *entire shape* of some functions rather than on a number of discrete variables: in physical terms, we would now have continuous systems, that is ones with an infinite number of degrees of freedom. Whereas the region of definition of a function is prescribed by definite values given to the variables, say a beam deflexion curve extending from $x = 0$ to $x = l$, the region of a functional is defined by a set or collection of admissible functions. A few examples will help to fix ideas.

EXAMPLE 4.6.1

The length L of a curve $y = f(x)$ over an interval $x = 0$ to $x = l$ is given by

$$L(y) = \int_0^l \sqrt{1 + (y')^2}\,dx .$$

This expression for $L(y)$ holds for any arbitrary $f(x)$ as long as it is a continuous function having a piecewise continuous first derivative. However, the *value* of $L(y)$ depends on the entire form of $f(x)$: we emphasise this by writing $L(y)$ and not just L.

EXAMPLE 4.6.2

The strain energy stored in a bent beam is given by

$$U(y) = \int_0^l \frac{EI}{2}(y'')^2 \, dx ,$$

where $y = f(x)$ is the deflexion function, l is the length and EI is the bending stiffness of the beam.

Here again, the functional $U(y)$ depends on the entire shape of $y = f(x)$.

EXAMPLE 4.6.3

The total potential energy for a beam supporting a uniformly distributed transverse load p/unit length together with end thrust T is given by

$$V(y) = \int_0^l \left[\frac{EI}{2}(y'')^2 - py - \frac{T}{2}(y')^2\right] dx$$

or

$$V(y) = \int_0^l F(y, y', y'') dx .$$

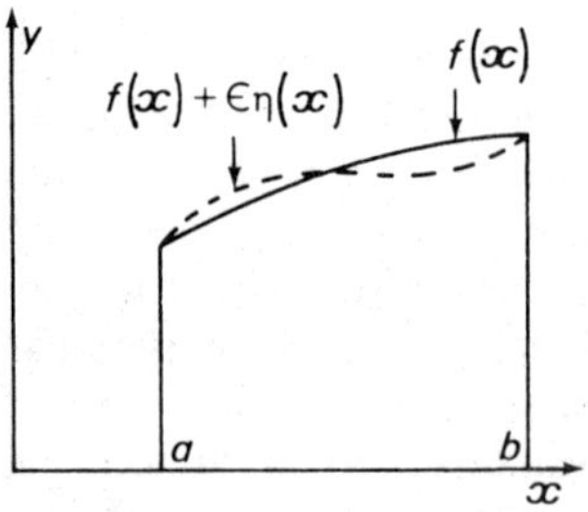

Fig. 4.6.1

To continue, suppose $y = f(x)$ is a function of interest, defined in the range $a \leqslant x \leqslant b$, and suppose $\eta(x)$ is some other function which is quite arbitrary except that it vanishes at $x = a$ and $x = b$. Then, if ϵ is an infinitesimal parameter, the curve defined by $f(x) + \epsilon\eta(x)$ will neighbour $f(x)$ as shown in Fig. 4.6.1. The function $\epsilon\eta(x)$, which represents a small change in the *overall shape* of the original function is called a **variation** in y: we shall later use the notation $\delta y = \epsilon\eta(x)$. If $I(y)$ is a functional of y then corresponding to a variation in y there will be a variation in I.

EXAMPLE 4.6.4

If

$$I(y) = \int_a^b F(x, y, y') dx$$

then
$$I(y + \epsilon\eta) = \int_a^b F(x, y + \epsilon\eta, y' + \epsilon\eta')\mathrm{d}x$$

so that

$$\Delta I = I(y + \epsilon\eta) - I(y) = \int_a^b [F(x, y + \epsilon\eta, y' + \epsilon\eta') - F(x, y, y')]\,\mathrm{d}x\,.$$

If the right hand side is expanded by Taylor's series we will find $\Delta I = (\delta I)\epsilon + 1/2!(\delta^2 I)\epsilon^2 + \ldots\ldots$ Then δI is called the first variation in I, $\delta^2 I$ is called the second variation in I etc. If we speak of the variation in I without qualification, we imly the first variation.

4.7 A typical problem in the variational calculus

Before proceeding to a more formal summary of the variational calculus it will be instuctive, and provide motivation for further study, if we follow the processes of the variational calculus in an application to the bending of a clamped-clamped (encastré) beam supporting a distributed load p per unit length.

In an attempt to arrive at the correct deflexion curve corresponding to a given load, we could imagine a mathematical experiment being carried out: we take several functions to see if they produce a configuration which implies equilibrium. In this process, the functions for the trials can be quite arbitrary except that we see immediately they must satisfy the 'forced' or 'kinematic' boundary conditions of $y = y' = 0$ at both ends: this is the requirement of compatibility with the end constraints. Such functions which satisfy the necessary continuity and differntiability requirements as well as the forced boundary conditions are referred to as **admissible**: all trial functions must belong to this admissible class. As described, the experiment is very cumbersome but in a refined way, this is what we do if we use the theorem of minimum potential energy in conjunction with the variational calculus.

Now the total potential energy of the beam is

$$V = \int_0^l \frac{EI}{2}(y'')^2\,\mathrm{d}x - \int_0^l py\mathrm{d}x\,. \tag{4.7.1}$$

It is worth noting that the concentrated forces and couples which constitute the reactions at the ends of the beam do no work since there is neither linear nor angular displacement corresponding to them. Accordingly, there is no potential energy term corresponding to them.

The theorem of minimum potential energy states that of all admissible trial functions $y = f(x)$, the one corresponding to the true equilibrium configuration is that which renders V a minimum. We will now carry out the mathematical experiment: if $f(x)$ is the true curve (as yet unkown) then a neighbouring curve is $f(x) + \epsilon\eta(x)$. Corresponding to this neighbouring curve, the potential energy is

$$V + \Delta V = \int_0^l \left[\frac{EI}{2}(y'' + \epsilon\eta'')^2 - p(y + \epsilon\eta)\right] dx$$

$$= \int_0^l \left[\frac{EI}{2}(y'')^2 - py\right] dx + \epsilon \int_0^l [EIy''\eta'' - p\eta]\, dx$$

$$+ \epsilon^2 \int_0^l \frac{EI}{2}(\eta'')^2\, dx\,, \qquad (4.7.2)$$

so that $$\Delta V = \delta V + \delta^2 V\,. \qquad (4.7.3)$$

δV is the part of ΔV linear in ϵ, $\delta^2 V$ is the part of ΔV quadratic in ϵ.

Equation (4.7.3) shows us that δV, the first variation in V, dominates the value of ΔV. Now if $f(x)$ is indeed the true deflexion curve, then V must be stationary so that δV must be zero. Thus, we can express the potential energy theorem alternatively by stating that the equilibrium configuration corresponds to a zero for the first variation of the potential energy. That is,

$$\delta V = 0 = \epsilon \int_0^l [EIy''\eta'' - p\eta]\, dx\,. \qquad (4.7.4)$$

The basic procedure of the calculus of variations is to transform the statement $\delta V = 0$ in such a way that it becomes independent of the arbitrary $\eta(x)$ leaving the conditions which are to be satisfied by the *correct* $f(x)$. Integrating by parts, we proceed as follows:

$$0 = \int_0^l [EIy''\eta'' - p\eta]\, dx = |EIy''\eta'|_0^l - \int_0^l \left[\frac{d}{dx}(EIy'')\eta' - p\eta\right] dx\,.$$

$$\text{ie. } 0 = |EIy''\eta'|_0^l - \left|\frac{d}{dx}(EIy'')\eta\right|_0^l + \int_0^l \left[\frac{d^2}{dx^2}(EIy'') - p\right]\eta\, dx\,. \qquad (4.7.5)$$

Now the geometric boundary conditions are that the slope and deflexion shall be zero at both ends for a clamped beam. This implies that for $\eta(x)$ to be an admissible variation, $\eta(x)$ and $\eta'(x)$ must vanish at both ends: when we see that the first two terms in equation (4.7.5) vanish. Equation (4.7.5) now shows that the integral term must also be zero. But from the **fundamental lemma of the calculus variations,**† if $\eta(x)$ is arbitrary, the integral can only be zero if

$$\frac{d^2}{dx^2}(EIy'') - p = 0\,. \qquad (4.7.6)$$

Thus of all the possible deflexion functions $f(x)$ which satisfy compatibility with the end constraints, the one which also implies stresses (and hence bending moments and shearing forces) in equilibrium with the applied loads is defined by

† See Section 4.8 below.

equation (4.7.6). This is the well known beam equation found in books on mechanics of materials.

We will now proceed to a more formal summary of the calculus of variations. Readers wishing a more extensive account should consult the list for further reading at the end of the chapter.

4.8 Euler's Equation†

The general statement of the simplest type of problem is as follows: given a function $F(x, y, y')$, which is explicit in x, y, y', we seek to determine that $y = f(x)$ which renders

$$I(y) = \int_a^b F(x, y, y')\mathrm{d}x \tag{4.8.1}$$

stationary. $f(x)$ must satisfy all the necessary continuity and boundary conditions: in addition, we require $f'(x)$ and $f''(x)$ to be continuous and that F can be differentiated with respect to x, y and y' up to and including the second order partial derivatives.

As outlined in the previous section, the procedure is to introduce a function

$$u(x) = f(x) + \epsilon\eta(x) \tag{4.8.2}$$

which neighbours on the true $f(x)$: $\eta'(x)$ and $\eta''(x)$ are continuous in the range and $\eta(a) = \eta(b) = 0$ so that $\epsilon\eta(x)$ is an admissible variation. Then

$$I(u) = \phi(\epsilon) = \int_a^b F(x, y + \epsilon\eta,\ y' + \epsilon\eta')\mathrm{d}x\,.$$

The notation $\phi(\epsilon)$ has been introduced, since for given $y(x)$ and $\eta(x)$ the integral is now simply a function of ϵ. Then, the proposition that $I(u)$ shall be stationary for $u(x) = f(x)$ implies that $\phi(\epsilon)$ shall be stationary at $\epsilon = 0$. For this latter condition, we have

$$\phi'(\epsilon)\Big|_{\epsilon=0} = \frac{\mathrm{d}\phi}{\mathrm{d}\epsilon}\Big|_{\epsilon=0} = \frac{\mathrm{d}}{\mathrm{d}\epsilon}\int_a^b F(x, y + \epsilon\eta, y' + \epsilon\eta')\mathrm{d}x\Big|_{\epsilon=0} = 0\,. \tag{4.8.3}$$

In the earlier example, we briefly introduced the δ notation and we defined δI, the coefficient of ϵ, as the first variation in $I(u)$. In the present case, then, for stationary $I(u)$,

$$\delta I = \epsilon\phi'(0) = 0 \tag{4.8.4}$$

so that the stationary character of I and the vanishing of its first variation mean exactly the same thing.

† Also called the **Euler – Lagrange equation.**

This vanishing of the first variation is a necessary condition for a stationary value of I, but it is not sufficient to identify whether a maximum or minimum has been found: such problems involve consideration of the second variation. It is, however, useful to recall that in engineering or physics, a minimum principle is usually established on physical grounds so that the existence of a solution is taken for granted. An important practical situation where consideration of the second variation would be crucial relates to buckling problems: here, the stability of equilibrium must be taken into account.†

As before, we now proceed to express $\delta I = 0$ in a form independent of the arbitrary function $\eta(x)$.

Returning to equation (4.8.3), we can differentiate with respect to ϵ under the integral sign, then

$$\phi'(\epsilon) = \int_a^b \frac{\mathrm{d}}{\mathrm{d}\epsilon}[F(x, u, u')]\,\mathrm{d}x$$

$$= \int_a^b \left[\frac{\partial F}{\partial u}\frac{\partial u}{\partial \epsilon} + \frac{\partial F}{\partial u'}\frac{\partial u'}{\partial x}\right]\mathrm{d}x$$

$$= \int_a^b \left[\frac{\partial F}{\partial u}\eta + \frac{\partial F}{\partial u'}\eta'\right]\mathrm{d}x \;.$$

Integrating the last term by parts, and noting that $\eta(a) = \eta(b) = 0$, we have

$$\phi'(\epsilon) = \int_a^b \left[\frac{\partial F}{\partial u} - \frac{\mathrm{d}}{\mathrm{d}x}\left(\frac{\partial F}{\partial u'}\right)\right]\eta\mathrm{d}x \;.$$

Now, we require $\phi'(\epsilon) = 0$, for $\epsilon = 0$; in this situation $u(x) = f(x)$ so that the condition becomes

$$\phi'(0) = \int_a^b \left[\frac{\partial F}{\partial y} - \frac{\mathrm{d}}{\mathrm{d}x}\left(\frac{\partial F}{\partial y'}\right)\right]\eta\mathrm{d}x = 0 \;. \tag{4.8.5}$$

If this result is to be true for any arbitrary $\eta(x)$, then from the fundamental lemma below, the necessary condition for the vanishing of the first variation is that

$$\frac{\partial F}{\partial y} - \frac{\mathrm{d}}{\mathrm{d}x}\left(\frac{\partial F}{\partial y'}\right) = 0 \;. \tag{4.8.6}$$

This is **Euler's differential equation** and its solution is said to be the extremal of the variational problem. In this case, it is an ordinary differential equation of the second order, the solution to which must satisfy the boundary conditions $y(a)$ and $y(b)$ prescribed.

† See Section 1.9.

We have previously referred to the **fundamental lemma of the variational calculus.** This asserts that, if $f(x)$ is a continuous function in an interval $a \leqslant x \leqslant b$ and

$$\int_a^b f(x)\eta(x)\mathrm{d}x = 0 \tag{4.8.7}$$

is to be true for all functions $\eta(x)$, which are continuous together with their first n derivatives, and $\eta(a) = \eta(b) = 0$, then $f(x) = 0$ identically.

The lemma may be proved by showing that if it is not true, then a contradiction arises. Thus, suppose that $f(x)$ is not zero but positive at some point ξ, where $a < \xi < b$, then because of the assumed continuity of $f(x)$ there must be some neighbourhood of $x = \xi$, say $\xi - \epsilon \leqslant x \leqslant \xi + \epsilon$, within which $f(x)$ is positive; ϵ is a positive constant. In the lemma, $\eta(x)$ is arbitrary so we can choose it to be

$$\eta(x) = \begin{cases} (x - \xi + \epsilon)^{2n}(x - \xi - \epsilon)^{2n} & \text{in } (\xi - \epsilon) \leqslant x \leqslant (\xi + \epsilon) \\ 0 & \text{elsewhere} \end{cases} \tag{4.8.8}$$

The integral of equation (4.8.7) is clearly positive under these circumstances and the lemma is contradicted. The supposition that $f(x)$ is not zero is thus untenable.

EXAMPLE 4.8.1

Find the equation of the curve joining two points in a plane such that the distance along the arc is a minimum.

Now $$L(y) = \int_a^b [1 + (y')^2]^{1/2}\mathrm{d}x, \quad a < b\,. \tag{a}$$

In our notation, $F = [1 + (y')^2]^{1/2}$, then

$$\left.\begin{aligned} &\frac{\partial F}{\partial y} \quad \text{does not appear} \\ &\frac{\partial F}{\partial y'} = \frac{1}{2\sqrt{1 + (y')^2}} \end{aligned}\right\} \tag{b}$$

so that Euler's equation beomes

$$\frac{\mathrm{d}}{\mathrm{d}x}\left[\frac{y'}{\sqrt{1 + (y')^2}}\right] = 0 \tag{c}$$

or

$$(y')^2 = C_1[1 + (y')^2] \tag{d}$$

giving $y' = C_2 =$ constant

or

$$y = C_2x + C_3\,. \tag{e}$$

That is, the required curve is a straight line.

Problems 4.8

4.8.1 Determine the Euler equation corresponding to equation (4.8.1) if
(i) $F(x, y, y') = p(x)(y')^2 - q(x)y^2 + 2r(x)y$.
(ii) $F(x, y, y') = A(x)(y')^2 + 2B(x)yy' + C(x)y^2 + D(x)y' + 2E(x)y$.

Ans:
(i) $(py')' + qy = r$.
(ii) $(Ay')' + (B' - C)y = E - D'$.

4.8.2 A smooth curve, joining two points (x_1, y_1) and (x_2, y_2), generates a surface of revolution by rotation about the x axis. Show that for the surface area to be a minimum, the curve must be a catenary

$$y = a\cosh(x/a + b) .$$

4.8.3 A smooth bead slides along a curved wire under its own weight and starting from rest. The time for descent from the starting point at the origin to a final point is given by

$$I = \int_1^2 \sqrt{\frac{1 + (y')^2}{2gy}}\, \mathrm{d}x$$

If the time of descent is to be a minimum, show that the wire must have a cycloidal shape given parametrically by

$$x = c(\theta - \sin\theta), \quad y = c(1 - \cos\theta) .$$

This is Johann Bernoulli's **problem of the Brachistochrone**. To effect a solution to the Euler equation, it will be helpful first to solve for y' and then use the substitution $y = 2c\sin^2(\theta/2)$.

4.9 Integrals involving higher derivatives of functions of one variable

The previous methods may be readily applied to functionals involving higher derivatives. Thus, suppose we wish to determine Euler's equation for the functional

$$I(y) = \int_a^b F(x, y, y^{(1)}, y^{(2)}, \ldots\ldots y^{(n)})\mathrm{d}x^{\dagger} \tag{4.9.1}$$

subject to the boundary conditions $y, y^{(1)}, \ldots\ldots y^{(n-1)}$ all prescribed at the end points or some linear combination so prescribed.

We shall suppose that, at least in the range of interest, $F(x, y, y^{(1)} \ldots\ldots y^{(n)})$ has continuous partial derivatives up to order $(n + 1)$ with respect to all the $(n + 2)$ arguments $x, y, y^{(1)} \ldots\ldots y^{(n)}$, and that $y \ldots\ldots y^{(2n)}$ shall be continuous.

† $y^{(n)}$ denotes $\mathrm{d}^n y/\mathrm{d}x^n$.

Then proceeding just as before by introducing a trial function $u(x) = f(x) + \epsilon\eta(x)$, etc., we find Euler's equation to be

$$\sum_{k=0}^{n} (-1)^k \frac{\mathrm{d}^k}{\mathrm{d}x^k}\left[\frac{\partial}{\partial y^{(k)}} F(x, y, y^{(1)} \ldots\ldots y^{(n)})\right] = 0 \,. \qquad (4.9.2)$$

This is an ordinary differential equation of order $2n$ which must satisfy $2n$ boundary conditions.

EXAMPLE 4.9.1

Looking again at the clamped beam of Section 4.7 we will use Euler's equation to obtain the governing differential equation directly.

Now $$V = \int_0^l \left[\frac{EI}{2}(y'')^2 - py\right] \mathrm{d}x \,, \qquad (a)$$

so that $$F(x, y, y^{(1)} \ldots\ldots y^{(n)}) = \frac{EI}{2}(y'')^2 - py \,, \qquad (b)$$

$$\frac{\partial F}{\partial y} = -p \,, \quad \frac{\partial F}{\partial y'} = 0 \,, \quad \frac{\partial F}{\partial y''} = EIy'' \,. \qquad (c)$$

Then substituting in Euler's equation (4.9.2), we have

$$\frac{\mathrm{d}^2}{\mathrm{d}x^2}\left(EI\frac{\mathrm{d}^2 y}{\mathrm{d}x^2}\right) - p = 0 \qquad (d)$$

as obtained previously.

Problems 4.9

4.9.1 The total potential energy for a simply supported beam carrying a distributed transverse load $p(x)$ per unit length and end thrust T is given by

$$V = \int_0^l \left\{\frac{EI}{2}(y'')^2 - \frac{T}{2}(y')^2 - py\right\} \mathrm{d}x \,.$$

Deduce the governing differential equation if EI = *constant.*

4.9.2 If the beam of Problem 4.9.1 is also supported by an elastic foundation having a foundation stiffness k/unit length, show that the governing differential equation is

$$EI\frac{\mathrm{d}^4 y}{\mathrm{d}x^4} - T\frac{\mathrm{d}^2 y}{\mathrm{d}x^2} + ky = p \,.$$

4.10 Integrals involving several unkown functions of one variable

Here, we seek the necessary conditions for a stationery value of, say,

$$I(y_1, y_2, \ldots\ldots y_n) = \int_a^b F(x, y_1, \ldots\ldots y_m; y_1' \ldots\ldots y_m')\mathrm{d}x \,. \tag{4.10.1}$$

For the purposes of illustration we have limited ourselves to include only first derivatives, but the discussion can be extended to include high order derivatives.

All the $y_i = f_i(x)$, $i = 1 \ldots\ldots m$ must satisfy the usual continuity and boundary conditions. We proceed as before, except that now, we have to introduce m arbitrary functions $\eta_i(x)$ so that we have $u_i(x) = f_i(x) + \epsilon_i\eta_i(x)$, with all the admissible $\eta_i(x)$ vanishing at the end points of the range: $\epsilon_i\eta_i(x) = \delta y_i(x)$ is the first variation in $y_i = f_i(x)$ etc. The result is that the necessary condition for stationary $I(y)$ is not just one Euler equation, but a set of them:

$$\frac{\partial F}{\partial y_i} - \frac{\mathrm{d}}{\mathrm{d}x}\left(\frac{\partial F}{\partial y_i'}\right) = 0\,. \quad i = 1, 2, \ldots\ldots m\,. \tag{4.10.2}$$

These are all ordinary differential equations (of second order in this illustration) and their solutions are the extremals of the variational problem.

EXAMPLE 4.10.1

Find the extremals of

$$I(y, z) = \int_0^{\pi/2} [(y')^2 + (z')^2 + 2yz]\,\mathrm{d}x \tag{a}$$

subject to $y(0) = 0\,,\quad y(\pi/2) = 1\,;\quad z(0) = 0\,,\quad z(\pi/2) = -1\,.$ (b)

The Euler equations become

$$y'' - z = 0\,,\quad z'' - y = 0\,. \tag{c}$$

Differentiating the first twice with respect to x and substituting from the second, we find

$$y^{\mathrm{iv}} - y = 0\,, \tag{d}$$

so that

$$y = Ae^x + Be^{-x} + C\cos x + D\sin x\,. \tag{e}$$

Then

$$z = y'' = Ae^x + Be^{-x} - C\cos x - D\sin x\,. \tag{f}$$

Adjusting $A, \ldots\ldots D$ to suit the boundary conditions, we find

$$y = \sin x\,,\quad z = -\sin x\,. \tag{g}$$

4.11 Integrals involving several independent variables

In stress analysis, we may well encounter problems formulated in two or three dimensions, so that we must now extend our previous knowledge to such problems.

To begin with, we will consider the functional

$$I(w) = \iint_D F(x, y, w, w_x, w_y)\mathrm{d}x\mathrm{d}y\,, \tag{4.11.1}$$

where for convenience w_x denotes $\partial w/\partial x$ etc. D is a region of the xy plane bounded by a curve C which has a piecewise continuously turning tangent. The function $F(x, y, w, w_x, w_y)$ is assumed to give continuous second derivatives with respect to its five arguments, and $w(x, y)$ is such that

$$w(x, y),\quad w_x(x, y),\quad w_y(x, y),\quad w_{xx}(x, y),\quad w_{yy}(x, y),\quad w_{xy}(x, y)$$

are all continuous in D. $w(x, y)$ is prescribed on C. Of all such $w(x, y)$'s, we seek that which renders $I(w)$ stationary. To this end we introduce a one parameter family of admissible functions

$$u(x, y) = w(x, y) + \epsilon\eta(x, y)\,, \tag{4.11.2}$$

where ϵ is an infinitesimal parameter and $\eta(x, y)$ is arbitrary, except that it has continuous derivatives up to and including second order and vanishes on C. [It is worth recalling that $\epsilon\eta(x, y) = \delta w(x, y)$, the variation in $w(x, y)$, so that if $w(x, y)$ is prescribed somewhere, as on C here, the corresponding $\delta w(x, y)$ must vanish there].

Proceeding as before, we again find that $I(w) = \phi(\epsilon)$ and that the necessary condition for a stationary value of $I(w)$ is that $\phi'(0) = 0$. This gives

$$0 = \iint_D \left[\frac{\partial F}{\partial w}\eta + \frac{\partial F}{\partial w_x}\eta_x + \frac{\partial F}{\partial w_y}\eta_y\right]\mathrm{d}x\,. \tag{4.11.3}$$

To get rid of η_x and η_y†, we can integrate by parts (or use Green's Theorem, see appendix) to give

$$0 = \iint_D \left[\frac{\partial F}{\partial w} - \frac{\partial}{\partial x}\left(\frac{\partial F}{\partial w_x}\right) - \frac{\partial}{\partial y}\left(\frac{\partial F}{\partial w_y}\right)\right]\eta\mathrm{d}x\mathrm{d}y$$
$$+ \int_C \left[\frac{\partial F}{\partial w_x}\frac{\mathrm{d}y}{\mathrm{d}s} - \frac{\partial F}{\partial w_y}\frac{\mathrm{d}x}{\mathrm{d}s}\right]\eta\mathrm{d}s\,.†† \tag{4.11.4}$$

† $\eta_x = \partial\eta/\partial x$, $\eta_y = \partial\eta/\partial y$.

†† It should be noted that since $\partial/\partial x$ and $\partial/\partial y$ arise from integrating by parts with respect to x and y, to evaluate the derivatives, $\partial F/\partial w_x$ and $\partial F/\partial w_y$ must be regarded as functions of x and y. Then the chain rule of differentiation gives

$$\frac{\partial}{\partial x}\left(\frac{\partial F}{\partial w_x}\right) = \frac{\partial G(x, y)}{\partial x} = \frac{\partial F_{w_x}}{\partial x} + \frac{\partial F_{w_x}}{\partial w}\frac{\partial w}{\partial x} + \frac{\partial F_{w_x}}{\partial w_x}\frac{\partial w_x}{\partial x} + \frac{\partial F_{w_x}}{\partial w_y}\frac{\partial w_y}{\partial x}\,.$$

$\eta = 0$ on C, so that the line integral vanishes. Furthermore, since η is arbitrary in the double integral, an extension of the fundamental lemma shows that the necessary condition for stationary $I(w)$ is that

$$\frac{\partial F}{\partial w} - \frac{\partial}{\partial x}\left(\frac{\partial F}{\partial w_x}\right) - \frac{\partial}{\partial y}\left(\frac{\partial F}{\partial w_y}\right) = 0 . \tag{4.11.5}$$

This is Euler's partial differential equation and it is of second order for this case.

EXAMPLE 4.11.1

Find the condition for $w(x, y)$ to extremise $I(w)$ where

$$I(w) = \iint_D [w_x^2 + w_y^2]\,\mathrm{d}x\mathrm{d}y$$

subject to $w(x, y)$ being prescribed on C.

Euler's equation becomes

$$-\frac{\partial}{\partial x}\left(2\frac{\partial w}{\partial x}\right) - \frac{\partial}{\partial y}\left(2\frac{\partial w}{\partial y}\right) = 0$$

or

$$\frac{\partial^2 w}{\partial x^2} + \frac{\partial^2 w}{\partial y^2} = \nabla^2 w = 0 .$$

Problems 4.11

4.11.1 A functional of considerable importance in problems of static equilibrium of solids and dynamics of plates and shells is

$$I(w) = \iiint_V F(x, y, z, w_x, w_y, w_z, w_{xx}, w_{yy}, w_{zz}, w_{xy}, w_{yz}, w_{zx})\mathrm{d}x\mathrm{d}y\mathrm{d}z .$$

Here V is the region of three dimensional space occupied by the solid.

By introducing the function $u(x, y, z) = w(x, y, z) + \epsilon\eta(x, y, z)$ and using the procedures developed above, show that the Euler equation becomes

$$\frac{\partial F}{\partial w} - \frac{\partial}{\partial x}\left(\frac{\partial F}{\partial w_x}\right) - \frac{\partial}{\partial y}\left(\frac{\partial F}{\partial w_y}\right) - \frac{\partial}{\partial z}\left(\frac{\partial F}{\partial w_z}\right) + \frac{\partial^2}{\partial x^2}\left(\frac{\partial F}{\partial w_{xx}}\right) + \frac{\partial^2}{\partial y^2}\left(\frac{\partial F}{\partial w_{yy}}\right)$$
$$+ \frac{\partial^2}{\partial z^2}\left(\frac{\partial F}{\partial w_{zz}}\right) + \frac{\partial^2}{\partial x \partial y}\left(\frac{\partial F}{\partial w_{xy}}\right) + \frac{\partial^2}{\partial y \partial z}\left(\frac{\partial F}{\partial w_{yz}}\right) + \frac{\partial^2}{\partial z \partial x}\left(\frac{\partial F}{\partial w_{zx}}\right) = 0 .$$

4.11.2 Show that the condition to extremise

$$I(w) = \iint_D [w_{xx}^2 + w_{yy}^2 + 2w_{xy}^2]\,\mathrm{d}x\mathrm{d}y$$

is

$$\nabla^4 w = \frac{\partial^4 w}{\partial x^4} + 2\frac{\partial^4 w}{\partial x^2 \partial y^2} + \frac{\partial^4 w}{\partial y^4} = 0$$

if w and $\partial w/\partial n$ vanish on the boundary of D.

4.12 Subsidiary conditions – Lagrange multipliers

In Section 4.4, we saw how the method of Lagrange multipliers provided an elegant treatment for problems of constrained maxima and minima of functions. An extension of the technique to variational calculus has even greater practical importance.

Typically, we now have the problem of extremising

$$I(y, z) = \int_a^b F(x, y, z, y', z')\mathrm{d}x \qquad (4.12.1)$$

with respect to the functions $y(x)$ and $z(x)$. These functions satisfy the boundary conditions, $y(a)$, $y(b)$, $z(a)$, $z(b)$ prescribed, but in addition they are constrained according to

$$G(x, y, z) = 0\ . \qquad (4.12.2)$$

Now it may be convenient to solve equation (4.12.2) for z, say, so that we are left with the task of extremising the functional of equation (4.12.1) with respect to only one independent function $y(x)$.

However, more often than not, such a solution of equation (4.12.2) would be most inconvenient and in any case seems to give unjustified emphasis to function $y(x)$. We can avoid these difficulties by proceeding as follows.

Introduce a new function

$$H(x, y, z) = F(x, y, z) + \lambda(x).\ G(x, y, z) \qquad (4.12.3)$$

and consider the **unconstrained** variation of the new functional

$$I(y, z, \lambda) = \int_a^b [F(x, y, x', y', z') + \lambda(x)G(x, y, z)]\,\mathrm{d}x\ .$$

Proceeding as before, we find the necessary conditions for $I(y, z, \lambda)$ to be stationary as

$$\begin{aligned} \frac{\partial H}{\partial y} &= \frac{\partial F}{\partial y} - \frac{\mathrm{d}}{\mathrm{d}x}\left(\frac{\partial F}{\partial y'}\right) + \lambda\frac{\partial G}{\partial y} = 0 \\ \frac{\partial H}{\partial z} &= \frac{\partial F}{\partial z} - \frac{\mathrm{d}}{\mathrm{d}x}\left(\frac{\partial F}{\partial z'}\right) + \lambda\frac{\partial G}{\partial z} = 0 \end{aligned} \qquad (4.12.4)$$

These two equations together with (4.12.2) allows the determination of $y(x)$, $z(x)$ and $\lambda(x)$. It is worth paying particular regard to the fact that, here, the Lagrange multiplier is a function of x and not a constant as it was in Section 4.4. When there are several unkown functions in the functional (as in Section 4.10) subject to several constraints, we can generalise the discussion in a manner analogous to Section 4.5.

The subsidiary condition may be in the form of an integral. The oldest example of such a variational problem is concerned with determining the shape of a closed curve of given length to enclose a maximum area. Another example is to find the shape taken up by a heavy chain of fixed length hung between the two points. Problems of this form are described as **isoperimetrical** and, typically, we require to determine the stationary value of

$$I = \int_a^b F(x, y, y')\,\mathrm{d}x \,, \tag{4.12.5}$$

subject to the constraint

$$J = \int_a^b G(x, y, y')\,\mathrm{d}x \tag{4.12.6}$$

where G is a given function and J is a constant.

We proceed in a manner similar to the above by forming a new functional

$$\bar{I} = \int_a^b H(x, y, y', \lambda)\,\mathrm{d}x \tag{4.12.7}$$

where

$$H = F + \lambda G \,. \tag{4.12.8}$$

λ is the Lagrange multiplier which is a *constant* here.

The constrained stationary condition is defined by

$$\frac{\partial H}{\partial y} - \frac{\mathrm{d}}{\mathrm{d}x}\left(\frac{\partial H}{\partial y'}\right) = 0 \,. \tag{4.12.9}$$

The integral of this differential equation involves λ and constants of integration. They must be determined from the boundary conditions and constraint (4.12.6).

EXAMPLE 4.12

A closed plane curve of length L is described by $x = x(t)$, $y = y(t)$ where t is a parameter. Determine that curve $y(x)$ which encloses the greatest area A.

The task is to extremise

$$A = \frac{1}{2}\int_{t_1}^{t_2} (x\dot{y} - y\dot{x})\,\mathrm{d}t \tag{a}$$

subject to

$$L = \int_{t_1}^{t_2} (\dot{x}^2 + \dot{y}^2)^{1/2}\,\mathrm{d}t \,, \tag{b}$$

where $\dot{x} = \mathrm{d}x/\mathrm{d}t$, $\dot{y} = \mathrm{d}y/\mathrm{d}t$. Using equation (4.12.8),

$$H = \tfrac{1}{2}(x\dot{y} - y\dot{x}) + \lambda(x^2 + y^2)^{1/2} \tag{c}$$

The procedure of Section 4.12 yields two Euler-Lagrange equations

$$-\frac{\dot{x}}{2} - \frac{\mathrm{d}}{\mathrm{d}t}\left[\frac{x}{2} + \frac{\lambda\dot{y}}{(\dot{x}^2 + \dot{y}^2)^{1/2}}\right] = 0 \,. \tag{d}$$

$$\frac{\dot{y}}{2} - \frac{\mathrm{d}}{\mathrm{d}t}\left[-\frac{y}{2} + \frac{\lambda\dot{x}}{(\dot{x}^2 + \dot{y}^2)^{1/2}}\right] = 0\ . \tag{e}$$

Integrating (d) and (e) once with respect to t and eliminating λ yields

$$(x - a)\mathrm{d}x + (y - b)\mathrm{d}y = 0\ . \tag{f}$$

Integrating again, we have

$$(x - a)^2 + (y - b)^2 = c^2 \tag{g}$$

where a, b and c are constants.

Thus, the curve of fixed length enclosing the greatest area is a circle radius c. The reader should demonstrate that $c = \lambda$.

Problems 4.12

4.12.1 Of all the curves, lying on a given surface, joining two points on the same surface, the shortest is called a **geodesic**. A circular cylinder has its axis along the z axis so that its equation is $x^2 + y^2 = a^2$. Two points on the surface are at $(a, 0, 0)$ and $(a \cos\alpha, a \sin\alpha, b)$ where α is an angle of rotation about the z axis. Show that the equations of the geodesics through the points are given by

$$x = a\cos t\,, \quad y = a\sin t\,, \quad z = bt/(2n\pi + \alpha)$$

t is a parameter and n an integer.

4.12.2 A heavy inextensible chain, length l mass per unit length μ, hangs under gravity between two points (x_1, y_1), (x_2, y_2). Using the potential energy principle (Chapter 1), show that the shape taken up by the chain is that which renders

$$V = \int_1^2 \mu g y\,[1 + (y')^2]^{1/2}\mathrm{d}x \tag{a}$$

stationary subject to the constraint

$$l = \int_1^2 [1 + (y')^2]^{1/2}\mathrm{d}x\ . \tag{b}$$

Show that the shape is a catenary

$$y = \lambda + a\cosh[(x + b)/a] \tag{c}$$

where λ, a, b are constants.

4.13 Variational notation

We have seen that the basic idea of the variational calculus is to take, say, the function $y = f(x)$ appearing in a functional, augment it by an infinitesimal function $\epsilon\eta(x)$ and to operate upon $f(x) + \epsilon\eta(x)$. In practical applications it is convenient to denote $\epsilon\eta(x)$ by δy. When we do this, it is important to be clear about the two kinds of increment which y may receive. First, due to an increment $\mathrm{d}x$, the function y will receive a differential increment $\mathrm{d}y$: that is, the change in y comes about from a small change in position along the given curve $y = f(x)$. On the other hand, the variation in y results from an infinitesimal function added to $f(x)$; that is, the change in y is due to a small overall change in the shape of the curve. In this example, where x is an independent variable, indicating position along a beam perhaps, $\delta x = 0$.

Now
$$\delta y = \epsilon\eta(x)\,, \tag{4.13.1}$$

so that
$$\delta y' = \delta\left(\frac{\mathrm{d}y}{\mathrm{d}x}\right) = \epsilon\eta'(x) = \epsilon\frac{\mathrm{d}\eta}{\mathrm{d}x} = \frac{\mathrm{d}}{\mathrm{d}x}(\epsilon\eta)$$

or
$$\delta\left(\frac{\mathrm{d}y}{\mathrm{d}x}\right) = \frac{\mathrm{d}}{\mathrm{d}x}(\delta y) \text{ etc.} \tag{4.13.2}$$

It can be readily verified by beginning with definition (4.13.1) that the rules for variation of sums, products, quotients etc. are completely analogous to the corresponding rules in ordinary differentiation. For example, if

$$y = f(x)\cdot g(x)\,, \tag{4.13.3}$$

$$\delta y = f(x)\cdot\delta\,[g(x)] + \delta\,[f(x)]\cdot g(x)\,, \tag{4.13.4}$$

or if
$$y = f(x)/g(x)\,, \tag{4.13.5}$$

$$\delta y = \frac{g(x)\delta\,[f(x)] - \delta\,[f(x)]\,g(x)}{[g(x)]^2} \tag{4.13.6}$$

As a more complicated example, suppose we have $F(x, y, u, v, u_x, u_y, v_x, v_y)$ where x and y are independent variables whilst u and v are functions of x and y. If u and v now receive variations $\delta u = \epsilon_1\xi(x, y)$ and $\delta v = \epsilon_2\eta(x, y)$, then $\delta x = \delta y = 0$, since they are independent. Recalling Section 4.3, the variation in F is given by terms linear in ϵ so that

$$\delta F = \frac{\partial F}{\partial u}\delta u + \frac{\partial F}{\partial v}\delta v + \frac{\partial F}{\partial u_x}\delta u_x + \frac{\partial F}{\partial u_y}\delta u_y + \frac{\partial F}{\partial v_x}\delta v_x + \frac{\partial F}{\partial v_y}\delta v_y\,, \tag{4.13.7}$$

where
$$\delta u_x = \delta\left(\frac{\partial u}{\partial x}\right) = \frac{\partial}{\partial x}(\delta u) \text{ etc.} \tag{4.13.8}$$

Here again, we have made use of the fact that the derivative of a variation with respect to an *independent* variable is the same as the variation of the derivative.

4.14 Natural boundary conditions

In all the previous work of this chapter, the various trial functions were 'forced' to satisfy certain boundary conditions. However, there are many practically important applications of the variational calculus in which there are no a-priory boundary conditions or where the behaviour of the function is restricted in some more general manner. Under such circumstances, the procedures of the variational calculus lead to the conditions which the trial functions must satisfy 'naturally'. For our present purpose, we shall only discuss natural boundary conditions in relation to free boundaries.

In Section 4.8 we examined the variation of

$$I(y) = \int_a^b F(x, y, y')\mathrm{d}x \,. \tag{4.14.1}$$

Forming $u(x) = f(x) + \epsilon\eta(x)$, we obtained

$$I(u) = \phi(\epsilon) = \int_a^b F(x, y + \epsilon\eta, y' + \epsilon\eta')\mathrm{d}x \,. \tag{4.14.2}$$

Then $$\left.\frac{\mathrm{d}\phi}{\mathrm{d}\epsilon}\right|_{\epsilon = 0} = \int_a^b \left[\frac{\partial F}{\partial y}\eta + \frac{\partial F}{\partial y'}\eta'\right]\mathrm{d}x = 0\,. \tag{4.14.3}$$

Integrating the second term by parts,

$$\left.\frac{\mathrm{d}\phi}{\mathrm{d}\epsilon}\right|_{\epsilon = 0} = \int_a^b \left[\frac{\partial F}{\partial y} - \frac{\mathrm{d}}{\mathrm{d}x}\left(\frac{\partial F}{\partial y'}\right)\right]\eta\mathrm{d}x + \left|\frac{\partial F}{\partial y'}\eta\right|_a^b = 0\,,$$

or recalling the notation $\epsilon\eta = \delta y$ and $\delta I = \epsilon\phi'(0)$,

$$\delta I = \int_a^b \left[\frac{\partial F}{\partial y} - \frac{\mathrm{d}}{\mathrm{d}x}\left(\frac{\partial F}{\partial y'}\right)\right]\delta y\mathrm{d}x + \left|\frac{\partial F}{\partial y'}\delta y\right|_a^b = 0\,. \tag{4.14.4}$$

Now in Section 4.8, y was prescribed at $x = a$ and $x = b$, so that δy vanished there. Accordingly the second term in equation (4.14.4) vanished and we were led to Lagrange's equation (4.8.6).

Now in this case, we can say that if $\delta I = 0$ for $y(x)$ free at the boundaries, then it is certainly true for the smaller class of problems for which $y(a) = y(b) = 0$, so that we conclude that Euler's equation must be required for equation (4.14.4) to be true as well.

However, we must also have

$$0 = \left|\frac{\partial F}{\partial y'}\delta y\right|_a^b = \left|\frac{\partial F}{\partial y'}\delta y\right|_{x = b} - \left|\frac{\partial F}{\partial y'}\delta y\right|_{x = a}\,.$$

But now, $\delta y(a)$ and $\delta y(b)$ are arbitrary so that we must have

$$\left|\frac{\partial F}{\partial y'}\right|_{x = a} = 0 \quad \text{and} \quad \left|\frac{\partial F}{\partial y'}\right|_{x = b} = 0\,. \tag{4.14.5}$$

Conditions (4.14.5) are the **natural boundary conditions** of the problem in contrast to the **forced boundary conditions** previously obtaining.

The concept of natural boundary conditions is also important in a more general type of variational problem in which the boundary values appear explicitly in the functional: it can be generalised also to functionals involving several dependent and independent variables and higher derivatives of the dependent variables.

Time and again in the pages that follow, we will see how important is the concept of setting up the natural boundary conditions for physical problems.

EXAMPLE 4.14.1

Determine the natural boundary conditions for a beam carrying a distributed load together with end couples and forces, given (Section 2.4), that the total potential energy is the functional

$$V = \int_o^l \left[\frac{EI}{2}(y'')^2 - py\right] \mathrm{d}x - M_o y_o' + M_l y_l' + Q_o y_o - Q_l y_l \,.$$

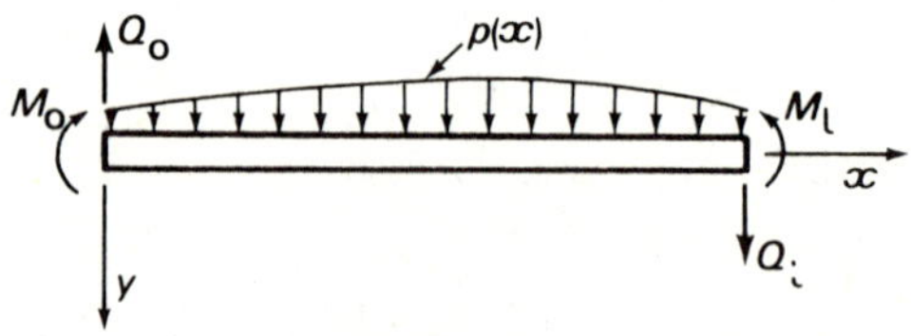

Fig. 4.14.1

For equilibrium

$$\delta V = \int_o^l EIy''\delta y''\mathrm{d}x - \int_o^l p\delta y \mathrm{d}x - M_o\delta y_o' + M_l\delta y_l' + Q_o\delta y_o - Q_l\delta y_l \,.$$

Integrating the first term by parts twice,

$$\delta V = \int_o^l \left[\frac{\mathrm{d}^2}{\mathrm{d}x^2}(EIy'') - p\right]\delta y\mathrm{d}x + [EIy_l'' + M_l]\,\delta y_l' - [EIy_o'' + M_o]\,\delta y_o'$$
$$- \left[\left|\frac{\mathrm{d}}{\mathrm{d}x}(EIy'')\right|_l + Q_l\right]\delta y_l + \left[\left|\frac{\mathrm{d}}{\mathrm{d}x}(EIy'')\right|_o + Q_o\right]\delta y_o \,.$$

Then, since δy is arbitrary in the interval $0 < x < l$ for $\delta V = 0$, we must have

$$\frac{\mathrm{d}^2}{\mathrm{d}x^2}\left(EI\frac{\mathrm{d}^2 y}{\mathrm{d}x^2}\right) - p = 0 \,.$$

In addition, we have the end conditions

(a) Either $EIy_l'' + M_l = 0$ or $\delta y_l' = 0$.

(b) Either $EIy_o'' + M_o = 0$ or $\delta y_o' = 0$.

(c) Either $\left|\frac{\mathrm{d}}{\mathrm{d}x}(EIy'')\right|_l + Q_l = 0$ or $\delta y_l = 0$.

(d) Either $\left|\frac{\mathrm{d}}{\mathrm{d}x}(EIy'')\right|_o + Q_o = 0$ or $\delta y_o = 0$.

Thus if we have the beam clamped at $x = 0$, we must have $\delta y_o = 0$ and $\delta y_o' = 0$: these are the **forced** or **rigid** or **kinematic boundary conditions.** If the end at $x = l$ supports a concentrated load and couple Q_l, M_l, δy_l and $\delta y_l'$ are arbitrary and we must have equilibrium between the shear force and bending moment and the applied load at the end of the beam.

In (a) to (d), the left hand equations represent the natural boundary conditions and the right hand ones represent the forced conditions.

A question may remain regarding the treatment of a beam for which there is, say, a concentrated force or moment somewhere along the span. Under these circumstances we have to consider each portion of the beam between such discontinuities in turn and the 'natural boundary conditions' at such points relate to the sudden change of shearing force or bending moment which occurs.

Problems 4.14

4.14.1 One dimensional heat conduction, in situations involving a distributed heat source and a radiation condition at one boundary, is described by $\delta I = 0$ where

$$I = \int_a^b \left[\frac{1}{2}T'^2 - F(x)T\right]\mathrm{d}x + \left|\frac{\alpha}{2}T^2\right|_{x=a}$$

Show that $T'' + F = 0$ in the region.

Either $T' = 0$ or $\delta T = 0$ at $x = b$ and either $\alpha T - T' = 0$ or $\delta T = 0$ at $x = a$.

4.15 Boundary value problems

Many problems are formulated directly in terms of differential equations and associated boundary conditions. The exact solution of these boundary value problems is often impossible and one method of evading such a difficulty is to generate a variational statement, corresponding to the given differential equation, and using one of the approximate methods of Chapter 5 to yield an approximate solution. We are thus faced with reversing the process with which we have been concerned in the earlier parts of this chapter: we must determine the functional for which the given differential equation is the Euler equation. To answer the question 'when is this possible', it is necessary to consider symmetric positive definite **linear differential operators** and **quadratic functionals.**

A useful notation for differential equations is

$$L(u) = f\,. \qquad (4.15.1)$$

Here, L is a linear differential operator which may be of the form

$$L = A(x) + B(x)\mathrm{d}/\mathrm{d}x + C(x)\mathrm{d}^2/\mathrm{d}x^2 + \ldots\ldots, \qquad (4.15.2)$$

for ordinary equations or perhaps

$$L = L(x) + M(x)\partial/\partial x + N(x)\partial/\partial y + \ldots\ldots, \tag{4.15.3}$$

for partial differential equations. In Chapter 3, we met ∇^2 and ∇^4 which are both linear operators; in contrast $L = \mathrm{d}/\mathrm{d}x + \ln$ is non-linear. The function f [$f(x)$ or $f(x, y)$ perhaps] is known and is sometimes called a forcing function. In this section, we shall limit ourselves to situations involving only one independent variable.

An operator L is said to be symmetric or **self-adjoint** if

$$\int_a^b uL(v)\,\mathrm{d}x = \int_b^a vL(u)\,\mathrm{d}x \tag{4.15.4}$$

where u and v are functions of x.

It is **positive definite** if

$$\int_a^b uL(u)\,\mathrm{d}x \geqslant 0 \tag{4.15.5}$$

for u's satisfying certain homogeneous boundary conditions the forms of which depend on L. (see Example 4.15.1 below).

A quadratic functional $I(u)$ corresponding to the differential equation $L(u) = f$ with homogeneous boundary conditions† is defined to be

$$I(u) = \int_a^b [uL(u) - 2uf]\,\mathrm{d}x\ . \tag{4.15.6}$$

We may show that if L is a self-adjoint positive definite operator, the solution u to equation (4.15.1) is that which renders $I(u)$ a minimum.

To proceed, we suppose that u is the function which minimises I and consider a family of trial functions $u + \epsilon\eta$ neighbouring on u: the parameter ϵ is small and η is a function which is arbitrary except that it satisfies the homogeneous boundary conditions. Then

$$I(u + \epsilon\eta) = \int_a^b [(u + \epsilon\eta)L(u + \epsilon\eta) - 2(u + \epsilon\eta)f]\,\mathrm{d}x\ . \tag{4.15.7}$$

Since u is the function minimising I,

$$\left|\frac{\mathrm{d}I}{\mathrm{d}\epsilon}\right|_{\epsilon=0} = 0 \tag{4.15.8}$$

or

$$\left|\frac{\mathrm{d}}{\mathrm{d}\epsilon}\int_a^b [(u + \epsilon\eta)L(u + \epsilon\eta) - 2(u + \epsilon\eta)f]\,\mathrm{d}x\right|_{\epsilon=0} = 0\ . \tag{4.15.9}$$

† If $L(u) = 0$, the equation is homogeneous. If $u = 0$ at $x = a$, say, the boundary condition is homogeneous there whereas if $u = c$ at $x = b$, the boundary condition is non-homogeneous.

Observing the linearity of L, carrying out the differentiation and setting $\epsilon = 0$, we have

$$\int_a^b [uL(\eta) + \eta L(u) - 2\eta f]\,dx = 0 . \tag{4.15.10}$$

But L is supposed self adjoint so that

$$\int_a^b uL(\eta)dx = \int_a^b \eta L(u)dx , \tag{4.15.11}$$

then
$$\int_a^b 2[L(u) - f]\eta dx = 0 . \tag{4.15.12}$$

The function η is arbitrary so that, according to the fundamental lemma of the variational calculus

$$L(u) - f = 0 . \tag{4.15.13}$$

To confirm that the solution to equation (4.15.13) corresponds to a minimum for $I(u)$, we note from above that

$$\delta I = I(u + \epsilon\eta) - I(u) = 2\epsilon\int_a^b [L(u) - f]\eta dx + \epsilon^2\int_a^b L(\eta)\eta dx$$

or
$$\delta I = \epsilon^2\int_a^b L(\eta)\eta dx . \tag{4.15.14}$$

Since L is positive definite, $\delta I \geqslant 0$ so that $I(u)$ is, in fact, a minimum.

We have seen that for a given differential equation with associated homogeneous boundary conditions to have a corresponding variational problem, L must be self adjoint. The form of the boundary conditions depends on L.

EXAMPLE 4.15.1

Given the bilinear form

$$B(u, v) = \int_a^b \left[p\frac{du}{dx}\frac{dv}{dx} + q\left(v\frac{du}{dx} + u\frac{dv}{dx}\right) + ruv\right]dx , \tag{a}$$

where u, v, p, q, r are functions of x, deduce the corresponding linear differential operator L and the boundary conditions necessary for it to be self adjoint.

Integrating (a) by parts and eliminating the derivatives of v, we find

$$B(u, v) = \int_a^b vL(u)dx + \left|\left(p\frac{du}{dx} + qu\right)v\right|_a^b \tag{b}$$

where
$$L \equiv \frac{d}{dx}\left(p\frac{d}{dx}\right) + \left(\frac{dq}{dx}\right) - r . \tag{c}$$

Alternatively, eliminating the derivatives of u,

$$B(u, v) = -\int_a^b uL(v)\,\mathrm{d}x + \left|\left(p\frac{\mathrm{d}v}{\mathrm{d}x} + qv\right)u\right|_a^b . \tag{d}$$

From (b) and (d)

$$\int_a^b [uL(v) - vL(u)]\,\mathrm{d}x = \left|p\left(v\frac{\mathrm{d}u}{\mathrm{d}x} - u\frac{\mathrm{d}v}{\mathrm{d}x}\right)\right|_a^b . \tag{e}$$

If L is to be self-adjoint, the left hand side of (e) must be zero; u and v must therefore satisfy the boundary conditions

$$\left|p\left(v\frac{\mathrm{d}u}{\mathrm{d}x} - u\frac{\mathrm{d}v}{\mathrm{d}x}\right)\right|_a^b = 0 . \tag{f}$$

Problems 4.15

4.15.1 The linear differential operator for beam bending problems is

$$L = \frac{\mathrm{d}^2}{\mathrm{d}x^2}\left(EI\frac{\mathrm{d}^2}{\mathrm{d}x^2}\right) , \tag{a}$$

where the bending stiffness EI is some function of x. By considering

$$\int_o^l L(u)v\mathrm{d}x \quad \text{and} \quad \int_o^l L(v)u\mathrm{d}x \tag{b}$$

where $u(x)$ and $v(x)$ are possible deflexion curves show that the proper homogeneous boundary conditions for L to be self-adjoint are

$$0 = \left|v\frac{\mathrm{d}}{\mathrm{d}x}\left(EI\frac{\mathrm{d}^2u}{\mathrm{d}x^2}\right) - u\frac{\mathrm{d}}{\mathrm{d}x}\left(EI\frac{\mathrm{d}^2v}{\mathrm{d}x^2}\right) + EI\left(\frac{\mathrm{d}^2v}{\mathrm{d}x^2}\frac{\mathrm{d}u}{\mathrm{d}x} - \frac{\mathrm{d}^2u}{\mathrm{d}x^2}\frac{\mathrm{d}v}{\mathrm{d}x}\right)\right|_a^b . \tag{c}$$

What is the physical significance of conditions (c)?

4.16 Further Reading

Courant, R. *and* Hilbert, D. — *Methods of Mathematical Physics*, Wiley Interscience (1953).

Forray, M. — *Variational Calculus in Science and Engineering*, McGraw Hill (1968).

Fox, C. — *Calculus of Variations*, Oxford University Press (1950).

Schechter, R.S. — *The Variational Method in Engineering*, McGraw Hill (1967).

Weinstock, R. — *Calculus of Variations*, McGraw Hill (1952).

Chapter 5

Variational Calculus - Direct Methods

5.1 Introduction

In Chapter 4 we saw that applying the formal processes of the calculus of variations to determining the stationary conditions for a functional yielded differential equations and appropriate boundary conditions. In contrast, substituting into the functional assumed solutions, involving adjustable parameters and determining the stationary conditions with respect to these parameters, provides **direct methods** of obtaining approximate solutions to otherwise intractable problems. In this connexion, we shall discuss, in this chapter, the **Rayleigh-Ritz**, **Kantorovitch**, **Euler**, **Galerkin** and **Trefftz methods**. Later, we shall discuss the more recently developed **finite element method** which, in its usual formulation for continua, has the character of a 'piece-wise' Rayleigh-Ritz technique.

Before proceeding further, it may be worth pondering the descriptions 'exact' and 'approximate' used here for solutions. In design work, we always have to idealise real engineering systems before we can apply our store of analytical techniques in order to obtain their performance characteristics. Therefore, when we talk of solution, 'exact' or not, we are talking about a solution to a mathematical model of the real situation. In an attempt to obtain a good representation we may well model our system in such a sophisticated way that methods of higher analysis are necessary and we obtain differential equations: it is to the happy situation where we have the means to obtain an analytic solution to these equations that we speak of an 'exact' solution. In contrast, the methods described in this chapter lead to 'approximate' solutions; they are, therefore, the means by which numerical values are obtained in the description of complex engineering systems.

5.2 The Rayleigh-Ritz method

A mechanical system which has its properties (mass, stiffness etc.) distributed in a continuous manner can take up any one of an infinite number of equilibrium configurations depending on the loading. Such a system is said to have an infinite number of degrees of freedom and its response to a given input is described by means of a differential equation (or even several such equations). It is in the solving of these equations to satisfy prescribed boundary conditions that we meet difficulties in engineering analysis. If, however, we make certain assumptions about the nature of the deformation, we approximate the real system by one having a finite number of degrees of freedom. This idea seems to have been first used in the 19th Century by Lord Rayleigh in his vibrations studies. Ritz refined this technique, and what has come to be called the Rayleigh-Ritz method has been very extensively used to obtain numerical solutions to complex engineering problems for which formal analytic methods leading to 'formulae' would have been impossible. Indeed, the finite element method which as stated can be formulated as a piecewise Rayleigh-Ritz process, is perhaps the most practically significant development in engineering analysis in recent times.

Suppose we wish to find a function $y(x)$ which is the extremal to some variational problem, then the Rayleigh-Ritz procedure is to assume that $y(x)$ can be approximated by a linear combination of suitably chosen 'coordinate functions' $\phi_1(x), \phi_2(x) \ldots\ldots \phi_n(x)$. That is

$$y_n(x) = c_1\phi_1(x) + c_2\phi_2(x) + \ldots\ldots c_n\phi_n(x)\,, \qquad (5.2.1)$$

where the c's are constants to be determined.

When $y(x)$ is approximated by $y_n(x)$ in this way, the quantity $I(y)$, which is to be extremised, becomes a function of the c's: we thus have an ordinary extremum problem where the task is to find the stationary value of a function of a finite number of parameters c_k which are found from the conditions

$$\frac{\partial I}{\partial c_k} = 0\,, \quad k = 1, 2 \ldots\ldots n\,. \qquad (5.2.2)$$

We therefore have a set of simultaneous algebraic equations in which the c's are the unknowns. In linear stress analysis work, $I(c_1, c_2 \ldots\ldots c_n)$ is a quadratic form in the c's so that the simultaneous equations are linear algebraic ones. Essentially the procedure is as follows: instead of testing *all possible* slightly different admissible functions to find the true one which gives a stationary $I(y)$, we are restricting the range of functions under comparison to the finite family $\phi_1, \phi_2, \ldots\ldots \phi_n$. Then, *within this limitation,* the 'best' solution is the one for which $I(c_1, c_2 \ldots\ldots c_n)$ has an ordinary stationary value.

The functions $\phi_i(x)$ are *chosen* in advance (they are *not* unkowns) and provided they satisfy the forced boundary conditions for any values of the c's, this choice of 'shapes' is arbitrary: having said this, we should notice that the *efficiency* of the procedure does depend on a judicious choice for the ϕ's and that it is an *advantage* to satisfy *all* the boundary conditions. In practical applications, however, one will have some idea of the general nature of the true solution $y(x)$ so that the question of using 'outrageous' shapes for the ϕ's rarely occurs.†

For problems where the boundary conditions are non-homogeneous it may be convenient to choose a series having the form

$$y_n(x) = \psi_o(x) + c_1\psi_1(x) + c_2\psi_2(x) + \ldots\ldots c_n\psi_n(x) \qquad (5.2.3)$$

† In fact, there are certain practical problems, such as determining the fundamental natural frequency or buckling load of a system, where even an outrageous shape has given good results. However, it is wise to give some thought to the choice of the ϕ's.

rather than (5.2.1). Here $\psi_0(x)$ is chosen to satisfy the forced boundary conditions exactly whilst the corresponding values of ψ_1, ψ_2 satisfy appropriate homogeneous conditions.

A more elaborate procedure consists in obtaining a *sequence* of approximations. Then, as a first approximation, we could take $y_1(x) = c_1\phi_1(x)$; as a second approximation, take $y_2(x) = \bar{c}_1\phi_1(x) + \bar{c}_2\phi(x)$ and so forth: comparison of successive solutions then gives some indication of how satisfactory the current solution may be. Here, it is important to realise that all the c's must be calculated afresh for each successive approximation: this means that the c_1 for the first approximation will not generally be the same as the $\bar{c}_1$ for the second approximation and so on. Before making a slightly more formal reference to the convergence of the procedure, it is worth noting a particular feature.

If $$y_n(x) = \sum_{k=1}^{n} c_k\phi_k(x) \text{ is the } n\text{'th approximation}$$

and $$y_{n+1}(x) = \sum_{k=1}^{n+1} \bar{c}_k\phi_k(x) \text{ is the } (n+1)\text{'th approximation},$$

then the latter solution will be better, or at least no worse, than the former.

We say that the process would converge uniformly if the functions $y_n(x)$ tend to the true required function $y(x)$ as n tends to infinity. Unfortunately, there is no assurance that the Rayleigh-Ritz process will converge in this sense. What Ritz required was that the sequence of approximations should converge in a mean square sense in the region of interest (a, b); that is he required

$$\lim_{n\to\infty} M_n = \lim_{n\to\infty}\int_a^b \left[y(x) - \sum_{k=1}^{n} c_k\phi_k(x)\right]^2 \mathrm{d}x = 0 \qquad (5.2.4)$$

We should notice that this is *not* the same as saying that $y_n(x)$ tends to $y(x)$ as n tends to ∞. It is in this connexion that we can introduce the idea of a 'complete set' of functions.

Suppose that the ϕ_k above are chosen to be orthonormal over the region (a, b). That is, suppose

$$\int_a^b \phi_i(x)\phi_j(x)\mathrm{d}x = \delta_{ij}, \qquad (5.2.5)$$

where $\delta_{ij} = 1$ if $i \neq j$, $\delta_{ij} = 0$ if $i = j$.

Then, the set of functions ϕ is said to be complete† if it is possible to choose the c's in such a way that equation (5.2.4) is satisfied.

It is fairly clear how the process can be extended. Suppose that the functional and coordinate functions involve more than one variable (which could involve

† Alternative definitions of completeness may be found in books on analysis.

time in a dynamical problem), then we would have the task of extremising, say, the functional

$$I(w) = \iint_D F(x, y, w, w_x, w_y)\mathrm{d}x\mathrm{d}y \tag{5.2.6}$$

subject to the boundary condition $w = w(s)$ on the contour c enclosing the region. Now, we could write

$$w_n(x, y) = \sum_{k=1}^{n} c_k\phi_k(x, y) \tag{5.2.7}$$

or

$$w_n(x, y) = \psi_0(x, y) + \sum_{k=1}^{n} \bar{c}_k\psi_k(x, y) , \tag{5.2.8}$$

where the ϕ's or ψ_0 satisfy $w = w(s)$ on C and the ψ_k's are zero there. Again, I becomes a function of the c's and the stationary value is determined by equations (5.2.2). Functionals involving more independent variables and higher derivatives are treated in just the same way, but we would have to notice that the forced boundary conditions could now involve specification of derivatives of w as well as w itself.

EXAMPLE 5.2.1

Determine, by the Rayleigh-Ritz method, an approximation to the function which extremises

$$I(y) = \int_0^l [y'^2 - y^2 - 2x^2y]\,\mathrm{d}x , \tag{a}$$

subject to the boundary conditions

$$y(0) = y(1) = 0 . \tag{b}$$

A polynomial satisfying conditions (b) is

$$y_N = x(1 - x)\sum_{n=1}^{N} c_n x^{n-1} . \tag{c}$$

As a first approximation, we take one term only so that

$$y_1 = c_1 x(1 - x) \tag{d}$$

and

$$y_1' = c_1(1 - 2x) . \tag{e}$$

Substituting from (d) and (e) in (a), carrying out the integration, differentiating with respect to c_1 and equating to zero gives

$$c_1 = 0.1667 . \tag{f}$$

A first approximation to y is thus

$$y_1 = 0.1667x(1 - x) \ . \tag{g}$$

To improve on this, we take two terms in (c) so that

$$y_2 = x(1 - x)[c_1 + c_2 x] \tag{h}$$

and

$$y_1' = c_1(1 - 2x) + c_2(2x - 3x^2) \ . \tag{j}$$

Substituting from (h) and (j) in (a), we find $I(y_2) = I(c_1, c_2)$. For extreme $I(y_2)$

$$\frac{\partial I}{\partial c_1} = \frac{\partial I}{\partial c_2} = 0 \ . \tag{k}$$

This leads to a pair of simultaneous equations in c_1 and c_2

$$0.6000c_1 + 0.3000c_2 = 0.1000 \ ,$$
$$0.3000c_1 + 0.2476c_2 = 0.0667 \ .$$

Solving,

$$c_1 = 0.0813 \ , \quad c_2 = 0.1703 \ . \tag{m}$$

Then, a second approximation to y is

$$y_2 = x(1 - x)[0.0813 + 0.1703x] \ . \tag{n}$$

We see that the value of c_1 in the second approximation is very different from its value in the first approximation.

It is instructive to develop these procedures in more general terms. Thus if

$$y_N = \sum c_n \phi_n(x) \tag{p}$$

then

$$y_N' = \sum c_n \phi_n'(x) \ . \tag{q}$$

Substituting in equation (a)

$$I(y_N) = \int_0^1 \left\{ \left(\sum c_n \phi_n' \right)^2 - \left(\sum c_n \phi_n \right)^2 - 2x^2 \sum c_n \phi_n \right\} \mathrm{d}x \ . \tag{r}$$

To determine $\partial I/\partial c_m$, it is an advantage to differentiate under the integral sign. Then

$$0 = \int_0^1 \left\{ 2\left(\sum c_n \phi_n' \right) \phi_m' - 2\left(\sum c_n \phi_n \right) \phi_m - 2x^2 \phi_m \right\} \mathrm{d}x$$

$$m, n = 1, 2 \ldots\ldots N. \tag{s}$$

That is,

$$\sum_n \alpha_{mn} c_n = \beta_m \ , \quad m, n = 1, 2 \ldots\ldots N \ , \tag{t}$$

where

$$\alpha_{mn} = \int_0^1 (2\phi_m'\phi_n' - 2\phi_m\phi_n)\mathrm{d}x \ , \tag{u}$$

$$\beta_m = \int_0^1 2x^2 \phi_m \, \mathrm{d}x \ . \tag{v}$$

Equation (u) shows immediately that the matrix of coefficients in the equations determining the c's is symmetric.

Further examples are given in Chapter 7.

Problems 5.2

5.2.1 The functional

$$I(y) = \int_1^2 \left[xy'^2 - [(x^2 - 1)/x]y^2 - 4x^2y \right] dx \qquad \text{(a)}$$

is to be extremised to satisfy $y(1) = 0$, $y(2) = 0$. Use the Rayleigh-Ritz technique to show that $c_1 = 1.622$ in a first approximation

$$y = c_1(x - 1)(x - 2) \ . \qquad \text{(b)}$$

5.2.2 The axial flow velocity $v(x, y)$ in a square section duct is given by the stationary value of

$$I(v) = \int_{-a}^{a}\int_{-a}^{a} [v_x{}^2 + v_y{}^2 + 4\alpha v]\, dx dy \ , \qquad \text{(a)}$$

subject to $v = 0$ at $x = \pm a$, $y = \pm a$. A cross section of the duct lies in the xOy plane with O at the centre, $v_x = \partial v/\partial x$, $v_y = \partial v/\partial y$ and α is a constant.

Assuming an approximation to the velocity profile to be

$$v(x, y) = (a^2 - x^2)(a^2 - y^2)[c_1 + c_2(x^2 + y^2) + c_3x^2y^2] \ , \qquad \text{(b)}$$

use the Ritz process to show that

$$c_1 = 0.5898\alpha/a^2 \ , \quad c_2 = 0.0820\alpha/a^4 \ , \quad c_3 = 0.2461\alpha/a^6 \ .$$

(This problem was treated by Sparrow and Siegel in the *Journal of Heat Transfer*, Vol. **81**, p. 157, 1959).

Further problems using the Rayleigh-Ritz method will be found in Chapter 7.

5.3 The Kantorovitch method

In the Rayleigh-Ritz method, the shapes of the approximating functions are completely specified at the start and extremal principles are used to determine the **constants of combination** c_k: these constants define what contribution each 'component' coordinate function makes to the *total* approximate solution and we have reduced the task of solving a partial differential equation to that of solving a set of algebraic equations. In the generalisation of this approach proposed by Kantorovitch, the problem of solving the partial differential equation is reduced to solving a set of *ordinary* differential equations with a prospect of improved accuracy. Evidently, whilst the Rayleigh-Ritz method can be used for functionals in one or more variables, the Kantorovitch method is appropriate for functionals in more than one variable.

To obtain a grasp of the ideas involved, suppose that $w(x, y)$ is an extremal to be determined in the region $-a \leqslant x \leqslant a$, $-b \leqslant y \leqslant b$. Then, if we were to use the Rayleigh-Ritz method, we would assume

$$w(x, y) \doteqdot w_n(x, y) = c_1\phi_1(x, y) + c_2\phi_2(x, y) + \ldots\ldots c_n\phi_n(x, y) \tag{5.3.1}$$

where the ϕ_i's, which must satisfy the forced boundary conditions, are completely specified at the beginning and the c_i's are constants to be determined.

Now in a particular situation, it may so happen that, although we have only a poor idea of how $w(x, y)$ varies with y, say, the general nature of the problem (symmetry of geometry or nature of loading etc.) gives a good indication of how $w(x, y)$ varies with x. Under these circumstances, we could afford only partially to specify the approximating functions and instead of using equation (5.3.1) we could write

$$w(x, y) \doteqdot w_n(x, y) = f_1(y)\phi_1(x) + f_2(y)\phi_2(x) + \ldots\ldots f_n(y)\phi_n(x) \tag{5.3.2}$$

Here, the ϕ's, which are functions of x alone, are completely specified and satisfy the forced boundary conditions at $x = -a$ and $x = a$: the f's, which are functions of y alone, are as yet unkown and so are to be determined. If to this end we apply variational principles, we have the task of finding the extremum of a functional of several functions of one variable when equation (5.3.2) is inserted into the original functional of two variables: then as shown in Section 4.10, the condition for an extremum is a set of ordinary differential equations. These ordinary differential equations must be solved and the constants of integration arising must be so adjusted that the f's satisfy the boundary conditions along $y = \pm b$.

In a more general form of Kantorovitch's method, the ϕ's can be completely specified coordinate functions in x *and* y so that we could take

$$w_n(x, y) = \phi_0(x, y) + \sum_{k=1}^{n} f_k(y)\phi_k(x, y) \ldots\ldots \tag{5.3.3}$$

as an approximation to $w(x, y)$.

As seen previously in Section 5.1, by using this form, non-homogeneous boundary conditions can be catered for by choosing $\phi_0(x, y)$ to satisfy them whilst the remaining $\phi_k(x, y)$ are zero on the boundary. Indeed, for restricted shapes of boundary, if we adjust the $f_k(y)$ to be zero along portions of the boundary, the $\phi_k(x, y)$ need not be zero there.

For a more extended account of this method, the interested reader should refer to the book by Kantorovitch and Krylov.

EXAMPLE 5.3.1

The deflexion $w(x, y)$ of a pre-tensioned membrane supporting a transverse pressure $p = T \sin x \sin y$, with T being the uniform pre-tension, is given by the stationary value of

$$I(w) = \int_0^\pi \int_0^1 [w_x^2 + w_y^2 + 2w \sin \pi x \sin y] \,\mathrm{d}x\mathrm{d}y \ . \qquad \text{(a)}$$

The boundary conditions are $w = 0$ at $x = 0, 1$ and $y = 0, \pi$. Assuming

$$w_1 = x(1 - x)f(y) \qquad \text{(b)}$$

determine an approximation to w.

From (b) in (a)

$$I(w_1) = \int_0^\pi \int_0^1 [(1-2x)^2 f + (x - x^2)^2 f' + 2(x - x^2) f \sin \pi x \sin y] \,\mathrm{d}x\mathrm{d}y. \qquad \text{(c)}$$

Integrating first with respect to x,

$$I(w_1) = \int_0^\pi [f'^2/30 + f^2/3 + (8f/\pi^3)\sin y] \,\mathrm{d}y \qquad \text{(d)}$$

or

$$I(w_1) = \int_0^\pi F(y, f, f')\,\mathrm{d}y \ . \qquad \text{(e)}$$

For $I(w_1)$ to be stationary, f is given by the Euler-Lagrange equation

$$\frac{\partial F}{\partial f} - \frac{\mathrm{d}}{\mathrm{d}y}\left(\frac{\partial F}{\partial y'}\right) = 0 \ , \qquad \text{(f)}$$

or

$$f'' - 10f = (120/\pi^3)\sin y \ . \qquad \text{(g)}$$

Integrating

$$y = A \exp(\sqrt{10}y) + B \exp(-\sqrt{10}y) - (120/11\pi^3) \sin y \ . \qquad \text{(h)}$$

Fitting the conditions $y(0) = y(\pi) = 0$, we find $A = B = 0$, then the first approximation to the deflexion of the membrane is

$$w_1 = -0.3518x(1 - x)\sin y \ . \qquad \text{(j)}$$

Further examples are given in Chapter 7.

Problems 5.3

5.3.1 Assuming, as a first approximation

$$u_1 = (b^2 - y^2)f(x) \ , \qquad \text{(a)}$$

deduce the required form of $f(x)$ to render

$$I(u) = \int_{-b}^{b}\int_{-a}^{a}[u_x^2 + u_y^2 - 2u]\,dxdy\,, \tag{b}$$

subject to $u = 0$ along $x = \pm a$, $y = \pm b$.

Ans:
$$u_1 = \tfrac{1}{2}(b^2 - y^2)\left[1 - \frac{\cosh(\sqrt{2.5}x/b)}{\cosh(\sqrt{2.5}a/b)}\right]. \tag{c}$$

5.4 The Euler finite difference method

In the Rayleigh-Ritz and Kantorovitch methods, we require that the admissible functions be continuously differentiable throughout the region of integration. We can extend the admissible range by admitting functions having piecewise continuous derivatives. In particular, if we proceed to divide the region of integration into sub-regions and assume linear functions within each sub-region, we have **Euler's method of finite differences.** The variational problem is again approximated by an ordinary stationary value problem.

Suppose, typically, we wish to find the extremal for

$$I(y) = \int_a^b F(x, y, y')\,dx\,, \tag{5.4.1}$$

subject to $y(x)$ prescribed at both ends. For our further discussion, it will help us if we change notation slightly: denote $a = x_0$ and $b = x_{n+1}$, then the end conditions become $y(a) = y_0$ and $y(b) = y_{n+1}$. Now divide the interval into $n + 1$ equal parts of length

$$\Delta x = \frac{x_{n+1} - x_0}{n + 1}.$$

Suppose that $y_1, y_2 \ldots\ldots y_n$ are the (as yet unknown) values of the wanted function at the points $x_1 = x_0 + \Delta x$, $x_2 = 2\Delta x, \ldots\ldots x_n = x_0 + n\Delta x$. Then a series of straight line segments joining (x_0, y_0), $(x_1 y_1)$, $(x_2 y_2)$ etc. would form a polygonal approximation to the required curve and an approximation to the integral of equation (5.4.1) would be

$$I_n(y_1, y_2 \ldots\ldots y_n) = \sum_{i=0}^{n} F\left(x_i, y_i, \frac{y_{i+1} - y_i}{\Delta x}\right)\Delta x\,. \tag{5.4.2}$$

Here, y_i' has been approximated by $(y_{i+1} - y_i)/\Delta x$ and the y_i's have to be chosen so that the summation in equation (5.4.2) attains a stationary value, that is, we require

$$\frac{\partial I_n}{\partial y_i} = 0\,, \quad i = 1, 2 \ldots\ldots n. \tag{5.4.3}$$

Now we notice that y_i is involved in only two terms in the series and these are adjacent to each other. Carrying out the differentiation of equation (5.4.3) gives

$$\frac{\partial I_n}{\partial y_i} = \frac{\partial}{\partial y_i}\left[F\left(x_i, y_i, \frac{y_{i+1} - y_i}{\Delta x}\right)\right]\Delta x + \frac{\partial}{\partial y_i{}'}\left[F\left(x_i, y_i, \frac{y_{i+1} - y_i}{\Delta x}\right)\right]\Delta x \frac{(-1)}{\Delta x}$$

$$+ \frac{\partial}{\partial y_{i-1}{}'}\left[F\left(x_{i-1}, y_{i-1}, \frac{y_i - y_{i-1}}{\Delta x}\right)\right]\Delta x \frac{1}{\Delta x} \ . \qquad (5.4.4)$$

Denoting $\Delta y_i = y_{i+1} - y_i$ and $\Delta F_i = F_{i+1} - F_i$ etc. the condition for a stationary value of I_n constitutes a set of n algebraic equations in n unkowns.

$$\frac{\partial F_i}{\partial y_i} - \frac{1}{\Delta x}\cdot\Delta\left(\frac{\partial F_{i-1}}{\partial y_{i-1}{}'}\right) = 0\,, \quad i = 1, 2 \ldots\ldots n\,. \qquad (5.4.5)$$

It is important to realise that in general these equations may be linear or non-linear. Increasing the number of sub-intervals leads to an improvement in accuracy, but this is at the expense of increased effort; indeed, we see that as $n \to \infty$, the system of equations (5.4.5) tend to the Euler equation for the originally desired extremum.

5.5 The Galerkin method

We pointed out in Section 4.15 that the solution of boundary value problems can very commonly be achieved by forming a corresponding variational problem;† under such circumstances, the Rayleigh-Ritz process is very effective for obtaining an approximate solution. In contrast Galerkin's method provides approximate solutions to differential equations directly and is applicable whether the transformation into a variational problem is possible or not. It therefore seems to have wider scope than the Rayleigh-Ritz technique and we would expect it to be more attractice in practice since there would be no need to evaluate the functional even in those situations where it existed. However, the two methods are closely related as explained in Section 5.6.

Galerkin's method proceeds as follows: suppose we wish to solve a linear (partial) differential equation

$$L[w(x, y)] = f(x, y) \qquad (5.5.1)$$

in a region over which $f(x, y)$ is prescribed and the boundary conditions are linear and homogeneous. L stands for a linear differential operator; examples having relevance to stress analysis being the Laplace or harmonic operator

$$\nabla^2 \equiv \frac{\partial^2}{\partial x^2} + \frac{\partial^2}{\partial y^2}$$

and the biharmonic operator

$$\nabla^4 \equiv \frac{\partial^4}{\partial x^4} + 2\frac{\partial^4}{\partial x^2 \partial y^2} + \frac{\partial^4}{\partial y^4} \ .$$

† Differential equations which have this property are 'self-adjoint'. See Section 4.15.

If by some means or other we secured an exact solution to equation (5.5.1), then we could write

$$L[w(x, y)] - f(x, y) = 0 . \tag{5.5.2}$$

If, further, the solution $w(x, y)$ were expressible in the form of a complete series of functions

$$w(x, y) = \sum_{k=1}^{\infty} c_k \phi_k(x, y) , \tag{5.5.3}$$

satisfying the required boundary conditions, then the 'exactness' of the solution could be expressed by the statement that in the region, the left hand side of equation (5.5.2) is orthogonal to every term in series (5.5.3). That is

$$\iint_D \{L[w(x, y)] - f(x, y)\} \phi_k(x, y) \mathrm{d}x\mathrm{d}y = 0 , \quad k = 1, 2, \ldots . . \tag{5.5.4}$$

Suppose we truncate series (5.5.3) to a finite number of terms n, then, we are in a position to use the above ideas to impose n conditions of orthogonality. Evidently, the requirements

$$\iint \left\{ L\left[\sum_{k=1}^{n} c_k \phi_k(x, y) \right] - f(x, y) \right\} \phi_i(x, y) \mathrm{d}x\mathrm{d}y = 0 , \tag{5.5.5}$$

with $i = 1, 2 \ldots . . n$, provides a kind of averaging basis for evaluating the n unknown c_k's such that

$$w_n(x, y) = \sum_{k=1}^{n} c_k \phi_k(x, y) \tag{5.5.6}$$

constitutes an approximate solution to the original differential equation. Equations (5.5.5) are called Galerkin's equations.

It is instructive to provide a physical interpretation to Galerkin's equations by arguing as follows. Suppose that equation (5.5.6) corresponds to an assumed displacement pattern for a system in equilibrium, then, the very acceptance of this pattern implies that the system has only a finite number of degrees of freedom with the c_k's as generalised coordinates. We saw earlier (Section 1.6) that when the principle of virtual work was used to define the equilibrium configuration, the result was

$$\delta w = 0 = Q_1 \delta q_1 + Q_2 \delta q_2 + \ldots . . Q_n \delta q_n . \tag{5.5.7}$$

Since the δq_i represented virtual displacements, the implication was that

$$Q_i = 0 \quad i = 1, 2, \ldots . . n . \tag{5.5.8}$$

The n equations (5.5.8) were, in fact, n equations of equilibrium. Suppose now that we interpret $L[w] - f = 0$ as a differential equation of equilibrium, then allowing virtual displacements δc_i from the equilibrium configuration and invoking the virtual work principle leads to equations (5.5.5).

This reasoning leads naturally to the utilisation of Galerkin's method for the direct solution of a variational problem. In Chapter 4 we saw that if, say $I(w) = \iint F(x, y, w, w_x, w_y)\mathrm{d}x\mathrm{d}y$ is to be extremised, then the methods of the variational calculus lead to the condition

$$0 = \iint \left[\frac{\partial F}{\partial w} - \frac{\partial}{\partial x}\left(\frac{\partial F}{\partial w_x}\right) - \frac{\partial}{\partial y}\left(\frac{\partial F}{\partial w_y}\right)\right] \eta(x, y)\mathrm{d}x\mathrm{d}y \ , \qquad (5.5.9)$$

where $\eta(x, y)$ is an arbitrary function: If we replace $\eta(x, y)$ by each of the functions $\phi_i(x, y)$ of series (5.5.6) in turn and impose the conditions

$$0 = \iint \left[\frac{\partial F}{\partial w_n} - \frac{\partial}{\partial x}\left(\frac{\partial F}{\partial w_{ny}}\right) - \frac{\partial}{\partial y}\left(\frac{\partial F}{\partial w_{ny}}\right)\right] \phi_i(x, y)\mathrm{d}x\mathrm{d}y \ ,$$

$$i = 1, 2 \ldots\ldots n, \qquad (5.5.10)$$

we produce n algebraic equations by means of which the 'best' values for the c_i's are obtained.

Substituting these values in series (5.5.6) then gives an approximate solution to the problem. Equations (5.5.10) correspond to equations (5.5.5) which were obtained directly.

Examples and problems on Galerkin's method are given in Chapter 7.

5.6 Equivalence of the Rayleigh-Ritz and Galerkin methods

In this section we present a brief discussion of the equivalence of the Rayleigh-Ritz and Galerkin methods. If he wishes to find a fuller account of this topic, the reader should consult the further reading list.

Suppose we wish to solve the self-adjoint second order differential equation

$$\frac{\mathrm{d}}{\mathrm{d}x}\left(p(x)\frac{\mathrm{d}y}{\mathrm{d}x}\right) - q(x)y - f(x) = 0 \qquad (5.6.1)$$

subject to homogeneous boundary restraints

$$y(0) = 0 \ , \quad y(l) = 0 \ . \qquad (5.6.2)$$

To approach this problem by the Rayleigh-Ritz method, the left hand side of the equation (5.6.1) is multiplied by δy and integrated with respect to x from $x = 0$ to $x = l$. That is

$$\int_0^l \left[\frac{\mathrm{d}}{\mathrm{d}x}\left(p\frac{\mathrm{d}y}{\mathrm{d}x}\right) - qy - f\right]\delta y \mathrm{d}x = 0 \ . \qquad (5.6.3)$$

Integrating the first term by parts and observing that $\delta y = 0$ at $x = 0$ and $x = l$, equation (5.6.3) becomes

$$\delta I = 0 \qquad (5.6.4)$$

where $$I = \int_0^l \left[p(x)\left(\frac{\mathrm{d}y}{\mathrm{d}x}\right)^2 + q(x)y^2 + 2f(x)y\right]\mathrm{d}x \ . \qquad (5.6.5)$$

For an approximate solution, assume

$$y_n(x) = \sum_{k=1}^{n} c_k \phi_k(x) . \tag{5.6.6}$$

Then substituting from equation (5.6.6) into equation (5.6.5) and carrying out the integrations would show that

$$I = I(c_1, c_2 \ldots\ldots c_n) . \tag{5.6.7}$$

On differentiating under the integral sign, the conditions $\partial I/\partial c_i = 0$, $i = 1, 2, \ldots\ldots n$ for I to be stationary become

$$0 = \int_0^l \left[2p(x)\, y_n{}'(x) \frac{\partial y_n{}'(x)}{\partial c_i} + 2q(x) y_n(x) \frac{\partial y_n(x)}{\partial c_i} + 2f(x) \frac{\partial y_n(x)}{\partial c_i} \right] \mathrm{d}x$$

$$= \int_0^l [p(x) y_n{}'(x)\, \phi_i{}'(x) + q(x) y_n(x) \phi_i(x) + f(x)\phi_i(x)]\, \mathrm{d}x ,$$

$$i = 1, 2 \ldots\ldots n.$$

Integrating the first term by parts and observing that $\phi_i(x) = 0$ at $x = 0$ and $x = l$ in accordance with the initially supposed homogeneous boundary conditions, we have

$$\int_0^l \left[\frac{\mathrm{d}}{\mathrm{d}x}\left(p(x) \frac{\mathrm{d}y_n(x)}{\mathrm{d}x} \right) - q(x) y_n(x) - f(x) \right] \phi_i(x) \mathrm{d}x = 0 , \quad i = 1, 2 \ldots\ldots n. \tag{5.6.8}$$

Equations (4.6.8) are precisely the ones which would be obtained directly by the Galerkin method.

EXAMPLE 5.6.1

The governing differential equation for a beam supporting a distributed load p is

$$\frac{\mathrm{d}^2}{\mathrm{d}x^2}\left(EI \frac{\mathrm{d}^2 w}{\mathrm{d}x^2} \right) - p = 0 . \tag{a}$$

To solve by the Rayleigh-Ritz process, we form

$$\int_0^l \left[\frac{\mathrm{d}^2}{\mathrm{d}x^2}\left(EI \frac{\mathrm{d}^2 w}{\mathrm{d}x^2} \right) - p \right] \delta w \mathrm{d}x = 0 . \tag{b}$$

For the purpose of this example, we shall suppose that the beam is clamped at both ends so that $w = w' = 0$ at $x = 0$ and $x = l$.

Integrating by parts as appropriate in equation (b) and taking the boundary conditions into account, we have

$$\delta \int_0^l \left[\frac{EI}{2} \left(\frac{\mathrm{d}^2 w}{\mathrm{d}x^2} \right)^2 - pw \right] \mathrm{d}x = 0 = \delta I . \tag{c}$$

To solve by the Rayleigh-Ritz method, we approximate w by w_n where

$$w_n = \sum_{k=1}^{n} c_k \phi_k(x) \tag{d}$$

and the ϕ_k's satisfy the boundary conditions.

Substituting from equation (d) in equation (c), we find, as previously described, that

$$I = I(c_1, c_2, \ldots\ldots), \tag{e}$$

so that $\delta I = 0$ requires

$$\frac{\partial I}{\partial c_i} = 0, \quad i = 1, 2 \ldots\ldots n. \tag{f}$$

That is

$$\int_0^l \left[EI \frac{\mathrm{d}^2 w_n}{\mathrm{d}x^2} \frac{\mathrm{d}^2 \phi_i}{\mathrm{d}x^2} - p\phi_i \right] \mathrm{d}x = 0, \quad i = 1, 2 \ldots\ldots n. \tag{g}$$

Integrating the first term by parts and again observing the boundary conditions, we find

$$\int_0^l \left[\frac{\mathrm{d}^2}{\mathrm{d}x^2} \left(EI \frac{\mathrm{d}^2 w_n}{\mathrm{d}x^2} \right) - p \right] \phi_i \mathrm{d}x = 0, \quad i = 1, 2 \ldots\ldots n. \tag{h}$$

These are precisely the Galerkin equations for the same problem so that, in this example, the Rayleigh-Ritz and Galerkin methods lead to exactly the same approximate solution.

We have seen above that, following Galerkin's method, the equations for determining the parameters in the approximating series are obtained in a very direct way which does not require the generation of a functional or even requires its existence. It would seem therefore that one would always choose this method in preference to that of Rayleigh and Ritz. However, in elastic stress analysis a functional, say the total potential energy V, is known to exist and when using the principle of stationary total potential energy, in conjunction with the Rayleigh-Ritz technique, one can form V directly by differentiations under the integral sign. The advantage of the Galerkin method is not nearly as dramatic now. Furthermore, whilst the Rayleigh-Ritz method only requires satisfaction of the geometric (forced) boundary conditions, the Galerkin method requires satisfaction of any traction (natural) boundary conditions as well. Example 5.6.2 below illustrates this point.

EXAMPLE 5.6.2

Consider again the beam of Example 5.6.1, but we will now let the end supports be pinned: this implies there will be one forced (geometric) condition and one natural (traction or load) condition to be satisfied at each end.

In this case, integrating equation (b) leads to

$$\delta\int_0^l\left[\frac{EI}{2}\left(\frac{d^2w}{dx^2}\right)^2 - p\right]\delta w dx - \left| EI\frac{d^2w}{dx^2}\delta\left(\frac{dw}{dx}\right)\right|_0^l . \tag{j}$$

Approximating w by $\overline{w}_n$ where

$$\overline{w}_n = \sum_{k=1}^{n} \overline{c}_k\overline{\phi}_k(x) \tag{k}$$

and the $\overline{\phi}_k = 0$ at $x = 0$ and $x = l$, we have, instead of equations (h)

$$\int_0^l\left[\frac{d^2}{dx^2}\left(EI\frac{d^2\overline{w}_n}{dx^2}\right) - p\right]\overline{\phi}_i\, dx + \left| EI\frac{d^2\overline{w}_n}{dx^2}\left[\frac{d\overline{\phi}_i}{dx} - \delta\left(\frac{d\overline{w}_n}{dx}\right)\right]\right|_0^l ,$$
$$i = 1, 2, \ldots\ldots n . \tag{l}$$

Clearly, equations (l) reduce to Galerkin type equations (h) only if

$$\left| EI\frac{d^2\overline{w}_n}{dx^2}\left(\frac{d\overline{\phi}_i}{dx} - \delta\left(\frac{d\overline{w}_n}{dx}\right)\right)\right|_0^l = 0 , \quad i = 1, 2 \ldots\ldots n . \tag{m}$$

Since the slope is not prescribed at the end points, $\delta(d\overline{w}_n/dx)$ is arbitrary so that if we are to have Galerkin type equations we must have $EI(d^2\overline{w}_n/dx^2) = 0$.

Now $EI(d^2\overline{w}_n/dx^2)$ is the bending moment so that requiring it to be zero at the end points is equivalent to requiring that the approximating function must satisfy the traction as well as the geometric boundary conditions for Galerkin's method.

Two dimensional problems may be tackled in the same way as will be shown in Example 7.5.2.

In connexion with problems of this nature there is a useful classification of functions due to Collatz. Thus if the governing differential equation for a system is of order $2p$ then

(i) **Admissible functions** are those which satisfy only the forced boundary conditions and are p times differentiable over the region.

(ii) **Comparison functions** are those which satisfy all the boundary conditions (forced and natural) and are $2p$ times differentiable over the region.

Approximate solution by Galerkin's method thus requires the use of comparison functions whilst the Rayleigh-Ritz technique only requires admissible functions.

5.7 The method of Trefftz

In the earlier sections of this chapter, we have described numerical processes utilising trial functions which were required to satisfy the boundary conditions but not the differential equations of the problem. A satisfactory approximation to the true solution was then obtained from the fact that the process, the Rayleigh-

Ritz technique say, arranges for the trial functions to satisfy the differential equations in some average way. An alternative approach is to use trial functions which satisfy the governing differential equation but not the boundary conditions, and then to arrange matters so that the boundary conditions are satisfied in some average way. One such method is due to Trefftz and it is of importance, since it provides a lower bound to the functional I in contrast to the Rayleigh-Ritz, Kantorovitch and Galerkin techniques which provide upper bounds. The procedure is as follows.

Suppose the true solution for a wanted function $w(x, y)$ satisfies

$$L[w(x, y)] = f(x, y), \tag{5.7.1}$$

where L is a linear differential operator and $f(x, y)$ is some prescribed forcing function. We choose as a trial solution

$$w_N = \psi_0 + \sum_{j=1}^{N} c_j\psi_j . \tag{5.7.2}$$

The ψ's do not satisfy the boundary conditions but

(i) ψ_0 is a particular solution of equation (5.7.1).

(ii) The ψ_j are solutions of $L[w(x, y)] = 0$; that is they are complementary functions for equation (5.7.1).

Failure to satisfy the boundary conditions implies an error in the solution: Trefft's method is to make this error as small as possible over the region of integration by requiring that

$$I(w_N - w) = \text{minimum}. \tag{5.7.3}$$

Since, from equations (5.7.2), $I(w_N - w)$ is a function of the c's, equation (4.7.3) requires that

$$\frac{\partial}{\partial c_i}[I(w_N - w)] = 0, \quad i = 1, 2, \ldots\ldots N. \tag{5.7.4}$$

This is a set of linear algebraic equations in the c's, the coefficients of which are integrals over the region of integration. Utilising Green's theorem (Appendix), these integrals may be transformed into new domain integrals, together with boundary integrals. The significance of this final move is that the new domain integrals actually vanish and since it is only they which involve the true unknown w of equation (5.7.4), we are left with a set of equations for the c's, independent of the unknown domain values of w. The coefficients of the c's in these equations derive from integrations of the prescribed boundary conditions which are thus satisfied in some average way.

By way of illustration, consider the determination of $\Psi(x, y)$ conjugate to the torsion warping function $\psi(x, y)$ introduced in Section 3.6. The task reduces to solution of the so-called Dirichlet problem wherein

$$\nabla^2\Psi = 0 \text{ in the region}, \tag{5.7.5}$$

$$\Psi = \tfrac{1}{2}(x^2 + y^2) \text{ on the boundary}. \tag{5.7.6}$$

Following the procedures of Chapter 4 it may be shown that the functional to be minimised in this case is

$$I(\Psi) = \iint [\Psi_x{}^2 + \Psi_y{}^2]\,dxdy\,. \qquad (5.7.7)$$

If a trial solution

$$\Psi_N = \sum_{i=1}^{N} c_i f_i(x, y) \qquad (5.7.8)$$

is chosen, the Trefftz procedure is to form

$$\frac{\partial}{\partial c_i}\left[I(\Psi_N - \Psi)\right] = \frac{\partial}{\partial c_i}\iint\left\{\left[\frac{\partial}{\partial x}(\Psi_N - \Psi)\right]^2 + \left[\frac{\partial}{\partial y}(\Psi_N - \Psi)\right]^2\right\}dxdy = 0 \qquad (5.7.9)$$

or

$$\iint\left[\frac{\partial}{\partial x}(\Psi_N - \Psi)\frac{\partial f_i}{\partial x} + \frac{\partial}{\partial y}(\Psi_N - \Psi)\frac{\partial f_i}{\partial y}\right]dxdy = 0\,. \qquad (5.7.10)$$

Transforming the right hand side of equation (5.7.10) by Green's theorem, we find

$$\int_C (\Psi_N - \Psi)\frac{\partial f_i}{\partial n}ds - \iint(\Psi_N - \Psi)\nabla^2 f_i dxdy = 0\,. \qquad (5.7.10)$$

But the f_i are chosen to satisfy equation (5.7.5) so that $\nabla^2 f_i = 0$. Then, in view of equation (5.7.6) and (5.7.8), the conditions for determining the c's become

$$\int_C\left[\sum_{j=1}^{N} c_j f_j - \frac{1}{2}(x^2 + y^2)\right]\frac{\partial f_i}{\partial n}ds = 0\,, \qquad i = 1, 2, \ldots\ldots N\,. \qquad (5.7.11)$$

or

$$\sum_{j=1}^{N} A_{ij}c_j = B_i \qquad i = 1, 2, \ldots\ldots N\,. \qquad (5.7.12)$$

where

$$A_{ij} = \int_C f_j\frac{\partial f_i}{\partial n}ds \qquad (5.7.13)$$

and

$$B_i = \frac{1}{2}\int_C (x^2 + y^2)\frac{\partial f_i}{\partial n}ds\,. \qquad (5.7.14)$$

In Section 7.3, we present an example in which the Trefftz and Rayleigh-Ritz techniques are employed to obtain upper and lower bounds to the torsion constant of a section.

5.8 Further reading

Courant, R. *and* Hilbert, D. — *Methods of Mathematical Physics,* Wiley Interscience (1953).

Dym, C.L. *and* Shames, I.H. — *Solid Mechanics: a Variational Approach,* McGraw (1973).

Forray, M. — *Variational Calculus in Science and Engineering,* McGraw Hill (1968).

Kantorovitch, L.V. *and* Krylov, V.I. — *Approximate Methods of Higher Analysis,* Noordhoff (1964).

Schechter, R.S. — *The Variational Method in Engineering,* McGraw Hill (1967).

Chapter 6

Variational Principles in Solid Continuum Mechanics

6.1 Introduction

When the principles of virtual work and virtual complementary work are applied to deformable solids, it is necessary to account for the work of the internal forces as well as that of the external forces. For the deformable elements of the lumped parameter representation of systems in Chapter 1, only their overall force-deformation characteristics are relevant. To understand the detailed behaviour throughout such elements, the higher level of idealisation incorporated in continuum mechanics is required. On the basis of the stress principle and the notion of the state of strain discussed in Chapter 3, we are in a position to formulate reasonably simple expressions for the work and complementary work of the internal forces and, in the case of elastic solids, to arrive at a strain energy and complementary strain energy density function. Then, we may formulate the principles of virtual work and virtual complementary work in the context of a continuum; this is a generalisation of the treatment of bars given in Chapter 2. We shall see how the principles yield the conditions of internal and surface equilibrium and compatibility and subsequently how we may derive approximate theories for structural systems.

6.2 Principle of virtual work for a solid continuum: strain energy density

Consider a solid body to be in equilibrium under the action of body forces F_i and surface tractions $\overset{n}{T}_i$. Over part of the surface, S_T, the tractions may be prescribed as $\overset{n}{T}_i{}^*$; over the remaining part of the surface, S_u, where the displacements are prescribed to be $u_i{}^*$, the tractions are the 'reactions' and are unknown. The total surface $S = S_u + S_T$.

The equilibrium configuration is described by functions $u_i(x_1, x_2, x_3)$ defining the displacement vector $\mathbf{u} = \mathbf{e}_i u_i$ for the material points in the solid: the corresponding stress and strain components are $\sigma_{ij}(x_1, x_2, x_3)$ and $\epsilon_{ij}(x_1, x_2, x_3)$. Any set of continuous u_i will define a possible configuration for the solid provided they satisfy the geometric constraints $u_i = u_i{}^*$ on S_u. Of all such geometrically possible configurations, the true one corresponding to equilibrium with the prescribed loads is identified by the principle of virtual work: to determine the conditions which the true u_i satisfy, we allow the solid to experience virtual displacements δu_i from the true position. As we saw in Chapters 1 and 2, the virtual displacements are quite arbitrary except that they must satisfy the constraints on the system. For a continuum, this implies that the δu_i must be continuous functions throughout the region of the solid† and that they must vanish on S_u. Then, following Chapters 1 and 2, equilibrium is identified by

$$\delta W_i + \delta W_e = 0 . \qquad (6.2.1)$$

In a virtual displacement, the distributed body force and surface tractions perform virtual work

$$\delta W_e = \int_{\text{vol}} F_i \delta u_i \, \text{dvol} + \int_{S_T} \overset{n}{T}_i{}^* \delta u_i \text{d}S . \qquad (6.2.2)$$

The surface integral has a contribution over S_T alone since $\delta u_i = 0$ on S_u.

Generalising the results of Sections 2.2 and 2.3 to a prismatic element such as that in Fig. 3.3.3(a), the virtual work of the internal forces $(\sigma_{11}\,\text{d}x_2\,\text{d}x_3)$ etc. through virtual strains $\delta\epsilon_{11}$ etc. corresponding to the δu_i is

$$\delta W_i = -\int (\sigma_{11}\delta\epsilon_{11} + \sigma_{22}\delta\epsilon_{22} + \ldots\ldots \sigma_{31}\delta\gamma_{31})\text{dvol} \qquad (6.2.3)$$

or

$$\delta W_i = -\int_{\text{vol}} \sigma_{ij}\delta\epsilon_{ij}\,\text{dvol}. \qquad (6.2.4)$$

Observing that for compatibility $\delta\epsilon_{ij} = \frac{1}{2}(\delta u_{i,j} + \delta u_{j,i})$, that $\sigma_{ij} = \sigma_{ji}$, and that i and j are dummy subscripts in equation (6.2.4), we can write

$$\delta W_i = -\int \sigma_{ij}\delta u_{i,j}\,\text{dvol}. \qquad (6.2.5)$$

Utilising equations (6.2.5) and (6.2.2) in (6.2.1), the condition for equilibrium becomes

$$\int_{\text{vol}} F_i\delta u_i \text{dvol} + \int_{S_T} \overset{n}{T}_i{}^*\delta u_i \text{dvol} = \int_{\text{vol}} \sigma_{ij}\delta\epsilon_{ij}\,\text{dvol} = \int_{\text{vol}} \sigma_{ij}\delta u_{i,j}\text{dvol} . \qquad (6.2.6)$$

The volume integral on the extreme right of (6.2.6) may be integrated by parts (see Appendix). Then observing that $\delta u_i = 0$ on S_u, equation (6.2.6) becomes

$$\int_{\text{vol}} (F_i + \sigma_{ji,j})\delta u_i \text{dvol} + \int_{S_T} (\overset{n}{T}_i{}^* - \sigma_{ji}n_j)\delta u_i \text{d}S = 0 . \qquad (6.2.7)$$

But the virtual displacements δu_i are arbitrary within the region and on S_T, so that equilibrium condition (6.2.7) implies.

$$\sigma_{ji,j} + F_i = 0 \text{ in the region} .$$

$$\overset{n}{T}_i{}^* = \sigma_{ji}n_j \text{ on the surface} . \qquad (6.2.9)$$

Thus the principle of virtual work has provided in one statement the conditions of equilibrium which must obtain in the interior and on the surface of the solid. We recognise equations (6.2.9) as precisely those obtained by direct statical principles in Chapter 3 and observe that they were obtained without reference

† Otherwise splits or overlappings are implied.

to any specific stress-strain law for the material. Returning to equation (6.2.4) if the material is elastic, linear or not, the internal forces have a potential so that

$$\delta W_i = -\delta U = -\int_{\text{vol}} \delta U_0 \, \mathrm{d}\,\text{vol}\,, \tag{6.2.10}$$

where δU_0 is a potential per unit volume. It is a function of the state of strain and is called the strain energy density function. Comparing (6.2.4) and (6.2.10)

$$\sigma_{ij}\delta\epsilon_{ij} = \delta U_0\,. \tag{6.2.11}$$

But since U_0 is a function of the strain components,

$$\delta U_0 = \frac{\partial U_0}{\partial \epsilon_{ij}}\delta\epsilon_{ij}\,, \tag{6.2.12}$$

and so

$$\sigma_{ij} = \frac{\partial U_0}{\partial \epsilon_{ij}}\,. \tag{6.2.13}$$

Thus the function $U_0(\epsilon_{ij})$ of the strain components defines the stress-strain law for an elastic material.

When the external forces are conservative, they too have a potential and we can write

$$F_i = -\frac{\partial \Omega_{0B}}{\partial u_i}\,, \quad \overset{n}{T}_i = -\frac{\partial \Omega_{0S}}{\partial u_i}\,. \tag{6.2.14}$$

The virtual work statement then becomes

$$\delta U + \int_{\text{vol}} \frac{\partial \Omega_{0B}}{\partial u_i}\delta u_i \mathrm{d}\,\text{vol} + \int_S \frac{\partial \Omega_{0S}}{\partial u_i}\delta u_i \mathrm{d}S = 0\,. \tag{6.2.15}$$

Now

$$\frac{\partial \Omega_{0B}}{\partial u_i}\delta u_i = \delta\Omega_{0B} \quad \text{and} \quad \frac{\partial \Omega_{0S}}{\partial u_i}\delta u_i = \delta\Omega_{0S}\,;$$

then denoting $\Omega_B = \int \Omega_{0B}\,\mathrm{d}\,\text{vol}$, $\Omega_S = \int \Omega_{0S}\,\mathrm{d}S$, equation (6.2.15) becomes

$$\delta[U + \Omega_B + \Omega_S] = 0\,. \tag{6.2.16}$$

The quantity in square brackets is the total potential energy V so that under these assumed conditions, the principle of virtual work becomes

$$\delta V = 0\,. \tag{6.2.17}$$

That is, at the equilibrium configuration, the total potential energy is stationary; this is the principle of stationary total potential energy.

When the F_i and $\overset{n}{T}_i{}^*$ are constant and so independent of displacements,

$$\Omega_{0B} = -F_i u_i \quad \text{and} \quad \Omega_{0S} = -\overset{n}{T}_i{}^* u_i\,, \tag{6.2.18}$$

then

$$\delta\Omega_{0B} = -\delta(F_i u_i) = -F_i \delta u_i\,.$$

$$\delta\Omega_{0S} = -\delta(\overset{n}{T}_i{}^* u_i) = -\overset{n}{T}_i{}^* \delta u_i\,. \tag{6.2.19}$$

T – prescribed surface traction F – prescribed body forces.

Then, we have the commonly expressed form of (6.2.16) as †

$$\delta\left[U - \int_{\text{vol}} F_i u_i \,\mathrm{d\,vol} - \int_{S_T} \overset{n}{T}_i u_i \,\mathrm{d}S\right] = 0 \,. \tag{6.2.20}$$

A characteristic of elastic bodies is that the deformation produced in them by some disturbance vanishes when the disturbance is removed. Thermodynamic considerations show that this feature endows the strain energy density function with the property of **positive definiteness**; that is, $U_0 = 0$ in the natural (unstrained) state and $U_0 > 0$ in the deformed state. This property allows us to strengthen condition (6.2.17); V is not only stationary, but a minimum at equilibrium. To prove this, we compare the potential energy V at the actual equilibrium configuration u_i with V' in a 'virtual' neighbouring position $u_i + \delta u_i$.

$$V' = \int_{\text{vol}} [U_0(\epsilon_{ij} + \delta\epsilon_{ij}) - F_i(u_i + \delta u_i)]\,\mathrm{d\,vol} - \int_S \overset{n}{T}_i(u_i + \delta u_i)\mathrm{d}S \tag{6.2.21}$$

From Taylor's expansion,

$$U_0(\epsilon_{ij} + \delta\epsilon_{ij}) = U_0 + \frac{\partial U_0}{\partial \epsilon_{ij}}\delta\epsilon_{ij} + \frac{1}{2}\frac{\partial^2 U_0}{\partial\epsilon_{ij}\partial\epsilon_{kl}}\delta\epsilon_{ij}\delta\epsilon_{kl} + \ldots\ldots, \tag{6.2.22}$$

so that

$$\begin{aligned} V' = {} & \int_{\text{vol}} [U_0(\epsilon_{ij}) - F_i u_i]\,\mathrm{d\,vol} - \int_S \overset{n}{T}_i u_i \,\mathrm{d}S \\ & + \int_{\text{vol}} \left(\frac{\partial U_0}{\partial\epsilon_{ij}}\delta\epsilon_{ij} - F_i \delta u_i\right)\mathrm{d\,vol} - \int_S \overset{n}{T}_i \delta u_i \mathrm{d}S \\ & + \frac{1}{2}\int_{\text{vol}} \frac{\partial^2 U_0}{\partial\epsilon_{ij}\partial\epsilon_{kl}}\delta\epsilon_{ij}\delta\epsilon_{kl}\,\mathrm{d\,vol} + \ldots\ldots \end{aligned} \tag{6.2.23}$$

But the first two integrals comprise V whilst the next two comprise δV which is zero for equilibrium. Further, if we refer to equation (6.2.22) and temporarily assume $\epsilon_{ij} = 0$,

$$U_0(\delta\epsilon_{ij}) = \frac{1}{2}\frac{\partial^2 U_0}{\partial\epsilon_{ij}\partial\epsilon_{kl}}\delta\epsilon_{ij}\delta\epsilon_{kl} + \ldots\ldots, \tag{6.2.24}$$

so that equation (6.2.23) becomes

$$V' - V = \int_{\text{vol}} U_0(\delta\epsilon_{ij})\mathrm{d\,vol} \,. \tag{6.2.25}$$

Now the right hand side of equation (6.2.25) is positive definite and so, therefore, must be $V' - V$. Thus, we have shown that if the body receives a virtual

† In applications, U is expressed as a function of the displacement components as we shall see in subsequent examples.

displacement from its actual position to some arbitrary neighbouring 'test' position, the potential energy increases: evidently, the potential energy is a minimum at the position of equilibrium. Thus we have the **principle of minimum total potential energy** which states:

> *Of all the geometrically possible configurations which a body can take up, the true one, corresponding to internal and external equilibrium between the stresses and applied loads, is identified by a minimum value for the total potential energy.*

The principle provides a means of formulating the correct equilibrium equations in the interior and on the surface of a continuum but, of course, makes no further contribution to the difficult task of solving the equations.

EXAMPLE 6.2.1

Deduce an expression for the strain energy density in a Hooke's law material.

From equation (6.2.11)
$$\delta U_0 = \sigma_{ij}\delta\epsilon_{ij} \; . \tag{a}$$

If we let the virtual strains $\delta\epsilon_{ij}$ correspond to actual strain increments $d\epsilon_{ij}$,

$$dU_0 = \sigma_{ij} d\epsilon_{ij} \; . \tag{b}$$

As the strains increase from zero to some current value ϵ_{ij},

$$U_0 = \int \sigma_{ij}\, d\epsilon_{ij} \; . \tag{c}$$

But from Hooke's law

$$\sigma_{ij} = \lambda\epsilon_{kk}\delta_{ij} + 2G\epsilon_{ij} \, , \tag{d}$$

so that

$$U_0 = \int_0^{\epsilon_{ij}} \lambda\epsilon_{kk}\delta_{ij}\, d\epsilon_{ij} + 2G\int_0^{\epsilon_{ij}} \epsilon_{ij}\, d\epsilon_{ij} \; . \tag{e}$$

Observing that $\delta_{ij} d\epsilon_{ij} = d\epsilon_{kk}$,† and integrating,

$$U_0 = \frac{\lambda}{2}(\epsilon_{kk})^2 + G\epsilon_{ij}\epsilon_{ij} \; . \tag{f}$$

Introducing the deviator,†† we may write

$$\epsilon_{ij} = \hat{\epsilon}_{ij} + \tfrac{1}{3}\epsilon_{kk}\delta_{ij} \; . \tag{g}$$

Substituting from (g) in (f), we find

$$U_0 = \frac{K}{2}(\epsilon_{kk})^2 + G\hat{\epsilon}_{ij}\hat{\epsilon}_{ij} \; . \tag{h}$$

The first term on the right hand side is the energy associated with volume change alone, whilst the second is associated with distortion. One criterion for the onset of yielding requires that the **distortion energy** reaches a critical value.

† See Appendix.

†† See example 3.2.1.

EXAMPLE 6.2.2

Use the principle of virtual work to formulate the mixed boundary value problem of elasticity in terms of displacements.

The starting point is equation (6.2.6). Now Hooke's law may be written as

$$\sigma_{ij} = \lambda\delta_{ij}u_{k,k} + G(u_{i,j} + u_{j,i}) \; . \tag{a}$$

If we wish, we may substitute from (a) in equation (6.2.6) and integrate by parts. However less manipulation is involved if equation (6.2.7) is first derived. The formulation is then given immediately as

$$(\lambda + G)u_{k,ki} + Gu_{i,kk} + F_i = 0 \tag{b}$$

for equilibrium throughout the interior of the solid and

$$\overset{n}{T}_i{}^* = [\lambda\delta_{ij}u_{k,k} + G(u_{i,j} + u_{j,i})]n_j \tag{c}$$

over portion S_T of the surface. Over S_u, we have

$$u_i = u_i{}^* \; . \tag{d}$$

Problems 6.2

6.2.1 A *linear elastic* solid, rigidly supported on part S_u of its surface, displaces u_i due to body forces F_i and surface tractions $\overset{n}{T}_i$. To calculate the work done by F_i and $\overset{n}{T}_i$, suppose that, at some intermediate stage in the loading process, the forces are μF_i and $\mu\overset{n}{T}_i$ where μ is some fraction $(0<\mu<1)$. In view of the linear elasticity, the corresponding displacement is μu_i. By considering a virtual displacement $\delta u_i = \Delta\mu u_i$, where $\Delta\mu$ ia an actual increment, write down an expression for virtual work. By allowing the forces to increase slowly in the same proportion as their final values, show that the total work done is

$$W = \frac{1}{2}\int_{\text{vol}} F_i u_i \,\text{d vol} + \frac{1}{2}\int_S \overset{n}{T}_i u_i \,\text{d}S \; .$$

This is **Clapeyron's theorem.**

The chosen loading path was a convenient one for calculating W. Why does the result depend only on the final values of F_i and $\overset{n}{T}_i$ for a linear elastic solid?

6.2.2 Consider that a set of forces ${}_1F_i$, ${}_1\overset{n}{T}_i$ produce stresses, strains and displacements ${}_1\sigma_{ij}$, ${}_1\epsilon_{ij}$, ${}_1u_i$ whilst ${}_2F_i$, ${}_2\overset{n}{T}_i$ produce ${}_2\sigma_{ij}$, ${}_2\epsilon_{ij}$ and ${}_2u_i$ in a given supported solid. Denote the work of the first set of forces in acting through the second set of displacements by

$$W_{12} = \int_{\text{vol}} {}_1F_i\,{}_2u_i \,\text{d vol} + \int_S {}_1\overset{n}{T}_i\,{}_2u_i \,\text{d}S \; . \tag{a}$$

Define W_{21} in an analogous manner.

Utilising Cauchy's formula, the divergence theorem and the equations of equilibrium show that

$$W_{12} = \int_{\text{vol}} {}_1\sigma_{ij\,2}\epsilon_{ij}\,\mathrm{d\,vol}\ . \tag{b}$$

Using Hooke's law, express the right hand side of (b) entirely in terms of strains to show that

$$W_{12} = W_{21} \tag{c}$$

This is the **Rayleigh-Betti reciprocal theorem** for linear elastic solids.

6.3 The engineering theory of beams

The formal application of the principles of continuum mechanics to practical problems usually leads to formidable difficulties in the solution of the appropriate equations. One possible way of making progress in design work is to make certain assumptions regarding the behaviour of the system with a view to establishing a simpler theory to describe its response to some disturbance. Hopefully, this simplified theory will be amenable to solution and will yield design predictions adequate for the purpose at hand. We have already pointed out in Section 4.1 that an important application of variational principles is to the formulation of approximate theories for structural systems. As a first example of this approach, we will establish the engineering theory of beams.

To make assumptions regarding the displacements in a system is equivalent to imposing constraints upon it. The forces necessary to maintain such constraints will generally be very complex, and the need to describe the precise nature of these forces can lead to difficulties in establishing an appropriate theory when vector methods are used. It is because we do not need to know these fine details in a formulation based on energy principles that such an approach is so valuable.

Treating a specific problem will not detract from showing the principles involved, but we will save ourselves a certain amount of distracting mathematical manipulation. Accordingly, we consider the slendar bar of Fig. 6.3.1. The ends of its centroidal axis are simply supported so that they are free to rotate but not to translate in the z direction: the left hand end cannot translate horizontally, but the right hand is free to do so.

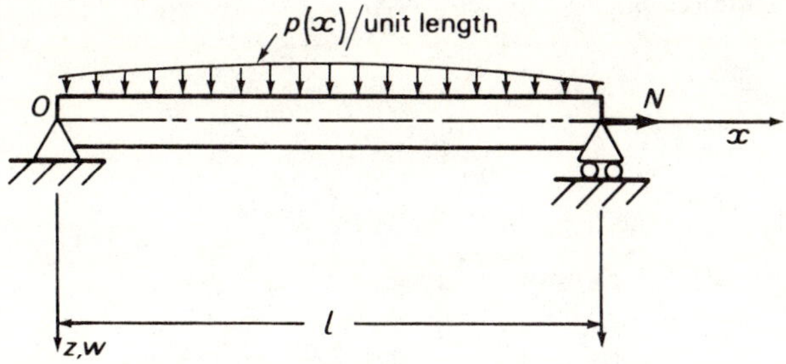

Fig. 6.3.1

A resultant axial force N acts at the centroidal axis at the right hand end, whilst the only non-zero component of body force is F_z/unit volume. This we take to be a function of x but independent of y and z, so that $p = F_z A$ is a transverse force having intensity p per unit length; A is the cross-sectional area. We will suppose that the section of the beam and the applied loading are symmetric about the place xOz so that displacements are symmetric with respect to this plane.

As a first step towards building up a plausible deformation picture, we assume that plane cross-sections remain plane after deformation and do not distort in their plane. This implies

$$\gamma_{yx} = \gamma_{xz} = \gamma_{zy} = 0 . \tag{6.3.1}$$

Fig. 6.3.2 shows the x and z components of displacement for a point distant z below the centroidal axis.

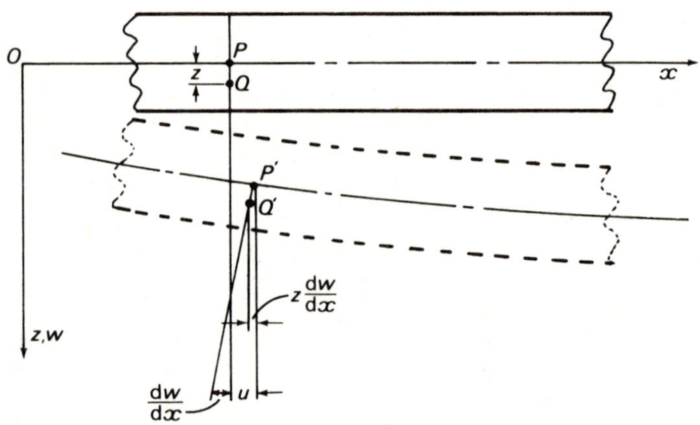

Fig. 6.3.2

Point P on the centroidal axis has displacement components $u(x)$, $v(x)$, $w(x)$ where $v = 0$ from the supposed symmetry. Then, point Q below P has displacement components u_Q, v_Q, w_Q where

$$u_Q = u - z\frac{dw}{dz} , \tag{6.3.2}$$

$$v_Q = 0 \tag{6.3.3}$$

and

$$w_Q = w + \overline{w} , \tag{6.3.4}$$

where, as yet, we do not know $\overline{w}$. Then

$$(\epsilon_x)_Q = \frac{du_Q}{dx} = \frac{du}{dx} - z\frac{d^2w}{dx^2} . \tag{6.3.5}$$

We are concerned with a *slender* bar, so we will tentatively assume that

$$\epsilon_y \;=\; \epsilon_z \;=\; -\nu\epsilon_x \;.^{\dagger} \tag{6.3.6}$$

We are now in a position to calculate the total strain energy stored in the bar. If the material of the beam obeys Hooke's law, substituting from equations (6.3.1) and (6.3.6) in equation (f) of Example 6.2.1 shows that the strain energy density $U_0 \;=\; E\epsilon_x{}^2/2$ so that

$$U \;=\; \int U_0 \,\mathrm{d\,vol} \;=\; \int_0^l \left[\int_A \frac{E}{2}\left(\frac{\mathrm{d}u}{\mathrm{d}x} - z\frac{\mathrm{d}^2 w}{\mathrm{d}x^2}\right)^2 \mathrm{d}A\right]\mathrm{d}x \tag{6.3.7}$$

or

$$U \;=\; \int_0^l \left[\frac{E}{2}\left(\frac{\mathrm{d}u}{\mathrm{d}x}\right)^2 \int_A \mathrm{d}A\right]\mathrm{d}x \;-\; \int_0^l \left[E\frac{\mathrm{d}u}{\mathrm{d}x}\frac{\mathrm{d}^2 w}{\mathrm{d}x^2}\int_A z\mathrm{d}A\right]\mathrm{d}x$$

$$+\; \int_0^l \left[\frac{E}{2}\left(\frac{\mathrm{d}^2 w}{\mathrm{d}x^2}\right)^2 \int_A z^2\mathrm{d}A\right]\mathrm{d}x \;. \tag{6.3.8}$$

But, since ox is the centroidal axis,

$$\int_A \mathrm{d}A \;=\; A\,, \qquad \int_A z\mathrm{d}A \;=\; 0\,, \qquad \int_A z^2\mathrm{d}A \;=\; I\,, \tag{6.3.9}$$

so that

$$U \;=\; \int_0^l \frac{EA}{2}\left(\frac{\mathrm{d}u}{\mathrm{d}x}\right)^2 \mathrm{d}x \;+\; \int_0^l \frac{EI}{2}\left(\frac{\mathrm{d}^2 w}{\mathrm{d}x^2}\right)^2 \mathrm{d}x \;. \tag{6.3.10}$$

The potential energy of the applied loads is

$$\Omega \;=\; -\int_{\mathrm{vol}} F_z(w + \overline{w})\mathrm{d\,vol} \;-\; Nu_l \;.^{\dagger\dagger} \tag{6.3.11}$$

Now, if a is the greatest distance between Q and the centroidal axis, $\overline{w}$ is of the order of $-\nu\epsilon_x a$, that is

$$\overline{w} \text{ is of the order } -\nu a\left(\frac{\mathrm{d}u}{\mathrm{d}x} \quad z\frac{\mathrm{d}^2 w}{\mathrm{d}x^2}\right), \tag{6.3.12}$$

so that

$$\Omega \;=\; -\int_0^l \left[\int_A F_z\left\{w \;-\; \nu a\left(\frac{\mathrm{d}u}{\mathrm{d}x} \;-\; z\frac{\mathrm{d}^2 w}{\mathrm{d}x^2}\right)\right\}\mathrm{d}A\right]\mathrm{d}x \;-\; Nu_l \,. \tag{6.3.13}$$

Integrating over the section and observing that F_z is a function of x alone,

$$\Omega \;=\; -\int_0^l p\left(w \;-\; \nu a\frac{\mathrm{d}u}{\mathrm{d}x}\right)\mathrm{d}x \;-\; Nu_l \,. \tag{6.3.14}$$

† Since the bar is *slender*, we tentatively assume that the lateral stresses σ_y and σ_z are negligible compared with σ_x. Longitudinal filaments of the beam thus behave like simple tension members and the direct strains are related by equation (6.3.6).

†† We suppose that the displacement w of the centre line of the bar is small, so that the axial travel of its end due to this 'bowing' is negligible compared with u_l.

The quantity $va\,du/dx$ being very small, then if we postulate that it is negligible compared with w, we have

$$\Omega = -\int_0^l pw\mathrm{d}x - Nu_l \,. \tag{6.3.15}$$

This result shows that the theory under development cannot discriminate between a transverse component of body force and a distributed traction along the top or bottom surfaces of the beam. It is the quantity p per unit length which is relevant.

The total potential energy of the system is thus given by

$$V = \int_0^l \frac{EI}{2}\left(\frac{\mathrm{d}^2w}{\mathrm{d}x^2}\right)^2 \mathrm{d}x + \int_0^l \frac{EA}{2}\left(\frac{\mathrm{d}u}{\mathrm{d}x}\right)^2 \mathrm{d}x - \int_0^l pw\,\mathrm{d}x - Nu_l \,. \tag{6.3.16}$$

The equilibrium configuration is given by $\delta V = 0$ which results from independent variations (virtual displacements) δw and δu. That is, we have

$$0 = \int_0^l EI\frac{\mathrm{d}^2w}{\mathrm{d}x^2}\frac{\mathrm{d}^2\delta w}{\mathrm{d}x^2}\mathrm{d}x + \int_0^l EA\frac{\mathrm{d}u}{\mathrm{d}x}\frac{\mathrm{d}\delta u}{\mathrm{d}x}\mathrm{d}x - \int_0^l p\delta w\mathrm{d}x - N\delta u_l \,. \tag{6.3.17}$$

Proceeding with integration by parts as described in Example 4.14.1, the stationary conditions become, with $\delta u_0 = 0$, $\delta w_0 = \delta w_l = 0$,

$$\begin{aligned}\delta V = 0 = {} & \int_0^l \left[\frac{\mathrm{d}^2}{\mathrm{d}x^2}\left(EI\frac{\mathrm{d}^2w}{\mathrm{d}x^2}\right) - p\right]\delta w\mathrm{d}x - \int_0^l \frac{\mathrm{d}}{\mathrm{d}x}\left(EA\frac{\mathrm{d}u}{\mathrm{d}x}\right)\delta u\,\mathrm{d}x \\ & + \left| EI\frac{\mathrm{d}^2w}{\mathrm{d}x^2}\delta\left(\frac{\mathrm{d}w}{\mathrm{d}x}\right)\right|_{x=l} - \left| EI\frac{\mathrm{d}^2w}{\mathrm{d}x^2}\delta\left(\frac{\mathrm{d}w}{\mathrm{d}x}\right)\right|_{x=0} \\ & + \left(EA\frac{\mathrm{d}u}{\mathrm{d}x} - N\right)\delta u_l \,.\end{aligned} \tag{6.3.18}$$

Since δw, δu, δu_l, $\delta(\mathrm{d}w/\mathrm{d}x)_l$, $\delta(\mathrm{d}w/\mathrm{d}x)_0$ are arbitrary, condition (6.3.18) implies

$$\frac{\mathrm{d}^2}{\mathrm{d}x^2}\left(EI\frac{\mathrm{d}^2w}{\mathrm{d}x^2}\right) = p \text{ along the beam}, \tag{6.3.19}$$

$$EI\frac{\mathrm{d}^2w}{\mathrm{d}x^2} = 0 \text{ at } x = 0 \text{ and } x = l, \tag{6.3.20}$$

$$\frac{\mathrm{d}}{\mathrm{d}x}\left(EA\frac{\mathrm{d}u}{\mathrm{d}x}\right) = 0 \text{ along the beam, and} \tag{6.3.21}$$

$$EA\frac{\mathrm{d}u}{\mathrm{d}x} = N \text{ at } x = l \,. \tag{6.3.22}$$

The previously prescribed geometric boundary conditions are

$$w = 0 \text{ at } x = 0 \text{ and } x = l, \tag{6.3.23}$$

$$u = 0 \text{ at } x = 0 . \tag{6.3.24}$$

Thus, on the basis of the various assumptions we have made, equations (6.3.19) to (6.3.24) give a complete and consistent statement of the boundary value problem for extension and bending of this bar; the theory is one dimensional, a considerable simplification, and the beam can be represented by a line. We notice that our assumptions imply *uncoupling* of the bending and extensional contributions to the deformation; the governing equations, which are conditions of internal and boundary equilibrium expressed in terms of displacements, can thus be treated independently for each mode. Relating these developments to the more formal discussion of variational calculus in Chapter 4†, we see that equations (6.3.19) and (6.3.21) are the Euler-Lagrange equations whilst equations (6.3.20) and (6.3.22) are the natural boundary conditions for the system.

The basic unknowns for this formulation are w and u so that we must now determine the physical significance of the various derivatives which have arisen. Equation (6.3.22) shows immediately that $EA\,\mathrm{d}u/\mathrm{d}x$ is the resultant extensional force over the cross section of the bar; this problem is straight forward and we will not discuss it further. The quantity $EI(\mathrm{d}^2w/\mathrm{d}x^2)$ can be rewritten as follows

$$EI\frac{\mathrm{d}^2w}{\mathrm{d}x^2} = \int_A E\frac{\mathrm{d}^2w}{\mathrm{d}x^2}z^2\,\mathrm{d}A = -\int E(\epsilon_x)_c z\mathrm{d}A = -\int(\sigma_x)_c z\mathrm{d}A , \tag{6.3.25}$$

where $(\epsilon_x)_c$ and $(\sigma_x)_c$ are the axial strain and stress components due to the curvature alone. Then with $M = \int_A(\sigma_x)_c z\mathrm{d}A$, the bending stress resultant or bending moment across a section,

$$EI\frac{\mathrm{d}^2w}{\mathrm{d}x^2} = -M . \tag{6.3.26}$$

Thus, the natural boundary condition (6.3.20) corresponds to the familiar condition of zero bending moment at the ends of a simply supported beam. Since p is a transverse force per unit length, equation (6.3.19) is a differential equation of transverse equilibrium, and since

$$\frac{\mathrm{d}^2}{\mathrm{d}x^2}\left(EI\frac{\mathrm{d}^2w}{\mathrm{d}x^2}\right) = \frac{\mathrm{d}}{\mathrm{d}x}\left[\frac{\mathrm{d}}{\mathrm{d}x}\left(EI\frac{\mathrm{d}^2w}{\mathrm{d}x^2}\right)\right] = -\frac{\mathrm{d}F}{\mathrm{d}x}\text{ , say ,} \tag{6.3.27}$$

the quantity F must be a transverse force. In fact, examining Fig. 6.3.3 equilibrium of transverse forces requires

$$\frac{\mathrm{d}F}{\mathrm{d}x} = -p , \tag{6.3.28}$$

† In particular, see example 4.14.1.

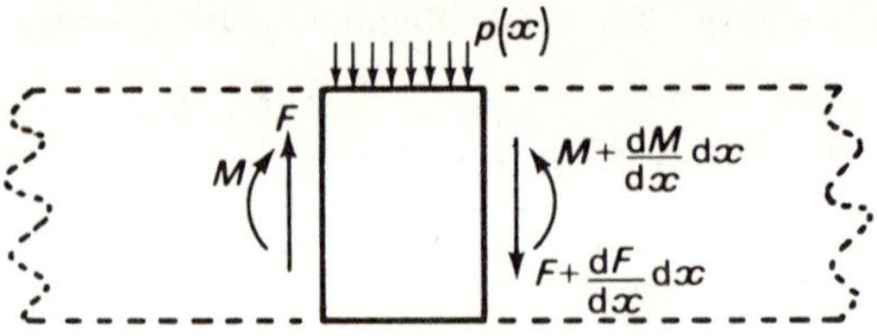

Fig. 6.3.3

so that
$$F = -\frac{d}{dx}\left(EI\frac{d^2w}{dx^2}\right) \tag{6.3.29}$$

is the familiar shearing force, or transverse stress resultant across a section.

A comparison of equations (6.3.26) and (6.3.29) yields a further familiar result,

$$\frac{dM}{dx} = F, \tag{6.3.30}$$

which represents equilibrium of moments for the beam element.

Our theory thus requires the existence of F which in turn implies the existence of a shearing stress component τ_{zx}. How do we reconcile this with our basic assumption that $\gamma_{zx} = 0$ as a consequence of plane sections remaining plane? Condition $\gamma_{zx} = 0$ implies the presence of some internal rigid constraints† so that the τ_{zx} must be the internal forces required to maintain them. In a virtual displacement (not violating the constraints), the τ_{zx} thus do not figure in any virtual work calculation; nevertheless the virtual work principle gave the necessary relationship between their resultant F and the applied transverse load intensity p.

It is also instructive to observe that this theory does not give any *direct* information concerning the reaction forces corresponding to the geometric constraints $u_0 = w_0 = w_l = 0$. Whilst in this particular situation of statically determinate supports they may be determined from considerations of overall equilibrium, (use of rigid body virtual displacements), in general they are given by the values of $EA(du/dx)$ and $-(d/dx)[EI(d^2w/dx^2)]$ at the supports: indeed, the latter treatment is mandatory when the supports are statically indeterminate. In applications, beams may well have end conditions other than those supposed here. The range of possibilities and their mathematical description are summarised in Example 4.14.1.

Our developments here were based on the principle of stationary total potential energy; elastic stress strain relations were thus implied and the net result was a set of equilibrium conditions in terms of displacements. Alternatively we may incorporate our assumptions in the principle of virtual work direct. Using equation (6.2.6), we then obtain the equilibrium relations between the stress resultants which obtain irrespective of the stress-strain law. The reader should carry out this analysis.

† We may view this as implying an infinite value for the shear modulus in the transverse direction.

The ultimate test of the adequacy of a theory, whether it purports to be 'exact' or 'approximate', is whether or not its predictions accord with reality to within some tolerance. Design experience over many years has given confidence in the engineering theory of beams, but the reader may care to compare its predictions with those of the plane theory of elasticity in Example 3.7.1.

Problems 6.3

6.3.1 A simply supported beam having a rectangular section carries a uniformly distributed load $p(x)$ per unit length. Proceeding as indicated below, develop an approximate theory for the deflexion $w(x)$ taking shear effects into account.

The slope of the deflexion curve may be written

$$\frac{\mathrm{d}w}{\mathrm{d}x} = \theta(x) + \psi(x) \,, \tag{a}$$

where $\theta(x)$ is the slope due to curvature alone and $\psi(x)$ is the transverse shearing strain ('shearing angle') at the centroidal axis.

Accepting the usual engineering assumption that the shearing stress and strain vary parabolically through the depth of the beam, show that the shearing strain energy may be written as

$$U_s = \int_0^l \frac{kGA}{2}\psi^2\,\mathrm{d}x \,, \tag{b}$$

where k is a constant. Then show that the total potential energy is

$$V = \int_0^l \left[\frac{EI}{2}\left(\frac{\mathrm{d}\theta}{\mathrm{d}x}\right)^2 + \frac{kGA}{2}\left(\frac{\mathrm{d}w}{\mathrm{d}x} - \theta\right)^2 - pw\right]\mathrm{d}x \,. \tag{c}$$

Determine the Euler-Lagrange equations corresponding to $\delta V = 0$ to be

$$\frac{\mathrm{d}}{\mathrm{d}x}\left[kGA\left(\frac{\mathrm{d}w}{\mathrm{d}x} - \theta\right)\right] + p = 0 \,, \tag{d}$$

$$\frac{\mathrm{d}}{\mathrm{d}x}\left(EI\frac{\mathrm{d}\theta}{\mathrm{d}x}\right) + kGA\left(\frac{\mathrm{d}w}{\mathrm{d}x} - \theta\right) = 0 \,. \tag{e}$$

Eliminating θ from (d) and (e), show that

$$EI\frac{\mathrm{d}^4 w}{\mathrm{d}x^4} = p - \frac{EI}{kGA}\frac{\mathrm{d}^2 p}{\mathrm{d}x^2} \tag{f}$$

when the beam is uniform.

6.4 The bending and stretching of plates

Plates are significant elements in a wide range of applications, from ship construction to delicate pressure sensing devices, so that an adequate theory for their structural behaviour is of considerable practical importance. The distinguishing geometric feature of a plate is that it is a solid body in which one characteristic dimension, the thickness, is very small compared with the other two. This contrasts with the bar or beam in which one dimension is very much greater than the other two. Midway between the faces of a plate is its **middle surface** which, for a flat plate in its undeformed configuration, is a plane. If the middle surface is initially slightly curved, it is usual still to speak of it as a plate whereas if the initial curvature is large, as in the wall of a boiler, it is usual to speak of a shell. Here, we are only concerned with flat plates.

As we saw in Section 6.3, the *slenderness* of the bar allowed us to make certain plausible assumptions about its behaviour, we will now take advantage of the *thinness* of a plate to make analogous assumptions. It will turn out that whilst beam theory is one dimensional and a deformed beam can be represented by a curve, plate theory is two dimensional and a deformed plate can be represented by a surface.

Suppose that a point P lying in the middle surface of a plate (Fig. 6.4.1) experiences displacement components $u(x, y)$, $v(x, y)$, $w(x, y)$.

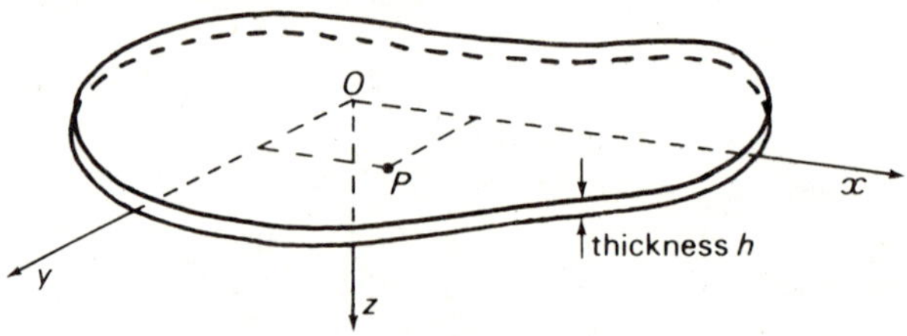

Fig. 6.4.1

We assume now that straight lines within the plate and normal to its middle surface remain straight and normal after deformation, then

$$\gamma_{xz} = \gamma_{yz} = 0 . \quad (6.4.1)$$

Visualising Fig. 6.3.2 as sections through the plate parallel to the xOz and yOz coordinate planes in turn, the horizontal displacements of point Q are given by

$$u_Q = u - z\frac{\partial w}{\partial x}, \qquad v_Q = v - z\frac{\partial w}{\partial y} . \quad (6.4.2)$$

It follows that

$$(\epsilon_x)_Q = \frac{\partial u_Q}{\partial x} = \frac{\partial u}{\partial x} - z\frac{\partial^2 w}{\partial x^2}, \qquad (\epsilon_y)_Q = \frac{\partial v_Q}{\partial y} = \frac{\partial v}{\partial y} - z\frac{\partial^2 w}{\partial y^2},$$

$$(\gamma_{xy})_Q = \frac{\partial v_Q}{\partial x} + \frac{\partial u_Q}{\partial y} = \frac{\partial v}{\partial x} + \frac{\partial u}{\partial y} - 2z\frac{\partial^2 w}{\partial x \partial y} . \quad (6.4.3)$$

In view of the thinness of the plate, if we further assume

$$\epsilon_z = \frac{-\nu}{1-\nu}(\epsilon_x + \epsilon_y)^\dagger , \qquad (6.4.4.)$$

then, as in the case of the beam we can reason that very nearly

$$w_Q = w . \qquad (6.4.5)$$

We can now calculate the strain energy stored in the plate. For a Hooke's law material, substitution from equations (6.4.1) and (6.4.4) in equation (f) of Example 6.2.1 shows that the strain energy density at point Q is

$$U_0 = \frac{E}{2(1-\nu^2)}\left[\epsilon_x{}^2 + \epsilon_y{}^2 + 2\nu\epsilon_x\epsilon_y + \frac{1-\nu}{2}\gamma_{xy}{}^2\right]_Q . \qquad (6.4.6)$$

Substituting from equations (6.4.3) and using the notation $u_x = \partial u/\partial x$, $w_{xx} = \partial^2 w/\partial x^2$ etc, we have

$$U_0 = \frac{E}{2(1-\nu^2)}\Big\{u_x{}^2 + v_y{}^2 + 2\nu u_x v_y + \frac{1-\nu}{2}(u_y + v_x)^2$$

$$+ z^2[w_{xx}{}^2 + w_{yy}{}^2 + 2\nu w_{xx}w_{yy} + 2(1-\nu)w_{xy}{}^2]$$

$$- 2z[u_x w_{xx} + v_y w_{yy} + \nu(u_x w_{yy} + v_y w_{xx})$$

$$+ (1-\nu)(v_x + u_y)w_{xy}]\Big\} . \qquad (6.4.7)$$

The total strain energy stored in the plate is given by

$$U = \int U_0 \, \mathrm{d}vol = \iiint U_0 \, \mathrm{d}x\mathrm{d}y\mathrm{d}z . \qquad (6.4.8)$$

We observe in equation (6.4.7) that the terms which couple u and v with w are linear in z so that if in equation (6.4.8) we integrate first with respect to z between the limits $\pm h/2$, such terms disappear leaving two uncoupled terms for strain energy. That is

$$U = \iint \frac{Eh}{2(1-\nu^2)}\{(u_x + v_y)^2 - 2(1-\nu)[u_x v_y - \tfrac{1}{4}(u_y + v_x)^2]\}\, \mathrm{d}x\mathrm{d}y$$

$$+ \iint \frac{1}{2}\frac{Eh^3}{12(1-\nu^2)}\{(w_{xx} + w_{yy})^2 - 2(1-\nu)[w_{xx}w_{yy} - w_{xy}{}^2]\}\, \mathrm{d}x\mathrm{d}y .$$

$$(6.4.9)$$

The stresses and strains associated with stretching of the middle surface are commonly called **membrane stresses and strains** whilst those arising from its

† Since the plate is *thin*, we tentatively assume that the lateral stress σ_z is negligible compared with σ_x and σ_y. Laminar elements parallel to the middle surface are thus assumed to be very nearly in a state of plane stress.

transverse curvature alone are the **bending stresses and strains.** Within the limitations of the present assumptions then,

$$U = U_m + U_b \tag{6.4.10}$$

where U_m and U_b are respectively the strain energy due to membrane and bending actions alone.

In engineering beam theory, the bending moment and shearing force were stress resultants which arose naturally. There is an analogous situation for plates except that, since the theory is two dimensional, it is necessary to speak of stress resultants per unit length of arc drawn in the middle surface. They are defined with reference to Fig. 6.4.2.

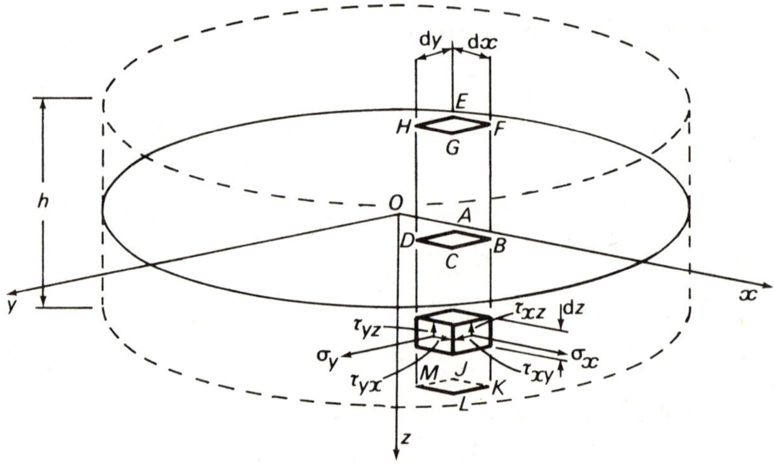

Fig. 6.4.2

Our assumed displacement mode implies that for a Hooke's law material the element $dxdydz$ at z below the middle surface supports stress components

$$\left.\begin{aligned} \sigma_x &= \frac{E}{1-\nu^2}[\epsilon_x + \nu\epsilon_y]\,, \qquad \sigma_y = \frac{E}{1-\nu^2}[\epsilon_y + \nu\epsilon_x]\,,\\ \sigma_z &= 0\,, \qquad \tau_{xy} = \frac{E}{2(1+\nu)}\gamma_{xy}\,. \end{aligned}\right\} \tag{6.4.11}$$

Strictly σ_z is not zero, after all it equals $-p(x, y)$, a distributed pressure, at $z = -h/2$. However, if on the basis of this theory we find that it is an order of magnitude smaller than σ_x, σ_y and τ_{xy}, σ_z is at least negligible and our assumption will have been proved reasonable.† Since we are assuming γ_{xz} and γ_{yz} to be zero, we imply rigid constraints against transverse shearing, but although we cannot describe the exact distribution of the τ_{xz} and τ_{yz} necessary to main-

† We demonstrate this in example 7.5.1.

tain these constraints, we do at least know that they are present. For face *GFKL* of element $hdxdy$ in Fig. 6.4.2 we have the following stress resultants

$$N_x\mathrm{d}y = \left(\int_{-(h/2)}^{h/2} \sigma_x \mathrm{d}z\right)\mathrm{d}y = \left\{\int_{-(h/2)}^{h/2} \frac{E}{(1-\nu^2)}[(u_x - zw_{xx}) + \nu(v_y - zw_{yy})]\,\mathrm{d}z\right\}\mathrm{d}y\ .$$

or

$$N_x = \frac{Eh}{1-\nu^2}\left(\frac{\partial u}{\partial x} + \nu\frac{\partial v}{\partial y}\right)\ . \tag{6.4.12}$$

Similarly

$$N_{xy} = \frac{Eh}{2(1+\nu)}\left(\frac{\partial v}{\partial x} + \frac{\partial u}{\partial y}\right)\ . \tag{6.4.13}$$

$$M_x\mathrm{d}y = \left(\int_{-(h/2)}^{h/2} \sigma_x z\mathrm{d}z\right)\mathrm{d}y = \left(\int_{-(h/2)}^{h/2} \frac{E}{(1-\nu^2)}[z(u_x - zw_{xx}) + \nu z(v_y - zw_{xx})]\,\mathrm{d}z\right)\mathrm{d}y\ ,$$

or

$$M_x = -D\left(\frac{\partial^2 w}{\partial x^2} + \nu\frac{\partial^2 w}{\partial y^2}\right)\ . \tag{6.4.14}$$

The quantity D is called the **bending stiffness** of the plate and plays a role similar to that of EI for beams.

$$D = \frac{Eh^3}{12(1-\nu^2)}\ . \tag{6.4.15}$$

$$M_{xy} = -\int_{-(h/2)}^{h/2} \tau_{xy} z\mathrm{d}z\mathrm{d}y = D(1-\nu)\frac{\partial^2 w}{\partial x \partial y}\ . \tag{6.4.16}$$

$$Q_x = \int_{-(h/2)}^{h/2} \tau_{xz}\mathrm{d}z\ . \tag{6.4.17}$$

For face *HGLM*

$$N_y = \frac{Eh}{1-\nu^2}\left(\frac{\partial v}{\partial y} + \nu\frac{\partial u}{\partial x}\right)\ . \tag{6.4.18}$$

$$N_{yx} = \frac{Eh}{2(1+\nu)}\left(\frac{\partial v}{\partial x} + \frac{\partial u}{\partial y}\right) = N_{xy}\ . \tag{6.4.19}$$

$$M_y = -D\left(\frac{\partial^2 w}{\partial y^2} + \nu\frac{\partial^2 w}{\partial x^2}\right)\ . \tag{6.4.20}$$

$$M_{yx} = D(1-\nu)\frac{\partial^2 w}{\partial x \partial y} = M_{yx}\ . \tag{6.4.21}$$

$$Q_y = \int_{-(h/2)}^{h/2} \tau_{yz}\mathrm{d}z\ . \tag{6.4.22}$$

In equations (6.4.12) to (6.4.22), N_x, N_y and N_{xy} are the **membrane forces**, M_x and M_y are **bending moments**, M_{xy} is a **twisting moment** and Q_x, Q_y are **transverse shearing forces.** All these quantities are expressed per unit length.

To determine the potential energy of the externally applied loads, we suppose that in addition to the transverse pressure $p(x, y)$, in-plane body force components F_x, F_y per unit plate surface and stress resultants $N_n{}^*, N_s{}^*, M_n{}^*, M_{ns}{}^*$, $Q_n{}^*$ act along the edge of the plate as shown in Fig. 6.4.3.

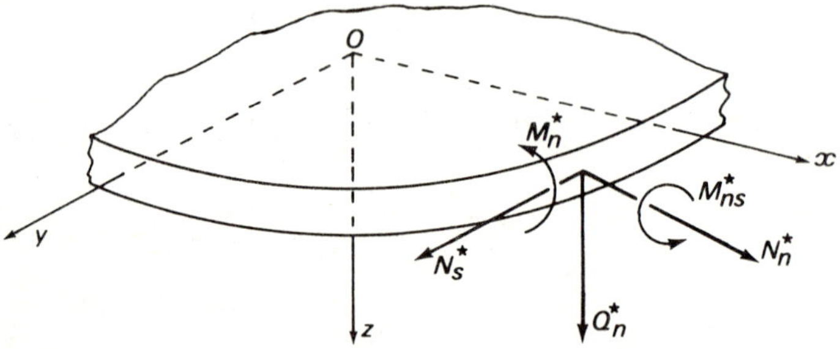

Fig. 6.4.3

Then
$$\Omega = -\iint(F_x u + F_y v)\mathrm{d}x\mathrm{d}y - \int(N_n{}^* n_x - N_s{}^* n_y)u\mathrm{d}s$$
$$-\int(N_n{}^* n_y + N_s{}^* n_x)v\mathrm{d}s - \iint pw\mathrm{d}x\mathrm{d}y$$
$$-\int Q_n{}^* w\mathrm{d}s + \int M_n{}^* \frac{\partial w}{\partial n}\mathrm{d}s - \int M_{ns}{}^* \frac{\partial w}{\partial s}\mathrm{d}s\,. \tag{6.4.23}$$

Before we proceed further, we re-write the last term of equation (6.4.23) as an integration by parts with respect to s. Then, proceeding from station 1 to station 2 on s,

$$\int M_{ns}{}^* \frac{\partial w}{\partial s}\mathrm{d}s = |M_{ns}{}^* w|_1^2 - \int \frac{\partial M_{ns}{}^*}{\partial s} w\mathrm{d}s\,. \tag{6.4.24}$$

If the integration extends right round the boundary curve s, the term $|M_{ns}{}^* w|_1^2$ is zero provided the tangent to s turns smoothly along s and $M_{ns}{}^*$ is continuous. Then

$$\Omega = -\iint pw\mathrm{d}x\mathrm{d}y + \int M_n{}^* \frac{\partial w}{\partial n}\mathrm{d}s - \int\left(Q_n{}^* - \frac{\partial M_{ns}{}^*}{\partial s}\right)w\mathrm{d}s$$
$$-\iint(F_x u + F_y v)\mathrm{d}x\mathrm{d}y - \int(N_n{}^* n_x - N_s{}^* n_y)u\mathrm{d}s$$
$$-\int(N_s{}^* n_x + N_n{}^* n_y)v\mathrm{d}s\,. \tag{6.4.25}$$

We notice that the potential energy of the stress resultants associated with in-plane and out-of-plane effects are uncoupled so that we can write

$$\Omega = \Omega_m + \Omega_b \ . \tag{6.4.26}$$

Then, the total potential energy of the system is

$$V = U_m + \Omega_m + U_b + \Omega_b \ . \tag{6.4.27}$$

or

$$V = V_m + V_b \ . \tag{6.4.27a}$$

Because of this uncoupled behaviour, we can determine the conditions for in-plane and bending equilibrium independently of each other by allowing variations δu, δv and δw corresponding to which,

$$\delta V = \delta V_m + \delta V_b \ . \tag{6.4.28}$$

For equilibrium,

$$\delta V_m = 0 \ , \quad \delta V_b = 0 \text{ independently} \ . \tag{6.4.29}$$

Considering first the bending problem, then

$$\begin{aligned} \delta V_b = & \frac{D}{2}\iint \{2(w_{xx} + w_{yy})(\delta w_{xx} + \delta w_{yy}) \\ & - 2(1 - \nu)[w_{xx}\delta w_{yy} + w_{yy}\delta w_{xx} - 2w_{xy}\delta w_{xy}]\} \, dxdy \\ & - \iint p\delta w dxdy + \int M_n{}^*\delta\left(\frac{\partial w}{\partial n}\right) ds - \int\left(Q_n{}^* - \frac{\partial M_{ns}}{\partial s}\right)\delta w ds \\ = & \ 0 \ . \end{aligned} \tag{6.4.30}$$

The δU term may be rewritten with the aid of integration by parts; after quite some manipulation and supposing D = constant we have

$$\begin{aligned} 0 = & \ D\iint(\nabla^4 w - p/D)\delta w dxdy - D\int\left[\frac{\partial}{\partial x}(\nabla^2 w)n_x + \frac{\partial}{\partial y}(\nabla^2 w)n_y\right]\delta w ds \\ & + D\int\{(w_{xx} + \nu w_{yy})n_x + (1 - \nu)w_{xy}n_y\}\delta w_x ds \\ & + D\int\{(w_{yy} + \nu w_{xx})n_y + (1 - \nu)w_{xy}n_x\}\delta w_y ds \\ & + \int M_n{}^*\delta\left(\frac{\partial w}{\partial n}\right) ds - \int\left(Q_n{}^* - \frac{\partial M_{ns}}{\partial s}\right)\delta w ds \ . \end{aligned} \tag{6.4.31}$$

The quantities δw_x and δw_y can be expressed in terms of $\delta(\partial w/\partial n)$ and $\delta(\partial w/\partial s)$. Thus from the chain rule of differentiation,

$$\frac{\partial w}{\partial n} = \frac{\partial w}{\partial x}\frac{\partial x}{\partial n} + \frac{\partial w}{\partial y}\frac{\partial y}{\partial n} \ , \qquad \frac{\partial w}{\partial s} = \frac{\partial w}{\partial x}\frac{\partial x}{\partial s} + \frac{\partial w}{\partial y}\frac{\partial y}{\partial s} \ . \tag{6.4.32a}$$

Referring to Fig. 6.4.4

$$\cos(\widehat{nx}) = n_x = \frac{\partial y}{\partial s} = \frac{\partial x}{\partial n} \ , \qquad \cos(\widehat{ny}) = n_y = -\frac{\partial x}{\partial s} = \frac{\partial y}{\partial n} \tag{6.4.32b}$$

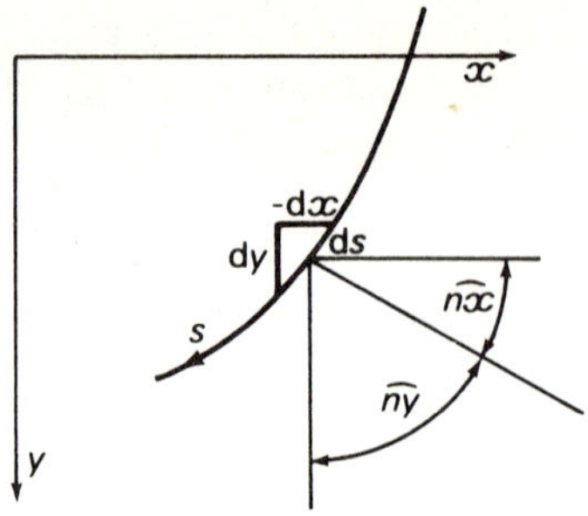

Fig. 6.4.4

and substitution in equation (6.4.32a) yields

$$\delta w_x = \delta w_n n_x - \delta w_s n_y , \quad \delta w_y = \delta w_n n_y + \delta w_s n_x , \quad (6.4.33)$$

where $$w_n = \frac{\partial w}{\partial n} \quad \text{and} \quad w_s = \frac{\partial w}{\partial s} .$$

These results may be substituted in equation (6.4.31), and after some rearrangement and integrating the term involving $\delta(\partial w/\partial s)$ with respect to s analogously to equation (6.4.24), we have

$$\begin{aligned} 0 = & \iint \left[\nabla^4 w - \frac{p}{D} \right] \delta w \mathrm{d}x\mathrm{d}y \\ & + \int \{ D[(w_{xx} + \nu w_{yy}) n_x^2 + (w_{yy} + \nu w_{xx}) n_y^2 + 2(1-\nu) w_{xy} n_x n_y] \\ & + M_n^* \} \delta \frac{\partial w}{\partial n} \mathrm{d}s + \int \{ D(1-\nu) \frac{\partial}{\partial s} [(w_{xx} - w_{yy}) n_x n_y - w_{xy}(n_x^2 - n_y^2)] \\ & - D \left[n_x \frac{\partial}{\partial x} (\nabla^2 w) + n_y \frac{\partial}{\partial y} (\nabla^2 w) \right] - \left(Q_n^* - \frac{\partial M_{ns}^*}{\partial s} \right) \} \delta w \mathrm{d}s . \end{aligned} \quad (6.4.34)$$

In arriving at this result, it is interesting to observe that the term $-2(1-\nu)[.....]$ in equation (6.4.30) makes no contribution to the term $\iint[\ ]\delta w\mathrm{d}x\mathrm{d}y$ above, whereas it does contribute to the line integrals. We shall return to this point later.

Now, over the entire surface of the plate, δw is arbitrary whilst, if along the edges the displacements are prescribed, $\delta(\partial w/\partial n)$ and δw must be zero there. Then, equation (6.4.34) can only be satisfied if

$$\nabla^4 w = p/D \text{ over the middle surface} \quad (6.4.35)$$

with w and $\partial w/\partial n$ prescribed along the boundary. Equation (6.4.35) is the Euler-Lagrange equation for the problem. If, in contrast, edge displacements are not prescribed, δw and $\delta(\partial w/\partial n)$ are arbitrary there. Again, if equation (6.4.34) is to be satisfied equation (6.4.35) must hold, but in addition we require along the plate edge,

$$M_n{}^* = -D(w_{xx} + \nu w_{yy})n_x{}^2 - D(w_{yy} + \nu w_{xx})n_y{}^2 - 2D(1-\nu)w_{xy}n_xn_y, \tag{6.4.36}$$

and
$$Q_n{}^* - \frac{\partial M_{ns}{}^*}{\partial s} = D(1-\nu)\frac{\partial}{\partial s}[(w_{xx} - w_{yy})n_xn_y - w_{xy}(n_x{}^2 - n_y{}^2)] -D\left[n_x\frac{\partial}{\partial x}(\nabla^2 w) + n_y\frac{\partial}{\partial y}(\nabla^2 w)\right] . \tag{6.4.37}$$

These two equations comprise the natural boundary conditions of the problem.

Equations (6.4.36) and (6.4.37) are equations of equilibrium along the edge of the plate; using the moment curvature relations (6.4.14) etc, equation (6.4.36) can be written as

$$M_n{}^* = M_xn_x{}^2 + M_yn_y{}^2 - 2M_{xy}n_xn_y . \tag{6.4.38}$$

The interesting feature of equation (6.4.37) is that the boundary conditions imposed on $Q_n{}^*$ and $M_{ns}{}^*$ appear as a single condition on the new quantity $(Q_n{}^* - \partial M_{ns}{}^*/\partial s)$. In the literature, this is referred to as the Kirchhoff Shear, whose physical significance was explained by Kelvin and Tait in Volume 1 in 1883.†

On the basis of our assumptions we have been able to express M_x, M_y, M_{xy} in terms of w very readily, equations (6.4.14) etc, but we could not do the same for Q_x and Q_y.†† Having established equation (6.4.35), we can correct this deficiency. That equation is for equilibrium of transverse forces and we can rewrite it as

$$D\left[\frac{\partial}{\partial x}\left(\frac{\partial}{\partial x}\nabla^2 w + \frac{\partial}{\partial y}\left(\frac{\partial}{\partial y}\nabla^2 w\right)\right] p - 0 . \tag{6.4.39}$$

Referring to the element $h\mathrm{d}x\mathrm{d}y$ of Fig. 6.4.5, we see that for equilibrium of transverse forces, we must have

$$\frac{\partial Q_x}{\partial x} + \frac{\partial Q_y}{\partial y} + p = 0 . \tag{6.4.40}$$

Comparing equations (6.4.39) and (6.4.40) we see that

$$Q_x = -D\frac{\partial}{\partial x}(\nabla^2 w), \quad Q_y = -D\frac{\partial}{\partial y}(\nabla^2 w) . \tag{6.4.41}$$

Thus, in a manner analogous to that used for the beam, we have obtained expressions for the transverse shear stress resultants in terms of w even though the shearing stresses τ_{xz} and τ_{yz} themselves were not explicitly taken into account in the formulation.

† See also Example 7.5.1.

†† Because we have no information on the distribution of τ_{zx} and τ_{zy}.

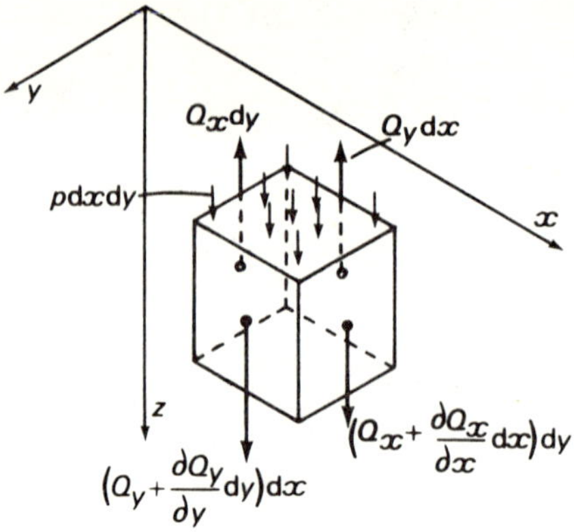

Fig. 6.4.5

Using equations (6.4.41) and (6.4.14) etc, equation (6.4.37) may be re-written as

$$Q_n{}^* - \frac{M_{ns}{}^*}{\partial s} = -\frac{\partial}{\partial s}\left[(M_x - M_y)n_x n_y + M_{xy}(n_x{}^2 - n_y{}^2)\right] + Q_x n_x + Q_y n_y .$$

It follows that $Q_n = Q_x n_x + Q_y n_y$ (6.4.42)

and $M_{ns}{}^* = (M_x - M_y)n_x n_y + M_{xy}(n_x{}^2 - n_y{}^2)$. (6.4.43)

Equations (6.4.38), (6.4.43) and (6.4.44) will, of course, hold for any sections through the plate even though we happen, here, to have established them for the edge. One other familiar result is obtained when the moment curvature relations are substituted in equation (6.4.35):

$$\frac{\partial^2 M_x}{\partial x^2} + \frac{\partial^2 M_y}{\partial y^2} - 2\frac{\partial^2 M_{xy}}{\partial x \partial y} + p = 0 . \tag{6.4.44}$$

Earlier, we observed that the [. . . . .] term in equation (6.4.30) made no contribution to the differential equation governing the problem but that it affected the natural boundary conditions. In numerical work, using the Rayleigh-Ritz of Kantorovitch methods, say, this fact can save significant computational effort in certain situations. Thus, if boundary displacements w and $\partial w/\partial n$ are prescribed all round the edge of the plate, there are no natural boundary conditions and the [. . . . .] term makes no contribution whatever to the solution. Under these circumstances, we can use the simpler expression

$$U - \frac{D}{2}\iint\left[\frac{\partial^2 w}{\partial x^2} + \frac{\partial^2 w}{\partial y^2}\right]^2 dxdy \tag{6.4.9a}$$

for the bending strain energy. It can also be shown that this expression may be used for polygonal plates when only the deflexion w is required to be zero along the edge. In particular, then, we can take advantage of this feature in the treatment of simply supported polygonal plates and clamped plates of any plan form.

To conclude this analysis of plate bending, it is useful to make two points. First, the principle of virtual work could be applied directly: incorporating the previous assumptions into equation (6.2.6), equilibrium equations (6.4.38), (6.4.42) and (6.4.44) are obtained independently of any stress-strain law.† The second point relates to the integration by parts of equation (6.2.24). If the edge of the plate has corners, or if there are discontinuities in M_{ns}, we must exercise care. Thus, suppose M_{ns} is discontinuous at some boundary point P, then we must write

$$\int M_{ns}{}^*\delta\frac{\partial w}{\partial s}\,\mathrm{d}s \;=\; -\;[\,|M_{ns}{}^*|_{p+} - |M_{ns}{}^*|_{p-}\,]\,\delta w_P - \int\frac{\partial M_{ns}{}^*}{\partial s}\,\delta w\mathrm{d}s\,. \tag{6.4.45}$$

We shall see the physical implications of this effect in Example 7.5.1.

The equilibrium equations in the interior and on the boundary of a stretched plate are derived in a manner similar to the above by considering the variation $\delta V_m = 0$ consequent upon variations δu and δv in the displacement components. We find that the condition for equilibrium becomes

$$0 = \iint\left\{\frac{Eh}{(1-\nu^2)}\{-(u_{xx}+v_{xy})+(1-\nu)[v_{xy}-\tfrac{1}{2}(u_{yy}+v_{xy})]\} - F_x\right\}\delta u\mathrm{d}A$$

$$+\iint\left\{\frac{Eh}{(1-\nu^2)}\{-(u_{xy}+v_{yy})+(1-\nu)[u_{xy}-\tfrac{1}{2}(u_{xy}+v_{xx})]\} - F_y\right\}\delta v\mathrm{d}A$$

$$+\int\left\{\frac{Eh}{1-\nu^2}\{(u_x+v_y)n_x-(1-\nu)[v_yn_x-\tfrac{1}{2}(u_y+v_x)n_y]\} - (N_n{}^*n_x - N_o{}^*n_y)\right\}\delta u\mathrm{d}s$$

$$+\int\left\{\frac{Eh}{1-\nu^2}\{(u_x+v_y)n_y-(1-\nu)[u_xn_y-\tfrac{1}{2}(u_y+v_x)n_x]\} - (N_n{}^*n_y + N_s{}^*n_x)\right\}\delta v\mathrm{d}s\,. \tag{6.4.46}$$

But the δu and δv are arbitrary and independent over the middle surface so that if the right hand side of equation (6.4.46) is to be zero, the coefficients of δu and δv in the integrands of the first two terms must separately be zero. Than, for equilibrium in the interior of the plate we have, after a slight re-arrangement,

$$\frac{\partial^2 u}{\partial x^2} + \frac{1-\nu}{2}\frac{\partial^2 u}{\partial y^2} + \frac{1+\nu}{2}\frac{\partial^2 v}{\partial x\partial y} + \frac{1-\nu^2}{Eh}F_x = 0\,. \tag{6.4.47}$$

† The reader should try this as an excercise.

$$\frac{\partial^2 v}{\partial y^2} + \frac{1-\nu}{2}\frac{\partial^2 v}{\partial x^2} + \frac{1+\nu}{2}\frac{\partial^2 v}{\partial x \partial y} + \frac{1-\nu^2}{Eh}F_y = 0 \,. \quad (6.4.48)$$

These are the two dimensional form of Navier's equations deduced in Section 3.5.

If in-plane displacements are prescribed all round the edge of the plate, δu and δv must be zero there and the line integrals of equation (6.4.46) are zero automatically. Equations (6.4.47) and (6.4.48) together with prescribed values for u and v along the edge thus define the boundary value problem for stretching of the plate. If in-plane edge tractions are prescribed, δu and δv are arbitrary along the edge so that for the line integrals to be zero we require

$$\frac{Eh}{1-\nu^2}\left\{\left(\frac{\partial u}{\partial x} + \frac{\partial v}{\partial y}\right)n_x - (1-\nu)\left[\frac{\partial v}{\partial y}n_x - \frac{1}{2}\left(\frac{\partial u}{\partial y} + \frac{\partial v}{\partial x}\right)n_y\right]\right\} = N_n{}^*n_x - N_s{}^*n_y \quad (6.4.49)$$

and

$$\frac{Eh}{1-\nu^2}\left\{\left(\frac{\partial u}{\partial x} + \frac{\partial v}{\partial y}\right)n_y - (1-\nu)\left[\frac{\partial u}{\partial x}n_y - \frac{1}{2}\left(\frac{\partial u}{\partial y} + \frac{\partial v}{\partial x}\right)n_x\right]\right\} = N_n{}^*n_y + N_s{}^*n_x \,. \quad (6.4.50)$$

These are the natural boundary conditions of the variational problem, whilst equations (6.4.47) and (6.4.48) constitute the Euler-Lagrange equations. Using the stress resultant definitions (6.4.12) etc, equilibrium throughout the interior of the plate is described by

$$\frac{\partial N_x}{\partial x} + \frac{\partial N_{xy}}{\partial y} + F_x = 0 \,.$$

$$\frac{\partial N_{xy}}{\partial x} + \frac{\partial N_y}{\partial y} + F_y = 0 \,. \quad (6.4.51)$$

On the edge of the plate, equations (6.4.49) and (6.4.50), slightly rearranged, become

$$N_n{}^* = N_x n_x{}^2 + N_y n_y{}^2 + 2N_{xy}n_x n_y \,.$$

$$N_s{}^* = -(N_x - N_y)n_x n_y + N_{xy}(n_x{}^2 - n_y{}^2) \,. \quad (6.4.52)$$

Again, had we used the principle of virtual work directly instead of the potential energy principle, we would have obtained these results irrespective of the stress-strain law for the material.

To conclude this section, we remind the reader that the application of variational principles to the formulation of an approximate theory for continuum systems yield the governing differential equations *together with* the correct boundary conditions: a consistent statement of the boundary value problem is thus obtained more directly and easily than by vector means. When the basis of the discussion is the virtual work principle, the procedure yields equations of equilibrium which are independent of the stress-strain relations for the material. If the behaviour is assumed to be elastic, a strain energy function exists and the equations are expressible in terms of the displacements which are the basic unkowns of the problem.

Problems 6.4

6.4.1 To obtain a theory for the deflexion of plates taking transverse shear effects into account, follow Mindlin's approach by proceeding as follows. Assuming plane sections remain plane, but not normal to the middle surface, a plausible displacement pattern is

$$u = -z\psi_x, \quad v = -z\psi_y, \quad w = w_0 . \tag{a}$$

Here $w_0(x, y)$ is the middle surface displacement whilst $\psi_x(x, y)$ and $\psi_y(x, y)$ are functions to be determined. The non-zero strain components are then

$$\epsilon_x = -z\frac{\partial \psi_x}{\partial x}, \quad \epsilon_y = -z\frac{\partial \psi_y}{\partial y}, \quad \gamma_{xy} = -z\left(\frac{\partial \psi_x}{\partial y} + \frac{\partial \psi_y}{\partial x}\right),$$

$$\gamma_{zx} = \frac{\partial w}{\partial x} - \psi_x, \quad \gamma_{zy} = \frac{\partial w}{\partial y} - \psi_y . \tag{b}$$

Further assume that σ_z is negligible compared with the other stress components.

Show that the total potential energy is

$$V = \iint\left\{\frac{D}{2}\left[\left(\frac{\partial \psi_x}{\partial x}\right)^2 + \left(\frac{\partial \psi_y}{\partial y}\right)^2 + 2\nu\frac{\partial \psi_x}{\partial x}\frac{\partial \psi_y}{\partial y}\right]\right.$$

$$\left. + \frac{G}{2}\left[\frac{h^3}{12}\left(\frac{\partial \psi_x}{\partial y} + \frac{\partial \psi_y}{\partial x}\right)^2 + h\left(\frac{\partial w}{\partial x} - \psi_x\right)^2 + h\left(\frac{\partial w}{\partial y} - \psi_y\right)^2\right] - pw\right\}$$

$$\times \, dxdy \tag{c}$$

or

$$V = \iint F\left(w, \frac{\partial w}{\partial x}, \frac{\partial w}{\partial y}, \psi_x, \frac{\partial \psi_x}{\partial x}, \frac{\partial \psi_x}{\partial y}, \psi_y, \frac{\partial \psi_y}{\partial x}, \frac{\partial \psi_y}{\partial y}\right) dA . \tag{d}$$

Show that the Euler-Lagrange equations for $\delta V = 0$ are

$$Gh\left(\frac{\partial^2 w}{\partial x^2} + \frac{\partial^2 w}{\partial y^2} - \frac{\partial \psi_x}{\partial x} - \frac{\partial \psi_y}{\partial y}\right) + p = 0, \tag{e}$$

$$Gh\left(\frac{\partial w}{\partial x} - \psi_x\right) + D\left(\frac{\partial^2 \psi_x}{\partial x^2} + \nu\frac{\partial^2 \psi_y}{\partial x \partial y}\right) + \frac{Gh^3}{12}\left(\frac{\partial^2 \psi_x}{\partial y^2} + \frac{\partial^2 \psi_y}{\partial x \partial y}\right) = 0, \tag{f}$$

$$Gh\left(\frac{\partial w}{\partial y} - \psi_y\right) + D\left(\frac{\partial^2 \psi_y}{\partial y^2} + \nu\frac{\partial^2 \psi_x}{\partial x \partial y}\right) + \frac{Gh^3}{12}\left(\frac{\partial^2 \psi_y}{\partial x^2} + \frac{\partial^2 \psi_x}{\partial x \partial y}\right) = 0. \tag{g}$$

After some considerable manipulation, ψ_x and ψ_y may be eliminated from equations (e), (f), (g) to yield

$$D\nabla^4 w = p - \frac{D}{Gh}\nabla^2 p . \tag{h}$$

G is the shear modulus in these developments.

6.5 Principle of virtual complementary work for a solid continuum: complementary strain energy density

In Chapter 1, we saw that of all the statically possible sets of forces induced in a system in equilibrium the true one, corresponding to the actual compatible configuration, is identified by the principle of virtual complentary work. To apply the principle to a continuum, for which the actual body forces, surface tractions, stresses, strains and displacements are F_i, $\overset{n}{T}_i$, σ_{ij}, ϵ_{ij} and u_i respectively, the virtual forces and stresses used for comparison are the variations δF_i, $\delta \overset{n}{T}_i$ and $\delta\sigma_{ij}$. These variations are not completely independent but *must* comprise equilibrate force and stress systems. That is, we must have

$$\frac{\partial \delta\sigma_{ji}}{\partial x_j} + \delta F_i = 0 \text{ in the interior;} \tag{6.5.1}$$

$$\delta \overset{n}{T}_i = \delta\sigma_{ji} n_j \text{ on the surface .} \tag{6.5.2}$$

Except for these restrictions, the δF_i etc, are quite arbitrary.

Now the principle is only valid when the actual forces produce small displacements. Then observing that the displacements remain fixed during the variation of forces and stresses, the virtual complementary work of the external forces is

$$\delta W_e{}^* = \int_{\text{vol}} u_i \delta F_i \text{d vol} + \int_S u_i \delta T_i \text{d}S \ . \tag{6.5.3}$$

Generalising the results of Section 2.2 to a prismatic element $\text{d}x_1 \,\text{d}x_2 \,\text{d}x_3$ such as that in Fig. 3.3.3(a), the virtual complementary work of the virtual stresses (virtual internal forces) $\delta\sigma_{11}\,\text{d}x_2\,\text{d}x_3$ through the actual strains, is

$$\delta W_i{}^* = -\int_{\text{vol}} [\epsilon_{11}\delta\sigma_{11} + \ldots\ldots \gamma_{31}\delta\sigma_{31}]\,\text{d vol} \tag{6.5.4}$$

or

$$\delta W_i{}^* = -\int_{\text{vol}} \epsilon_{ij}\delta\sigma_{ij}\text{d vol} \ . \tag{6.5.5}$$

The principle then becomes

$$\delta W_e{}^* + \delta W_i{}^* = 0 = -\int_{\text{vol}} (\epsilon_{ij}\delta\sigma_{ij} - u_i\delta F_i)\text{d vol} + \int_S u_i \delta \overset{n}{T}_i \,\text{d}S \tag{6.5.6}$$

as the condition for compatible displacements.

Taking account of equation (6.5.1), we have

$$-\int_{\text{vol}} \left(\epsilon_{ij}\delta\sigma_{ij} + u_i \frac{\partial\delta\sigma_{ji}}{\partial x_j}\right)\text{d vol} + \int_S u_i \delta \overset{n}{T}_i \text{d}S = 0 \ . \tag{6.5.7}$$

Integrating by parts,

$$0 = \int_{\text{vol}} \left(\delta\sigma_{ji}\frac{\partial u_i}{\partial x_j} - \epsilon_{ij}\delta\sigma_{ij}\right)\text{d vol} + \int_S (u_i \delta \overset{n}{T}_i - u_i\delta\sigma_{ji}n_j)\text{d}S \ . \tag{6.5.8}$$

Since i and j are dummy subscripts and $\sigma_{ij} = \sigma_{ji}$, we can write

$$0 = \int\left[\frac{1}{2}\left(\frac{\partial u_i}{\partial x_j} + \frac{\partial u_j}{\partial x_i}\right) - \epsilon_{ij}\right]\delta\sigma_{ij}\,\mathrm{dvol} + \int_S(\delta\overset{n}{T}_i u_i - \delta\sigma_{ji}n_ju_i)\mathrm{d}S. \tag{6.5.9}$$

Throughout the interior of the solid, $\delta\sigma_{ij}$ is arbitrary so that, if equation (6.5.9) is to hold, both integrals must be zero independently. We have first that

$$\epsilon_{ij} = \frac{1}{2}\left(\frac{\partial u_i}{\partial x_j} + \frac{\partial u_j}{\partial x_i}\right). \tag{6.5.10}$$

throughout the interior. On part S_u of the surface, we suppose the displacements $u_i = u_i{}^*$, prescribed, whilst on part S_T, $\overset{n}{T}_i = \overset{n}{T}_i{}^*$, prescribed. Then, the line integral in equation (6.5.9) can be written as

$$\int_{S_u}(\delta\overset{n}{T}_i u_i{}^* - \delta\sigma_{ij}n_ju_i)\mathrm{d}S + \int_{S_T}(\delta\overset{n}{T}_i{}^*u_i{}^{**} - \delta\sigma_{ji}n_ju_i)\mathrm{d}S\ \dagger$$

which, in view of equation (6.5.2), becomes

$$\int_{S_u}(u_i{}^* - u_i)\delta\overset{n}{T}_i\,\mathrm{d}S + \int_{S_T}(u_i{}^{**} - u_i)\delta\overset{n}{T}_i{}^*\mathrm{d}S.$$

But since $\overset{n}{T}_i$ is prescribed on S_T, $\delta\overset{n}{T}_i{}^* = 0$, so that for equation (6.5.9) to be satisfied we have, secondly,

$$u_i{}^* = u_i \text{ on } S_u. \tag{6.5.11}$$

Thus we see that the principle of virtual complementary work has yielded the conditions of compatibility which must be satisfied both in the interior and at the boundary of the solid. This is a purely geometric result, independent of any stress-strain law, as we would expect.

For an elastic solid, linear or not, thermodynamic considerations show that a function $U_0{}^*(\sigma_{ij})$ exists such that

$$\epsilon_{ij} = \frac{\partial U_0{}^*}{\partial\sigma_{ij}}. \tag{6.5.12}$$

Then $$\delta W_i{}^* = -\int_{\mathrm{vol}}\epsilon_{ij}\delta\sigma_{ij}\mathrm{dvol} = -\int_{\mathrm{vol}}\frac{\partial U_0{}^*}{\partial\sigma_{ij}}\delta\sigma_{ij}\,\mathrm{dvol}$$

or $$\delta W_i{}^* = -\int_{\mathrm{vol}}\delta U_0{}^*\mathrm{dvol} = -\delta U^*. \tag{6.5.13}$$

$U_0{}^*$ is called the complementary strain energy density.

The principle of virtual complementary work now takes the form

$$\delta U^* = \int u_i\delta F_i\,\mathrm{dvol} + \int u_i\delta\overset{n}{T}_i\,\mathrm{d}S.$$

† $u_i{}^{**}$ is the displacement corresponding to $\overset{n}{T}_i{}^*$.

For the usual situation where body forces and tractions over S_T are prescribed, we must have $\delta F_i = 0$ and $\delta \overset{n}{T}_i = 0$ on S_T, the principle of virtual complementary work then becoming

$$0 = \int \delta U_0{}^* \mathrm{d}\,\mathrm{vol} - \int_{S_u} u_i{}^* \delta \overset{n}{T}_i \mathrm{d}S \ . \qquad (6.5.14)$$

Since the u_i remain fixed whilst the variations $\delta \overset{n}{T}_i$ etc. are imposed, equation (6.5.14) can be written

$$\delta V^* = 0 \qquad (6.5.15)$$

where
$$V^* = \int_{\mathrm{vol}} U_0{}^*(\sigma_{ij})\mathrm{d}\,\mathrm{vol} - \int_{S_u} u_i{}^* \overset{n}{T}_i \ \mathrm{d}S \ . \qquad (6.5.16)$$

V^*, which is a function of F_i, $\overset{n}{T}_i$ and σ_{ij} is called the total complementary energy. Equation (6.5.15) represents the **principle of stationary total complementary energy** which asserts that:

> *Of all the possible equilibrium stress fields for a solid body subjected to prescribed loadings and boundary displacements the true one, corresponding to a compatible strain field, renders the total complementary energy stationary.*

It may be shown on thermodynamic grounds that $U_0{}^*$ is positive definite so that this principle can be strengthened: in fact V^* is a *minimum* for compatibility. The proof is similar to that used to show that the potential energy is a minimum for equilibrium.

The assumption that the supports of a structure are *rigid* is frequently and quite reasonably made in design. Under these circumstances, $u_i{}^* = 0$ on S_u and equation (6.5.15) reduces to the important special case

$$\delta U^* = 0 \qquad (6.5.17)$$

This is the famous **Castigliano's theorem of least work,** but generalised for non-linear elastic behaviour. When Hooke's law obtains we find that $U^*(\sigma_{ij})$ is numerically equal to $U(\sigma_{ij})$. Then, we have the usual form of Castigliano's theorem

$$\delta U = 0 \qquad (6.5.17\mathrm{a})$$

where U must be expressed as a function of the stresses.

For structures which are mounted on *rigid* supports (those for which $u_i{}^* = 0$ on S_u), Castigliano developed a second famous theorem for calculating displacements corresponding to point loads and couples: we saw in Chapter 2 how valuable this theorem is for frames. We now generalise it for an arbitrary solid.

Returning to equations (6.5.6) let us suppose the material to be elastic and that in addition to $\overset{n}{T}_i{}^*$ there act on the part S_T of the surface of the solid, several point loads $P_1, P_2 \ldots\ldots P_n$ having corresponding displacements

$\bar{u}_1, \bar{u}_2 \ldots\ldots \bar{u}_n$. Then with $u_i^* = 0$ on S_u, the principle of stationary complementary energy becomes

$$\delta U^* = \int_{\text{vol}} u_i \delta F_i \mathrm{d}\,\text{vol} + \int_{S_T} u_i \delta \overset{n}{T}_i \mathrm{d}S + \sum_{k=1}^{n} \bar{u}_k \delta P_k \,. \quad (6.5.18)$$

Let us now allow variations for which $\delta F_i = \delta \overset{n}{T}_i{}^* = 0$ and $\delta P_k = \Delta P_k$, *actual independent increments* in the forces P_k, then equation (6.5.18) becomes

$$\Delta U^* = \sum_{k=1}^{n} \bar{u}_k \Delta P_k \,. \quad (6.5.19)$$

But

$$\Delta U^* = \sum_{k=1}^{n} \frac{\partial U^*}{\partial P_k} \Delta P_k \,, \quad (6.5.20)$$

so comparing equations (6.5.19) and (6.5.20),

$$\sum_{k=1}^{n} \left(\frac{\partial U^*}{\partial P_k} - \bar{u}_k \right) \Delta P_k = 0 \,. \quad (6.5.21)$$

Since the ΔP_k are independent, it follows that

$$\bar{u}_k = \frac{\partial U^*}{\partial P_k} \,. \quad (6.5.22)$$

This is **Castigliano's theorem on displacements** generalised for non-linear elasticity. For a Hooke's law material $U(\epsilon_{ij}) = U^*(\sigma_{ij})$ numerically, and we have then,

$$\bar{u}_k = \frac{\partial U}{\partial P_k} \quad (6.5.23)$$

where U must be expressed as a function of the loads.

It follows from the method of deriving equation (6.5.22) that P_k may be either a point force or a couple. In the former case, $\bar{u}_k$ is a linear displacement in the direction of P_k; in the latter case, $\bar{u}_k$ is an angular displacement.

EXAMPLE 6.5.1

Deduce an expression for the complementary energy density for a Hooke's law material and show that it is numerically equal to the strain energy density.

Since

$$\delta U_0^* = \epsilon_{ij} \delta \sigma_{ij} \,, \quad \text{(a)}$$

$$U_0^* = \int_0^{\sigma_{ij}} \epsilon_{ij} \mathrm{d}\sigma_{ij} \,. \quad \text{(b)}$$

From Hooke's law

$$\epsilon_{ij} = \frac{1+\nu}{E} \sigma_{ij} - \frac{\nu}{E} \delta_{ij} \sigma_{kk} \,. \quad \text{(c)}$$

Then
$$U_0^* = \int_0^{\sigma_{ij}} \left(\frac{1+\nu}{E}\right) \sigma_{ij} d\sigma_{ij} - \int_0^{\sigma_{ij}} \frac{\nu}{E} \sigma_{kk} \delta_{ij} d\sigma_{ij}$$

$$= \frac{1+\nu}{E} \int_0^{\sigma_{ij}} \sigma_{ij} d\sigma_{ij} - \frac{\nu}{E} \int_0^{\sigma_{kk}} \sigma_{kk} d\sigma_{kk}$$

or
$$U_0^* = \frac{1+\nu}{2E} \sigma_{ij}\sigma_{ij} - \frac{\nu}{2E} \sigma_{kk}^2 . \tag{d}$$

Also from Hooke's law
$$\sigma_{ij} = \lambda \delta_{ij} \epsilon_{kk} + 2G\epsilon_{ij} , \tag{e}$$

so that
$$d\sigma_{ij} = \lambda \delta_{ij} d\epsilon_{kk} + 2G d\epsilon_{ij} .$$

Substituting in equation (b)

$$U_0^* = \lambda \int_0^{\epsilon_{ij}} \delta_{ij} \epsilon_{ij} d\epsilon_{kk} + 2G \int_0^{\epsilon_{ij}} \epsilon_{ij} \, d\epsilon_{ij}$$

or
$$U_0^* = \frac{\lambda}{2} (\epsilon_{kk})^2 + G\epsilon_{ij}\epsilon_{ij} \tag{f}$$

which is precisely the expression for U_0.

Problems 6.5

6.5.1 Accepting that U_0^* is positive-definite, proceed in a manner analogous to that used in Section 6.2 to prove that the total complementary energy is a minimum for a solid to be in a compatible equilibrium state.

6.6 Principle of complementary energy applied to the plane stress problem

This principle is concerned with the relative merits of equilibriant stress systems. A convenient way in which to ensure that we are, indeed, working on stresses which are in equilibrium is to carry out formulations in terms of stress functions. As we saw in Section 3.7, if a function $\chi(x, y)$ is introduced such that

$$\sigma_x = \frac{\partial^2 \chi}{\partial y^2} , \quad \sigma_y = \frac{\partial^2 \chi}{\partial x^2} , \quad \tau_{xy} = -\frac{\partial^2 \chi}{\partial x \partial y} , \tag{6.6.1}$$

the equations of equilibrium are satisfied automatically when body forces are negligible. Furthermore, specification of tractions on the surface implies specification of χ, $\partial\chi/\partial x$ and $\partial\chi/\partial y$ there. [Equations (3.7.23), (3.7.24), and (3.7.26).]

Let us suppose that we have a plane stress problem in which tractions are prescribed over the entire boundary of the region and that body forces are negligible. Then, the principle of minimum complementary energy reduces to

$$\delta V^* = \delta U^* = 0 . \tag{6.6.2}$$

In a state of plane stress with respect to the xOy plane, the significant stress components are σ_x, σ_y and τ_{xy}. Then using equation (d) of Example 6.5.1 we have for a Hooke's law material

$$U_0{}^* = \frac{1}{2E}[\sigma_x{}^2 + \sigma_y{}^2 - 2\nu\sigma_x\sigma_y] + \frac{1}{2G}\tau_{xy}{}^2 \tag{6.6.3}$$

Substituting from equation (6.6.1) and (6.6.3), in (6.5.13), we find

$$U^* = \frac{1}{2E}\iint\left\{\left(\frac{\partial^2\chi}{\partial x^2} + \frac{\partial^2\chi}{\partial y^2}\right)^2 - 2(1+\nu)\left[\frac{\partial^2\chi}{\partial x^2}\frac{\partial^2\chi}{\partial y^2} - \left(\frac{\partial^2\chi}{\partial x\partial y}\right)^2\right]\right\}\mathrm{d}x\mathrm{d}y\dagger \tag{6.6.4}$$

Allowing a variation $\delta\chi$ in the stress function implies the imposition of equilibriant virtual forces so that

$$\delta U^* = \frac{1}{2E}\iint\{2\nabla^2\chi\,(\delta\chi_{xx} + \delta\chi_{yy}) - 2(1+\nu)[\chi_{xx}\delta\chi_{yy} + \chi_{yy}\delta\chi_{xx} - 2\chi_{xy}\delta\chi_{xy}]\}\,\mathrm{d}x\mathrm{d}y\ . \tag{6.6.5}$$

Integrating parts, we find

$$\begin{aligned}\delta U^* = \frac{1}{E}\iint\nabla^4\chi\delta\chi\mathrm{d}x\mathrm{d}y &+ \int[n_x\nabla^2\chi - (1+\nu)(\chi_{yy}n_x - \chi_{xy}n_y)]\,\delta\chi_x\mathrm{d}S\\ &+ \int[n_y\nabla^2\chi - (1+\nu)(\chi_{xx}n_y - \chi_{xy}n_x)]\,\delta\chi_y\mathrm{d}S\\ &- \int\left[n_x\frac{\partial}{\partial x}(\nabla^2\chi) + n_y\frac{\partial}{\partial y}(\nabla^2\chi)\right]\delta\chi\mathrm{d}S\ .\end{aligned} \tag{6.6.6}$$

But $\delta\chi$ is arbitrary over the region and, since tractions are supposed prescribed along the entire boundary, $\delta\chi = \delta\chi_x = \delta\chi_y = 0$ there. Then, the condition $\delta U^* = 0$ leads to

$$\nabla^4\chi = 0 \quad \text{over the region}\ . \tag{6.6.7}$$

This is an equation of compatibility and since, from the definition of χ, the stresses implied are in equilibrium, solutions to equation (6.6.7) represent the true solution to an elastic stressing problem.

A stress function may be used directly in the virtual complementary work principle using equation (6.5.6) as the starting point with $\delta F_i = 0$ in the region and $\overset{n}{T_i} = 0$ on the boundary. The reader should show that integration by parts then yields the familiar compatibility condition

$$\frac{\partial^2\epsilon_y}{\partial x^2} + \frac{\partial^2\epsilon_x}{\partial y^2} = \frac{\partial^2\gamma_{xy}}{\partial x\partial y} \tag{6.6.8}$$

as a necessary condition for zero virtual complementary work.

† The reader should notice the similarity between this expression for U^* and that for bending strain energy in a plate, Equation (6.4.9). Again, the term in square brackets makes a contribution to the natural boundary conditions alone and not to the Euler Equation (6.6.7).

Comparing the work of this section with that of Section 6.4, we see the foundations of the so-called **slab-slice analogy**: the form of the governing differential equations and boundary conditions are the same for plate bending and stretching. Thus, the *plate bending* (slab) problem, where *edge displacements* are prescribed, is mathematically equivalent to the *plate stretching* (slice) problem, where *edge tractions* are specified. This is just one instance of a number of analogous situations in mathematical physics and engineering which are practically useful: whatever solution process is developed for the one situation is immediately applicable to another; the physically simpler system may be used as an analogue computing device for the other, or the more familiar situation may prove helpful in visualising the solution to an unfamiliar one.

Problem 6.6

6.6.1 Fig. 6.6.1 shows the details of a bonded joint. Each cover plate has thickness t_1 and material having Young's modulus E_1. The main plate is t_2 thick and its material has Young's modulus E_2. The adhesive is t_a thick with a shear modulus G_a. The assembly is b wide. Establish an approximate theory for the shearing stress in the adhesive on the basis of the following assumptions:

(i) The adhesive is so thin that its shearing stress τ_a does not change through the thickness although it is a function of x. After curing, the adhesive behaves in a linear elastic manner.

(ii) The main and cover plates are so stiff compared with the adhesive that their plane normal sections remain plane and normal when the load is applied.

Show that for equilibrium in the region $0 < x < l$

$$\tau_a = -t_1\frac{d\sigma_1}{dx}, \qquad \tau_a = \frac{t_2}{2}\frac{d\sigma_2}{dx}. \tag{a}$$

Introducing constraints (a) by means of Lagrange multipliers λ_1, λ_2, show that a suitable modified complementary energy functional is

$$\bar{U}^* = \frac{b}{2}\left\{\int_0^l\left[\frac{2t_1\sigma_1^2}{E_1} + \frac{t_2\sigma_2^2}{E_2} + \frac{2t_a\tau_a^2}{G_a}\right]dx + \frac{\sigma_{20}^2 at_2}{E_2}\right\}$$
$$- b\left\{\int_0^l\lambda_1\left(\tau_a + t_1\frac{d\sigma_1}{dx}\right)dx + \int_0^l\lambda_2\left(\tau_a - \frac{t_2}{2}\frac{d\sigma_2}{dx}\right)dx\right\}.$$

Show that $\delta\bar{U}^* = 0$ leads to

$$\frac{d^2\tau_a}{dx^2} - \mu^2\tau_a = 0$$

where μ is a constant for the system.

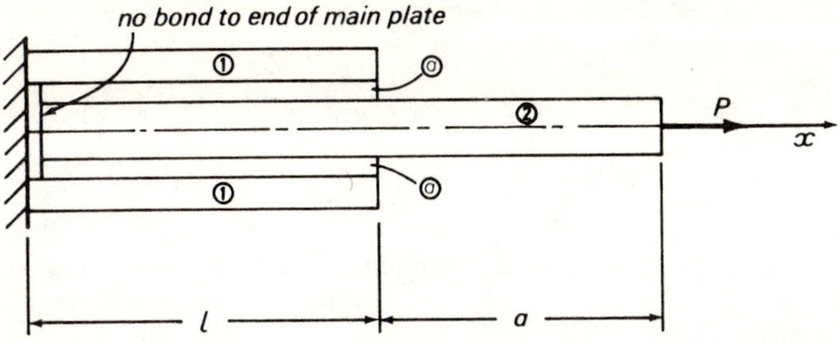

Fig. 6.6.1

6.7 Principle of complementary energy applied to the torsion problem

Following Section 3.6, we assume that the only non-zero stress components are τ_{zx} and τ_{zy}. A stress function then exists such that

$$\tau_{zx} = \frac{\partial \phi}{\partial y}, \qquad \tau_{zy} = -\frac{\partial \phi}{\partial x}. \tag{6.7.1}$$

We suppose that end $z = 0$ of the bar, Fig. 3.6.1 is prevented from rotating so that u and v are zero there. At $z = l$, a total angle of twist αl is imposed. The part S_u of the bar surface thus consists of the two end faces, but since $u = v = 0$ at $z = 0$, there is a non-zero contribution to the surface integral of equation (6.5.16) at $z = l$ alone. Then, the principle of virtual complementary energy becomes:

$$\delta V^* = 0 = \delta\left\{\frac{l}{2G}\iint\left[\left(\frac{\partial\phi}{\partial x}\right)^2 + \left(\frac{\partial\phi}{\partial y}\right)^2\right]\mathrm{d}x\mathrm{d}y - \iint_{S_u} u_i\overset{n}{T}_i\,\mathrm{d}x\mathrm{d}y\right\}. \tag{6.7.2}$$

The second term in equation (6.7.2) must be expressed in terms of α and ϕ. Now, for torsion by end tractions at $x = l$

$$\left|\overset{n}{T}_x\right|_l = \left|\tau_{zx}\right|_l = \left|\frac{\partial\phi}{\partial y}\right|_l; \quad \left|\overset{n}{T}_y\right|_l = \left|\tau_{zy}\right|_l = -\left|\frac{\partial\phi}{\partial x}\right|_l; \quad \overset{n}{T}_z = 0. \tag{6.7.3}$$

Again following Section 3.6

$$u_l = -\alpha l y, \qquad v_l = \alpha l x, \qquad w_l = \alpha\psi(x, y) \tag{6.7.4}$$

then
$$\iint_{S_u} u_i\overset{n}{T}_i\mathrm{d}x\mathrm{d}y = -\alpha l\iint\left[y\frac{\partial\phi}{\partial y} + x\frac{\partial\phi}{\partial x}\right]\mathrm{d}x\mathrm{d}y$$
$$= -\alpha l\left\{\int[y\phi n_y + x\phi n_x]\,\mathrm{d}s - \iint[1\cdot\phi + 1\cdot\phi]\,\mathrm{d}x\mathrm{d}y\right\}.$$

But for equilibrium with the prescribed zero traction along the lateral surface of the bar, $\phi = 0$ along the boundary of the section (see Section 3.6) so that

$$\iint_{S_u} u_i\overset{n}{T}_i\mathrm{d}x\mathrm{d}y = 2\alpha l\iint\phi\,\mathrm{d}x\mathrm{d}y. \tag{6.7.5}$$

The condition for stationary complementary energy thus becomes

$$\delta V^* = 0 = \delta\left\{\frac{l}{2G}\iint\left[\left(\frac{\partial\phi}{\partial x}\right)^2 + \left(\frac{\partial\phi}{\partial y}\right)^2 - 4G\alpha\phi\right]\mathrm{d}x\mathrm{d}y\right\} \quad (6.7.6)$$

or

$$0 = \frac{l}{2G}\iint[2\phi_x\delta\phi_x + 2\phi_y\delta\phi_y - 4G\alpha\delta\phi]\,\mathrm{d}x\mathrm{d}y\ . \quad (6.7.7)$$

Integrating the first two terms and observing that $\phi = 0$ on the boundary, we have

$$0 = \iint[\nabla^2\phi + 2G\alpha]\,\delta\phi\mathrm{d}x\mathrm{d}y\ . \quad (6.7.8)$$

But $\delta\phi$ is arbitrary over the region so that

$$\nabla^2\phi + 2G\alpha = 0 \quad \text{in the region}\ . \quad (6.7.9)$$

This is the Euler-Lagrange equation for the problem.

Since the formulation is based on complementary energy, we expect that (6.7.9) is an equation of compatibility: we readily demonstrate this as follows. For the axial displacement $w(x, y)$ to be single valued, we require

$$\oint_{\mathscr{C}} \mathrm{d}w = 0 \quad (6.7.10)$$

where $\mathscr{C}$ is *any* closed contour lying in a cross section. But from the calculus,

$$\mathrm{d}w = \frac{\partial w}{\partial x}\mathrm{d}x + \frac{\partial w}{\partial y}\mathrm{d}y\ . \quad (6.7.11)$$

Observing that eventually, we wish to work in terms of ϕ, we write

$$\frac{\partial w}{\partial x} = \frac{\partial w}{\partial x} + \frac{\partial u}{\partial z} - \frac{\partial u}{\partial z} = \gamma_{zx} + \alpha y = \frac{1}{G}\frac{\partial\phi}{\partial y} + \alpha y$$

$$\frac{\partial w}{\partial y} = \frac{\partial w}{\partial y} + \frac{\partial v}{\partial z} - \frac{\partial v}{\partial z} = \gamma_{zy} - \alpha x = -\frac{1}{G}\frac{\partial\phi}{\partial x} - \alpha x\ . \quad (6.7.12)$$

Furthermore, from Fig. 6.4.4,

$$\mathrm{d}x = -n_y\mathrm{d}s\,, \qquad \mathrm{d}y = n_x\mathrm{d}s\ . \quad (6.7.13)$$

Substituting from equations (6.7.11), (6.7.12) and (6.7.13) in (6.7.10),

$$\oint_{\mathscr{C}}\mathrm{d}w = -\frac{1}{G}\oint_{\mathscr{C}}\left\{\frac{\partial\phi}{\partial y}n_y + \frac{\partial\phi}{\partial x}n_x\right\}\mathrm{d}s - \alpha\oint_{\mathscr{C}}\left\{yn_y + xn_x\right\}\mathrm{d}s = 0\ .$$

These line integrals may be transformed by Gauss's theorem† (Appendix) giving

$$-\frac{1}{G}\left\{\iint\left[\frac{\partial^2\phi}{\partial x^2} + \frac{\partial^2\phi}{\partial y^2} + 2G\alpha\right]\mathrm{d}x\mathrm{d}y\right\} = 0\ . \quad (6.7.14)$$

† Recall that we are only concerned with singly connected sections here.

But the region enclosed by $\mathscr{C}$ is arbitrary, so for equation (6.7.14) to be true it is required that the integrand be zero. That is, equation (6.7.9) must be satisfied and we see that it does, indeed, imply compatibility.

In this reasoning, a total angle of twist αl was specified, so that a solution of equation (6.7.9) corresponds to this. We are now in a position to calculate the twisting moment called for to produce this angle of twist.

Now,
$$M_t = \iint [\tau_{zy}x - \tau_{zx}y]\,\mathrm{d}x\mathrm{d}y \ .$$

Then substituting from (6.7.1), integrating by parts and observing that $\phi = 0$ on the section boundary

$$M_t = 2\iint \phi \,\mathrm{d}x\mathrm{d}y \ . \qquad (6.7.15)$$

6.8 The Hellinger-Reissner and other derived variational principles

In Chapters 1 and 4 we saw that the method of Lagrange multipliers provided a very convenient means of explicitly taking account of constraint conditions. We can exploit this procedure to deduce new variational principles: one of these, due to Hellinger and Reissner, is becoming increasingly important in structural mechanics especially in connexion with the finite element treatment of plate and shell problems.

In the earlier sections of this chapter, we developed two *minimum* principles. However, if we are to modify the functionals involved by means of Lagrange multipliers, it is important to recall (Chapter 4) that we can deduce conditions for *stationary* values of these modified functionals but that we cannot assert that they are either a maximum or a minimum.

Our approach here will be successively to modify the potential energy functional so that it will be helpful to begin by summarising the essentials of the principle of minimum total potential energy.

(a) Assuming elastic behaviour, we imply that internal forces have a potential so that we may write a stress-strain law as

$$\sigma_{ij} = \frac{\partial U_0(\epsilon_{ij})}{\partial \epsilon_{ij}} \ . \qquad (6.8.1)$$

(b) In testing the relative merits of various displacement patterns, with a view to identifying the true one corresponding to the actual loads, we are only concerned with those which imply compatibility with the internal and external constraints. That is, throughour the interior of the solid, the strain and displacement components are related by

$$\epsilon_{ij} = \tfrac{1}{2}(u_{i,j} + u_{j,i}) \ . \qquad (6.8.2)$$

On the surface, the displacements must be compatible with the external constraints so that if the displacement components are prescribed functions $u_i{}^*$ on a part S_u of the surface,

$$u_i = u_i^* \text{ on } S_u \,. \tag{6.8.3}$$

(c) Under the above circumstances, the true deformed configuration of the solid, which corresponds to equilibrium stresses in the interior and on S_T, is identified by a minimum value for

Standard min PE representation

$$V = \int_{\text{vol}} U_0 \,\text{d vol} - \int_{\text{vol}} F_i u_i \text{d vol} - \int_{S_T} u_i \overset{n}{T}_i{}^* \,\text{d}S \,. \tag{6.8.4}$$

In applications, U_0 must be expressed in terms of the displacements by means of (6.8.2). However, in view of equation (6.8.1), it might be considered more rational to have U_0 as a function of the strains and to introduce the nine constraints (6.8.2) by means of Lagrange multipliers $\lambda_{ij}(x_i)$, defined throughout the interior of the solid. Furthermore, constraints (6.8.3) may be introduced by means of Lagrange multipliers $\mu_i(x_i)$ defined over S_u. Then, a modified potential energy V_1 may be defined by

Constraints imposed by use of Lagrange multipliers.

$$V_1 = \int_{\text{vol}} \{U_0(\epsilon_{ij}) - F_i u_i + \lambda_{ij}[\epsilon_{ij} - \tfrac{1}{2}(u_{i,j} + u_{j,i})]\} \,\text{d vol}$$
$$- \int_{S_T} u_i \overset{n}{T}_i{}^* \text{d}S + \int_{S_u} \mu_i (u_i - u_i^*) \text{d}S \,. \tag{6.8.5}$$

We observe that since the ϵ_{ij} are symmetric in i and j, so also must be the λ_{ij}.

Following Lagrange's method, the eighteen functions ϵ_{ij}, u_i, λ_{ij}, μ_i are to be treated as independent in functional V_1 and capable of receiving independent variations $\delta\epsilon_{ij} \ldots\ldots \delta\mu_i$. Then,

$$\delta V_1 = \int_{\text{vol}} \{\frac{\partial U_0}{\partial \epsilon_{ij}} \delta\epsilon_{ij} - F_i \delta u_i + \delta\lambda_{ij}[\epsilon_{ij} - \tfrac{1}{2}(u_{i,j} + u_{j,i})]$$
$$+ \lambda_{ij}[\delta\epsilon_{ij} - \tfrac{1}{2}(\delta u_{i,j} + \delta u_{j,i})]\} \,\text{d vol} - \int_{S_T} \overset{n}{T}_i{}^* \delta u_i \text{d}S$$
$$+ \int_{S_u} \{\mu_i(\delta u_i - \delta u_i^*) + (u_i - u_i^*)\delta\mu_i\} \text{d}S \,.$$

Now $\quad -\int \lambda_{ij} \tfrac{1}{2}(\delta u_{i,j} + \delta u_{j,i}) \text{d vol} = -\int \lambda_{ij} \delta u_{i,j} \text{d vol}$

$$= -\int \{(\lambda_{ij}\delta u_i)_{,j} - \lambda_{ij,j}\delta u_i\} \,\text{d vol} \,.$$

Also, using Gauss's theorem

$$\int_{\text{vol}} (\lambda_{ij}\delta u_i)_{,j} \text{d vol} = \int_S \lambda_{ij} n_j \delta u_i \,\text{d}S$$

where, $S = S_u + S_T$. Then,

$$\delta V_1 = \int_{\text{vol}} \left\{ \left(\frac{\partial U_0}{\partial \epsilon_{ij}} + \lambda_{ij} \right) \delta\epsilon_{ij} + (\lambda_{ij,j} - F_i)\delta u_i + [\epsilon_{ij} - \tfrac{1}{2}(u_{i,j} + u_{j,i})]\,\delta\lambda_{ij} \right\} \text{d vol} - \int_{S_T} (\lambda_{ij}n_j + \overset{n}{T_i}{}^*)\delta u_i \text{d}S$$
$$+ \int_{S_u} (u_i - u_i^*)\delta\mu_i \text{d}S + \int_{S_u} (\mu_i - \lambda_{ij}n_j)\delta u_i \,\text{d}S\,. \qquad (6.8.6)$$

Here, $\delta u_i^* = 0$ since u_i^* is prescribed.

For $\delta V_1 = 0$ with $\delta\epsilon_{ij}$, δu_i, $\delta\lambda_{ij}$, $\delta\mu_i$ all arbitrary, equation (6.8.6) implies

$$\left.\begin{aligned}
\frac{\partial U_0}{\partial \epsilon_{ij}} &= -\lambda_{ij} && \text{throughout the interior}, && \text{(a)}\\
\frac{\partial \lambda_{ij}}{\partial x_j} - F_i &= 0 && \text{,,} \quad \text{,,} \quad \text{,,} \quad , && \text{(b)}\\
\epsilon_{ij} &= \frac{1}{2}\left(\frac{\partial u_i}{\partial x_j} + \frac{\partial u_j}{\partial x_i}\right) && \text{,,} \quad \text{,,} \quad \text{,,} \quad , && \text{(c)}\\
\lambda_{ij}n_j &= \overset{n}{T_i}{}^* && \text{on } S_T\,, && \text{(d)}\\
\lambda_{ij}n_j &= \mu_i && \text{on } S_u\,, && \text{(e)}\\
u_i &= u_i^* && \text{on } S_u\,. && \text{(f)}
\end{aligned}\right\} \qquad (6.8.7)$$

Equations (6.8.7a) identify the λ_{ij}'s with the stress components; in fact $\lambda_{ij} = -\sigma_{ij}$. Referring to equations (6.8.7e) we see that the μ_i are identifiable with the tractions ('reactions') $\overset{n}{T_i}$ on S_u.

Thus we see that $\delta V_1 = 0$ implies a complete specification of the elastic small displacement theory of stress analysis. It is interesting to observe that if equations (6.8.7c) and (6.8.7f) are accepted implicitly, $\delta V_1 = 0$ reduces to $\delta V = 0$, which is the principle of stationary total potential energy.

Recognising now that $\lambda_{ij} = -\sigma_{ij}$, and $\mu_i = -\overset{n}{T_i} = -\sigma_{ij}n_j$, we can write equation (6.8.5) in a slightly different form

$$V_2 = \int_{\text{vol}} \{U_0(\epsilon_{ij}) - \sigma_{ij}\epsilon_{ij} + \tfrac{1}{2}\sigma_{ij}(u_{i,j} + u_{j,i}) - F_i u_i\}\text{d vol}$$
$$- \int_{S_T} u_i \overset{n}{T_i}{}^* \text{d}S - \int_{S_u} (u_i - u_i^*)\sigma_{ij}n_j \text{d}S\,. \qquad (6.8.8)$$

In effect, we have *accepted* equation (6.8.1) as a constitutive equation expressing the stresses in terms of the strains. Now $U_0(\epsilon_{ij})$ is positive definite so that $\det(\partial^2 U_0/\partial\epsilon_{ij}\partial\epsilon_{mn})$ does not vanish; an inverse relation $\epsilon_{ij} = f_{ij}(\sigma_{ij})$ thus exists so that the function

$$U_0^* = \sigma_{ij}\epsilon_{ij} - U_0(\epsilon_{ij}) \qquad (6.8.9)$$

can be written in terms of the σ_{ij} alone and

$$\delta U_0{}^* = \frac{\partial U_0{}^*(\sigma_{ij})}{\partial \sigma_{ij}} \delta \sigma_{ij} . \tag{6.8.10}$$

But from equaton (6.8.9)

$$\delta U_0{}^* = \sigma_{ij}\delta\epsilon_{ij} + \epsilon_{ij}\delta\sigma_{ij} - \frac{\partial U_0}{\partial \epsilon_{ij}} \delta\epsilon_{ij} = \epsilon_{ij}\delta\sigma_{ij} . \tag{6.8.11}$$

Comparing equations (6.8.10) and (6.8.11), we see that

$$\epsilon_{ij} = \frac{\partial U_0{}^*(\sigma_{ij})}{\partial \sigma_{ij}} \tag{6.8.12}$$

so that $U_0{}^*$, as defined in equation (6.8.9), is nothing more than the complementary strain energy density function. Equation (6.8.8) can thus be written as

$$V_R = \int_{\text{vol}} \{\tfrac{1}{2}\sigma_{ij}(u_{i,j} + u_{j,i}) - F_i u_i - U_0{}^*(\sigma_{ij})\}\text{d vol}$$
$$- \int_{S_T} u_i \overset{n}{T}_i{}^* \text{d}S - \int_{S_u} (u_i - u_i{}^*)\sigma_{ij} n_j \text{d}S . \tag{6.8.13}$$

This functional, in which the $\int_{S_u}$ term is sometimes left out by accepting $u_i = u_i{}^*$ on S_u, is the Hellinger-Reissner functional: the u_i and σ_{ij} are treated as the independent quantities capable of taking independent variations δu_i and $\delta\sigma_{ij}$.

The condition $\delta V_R = 0$ gives

$$0 = \int_{\text{vol}} \{\tfrac{1}{2}\sigma_{ij}(\delta u_{i,j} + \delta u_{j,i}) + \tfrac{1}{2}(u_{i,j} + u_{j,i})\delta\sigma_{ij} - F_i\delta u_i - \frac{\partial U_0{}^*}{\partial \sigma_{ij}} \delta\sigma_{ij}\}\text{d vol}$$
$$- \int_{S_T} \overset{n}{T}_i{}^* \delta u_i \text{d}S - \int_{S_u} \{\sigma_{ij} n_j \delta u_i + (u_i - u_i{}^*)\delta\sigma_{ij} n_j\}\text{d}S . \tag{6.8.14}$$

Employing the manipulative device introduced above,

$$\int_{\text{vol}} \tfrac{1}{2}\sigma_{ij}(\delta u_{i,j} + \delta u_{j,i})\text{d vol} = -\int_{\text{vol}} \sigma_{ij,j}\delta u_i \text{d vol} + \int_S \sigma_{ij} n_j \delta u_i \text{d}S ,$$

and noting that $S = S_u + S_T$, we have

$$0 = -\int_{\text{vol}} \left\{(\sigma_{ij,j} + F_i)\delta u_i - \left[\frac{\partial U_0{}^*}{\partial \sigma_{ij}} - \tfrac{1}{2}(u_{i,j} + u_{j,i})\right]\delta\sigma_{ij}\right\}\text{d vol}$$
$$- \int_{S_T} (\overset{n}{T}_i{}^* - \sigma_{ij} n_j)\delta u_i \text{d}S - \int_{S_u} (u_i - u_i{}^*)\delta\sigma_{ij} n_j \text{d}S . \tag{6.8.15}$$

With the δu_i and $\delta \sigma_{ij}$ arbitrary the conditions for stationary V_R are therefore

$$\left.\begin{aligned}
\frac{\partial \sigma_{ij}}{\partial x_j} + F_i &= 0 && \text{throughout the interior}, && \text{(a)}\\
\frac{\partial U_0^*}{\partial \sigma_{ij}} &= \frac{1}{2}\left(\frac{\partial u_i}{\partial x_j} + \frac{\partial u_j}{\partial x_i}\right) && \text{"} \quad \text{"} \quad \text{"} , && \text{(b)}\\
\sigma_{ij} n_j &= \overset{n}{T_i}{}^* && \text{on } S_T , && \text{(c)}\\
u_i &= u_i^* && \text{on } S_u . && \text{(d)}
\end{aligned}\right\} \quad (6.8.16)$$

Thus, accepting a stress-strain law for the material in the form (6.8.12), a stationary value for the Hellinger-Reissner functional yields the equations of equilibrium (6.8.16a) and the equations of compatibility (6.8.16b) in the interior, together with the conditions of equilibrium (6.8.16c) and the conditions of compatibility with the external constraints (6.8.16d), on the appropriate portions S_T and S_u of the surface of the solid.

We notice that, in contrast to the principles of potential energy or complementary energy, the Hellinger-Reissner principle does not place any preferential emphasis on establishing the conditions of equilibrium or compatibility: *both* conditions arise naturally.

In view of the symmetry of σ_{ij} and the repeated subscripts, the first term in V_R, equation (6.8.13), can be re-written as

$$\tfrac{1}{2}\sigma_{ij}(u_{i,j} + u_{j,i}) = \sigma_{ij} u_{i,j} = (\sigma_{ij} u_i)_{,j} - u_i \sigma_{ij,j} .$$

Transforming the integral by Gauss's theorem and noting that $S = S_u + S_T$, V_R can be written as

$$V_{R1} = -\int_{\text{vol}} \{(\sigma_{ij,j} + F_i)u_i + U_0^*(\sigma_{ij})\} \, \mathrm{d}\text{vol}$$

$$- \int_{S_T} (\overset{n}{T_i}{}^* - \sigma_{ij} n_j) u_i \, \mathrm{d}S + \int_{S_u} u_i^* \sigma_{ij} n_j \, \mathrm{d}S . \quad (6.8.17)$$

If we now consider only equilibriant stress systems, $\sigma_{ij,j} + F_i = 0$ in the interior, $\sigma_{ij} n_j = \overset{n}{T_i}{}^*$ on S_T and $\sigma_{ij} n_j = \overset{n}{T_i}$ on S_u, and V_{R1} becomes

$$-V_{R2} = \int_{\text{vol}} U_0^*(\sigma_{ij}) \, \mathrm{d}\text{vol} - \int_{S_u} u_i^* \overset{n}{T_i} \, \mathrm{d}S = V^* . \quad (6.8.18)$$

That is, we have arrived at the complementary energy functional (6.5.16), and variations in $-V_{R2}$ can only arise from variations in equilibriant stresses.

The above developments show that if geometric constraints are introduced by means of Lagrange multipliers, the principle of virtual work can be regarded as *the* basic principle of small displacement solid mechanics. Not only can we derive the complementary energy principle from it (and not regard it as a separate principle), but other variational principles, such as that of Hellinger and Reissner can also be deduced. A little reflection will show that we may, alternatively, accept the complementary energy principle and carry out the above developments in reverse: indeed, this was the route chosen by Reissner in his original work in 1950.

It is reasonable to ask if these discussions have any significance for practical numerical work in design: they do indeed. The so-called direct methods of the variational calculus were discussed in Chapter 5 and we shall apply them further in Chapters 7 and 8. We shall see that, if we have obtained an approximate solution to the deformed configuration, by the potential energy principle say, the calculation of stresses involves differentiation of these approximating functions. The accuracy of the stresses will thus be inferior to that of the displacements. On the other hand one might hope that the use of Reissner's principle, with its balanced emphasis on equilibrium and compatibility, would lead to comparable accuracy in stresses and displacements.

EXAMPLE 6.8.1

Apply the Hellinger-Reissner principle to formulate the problem of bending of the cantilever beam of Fig. 6.8.1 Hooke's law material.

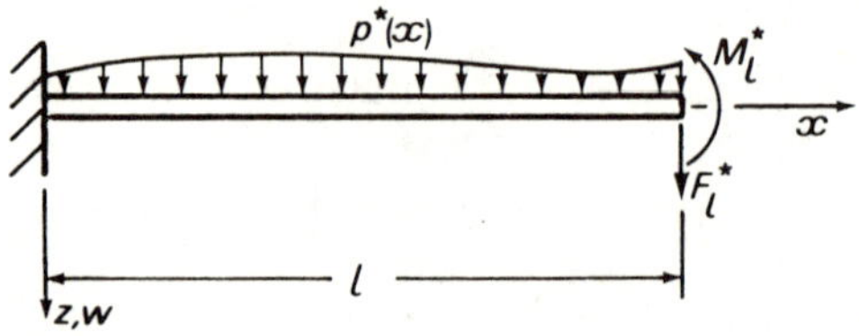

Fig. 6.8.1

Employing the usual assumptions of beam theory, see Section 6.3, equation (6.8.13) becomes

$$V_R = \int_0^l \{\int_A [\sigma_x \epsilon_x - U_0^*(\sigma_x)]\,\mathrm{d}A\}\,\mathrm{d}x - \int_0^l p^* w \mathrm{d}x - F_l^* w_l + M_l^* w_l' \quad . \qquad \text{(a)}$$

But, $\epsilon_x = -zw''$, $U_0^*(\sigma_x) = \dfrac{{\sigma_x}^2}{2E}$, $\sigma_x = \dfrac{Mz}{I}$, $M = \int_A \sigma_x z \mathrm{d}A$.

Integrating over the section A yields

$$V_R = -\int_0^l \left[\frac{M^2}{2EI} + Mw''\right]\mathrm{d}x - \int_0^l p^* w \mathrm{d}x - F_l^* w_l + M_l^* w_l' \ . \qquad \text{(b)}$$

Then, taking variations and equating to zero

$$\delta V_R = 0 = -\int_0^l \left[\left(\frac{M}{EI} + w''\right)\delta M + M\delta w''\right]\mathrm{d}x - \int_0^l p^* \delta w \mathrm{d}x$$
$$- F_l^* \delta w_l + M_l^* \delta w_l' \ . \qquad \text{(c)}$$

Integrating by parts

$$-\int_0^l M\delta w''\mathrm{d}x = -\{|M\delta w'|_0^l - \int_0^l M'\delta w'\mathrm{d}x\}$$

$$= -\{|M\delta w'|_0^l - |M'\delta w|_0^l + \int_0^l M''\delta w\mathrm{d}x\}. \qquad \text{(d)}$$

Then, $$\delta V_R = 0 = -\int_0^l \left[\left(\frac{M}{EI} + w''\right)\delta M + (M'' + p^*)\delta w\right]\mathrm{d}x$$

$$+ (M_l' - F_l^*)\delta w_l + (M_l - M_l^*)\delta w_l' \qquad \text{(d)}$$

where $\delta w_0 = \delta w_0' = 0$ since deflexion and slope are prescribed (zero) at a clamped end.

But δM and δw are arbitrary along the beam as are δw_l and $\delta w_l'$ at $x = l$. Equation (d) thus implies

$$\frac{\mathrm{d}^2M}{\mathrm{d}x^2} = -p \qquad \text{along beam} \qquad \text{(e)}$$

$$M = -EI\frac{\mathrm{d}^2w}{\mathrm{d}x^2} \qquad \text{,,} \quad \text{,,} \qquad \text{(f)}$$

$$\frac{\mathrm{d}M}{\mathrm{d}x} = F^* \qquad \text{at } x = l \qquad \text{(g)}$$

$$M = M^* \qquad \text{at } x = l\,. \qquad \text{(h)}$$

These results together with

$$w = \frac{\mathrm{d}w}{\mathrm{d}x} = 0 \text{ at } x = 0 \qquad \text{(j)}$$

thus define the cantilever beam boundary value problem.

Problem 6.8

6.8.1 Accepting the displacement pattern suggested in Problem 6.4.1 and assuming the following stress distribution, deduce Reissner's functional for plate bending incorporating shear effects.

$$\sigma_x = \frac{12M_x}{h^3}z\,, \quad \sigma_y = \frac{12M_y}{h^3}z\,, \quad \tau_{xy} = -\frac{12M_{xy}}{h^3}z\,,$$

$$\tau_{zx} = \frac{3Q_x}{2h}\left[1 - \left(\frac{2z}{h}\right)^2\right], \quad \tau_{zy} = \frac{3Q_y}{2h}\left[1 - \left(\frac{2z}{h}\right)^2\right],$$

$$\sigma_z = \text{negligible.}$$

Ans: $$V_R = \iint \{-M_x \frac{\partial \psi_x}{\partial x} - M_y \frac{\partial \psi_y}{\partial y} + M_{xy}\left(\frac{\partial \psi_x}{\partial y} + \frac{\partial \psi_y}{\partial x}\right) + Q_x\left(\frac{\partial w}{\partial x} - \psi_x\right)$$
$$+ Q_y\left(\frac{\partial w}{\partial y} - \psi_y\right) - \frac{12}{2Eh^3}[M_x{}^2 + M_y{}^2 - 2\nu M_x M_y]$$
$$- \frac{12}{2Gh^3} M_{xy} - \frac{3}{5Gh}(Q_x{}^2 + Q_y{}^2) - pw\}\mathrm{d}x\mathrm{d}y\,.$$

6.8.2 Treating w, ψ_x, ψ_y, M_x, M_y, M_{xy}, Q_x and Q_y as free variables, show that the Euler-Lagrange equations corresponding to $\delta V_R = 0$ in problem 6.8.1 are

$$\frac{\partial Q_x}{\partial x} + \frac{\partial Q_y}{\partial y} + p = 0\,, \qquad \frac{\partial M_x}{\partial x} - \frac{\partial M_{xy}}{\partial y} - Q_x = 0\,,$$
$$\frac{\partial M_y}{\partial y} - \frac{\partial M_{xy}}{\partial x} - Q_y = 0\,, \qquad \frac{\partial w}{\partial x} - \psi_x - \frac{6}{5Gh} Q_x = 0\,,$$
$$\frac{\partial w}{\partial y} - \psi_y - \frac{6}{5Gh} Q_y = 0\,.$$

6.9 Further Reading

Argyris, J.H. *and* Kelsey, S. — *Energy Theorems and Structural Analysis,* Butterworth (1960).

Biezeno, C.B. *and* Grammel, R. — *Engineering Dynamics,* Blackie (1955).

Fung, Y.C. — *Foundations of Solid Mechanics,* Prentice Hall, (1965).

Kelvin, W.T. *and* Tait, P.G. — *Elements of Natural Philosophy,* 2nd Edition, Vol 1, Cambridge University Press (1885).

Lanczos, C. — *The Variational Principles of Mechanics,* 4th Edition, University of Toronto Press (1970).

Langhaar, H.L. — *Energy Methods in Applied Mechanics,* John Wiley (1962).

Mansfield, E.H. — *The Bending and stretching of Plates,* Pergamon (1964).

Reissner, E. — *On Variational Principles in Elasticity,* Proceedings of Symposia in Applied Mathematics, Vol 8 pp 1–6, McGraw Hill (1958).

Sokolnikoff, I.S. — *Mathematical Theory of Elasticity,* McGraw Hill (1956).

Southwell, R.V. — On the analogues relating the flexure and extension of flat plates; *Quart. Jnl. Mech. & Appl. Math.* Vol **3** pp 257–270 (1950).

Timoshenko, S. *and* Goodier, J.N. — *Theory of Elasticity,* 3rd Edition, McGraw Hill (1970).

Timoshenko, S. *and* Woinowsky-Kreiger, S. — *Theory of Plates and Shells,* 2nd Edition, McGraw Hill, (1959).

Washizu, K. — *Variational Methods in Elasticity and Plasticity,* 2nd Edition, Pergamon Press (1973)

Chapter 7

Applications of the Direct Methods of the Calculus of Variations

7.1 Introduction

In attempting to provide a mathematical description of all aspects of the mechanical behaviour of a real engineering system one meets insuperable difficulties so that a central activity in design work is the idealisation of a real system into something which can be handled. Increased precision in design calls for evermore sophisticated idealisation, but unfortunately this entails increasing difficulties with the mathematics. Thus, as we have seen in Chapter 3, to treat all stressing problems by means of the formal methods of continuum mechanics is an impossible task. The purpose of the so-called 'approximate' theories is to produce a mathematical description of a situation which is adequate for the purpose at hand. One can then (hopefully) cope with a number of situations of practical significance without too much mathematical difficulty. As design requirements become more demanding, problem areas arise in which the traditional methods of mechanics of materials and structural analysis prove to be inadequate; and we can no longer side step the difficulties of continuum mechanics, but have to deal with them directly. One method is by the use of one of the direct methods of the variational calculus described in Chapter 5. Here, we shall work through a number of examples to illustrate the 'classical' application of these methods. These techniques are still of considerable importance in their own right, and the examples also serve as valuable preparation for a study of the finite element method in Chapter 8. We shall see that the net effect of applying these processes is generally to transform the continuum system into a substitute finite degree of freedom, one whose characteristics approximate those of the original. The basis for the transformation is a variational one and when, say, the potential energy principle is appropriate, generalised forces are generated to correspond, in a rational manner to generalised coordinates: situations analogous to those treated in Chapter 1 are thus created.

7.2 Beam problems treated by the Rayleigh-Ritz method

Beam problems provide a very good starting point from which to gain some experience of the direct methods of the variational calculus. We saw in Chapter 5 that the Rayleigh-Ritz procedure can yield a sequence of approximate solutions which converge in a mean square sense when complete sets of functions are used, but that if they are chosen judiciously very good accuracy can be attained with relatively few functions. When the principle of minimum potential energy governs the formulation, trial functions are required to satisfy only the forced (or geometric) boundary conditions. However, the efficiency of the Rayleigh-Ritz process is considerably enhanced if the natural (or mechanical) boundary conditions are also satisfied; this is an important consideration when only few functions are to be used.

EXAMPLE 7.2.1

Determine the deflexion curve for the symmetrically loaded simply supported uniform beam of Fig. 7.2.1.

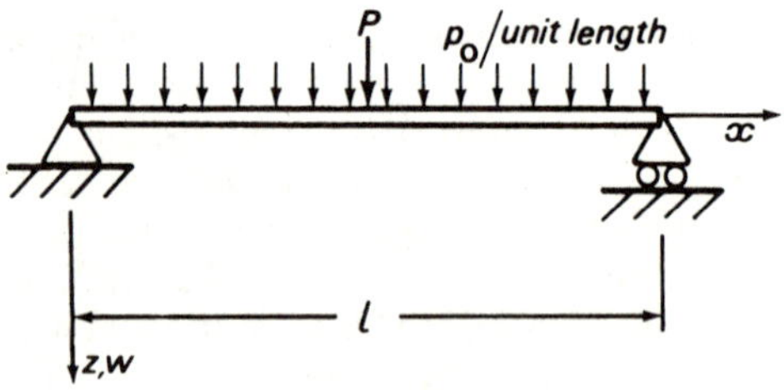

Fig. 7.2.1

Using the principle of minimum potential energy, the basic unkown is the deflexion curve $w(x)$. Then, the total potential is given by

$$V = \int_0^l \frac{EI}{2}\left(\frac{d^2w}{dx^2}\right)^2 dx - Pw(l/2) - \int_0^l p_0 w dx \; , \tag{a}$$

and for equilibrium,

$$\delta V = 0 \; . \tag{b}$$

To solve by the Rayleigh-Ritz method, we assume $w(x)$ to be approximated by

$$w_n(x) = \sum_{i=1}^{n} c_i \phi_i(x) \tag{c}$$

so that, in effect, we are replacing the beam (which has an infinite number of degrees of freedom), by a system having a finite number of degrees of freedom. The c's are thus generalised coordinates, being the amplitudes of the prescribed ϕ's contributing to the total deflexion curve.

As a first approximation, we choose a one-term series so that $n = 1$. Being guided by the need for symmetry about midspan and for $w_n = 0$ at $x = 0$ and $x = l$, we choose

$$w_1 = c_1 \sin \frac{\pi x}{l} \; . \tag{d}$$

Then,

$$\frac{dw_1}{dx} = c_1 \frac{\pi}{l} \cos \frac{\pi x}{l} \; , \quad \frac{d^2 w_1}{dx^2} = -c_1 \frac{\pi^2}{l^2} \sin \frac{\pi x}{l} \; . \tag{e}$$

We notice that with this assumed shape for the deflexion curve, $M = EIw'' = 0$ at both ends of the beam; with both the geometric and natural boundary conditions satisfied, equation (d) is thus a reasonable choice. Substituting in equation (a)

$$V = \int_0^l \frac{EI}{2} c_1^2 \frac{\pi^4}{l^4}\left(\sin \frac{\pi x}{l}\right)^2 \mathrm{d}x - Pc_1 - \int_0^l p_0 c_1 \sin \frac{\pi x}{l} \mathrm{d}x \,. \tag{f}$$

Integrating

$$V = \frac{\pi^4 EI}{4l^3} c_1^2 - Pc_1 - \frac{2p_0 l}{\pi} c_1 \,. \tag{g}$$

Here, c_1, is the generalised coordinate and for equation (b) to be satisfied, we need $\partial V/\partial c_1 = 0$ so that

$$c_1 = \frac{2l^3}{\pi^4 EI}\left(P + \frac{2p_0 l}{\pi}\right) . \tag{h}$$

At mid-span, this gives

$$w = 0.9855 \frac{Pl^3}{48EI} + 1.0039 \frac{5p_0 l^4}{384EI} \tag{j}$$

which compares extremely well with the known result

$$w = \frac{Pl^3}{48EI} + \frac{5p_0 l^4}{384EI} \,. \tag{k}$$

We record the approximate and 'exact' values of w for $x = l/2$ and $x = l/4$ in Table 7.2.1.

Table 7.2.1

	$x = l/4$	$x = l/2$
w_{ex}	$0.01432\frac{Pl^3}{EI} + 0.00928\frac{p_0 l^4}{EI}$	$0.02083\frac{Pl^3}{EI} + 0.01302\frac{p_0 l^4}{EI}$
w_{app}	$0.01456\frac{Pl^3}{EI} + 0.00924\frac{p_0 l^4}{EI}$	$0.02059\frac{Pl^3}{EI} + 0.01307\frac{p_0 l^4}{EI}$

We have computed these two sets of figures for the following reason. Assuming the form equation (d) for the deflexion curve implies the imposition of constraints. The beam is thus stiffer than it should be, so that we would expect the deflexions to be smaller than the 'exact' ones. Table 7.2.1 shows that this is not literally the case everywhere: the assumed shape (d) implies that the beam is stiffer than it should be in some average way so that the deflexions are too small in some average way.

The agreement between the exact and approximate solutions is remarkably good, but if we compute the bending moment (and hence bending stress) distribution from $M = -EIw''$, the situation is not so happy. Table 7.2.2 shows a comparison of the exact and approximate values of the bending moments.

Table 7.2.2

	$x = l/4$	$x = l/2$
M_{ex}	$0.125Pl + 0.09375p_0l^2$	$0.25Pl + 0.125p_0l^2$
M_{app}	$0.14329Pl + 0.09122p_0l^2$	$0.20264Pl + 0.12901p_0l^2$

Examining the contribution from the distributed load alone, the agreement between exact and approximate bending moments is still quite good. For the point load, however, there is an error of about 19% in the midspan moment. We can understand this state of affairs since, from statics, the exact bending moment diagram for the point load alone is curve (i) in Fig. 7.2.2 whilst (ii) shows our approximation.

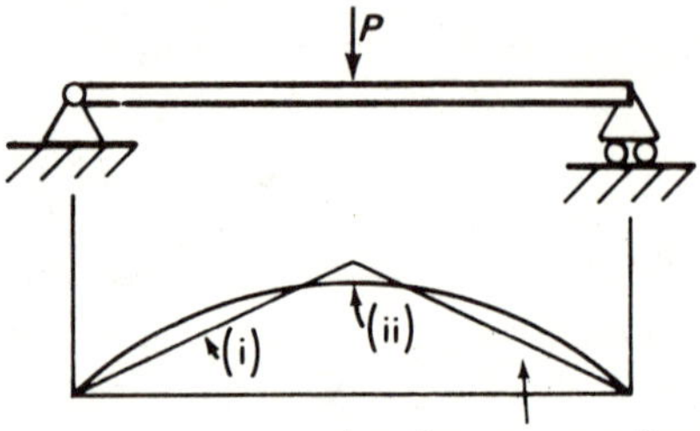

Fig. 7.2.2

For the distributed load contribution, the reader can demonstrate the similarity in shape between the exact and approximate bending moment diagrams.

The above example illustrates a characteristic feature of the use of the principle of minimum potential energy in conjunction with the Rayleigh-Ritz method: if stresses are required with good accuracy, displacements must be determined to a very high accuracy. Stresses are obtained from displacements by differentiation and this is essentially an inaccurate process when applied to approximating functions.

To improve accuracy, we can take more terms in the series. Now it can be shown that the set $\sin(\pi x/l)$, $\sin(2\pi x/l)$, ... is complete in the region $x = 0$, $x = l$ so that we would expect solutions based on

$$w = \sum_{1}^{n} c_m \sin \frac{m\pi x}{l} \tag{7.2.1}$$

to converge to the exact answer as n is increased indefinitely. In the next example we employ this series to treat a simply supported beam loaded as above but, in addition, we will impose an end thrust too. Solving by the differential equation approach this makes quite a difference to the treatment of the problem: a formal knowledge of differential equations is necessary, whereas with no end thrust only successive integration is involved. As we shall see treating the problem by the potential energy method, the procedure is exactly as before: we merely have to incorporate the potential energy of the end thrust.

EXAMPLE 7.2.2

Determine the deflexion curve for the beam-column of Fig. 7.2.3.

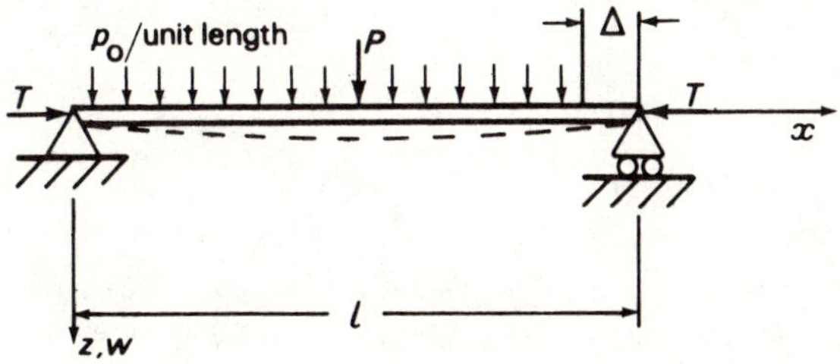

Fig. 7.2.3

The constraints chosen are such that the left hand pinned support does not move whilst the right hand one may move axially. Now this axial travel of T arises from straight forward compression and bending of the beam. We are only interested in bending deflexions here so that we only have to concern ourselves with the axial travel Δ due to the bow in the beam.

We first obtain an expression for Δ in terms of w. Since we are ignoring the effects of direct compression, the curved length of the beam is still l and

$$l = \int_{x=0}^{x=(l-\Delta)} \mathrm{d}s = \int_{x=0}^{x=(l-\Delta)} \left[1 + \left(\frac{\mathrm{d}w}{\mathrm{d}x}\right)^2\right]^{1/2} \mathrm{d}x \,. \tag{a}$$

We suppose that slopes and deflexions are small so that $\mathrm{d}w/\mathrm{d}x$ is small compared with unity. Then, expanding the square bracket term by the binomial theorem, we can write (approximately)

$$l = \int_0^{l-\Delta} \left[1 + \frac{1}{2}\left(\frac{\mathrm{d}w}{\mathrm{d}x}\right)^2\right] \mathrm{d}x = (l - \Delta) + \frac{1}{2}\int_0^{l-\Delta} \left(\frac{\mathrm{d}w}{\mathrm{d}x}\right)^2 \mathrm{d}x \,. \tag{b}$$

Observing that Δ is small compared with l, equation (b) becomes

$$\Delta = \frac{1}{2}\int_0^{l} \left(\frac{\mathrm{d}w}{\mathrm{d}x}\right)^2 \mathrm{d}x \,. \tag{c}$$

The total potential energy for the system can thus be written

$$V = \int_0^l \frac{EI}{2}\left(\frac{d^2w}{dx^2}\right)^2 dx - \frac{T}{2}\int_0^l \left(\frac{dw}{dx}\right)^2 dx - \int_0^l pw\,dx - Pw(l/2)\,. \qquad \text{(d)}$$

To proceed, we assume
$$w = \sum_1^n c_m \sin\frac{m\pi x}{l} \qquad \text{(e)}$$

so that
$$\frac{dw}{dx} = \sum_1^n c_m\left(\frac{m\pi}{l}\right)\cos\frac{m\pi x}{l},\quad \frac{d^2w}{dx^2} = -\sum_1^n c_m\left(\frac{m\pi}{l}\right)^2 \sin\frac{m\pi x}{l}\,. \qquad \text{(f)}$$

Then,

$$V = \int_0^l \frac{EI}{2}\left(\sum_1^n c_m\left(\frac{m\pi}{l}\right)^2 \sin\frac{m\pi x}{l}\right)^2 dx - \frac{T}{2}\int_0^l\left(\sum_1^n c_m\left(\frac{m\pi}{l}\right)\cos\frac{m\pi x}{l}\right)^2 dx$$

$$-\int_0^l p_0 \sum_1^n c_m \sin\frac{m\pi x}{l}\,dx - P\sum_1^n c_m \sin\frac{m\pi}{l}\,. \qquad \text{(g)}$$

For equilibrium, $\delta V = 0$ and since after the integrations are carried out, we see that $V = V(c_1, c_2 \ldots\ldots c_n)$; this requires $\partial V/\partial c_r = 0$, $r = 1, 2, \ldots\ldots n$. Differentiating under the integral signs,

$$\frac{\partial V}{\partial c_r} = 0 = EI\int_0^l\left\{\left(\frac{r\pi}{l}\right)^2 \sin\frac{r\pi x}{l}\sum_1^n c_m\left(\frac{m\pi}{l}\right)^2 \sin\frac{m\pi x}{l}\right\}dx$$

$$- T\int_0^l\left\{\left(\frac{r\pi}{l}\right)\cos\frac{r\pi x}{l}\sum_1^n c_m\frac{m\pi}{l}\cos\frac{m\pi x}{l}\right\}dx$$

$$- p_0\int_0^l \sin\frac{r\pi x}{l}\,dx - P\sin\frac{r\pi}{2}\,. \qquad \text{(h)}$$

Changing the order of summation and integration in equation (h), there arise integrals of the type

$$I_{mr} = \int_0^l \sin\frac{r\pi x}{l}\sin\frac{m\pi x}{l}\,dx = \begin{cases} 0 \text{ if } m \neq r \\ \dfrac{l}{2} \text{ if } m = r \end{cases} \qquad \text{(j)}$$

$$J_{mr} = \int_0^l \cos\frac{r\pi x}{l}\cos\frac{m\pi x}{l}\,dx = \begin{cases} 0 \text{ if } m \neq r. \\ \dfrac{l}{2} \text{ if } m = r. \end{cases} \qquad \text{(k)}$$

These results correspond to the orthogonality properties of sets of sines and cosines. Equation (h) thus simplifies to

$$\frac{\partial V}{\partial c_r} = 0 = EI\left(\frac{r\pi}{l}\right)^4 \frac{l}{2}c_r - T\left(\frac{r\pi}{l}\right)^2 \frac{l}{2}c_r - \frac{p_0 l}{r\pi}(\cos r\pi - 1) - P\sin\frac{r\pi}{2} . \quad (1)$$

Here, we have the remarkable situation that in spite of the complexity of equation (g), $\partial V/\partial c_r = 0$ leads to an explicit expression for c_r, independently of all the other c_m's. This means that the value of c_r is final and does not change if more terms are taken in the series to improve accuracy: this is a consequence of the orthogonality of the coordinate functions ϕ_n chosen here. If they did not enjoy this property, the c's would be determined by a set of simultaneous algebraic equations; and if a sequence of approximations with an increasing number of terms were chosen, the c's would have to be evaluated completely afresh each time as pointed out in Section 5.2. From Equation (1)

$$c_r = \frac{2l^3}{\pi^4 EIr^4}\left[\frac{p_0 l}{r\pi}(\cos r\pi - 1) + P\sin\frac{r\pi}{2}\right]\left[1 - \frac{Tl^2}{\pi^2 EIr^2}\right]^{-1} . \quad (m)$$

We notice that $c_r = 0$ for $r =$ even. Indeed, since we require w to be non-zero at mid-span, we could have decided at the outset that m need only be odd in equation (7.2.1). Either way,

$$w = \sum_{m=1,3..}^{n} c_m \sin\frac{r\pi x}{l} \quad (n)$$

Equation (m) shows that c_r tends to infinity as T tends to $r^2\pi^2 EI/l^2$. For $r = 1$, this corresponds to the familiar Euler buckling load for a pin-ended strut so that, if the end thrust approaches this value, very large deflexions will result from the slightest transverse disturbance. For the case where $p_0 = T = 0$, we have a simply supported beam with a central point load and equation (e) becomes

$$w = \sum_{r=1,3..}^{n} \left(\frac{2Pl^3}{\pi^4 EI}\frac{1}{r^4}\sin\frac{r\pi}{2}\sin\frac{r\pi x}{l}\right) \quad (p)$$

and

$$M = -EI\frac{d^2w}{dx^2} = \sum_{r=1,3..}^{n}\left(\frac{2Pl}{\pi^2}\frac{1}{r^2}\sin\frac{r\pi}{2}\sin\frac{r\pi x}{l}\right)^\dagger \quad (q)$$

We notice that both these series converge rapidly, but that for M not as rapidly as that for w. The reader should use a sequence of n's to calculate the deflexion and bending moment at midspan to see just how rapid is the convergence.

Polynomial expansions are often valuable in the Rayleigh-Ritz method. We now present one example to illustrate how, in general, the Ritz coefficients must be evaluated afresh as more terms are added into the series.

† For a series to be differentiable term by term, certain conditions have to be met. They are met here as reference to a standard text on the calculus will show.

EXAMPLE 7.2.3

Determine approximations to the deflexion curve of the cantilever of Fig. 7.2.4. EI and $p(x)$ are constant.

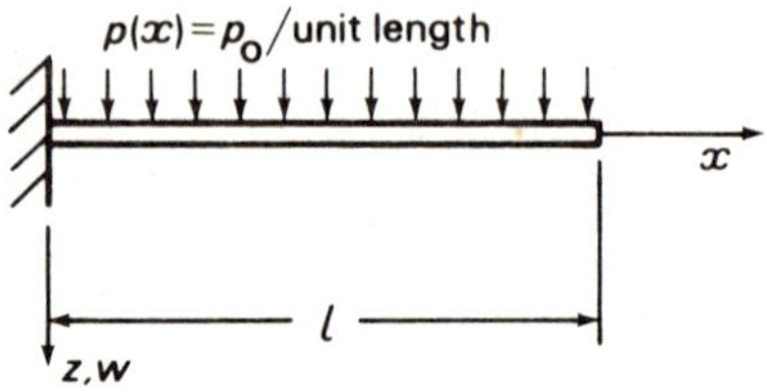

Fig. 7.2.4

The geometric boundary conditions are $w = w' = 0$ at $x = 0$, then a series satisfying these conditions is

$$w = x^2 \sum_{m=1}^{n} c_m x^{m-1} . \tag{a}$$

As a first approximation, we take $n = 1$ so that

$$w = c_1 x^2 . \tag{b}$$

Then, denoting $\mathrm{d}w/\mathrm{d}x$ and $\mathrm{d}^2w/\mathrm{d}x^2$ by w' and w'' respectively

$$w' = 2c_1 x , \quad w'' = 2c_1 . \tag{c}$$

Now the purpose of using the potential energy principle in conjunction with equation (b) is to find the 'best' c_1 which allows equilibrium to be satisfied in some average way. Equation (c) implies a constant bending moment which is very different from the true parabolic distribution; see curves (i) and (ii) in Fig. 7.2.5. However, in spite of this, it is instructive to proceed.

$$V = \int_0^l \frac{EI}{2} (w'')^2 \,\mathrm{d}x - \int_0^l pw \,\mathrm{d}x \tag{d}$$

and for $\delta V = 0$,

$$\frac{\partial V}{\partial c_1} = \int_0^l EIw'' \frac{\partial w''}{\partial c_1} \,\mathrm{d}x - \int^l p_0 \frac{\partial w}{\partial c_1} \,\mathrm{d}x = 0 . \tag{e}$$

Substituting from equations (b) and (c) into (e) yields

$$c_1 = p_0 l^2/12EI , \tag{f}$$

so that the tip deflection is $p_0 l^4/12EI$ compared with the exact value of $p_0 l^4/8EI$: not too good!

We now seek to improve the solution and take

$$w = c_1 x^2 + c_2 x^3 . \tag{g}$$

Then
$$w' = 2c_1 x + 3c_2 x^2 , \quad w'' = 2c_1 + 6c_2 x . \tag{h}$$

Substituting from equations (g) and (h) in (d) we would find $V = V(c_1, c_2)$, c_1 and c_2 are generalised coordinates and for $\delta V = 0$,

$$\frac{\partial V}{\partial c_1} = 0 , \quad \frac{\partial V}{\partial c_2} = 0 . \tag{j}$$

This leads to

$$2c_1 + 3lc_2 = p_0 l^2/6EI$$
$$3lc_1 + 6l^2 c_2 = p_0 l^3/8EI$$

or

$$\begin{bmatrix} 2l & 3l \\ 3l & 6l^2 \end{bmatrix} \begin{Bmatrix} c_1 \\ c_2 \end{Bmatrix} = \frac{p_0}{EI} \begin{Bmatrix} l^2/3 \\ l^3/8 \end{Bmatrix} \tag{k}$$

Notice that the coefficient matrix is symmetric.† Solving,

$$c_1 = \frac{5p_0 l^2}{24EI} , \quad c_2 = -\frac{p_0 l}{12EI} . \tag{l}$$

We observe that c_1 has taken on a new value different from that in the first approximation.

The deflexion curve is then

$$w = \frac{5p_0 l^2}{24EI} x^2 - \frac{p_0 l}{12EI} x^3 . \tag{m}$$

At $x = l$, $w = p_0 l^4/8EI$, so that the approximate solution agrees precisely with the exact one at this position. Proceeding to two terms has improved the solution considerably, but the bending moment diagram is still significantly in error as shown by curve (iii) in Fig. 7.2.5.

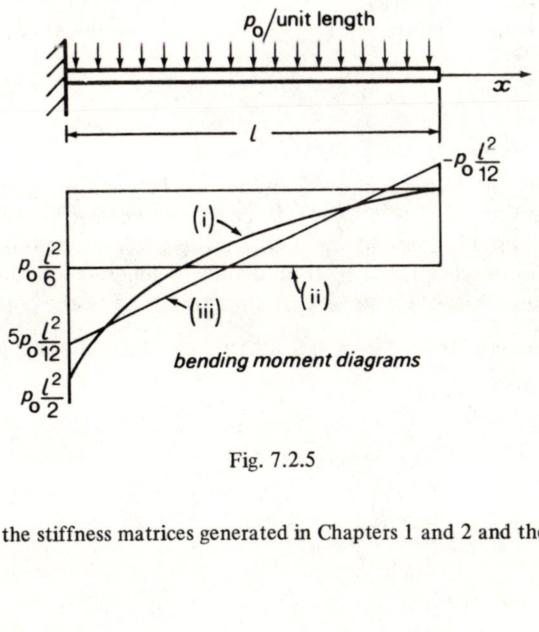

Fig. 7.2.5

† Compare with the stiffness matrices generated in Chapters 1 and 2 and the generalisations below.

In Problem 7.2.1, the reader is asked to re-solve this case assuming $w = c_1x^2 + c_2x^3$, but with c_1 and c_2 related such that a zero bending moment is obtained at $x = l$ before substituting in the potential energy functional. In this way we revert to a one parameter family of functions but now, in addition to the geometric boundary conditions, one of the mechanical boundary conditions is also satisfied.

It is instructive to generalise this treatment, and to this end we suppose the deflexion curve to be represented by

$$w = \sum_{i=1}^{n} c_i\phi_i(x) \tag{n}$$

where the $\phi_i(x)$ are prescribed coordinate functions and the c_i serve as generalised coordinates as before (see also Section 1.3). From equation (n),

$$\frac{\mathrm{d}^2 w}{\mathrm{d}x^2} = w'' = \sum_{i=1}^{n} c_i\phi_i'' . \tag{p}$$

Then, for a beam carrying a distributed load $p(x)$, the total potential energy is given by

$$V = \int_0^l \frac{EI}{2}\left(\sum_{i=1}^{n} c_i\phi_i''\right)^2 \mathrm{d}x - \int_0^l p\sum_{i=1} c_i\phi_i \mathrm{d}x \tag{q}$$

or
$$V = \frac{1}{2}\sum_i\sum_j\left(c_ic_j\int_0^l EI\phi_i''\phi_j''\mathrm{d}x\right) - \sum_i\left(c_i\int_0^l p\phi_i\mathrm{d}x\right). \tag{r}$$

Denoting
$$\int_0^l EI\phi_i''\phi_j''\mathrm{d}x = k_{ij} = k_{ji}, \quad \int_0^l p\phi_i\mathrm{d}x = P_i , \tag{s}$$

$$V = \frac{1}{2}\sum_i\sum_j k_{ij}c_ic_j - \sum_i P_ic_i . \tag{t}$$

This is precisely the form taken by V in the case of the finite degree of freedom systems met in Chapter 1; the k_{ij} can be interpreted as stiffness coefficients and the P_i as generalised forces corresponding to the generalised coordinates c_i. The net effect of the Rayleigh-Ritz method is thus to generate a discrete system having mechanical characteristics approximating those of the continuum system.

The equilibrium condition $\delta V = 0$ leads to a set of simultaneous linear algebraic equations

$$\frac{\partial V}{\partial c_r} = 0, \quad r = 1, 2 \ldots\ldots n . \tag{u}$$

That is
$$\frac{1}{2}\left[\sum_i k_{ir}c_i + \sum_j k_{rj}c_j\right] - P_r = 0 . \tag{v}$$

We observe that i and j merely indicate summations over the same range n so that we can replace i by j (see 'dummy indices' in the Appendix). Then, since we have already seen that $k_{ir} = k_{ri}$ from equation (s), we can write equation (v) as

$$\sum_{j=1}^{n} k_{rj}c_j = P_r, \quad r = 1, 2, \ldots\ldots n. \tag{w}$$

Written in matrix form, equation (t) becomes

$$V = \tfrac{1}{2}\{c\}^t[K]\{c\} - \{c\}^t\{P\} \tag{x}$$

and $\delta V = 0$ leads to

$$[K]\{c\} = \{P\}. \tag{y}$$

Here, $[K]$ is a stiffness matrix, which is symmetric, whilst $\{c\}$ and $\{P\}$ are vectors of generalised coordinates and generalised forces respectively.

When interest centres on stress levels rather than in the displacements in a structure, we may formulate the problem in terms of the complementary energy principle. We now treat a beam as a one dimensional continuum to utilise this principle. Since all trial stress systems must satisfy equilibrium requirements, the principle of complementary energy is only practical in treating statically indeterminate systems: after all, in a statically determinate situation, the equilibriant stress system is unique. We thus illustrate the process by means of the propped cantilever below.

EXAMPLE 7.2.4

Determine the bending moment diagram for the uniform cantilever of Fig. 7.2.6. The right hand support does not sink and Hooke's law obtains.

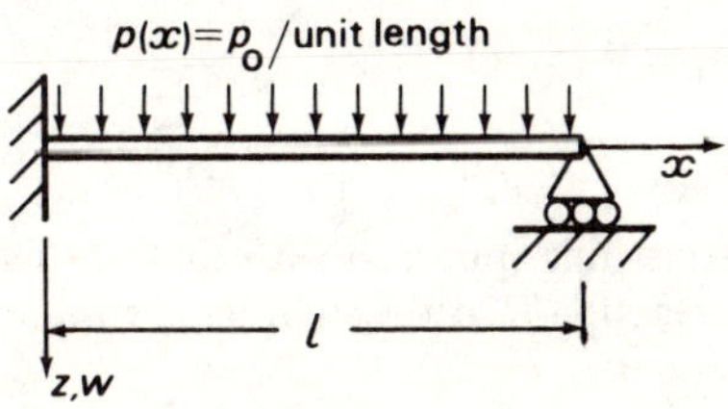

Fig. 7.2.6

All trial stress systems must be equilibriant and so must satisfy

$$\frac{d^2M}{dx^2} = -p_0. \tag{a}$$

Integrating twice

$$M = -\frac{p_0x^2}{2} + C_1x + C_2. \tag{b}$$

Furthermore, equilibrium at the propped end requires $M = 0$ at $x = l$; adjusting the values of C_1 and C_2, we find

$$M = \frac{p_0 l^2}{2}\left[1 - \left(\frac{x}{l}\right)^2\right] + C_1 l\left[\left(\frac{x}{l}\right) - 1\right] \tag{c}$$

as an admissible trial function for M.

The geometric boundary conditions are $w = w' = 0$ at $x = 0$ and $w = 0$ at $x = l$. Then, expression (6.5.16) for V^* reduces to

$$V^* = U^* \tag{d}$$

so that for compatibility $\delta V^* = \delta U^* = 0$. (e)

For a Hooke's law material

$$U^* = U = \int_0^l \frac{M^2}{2EI}\,dx \tag{f}$$

and in view of equation (c), substitution in equation (f) yields

$$U = U(C_1)\,. \tag{g}$$

Then, for $\delta U = 0$, we have

$$\frac{\partial U}{\partial C_1} = \int_0^l \frac{M}{EI}\frac{\partial M}{\partial C_1}\,dx = 0\,. \tag{h}$$

Substituting from equation (c) in (h) we find

$$C_1 = \frac{5p_0 l}{8} \tag{j}$$

so that

$$M = \frac{p_0 l^2}{2}\left[1 - \left(\frac{x}{l}\right)^2\right] - \frac{5p_0 l^2}{8}\left[1 - \left(\frac{x}{l}\right)\right]. \tag{k}$$

At

$$x = 0, \quad M = -\frac{p_0 l^2}{8} \tag{l}$$

which is the correct result.

The reader should contrast this approach with that of Chapter 2. There, we could have chosen the reaction R at the right hand support as the redundant quantity. It turns out that $C_1 = 5R/3$.

As a final example of a one dimensional problem, we treat a clamped-clamped beam by the Hellinger-Reissner principle.

EXAMPLE 7.2.5

Determine the bending moment diagram and deflexion curve for the beam of Fig. 7.2.7. Hooke's law material.

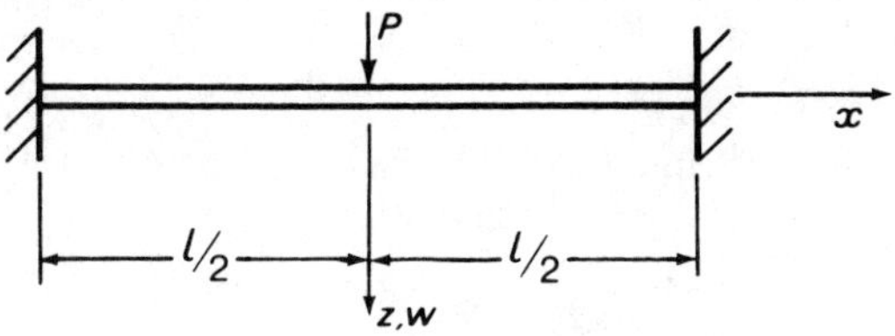

Fig. 7.2.7

In view of the symmetry involved, we choose the origin at the centre. Then, for $0 \leqslant x \leqslant l/2$, we choose as trial functions†

$$M = Ax + B \tag{a}$$

and

$$w = C\left(x^2 - \frac{l^2}{4}\right)^2 . \tag{b}$$

The prescribed end conditions are $w = w' = 0$ at $x = l/2$, whilst from symmetry $w' = 0$ at $x = 0$: Equation (b) meets these requirements. M is not prescribed at the ends so that Equation (a) is acceptable as a trial function for it.

The Reissner functional is

$$V_R = -2\int_0^{l/2}\left[\frac{M^2}{2EI} + Mw''\right]\mathrm{d}x - Pw(0) . \tag{c}$$

Substituting from equations (a) and (b) in (c),

$$V_R = -2\int_0^{l/2}\left[\frac{(Ax + B)^2}{2EI} + 4C(Ax + B)\left(3x^2 - \frac{l^2}{4}\right)\right]\mathrm{d}x - \frac{PCl^4}{16}, \tag{d}$$

or $V_R = V_R(A, B, C)$. (e)

Then, since for equilibrium and compatibility $\delta V_R = 0$, we require

$$\frac{\partial V_R}{\partial A} = \frac{\partial V_R}{\partial B} = \frac{\partial V_R}{\partial C} = 0 . \tag{f}$$

Carrying out the integrations and differentiations we find

$$\begin{aligned}\frac{\partial V_R}{\partial A} &= 0 = Al + 3B + \tfrac{3}{2}EIl^2C, \\ \frac{\partial V_R}{\partial B} &= 0 = \frac{Al}{4} + B , \\ \frac{\partial V_R}{\partial C} &= 0 = A + P/2 .\end{aligned} \tag{g}$$

From which $A = -P/2$, $B = Pl/8$, $C = P/12EIl$ (h)

† Notice that the Euler equations for the Reissner functional are the equations of equilibrium and the stress-strain relations whilst all boundary conditions are natural ones. Approximate solution can thus imply approximate satisfaction of all these requirements.

and $$w = \frac{P}{12EIl}\left(x^2 - \frac{l^2}{4}\right)^2, \quad M = \frac{Pl}{8} - \frac{Px}{2}, \quad 0 \leqslant x \leqslant l/2 . \qquad \text{(j)}$$

The bending moment distribution is exact here, and whilst the *shape* of w is not exact (it should be a cubic), we find that at $x = 0$,

$$w(0) = \frac{Pl^4}{192EI} \qquad \text{(k)}$$

which is the correct solution.

Thus, we see that the Reissner principle is a very powerful tool indeed, yielding displacements and stresses of comparable accuracy. We would however point out that whilst stiffness and flexibility matrices are symmetric and positive definite, the coefficient matrix for determining the stationary parameters in the Reissner functional is not positive definite: from a computational point of view, this is a disadvantage.

Problems 7.2

7.2.1 Treat the cantilever of Example 7.2.3 assuming $w = c_1x^2 + c_2x^3$, but with c_1 and c_2 related such that the bending moment is zero at $x = l$. Show that an approximation to the deflexion is

$$w = \frac{3p_0l^2}{16EI}\left(x^2 - \frac{x^3}{3l}\right).$$

Is this better than the result of Example 7.2.3?

7.2.2 A simply supported beam, length l, carries a distributed load the intensity of which varies linearly from zero at mid-span to a maximum at each support. If the *total* load is W, use a trigonometric series and the potential energy principle to show that the deflexion at mid-span is

$$w_{ms} = 9.41 \times 10^{-3}\, Wl^3/EI .$$

7.2.3 A uniform beam of length l, simply supported at both ends, carries a uniformly distributed load p_0/unit length over its entire span and an end thrust T. Additional support is provided by a linear spring, stiffness k, at mid span. With the origin at one support, and assuming a deflexion mode

$$w = \sum_{n=1}^{N} b_n \sin\frac{n\pi x}{l}, \qquad \text{(a)}$$

use the potential energy principle to show that the b_n's are given by simultaneous equations of the type

$$\left[\frac{\pi^4n^4EI}{2l^3} - \frac{\pi^2n^2T}{2l}\right]b_n + k\sin\frac{n\pi}{2}\sum_{n=1}^{N} b_n \sin\frac{n\pi}{2} = \frac{p_0l}{\pi}\left(1 - \frac{\cos n\pi}{n}\right).$$

7.2.4 Use the principle of complementary energy to determine the spring force in the system of Problem 7.2.3 above with the end thrust T removed.

7.2.5 The beam of Example 7.2.5 is given extra support by a linear spring, stiffness k at mid-span. Formulate Reissner's functional for the new system and show that the spring force is given by

$$T = P[1 + 192EI/kl^3]^{-1} .$$

7.2.6 A cantilever beam is tapered such that

$$EI = EI_0\left[1 - \frac{1}{2}\left(\frac{x}{l}\right)^2\right] \quad \text{(a)}$$

where x is measured from the clamped end. The beam carries a distributed load having intensity

$$p = p_0(1 - x) . \quad \text{(b)}$$

Assuming that the deflexion curve can be approximated by

$$w = c_1x^2 + c_2x^3 , \quad \text{(c)}$$

use the Rayleigh-Ritz technique to obtain stiffness equilibrium equations in the form

$$\begin{bmatrix} k_{11} & k_{12} \\ k_{21} & k_{22} \end{bmatrix} \begin{Bmatrix} c_1 \\ c_2 \end{Bmatrix} = \begin{Bmatrix} F_1 \\ F_2 \end{Bmatrix} . \quad \text{(d)}$$

7.3 Torsion problems

In Chapter 3 we saw that formulating the torsion problem in terms of a stress function led to a simpler boundary condition than for the warping function. Since the more awkward part of obtaining numerical solutions to practical design problems is usually satisfying the boundary conditions, most treatments have been based on the stress function. Such formulations are, of course, effectively in terms of stresses so that it is appropriate to base one of the direct methods of the variational calculus on the complementary energy principle. The attraction of working in terms of a stress function is that internal equilibrium conditions are automatically satisfied by all trial functions: a central requirement of the complementary energy principle is thus easily met.

EXAMPLE 7.3.1

Treat the torsion of a rectangular section bar by the Rayleigh-Ritz method. Hooke's law material. The section is shown in Fig. 7.3.1 below.

We assume trial functions in the form

$$\phi = \sum_{i=1}^{n} c_i\phi_i(x, y) . \quad \text{(a)}$$

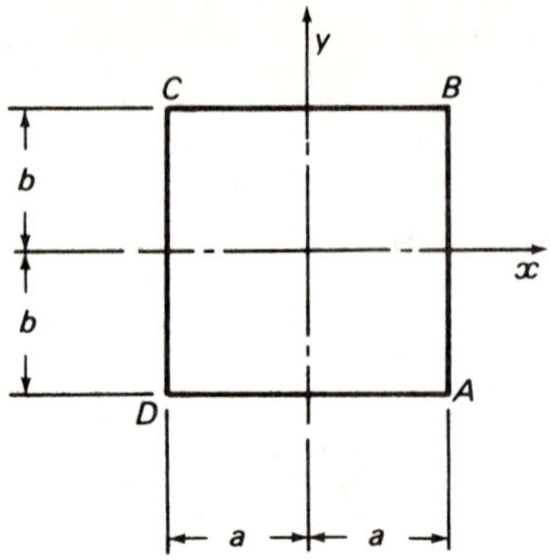

Fig. 7.3.1

To satisfy equilibrium on the lateral surface of the bar, we require $\phi = 0$ along the boundary of the section. To provide a guide in forming suitable ϕ_i's, we observe that the boundary of the section is defined by

$$x^2 - a^2 = 0, \quad y^2 - b^2 = 0, \tag{b}$$

so that, by choosing

$$\phi = (x^2 - a^2)(y^2 - b^2) \sum_{m=0}^{M} \sum_{n=0}^{N} c_{mn} x^m y^n, \tag{c}$$

we ensure $\phi = 0$ along the boundary. Further, in view of the symmetry of the situation, we conclude that m and n must be even.†

As a first approximation, we choose one term only, then

$$\phi = c_1 (x^2 - a^2)(y^2 - b^2). \tag{d}$$

$$V^* = \frac{l}{2G} \iint \{\phi_x{}^2 + \phi_y{}^2 - 4G\alpha\phi\} \, dx dy = V^*(c_1).^{\dagger\dagger} \tag{e}$$

For compatibility, $\delta V^* = 0$, so that

$$\frac{\partial V^*}{\partial c_1} = 0. \tag{f}$$

Substituting from equation (d) in (e) and carrying out the necessary operations, we find

$$c_1 = \frac{5G\alpha}{4(a^2 + b^2)}. \tag{g}$$

† We notice that $\phi_i(x, y) = \cos(m\pi x/2a) \cos(n\pi y/2a)$, m, n odd, would also be suitable.
†† ϕ_x and ϕ_y denote $\partial\phi/\partial x$ and $\partial\phi/\partial y$ respectively.

The first approximation solution is thus

$$\phi = \frac{5G\alpha}{4(a^2 + b^2)}(x^2 - a^2)(y^2 - b^2) . \tag{h}$$

The maximum shearing stress occurs at the centre of the longest side: this can be seen by calculation or directly by visualising the analogous soap film. Suppose that $b > a$, then the maximum shearing stress is

$$|\tau_{zy}|_{\substack{x = a\\ y = 0}} = \frac{5}{2}\frac{G\alpha a b^2}{a^2 + b^2} . \tag{j}$$

The torsion constant for the section is defined by $K = M_t/G\alpha$. Then substituting in equation (3.6.19),

$$K = \frac{40}{9}\frac{a^3 b^3}{(a^2 + b^2)} . \tag{k}$$

For a square section $a = b$, then

$$(\tau_{\max})_{\text{app}} = 1.2500 G\alpha a , \quad K_{app} = 2.2222a^4 . \tag{l}$$

The corresponding exact values are, from Timoshenko and Goodier,

$$(\tau_{\max})_{\text{ex}} = 1.3500 G\alpha a , \quad K_{\text{ex}} = 2.2496a^4 , \tag{m}$$

so that the corresponding errors are 7.4% and 1.2%.

For a thin rectangular section blade, $b \gg a$ and we have, from above,

$$(\tau_{\max})_{\text{app}} = 2.500 G\alpha a , \quad K_{\text{app}} = 0.8333[\tfrac{1}{3}(2b)(2a)^3] \tag{n}$$

in contrast to the exact values

$$(\tau_{\max})_{\text{ex}} = 2.000 G\alpha a , \quad K_{\text{ex}} = \tfrac{1}{3}(2b)(2a)^3 . \tag{p}$$

Here, the errors are 25% and 16.67%: to improve on this, we can take more terms in the series, say

$$\phi = (x^2 - a^2)(y^2 - b^2)(c_1 + c_2 x^2 + c_3 y^2) . \tag{q}$$

Then the equations for determining the c's are found to be

$$\begin{aligned} 140(a^2 + b^2)c_1 + (20a^4 + 28a^2b^2)c_2 + (28a^2b^2 + 20b^4)c_3 &= 175G\alpha . \\ (60a^2 + 84b^2)c_1 + (20a^4 + 132a^2b^2)c_2 + 12(a^2b^2 + b^4)c_3 &= 105G\alpha . \\ (84a^2 + 60b^2)c_1 + 12(a^4 + a^2b^2)c_2 + (132a^2b^2 + 20b^4)c_3 &= 105G\alpha . \end{aligned} \tag{r}$$

With a very simple little computer programme, equations (r) can be solved for a^2c_1, a^4c_2 and a^4c_3 for a whole range of b/a. We tabulate just two sets of results below in Table 7.3.1.

Table 7.3.1

b/a	$(k_1)_{ex}$	$(k_1)_{app}$	% error k_1	$(k_2)_{ex}$	$(k_2)_{app}$	% error k_2
1	1.35	1.046	4.1	0.1406	0.1404	0.14
10	2.00	1.80	10	0.312	0.305	2.3

Here, we have quoted exact values from Timoshenko and Goodier with

$$\tau_{max} = k_1 G\alpha a\,, \qquad K = k_2(2a)^3(2b)\,. \tag{s}$$

We note that for $a = b$, $c_1 = 0.625G\alpha/a^2$ for a one parameter family whilst from equation (r) $c_1 = 0.584G\alpha/a^2$: this illustrates the change in the Ritz coefficients as more terms are used. It also turns out from equation (r) that when $a = b$, $c_2 = c_3 = 0.1185G\alpha/a^4$: we would expect this equality of c_1 and c_2 from the symmetry of the situation.

The torsion constant has been obtained with good accuracy even in the case of $b/a = 10$, but the maximum shearing stress is still quite significantly in error (although for many design purposes it may be adequate). We can explain the less accurate results for $b/a = 10$ as follows. The 'shape' which equation (q) proposed for ϕ is the same in the x and y directions, whereas, if we visualise the analogous soap film, we see that as b/a becomes larger, the soap film (and hence the ϕ surface) should become 'flatter' in the y direction except near the short sides of the section.

We now attempt to improve on the above treatment by using Kantorovitch's method, (Section 5.3).

EXAMPLE 7.3.2

Treat Example 7.3.1 by Kantorovitch's method.

We assume a one term approximating function

$$\phi = f(y)(x^2 - a^2)\,. \tag{a}$$

Then

$$\phi_x = 2xf(y)\,, \quad \phi_y = f'(y)(x^2 - a^2) \tag{b}$$

and

$$V^* = \frac{l}{2G}\int_{-b}^{b}\int_{-a}^{a}\{4x^2f^2 + (x^2 - a^2)^2(f')^2 - 4G\alpha f(x^2 - a^2)\}\,dx\,dy\,. \tag{c}$$

Integrating with respect to x first,

$$V^* = \frac{8la^3}{6G}\int_{-b}^{b}[\tfrac{2}{5}a^2(f')^2 + f^2 + 2G\alpha f]\,dy\,, \tag{d}$$

or

$$V^* = \int_{-b}^{b} F(f, f')dy\,. \tag{e}$$

For compatibility, $\delta V^* = 0$ so that, following Section 4.8, the Euler-Lagrange equation is

$$\frac{\partial F}{\partial f} - \frac{\mathrm{d}}{\mathrm{d}y}\left(\frac{\partial F}{\partial f'}\right) = 0 \tag{f}$$

which yields

$$\frac{\mathrm{d}^2 f}{\mathrm{d}y^2} - \mu^2 f = \mu^2 G\alpha \tag{g}$$

as the condition to determine f with the notation $\mu^2 = 5/(2a^2)$. Equation (g) is an ordinary differential equation with constant coefficients and its solution consists of a particular integral and a complementary function

$$f = C_1 \sinh \mu y + C_2 \cosh \mu y - G\alpha\,. \tag{h}$$

Our trial ϕ thus becomes

$$\phi = [C_1 \sinh \mu y + C_2 \cosh \mu y - G\alpha](x^2 - a^2)\,. \tag{j}$$

Now ϕ must be symmetric in y so that $C_1 = 0$.

Furthermore, with $\phi = 0$ at $y = \pm b$, equation (j) becomes

$$\phi = G\alpha\left(1 - \frac{\cosh \mu y}{\cosh \mu b}\right)(a^2 - x^2)\,. \tag{k}$$

The maximum shearing stress is at $x = a$, $y = 0$ (for $b > a$) with

$$\tau_{\max} = 2G\alpha a\left(1 - \frac{1}{\cosh \mu b}\right). \tag{l}$$

The torsion constant is

$$K = \frac{16a^3}{3}\left[b - \frac{1}{\mu}\tanh \mu b\right]. \tag{m}$$

Table 7.3.2 shows two typical results for the coefficients k_1 and k_2 of equations (s) of example 7.3.1.

Table 7.3.2

b/a	$(k_1)_{ex}$	$(k_1)_{app}$	% error k_1	$(k_2)_{ex}$	$(k_2)_{app}$	% error k_2
1	1.35	1.21	10.4	0.1406	0.1396	0.68
10	2.00	2.00	0	0.312	0.312	0

We see that the improvement in these results compared with those for the one term Rayleigh-Ritz treatment of Example 7.3.1 is dramatic for the case $b/a = 10$. For the case of $a = b$, the approximation to K is better by Kantorovitch, but the maximum stress value is not so good. We can again explain this with the aid of the soap film analogy. Along $x = 0$, the 'height' of ϕ above the xOy plane is given, for $a = b$, by

$$\bar{\phi}_{RR} = \phi_{RR}/G\alpha a^2 = 0.625(1 - r^2). \tag{n}$$

$$\bar{\phi}_K = \tilde{\phi}_K/G\alpha a^2 = \left[1 - \frac{\cosh(\sqrt{2.5}r)}{\cosh(\sqrt{2.5})}\right]. \tag{p}$$

Here, $r = y/a$ whilst ϕ_{RR} and ϕ_K are the one term Rayleigh-Ritz and Kantorovitch solutions respectively. Computing values of ϕ_{RR} and ϕ_K from equations (n) and (p), we obtain Table 7.3.3.

Table 7.3.3

r	0	0.2	0.4	0.6	0.8	0.9	0.95	1.0
$\bar{\phi}_{RR}$	0.6250	0.6000	0.5250	0.4000	0.2250	0.1188	0.0609	0
$\bar{\phi}_K$	0.6052	0.5853	0.5236	0.4139	0.2450	0.1334	0.0696	0

Thus, the Kantorovitch curve is 'flatter' in the y direction than the Rayleigh-Ritz one. Since the same angle of twist is implied for each, the torque transmitted must be the same and, in order that $2\int\int\phi\, dx\, dy$† shall be the same for both, $(\bar{\phi}_K)_{max}$ must be less than $(\bar{\phi}_{RR})_{max}$ as confirmed by the table. The slope component ϕ_x at $y = r = 0$ and $x = a$ is thus smaller for the Kantorovitch surface than the Rayleigh-Ritz one and we see why the Kantorovitch shearing stress is smaller than the Rayleigh-Ritz one in this case.

For a square section bar we know that τ_{xy} at $x = a, y = 0$ should be the same as τ_{zx} at $x = 0, y = a$. The above discussion also explains why the Kantorovitch τ_{zx} at $x = 0, y = a$ is not as accurate as the τ_{zy} at $x = a$, $y = 0$. In fact

$$|(\tau_{zx})_K|_{\substack{x=0\\y=a}} = 1.45G\alpha a. \tag{q}$$

This example demonstrates an essential point regarding the use of any approximation method. To be successful, a certain amount of engineering judgement is required in deciding how the discretisation†† is to be carried out. A further point to observe is that, for the Ritz and Kantorovitch methods, the correct torsion stiffness for the bar is approached from *below:* to prove that this is always the case, define an error function by

$$\eta(x, y) = \phi(x, y) - \phi_n(x, y) \tag{7.3.1}$$

† This is twice the volume under the analogous membrane or the stress function 'hill'.

†† That is, the reduction to a finite number of degrees of freedom.

so that $\frac{2G}{l}V^*(\phi_n) = I_n$

$$= \iint\{(\phi_x - \eta_x)^2 + (\phi_y - \eta_y)^2 - 4G\alpha(\phi - \eta)\}\mathrm{d}x\mathrm{d}y^{\dagger} \qquad (7.3.2)$$

or

$$I_n = \iint\{\phi_x{}^2 + \phi_y{}^2 - 4G\alpha\phi\}\mathrm{d}x\mathrm{d}y$$
$$- 2\iint\{\phi_x\eta_x + \phi_y\eta_y\}\mathrm{d}x\mathrm{d}y$$
$$+ \iint\{\eta_x{}^2 + \eta_y{}^2\}\mathrm{d}x\mathrm{d}y + 4G\alpha\iint\eta\mathrm{d}x\mathrm{d}y. \qquad (7.3.3)$$

The second integral, call it I_{n_2}, may be transformed by Green's theorem (see Appendix) to give

$$I_{n2} = -2\{\int\eta\frac{\partial\phi}{\partial n}\mathrm{d}S - \iint\eta\nabla^2\phi\mathrm{d}x\mathrm{d}y\}. \qquad (7.3.4)$$

But $\eta = 0$ on the boundary and $\nabla^2\phi = -2G\alpha$ so that

$$I_n = I + \iint\{\eta_x{}^2 + \eta_y{}^2\}\mathrm{d}x\mathrm{d}y. \qquad (7.3.5)$$

The second integral cannot be negative so

$$I_n \geqslant I \text{ and } V_n{}^* \geqslant V^* \qquad (7.3.6)$$

showing that the complementary energy is approached from above. Green's theorem may also be used to transform I. Thus,

$$I = \iint\{\phi_x{}^2 + \phi_y{}^2 - 4G\alpha\phi\}\mathrm{d}x\mathrm{d}y$$
$$= \int_{\mathscr{C}}\phi\frac{\partial\phi}{\partial n}\mathrm{d}S - \iint\phi\nabla^2\phi\mathrm{d}x\mathrm{d}y - \iint 4G\alpha\phi\mathrm{d}x\mathrm{d}y. \qquad (7.3.7)$$

Then recalling that $\nabla^2\phi = -2G\alpha$ in the region and $\phi = 0$ on the boundary,

$$I = -2G\alpha\iint\phi\mathrm{d}x\mathrm{d}y. \qquad (7.3.8)$$

Denoting $M_t = KG\alpha = 2\iint\phi\mathrm{d}x\mathrm{d}y$, where K is the torsion constant for the section,

$$I = -(G\alpha)K. \qquad (7.3.9)$$

Thus, since I is approached from above, equation (7.3.9) shows that the torsion constant for the section, measuring stiffness, is approached from below. The Ritz, Kantorovitch and Galerkin methods thus give a lower bound to the stiffness.

† Note that $\phi_x \equiv \partial\phi/\partial x$ etc.

We have seen that, in the Rayleigh-Ritz process, taking more terms in a series of coordinate functions generally improves the solution, but that we cannot compute the error involved at any particular stage in the process. However, if we have a means of approximating the stiffness from above, this, together with the Rayleigh-Ritz solution, will straddle the true value. In this way we can estimate the maximum error involved and, if this is acceptable for engineering purposes, it does not matter if the chosen sequence of coordinate functions yields a solution converging to the true one or not. We now show that the Trefftz method provides an upper bound to the stiffness. For this purpose it proves slightly more convenient to work in terms of a function $\Psi(x, y)$† obtained from $\phi(x, y)$ by

$$\phi = G\alpha[\Psi - \tfrac{1}{2}(x^2 + y^2)] \quad . \tag{7.3.10}$$

Then, since $\nabla^2\phi + 2G\alpha = 0$ in the region and $\phi = 0$ on the boundary, we see that

$$\nabla^2\Psi = 0 \qquad \text{in the region ,} \tag{7.3.11}$$

$$\Psi = \tfrac{1}{2}(x^2 + y^2) \quad \text{on the boundary .} \tag{7.3.12}$$

Using the method outlined in Section 4.15, we may show that the variational statement

$$J(\Psi) = \iint [\Psi_x{}^2 + \Psi_y{}^2]\,\mathrm{d}x\mathrm{d}y = \text{stationary} \tag{7.3.13}$$

corresponds to equation (7.3.11) subject to condition (7.3.12). Let a trial solution for equation (7.3.11) be Ψ_N; then, we may define an error function

$$\mu(x, y) = \Psi(x, y) - \Psi_N(x, y) \tag{7.3.14}$$

so that
$$J(\Psi) = \iint \{(\Psi_{Nx} + \mu_x)^2 + (\Psi_{Ny} + \mu_y)^2\}\,\mathrm{d}x\mathrm{d}y \tag{7.3.15}$$

$$= \iint [\Psi_{Nx}{}^2 + \Psi_{Ny}{}^2]\,\mathrm{d}x\mathrm{d}y$$
$$+ 2\iint [\Psi_{Nx}\mu_x + \Psi_{Ny}\mu_y]\,\mathrm{d}x\mathrm{d}y + \iint [\mu_x{}^2 + \mu_y{}^2]\,\mathrm{d}x\mathrm{d}y. \tag{7.3.16}$$

The second integral may be transformed by Green's theorem so that

$$J(\Psi) = J(\Psi_N) + J(\mu) + 2\{\int_{\mathscr{C}} \frac{\partial\Psi_N}{\partial n}\,\mathrm{d}S - \iint \mu\nabla^2\Psi_N\mathrm{d}x\mathrm{d}y\}. \tag{7.3.17}$$

Now $\nabla^2\Psi_N = 0$ and $J(\mu)$ cannot be negative so that

$$J(\Psi_N) \leqslant J(\Psi) \tag{7.3.18}$$

provided
$$\int_{\mathscr{C}} \mu\frac{\partial\Psi_N}{\partial n}\,\mathrm{d}S = \int_{\mathscr{C}} (\Psi - \Psi_N)\frac{\partial\Psi_N}{\partial n}\,\mathrm{d}S = 0\,. \tag{7.3.19}$$

† In fact Ψ is conjugate to the warping function ψ in Section 3.6. They are related by the Cauchy-Riemann relations

$$\frac{\partial\psi}{\partial x} = \frac{\partial\Psi}{\partial y}, \quad \frac{\partial\psi}{\partial y} = -\frac{\partial\Psi}{\partial x}\,.$$

But achieving equation (7.3.19) is precisely Trefftz's method so that it yields a lower bound to $J(\Psi)$. We now show that this corresponds to an upper bound to the torsion stiffness of a bar.

Substituting from equation (7.3.10) in (7.3.13)

$$J(\Psi) = \frac{1}{(G\alpha)^2}\iint(\phi_x{}^2 + \phi_y{}^2)\mathrm{d}x\mathrm{d}y + \frac{2}{G\alpha}\iint(x\phi_x + y\phi_y)\mathrm{d}x\mathrm{d}y + \iint(x^2 + y^2)\mathrm{d}x\mathrm{d}y. \tag{7.3.20}$$

The second integral may be transformed by Green's theorem to yield

$$\frac{2}{G\alpha}\left\{\int_{\mathscr{C}} \phi\frac{\partial}{\partial n}\left(\frac{x^2 + y^2}{2}\right)\mathrm{d}S - \iint \phi\nabla^2\left(\frac{x^2 + y^2}{2}\right)\mathrm{d}x\mathrm{d}y\right\} = -\frac{4}{G\alpha}\iint\phi\mathrm{d}x\mathrm{d}y \tag{7.3.21}$$

since ϕ is zero on the boundary.

Then $$J(\Psi) = \frac{1}{(G\alpha)^2}\iint\{\phi_x{}^2 + \phi_y{}^2 - 4\phi G\alpha\}\mathrm{d}x\mathrm{d}y + J_0\,. \tag{7.3.21}$$

Here J_0 is the polar second moment of area of the section so that, recalling equations (7.3.8) and (7.3.9),

$$J(\Psi) = -K + J_0\,. \tag{7.3.22}$$

Thus, since in the Trefftz process we approach $J(\Psi)$ from below, we approach the torsion constant K from *above*.

EXAMPLE 7.3.3

Use Trefftz's method to determine the torsion constant for the section of Fig. 7.3.1.

Assuming a trial function

$$\Psi_N = \sum_{i=1}^{N} c_i f_i(x, y)\,, \tag{a}$$

where the f_i are harmonic, we form (Section 5.7)

$$\sum_j A_{ij}c_j = B_i\,, \quad i, j = 1\,.\,.\,.\,.\,.N \tag{b}$$

where $$A_{ij} = \int_{\mathscr{C}} \frac{\partial f_i}{\partial n} f_j\mathrm{d}S\,, \quad B_i = \frac{1}{2}\int_{\mathscr{C}}(x^2 + y^2)\frac{\partial f_i}{\partial n}\,\mathrm{d}S\,. \tag{c}$$

We may generate harmonic functions by taking the real and imaginary parts of the complex function $z = (x + iy)^n$ for various n. For this particular problem, we can visualise (from the soap film analogy) that ϕ must be even in x and y and so, then, must Ψ. Choosing $n = 2$ yields the lowest degree even polynomial so that

$$\Psi_1 = c_1(x^2 - y^2)\,. \tag{d}$$

Observing that $$\oint_{\mathscr{C}} = \int_A^B + \int_B^C + \int_C^D + \int_D^A\,, \tag{e}$$

$$A_{11} = \int_{-b}^{b} 2a(a^2 - y^2)\mathrm{d}y + \int_{a}^{-a} 2b(x^2 - b^2)\mathrm{d}x - \int_{b}^{-b} 2a(a^2 - y^2)\mathrm{d}y$$
$$- \int_{-a}^{a} 2b(x^2 - b^2)\mathrm{d}x \tag{f}$$

$$\text{or } A_{11} = \frac{16ab}{3}(a^2 + b^2)\,. \tag{g}$$

Similarly,

$$B_1 = \frac{1}{2}\int_{-b}^{b}(a^2 + y^2)2a\mathrm{d}y + \frac{1}{2}\int_{a}^{-a}(x^2 + b^2)2b\mathrm{d}x - \frac{1}{2}\int_{b}^{-b}(a^2 + y^2)2a\mathrm{d}y$$
$$- \frac{1}{2}\int_{-a}^{a}(x^2 + b^2)2b\mathrm{d}x\,, \tag{h}$$

$$\text{or } B_1 = \frac{8ab}{3}(a^2 - b^2)\,. \tag{j}$$

It follows that $$c_1 = \frac{1}{2}\left(\frac{a^2 - b^2}{a^2 + b^2}\right) \tag{k}$$

and $$\Psi_1 = \frac{1}{2}\left(\frac{a^2 - b^2}{a^2 + b^2}\right)(x^2 - y^2)\,. \tag{l}$$

The torsion constant is given by equation (7.3.22) so that

$$K = \int_{-b}^{b}\int_{-a}^{a}(x^2 + y^2)\mathrm{d}x\mathrm{d}y - 4c_1{}^2\int_{-b}^{b}\int_{-a}^{a}(x^2 + y^2)\mathrm{d}x\mathrm{d}y$$

$$\text{or} \quad K = \frac{16}{3}\frac{a^3b^3}{(a^2 + b^2)} = \frac{1}{3}\frac{(2a)^3(2b)}{1 + (a/b)^2} = k_2(2a)^3(2b)\,. \tag{m}$$

In Table 7.3.4, we contrast the k_2 values for the one term Rayleigh-Ritz solution of example 7.3.1 the one term Trefftz solution here and that quoted by Timoshenko and Goodier.

Table 7.3.4

b/a	R.R.	T & G	T
1	0.1389	0.1406	0.1667
10	0.2750	0.312	0.3300

We observe that neither the Rayleigh-Ritz nor the Trefftz method gives a consistently better solution: the details of the structure (i.e. the ratio b/a) influence the situation. However, the two methods give solutions straddling the true one.

Problems 7.3

7.3.1 The torsion problem may be solved by applying the potential energy principle to the analogous membrane. Taking the origin of coordinates at the centre, the height of the membrane can be written

$$w = \sum_{m,\, n\, =\, \text{odd}} \sum a_{mn} \cos \frac{m\pi x}{2a} \cos \frac{n\pi y}{2b} \qquad \text{(a)}$$

for a rectangular opening $2a \times 2b$.

Accepting that the change in the membrane strain energy due to its being distended is

$$U = \frac{T}{2} \iint \left[\left(\frac{\partial w}{\partial x} \right)^2 + \left(\frac{\partial w}{\partial y} \right)^2 \right] \mathrm{d}A\,, \qquad \text{(b)}$$

determine the a_{mn}'s for equilibrium with the transverse pressure p.

Reasoning by analogy, deduce the torsion constant and maximum shearing stress for a square section bar.

Ans: Using only the first term in expansion (a), $K = 0.139(2a)^4$, $\tau_{\max} = 1.252G\alpha a$.

7.3.2 Adopting a one term approximating function

$$\phi = f(y) \cdot \cos \frac{\pi x}{2a}\,, \qquad \text{(a)}$$

use the Kantorovitch method to solve for the torsion of a rectangular section $2a \times 2b$. Compare the results with those of Example 7.3.2.

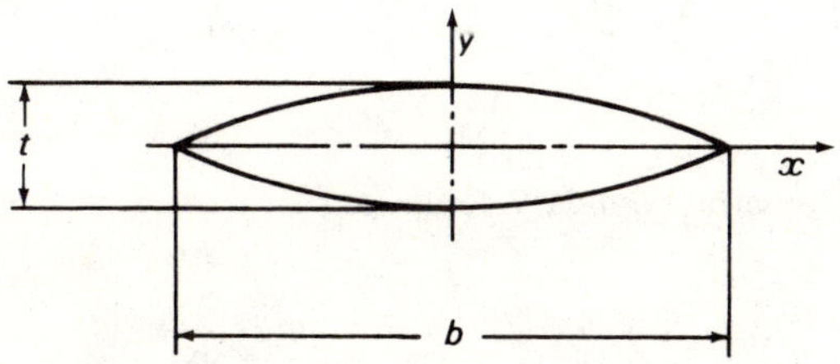

Fig. 7.3.2

7.3.3 The section of Fig. 7.3.2 is bounded by parabolic curves. Show that

$$\phi = A\left[y^2 - \frac{t^2}{4}\left(1 - \frac{4x^2}{b^2}\right)^2 \right] \qquad \text{(a)}$$

satisfies the torsion boundary condition $\phi = 0$ on the boundary. Determine a value for A, by means of the complementary energy principle, and show that the torsion constant for the section is given approximately by

$$K = 0.1524bt^3[1 + 1.333(t/b)^2]^{-1} .$$

7.4 Plate stretching problems

Depending on the situation at hand, it is convenient to formulate the problem either in terms of displacements or stresses (see Section 3.5). We present examples for each of these approaches.

EXAMPLE 7.4.1

The rectangular plate of Fig. 7.4.1 has all its edges rigidly clamped and carries a patch of uniformly distributed load, components F_x and F_y per unit surface, tangential to the plate surface and symmetric with respect to the middle surface. Use the potential energy principle to obtain the in-plane displacements: the plate is of unit thickness and the material obeys Hooke's law.

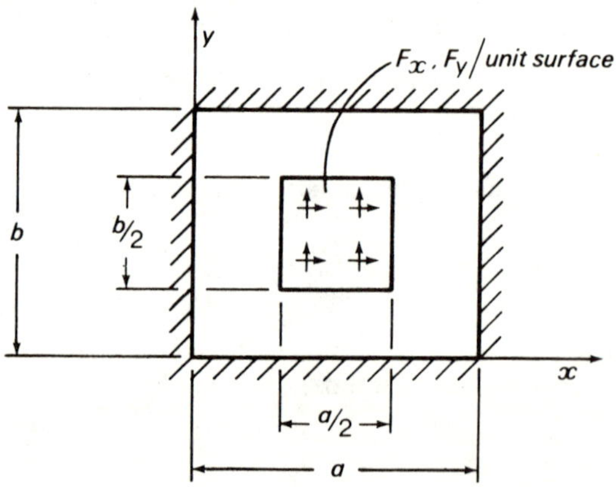

Fig. 7.4.1

The prescribed geometric boundary conditions are $u = v = 0$ all round. The functions

$$u = \sum_{m,n=1,3\,..}\sum A_{mn} \sin\frac{m\pi x}{a} \sin\frac{n\pi y}{b} ,$$

$$v = \sum_{m,n=1,3\,..}\sum B_{mn} \sin\frac{m\pi x}{a} \sin\frac{n\pi y}{b} , \qquad \text{(a)}$$

clearly satisfy these conditions. Only odd values of m and n are required to allow non-zero u and v at the centre.

The strain energy stored in the plate is given by the first integral in equation (6.4.9) with

$$\frac{\partial u}{\partial x} = \sum\sum A_{mn} \frac{m\pi}{a} \cos \frac{m\pi x}{a} \sin \frac{n\pi y}{b},$$

$$\frac{\partial u}{\partial y} = \sum\sum A_{mn} \frac{n\pi}{b} \sin \frac{m\pi x}{a} \cos \frac{n\pi y}{b};$$

$$\frac{\partial v}{\partial x} = \sum\sum B_{mn} \frac{m\pi}{a} \cos \frac{m\pi x}{a} \sin \frac{n\pi y}{b},$$

$$\frac{\partial v}{\partial y} = \sum\sum B_{mn} \frac{n\pi}{b} \sin \frac{m\pi x}{a} \cos \frac{n\pi y}{b}. \tag{b}$$

The potential energy of the load is

$$\Omega = -\int_{b/4}^{3b/4}\int_{a/4}^{3a/4} [F_x u + F_y v]\,\mathrm{d}x\mathrm{d}y, \tag{c}$$

so that, after carrying out the integrations,

$$V = V(A_{mn}\text{'s}, \quad B_{mn}\text{'s}). \tag{d}$$

For equilibrium, $\delta V = 0$ or

$$\frac{\partial V}{\partial A_{pq}} = \frac{\partial U}{\partial A_{pq}} + \frac{\partial \Omega}{\partial A_{pq}} = 0. \tag{e}$$

$$\frac{\partial V}{\partial B_{pq}} = \frac{\partial U}{\partial B_{pq}} + \frac{\partial \Omega}{\partial B_{pq}} = 0. \tag{f}$$

After quite some manipulation, in which differentiating partially with respect to A_{pq} and B_{pq} under the integral signs is a help, we have

$$\frac{\partial V}{\partial A_{pq}} = 0 = \frac{\pi^2 Eab}{4(1-\nu^2)}\left[\frac{p^2}{a^2} + \frac{1-\nu}{2}\frac{q^2}{b^2}\right]A_{pq} + \frac{4(-1)^{(p+q)/2}F_x ab}{\pi^2 pq} \sin\frac{p\pi}{4} \sin\frac{q\pi}{4}. \tag{g}$$

$$\frac{\partial V}{\partial B_{pq}} = 0 = \frac{\pi^2 Eab}{4(1-\nu^2)}\left[\frac{1-\nu}{2}\frac{p^2}{a^2} + \frac{q^2}{b^2}\right]B_{pq} + \frac{4(-1)^{(p+q)/2}F_y ab}{\pi^2 pq} \sin\frac{p\pi}{4} \sin\frac{q\pi}{4}. \tag{h}$$

Thus in spite of the complicated form of equation (d), we actually have two uncoupled equations yielding the coefficients A_{pq} and B_{pq} explicitly.

As an example of a situation wherein tractions are specified over the entire boundary, we present a problem first tackled by Timoshenko in 1924.

EXAMPLE 7.4.2

A rectangular plate is subjected to a parabolic distribution of normal traction over edges $x = \pm a$ whilst edges $y = \pm b$ are free of load as shown in Fig. 7.4.2. Use the complementary energy principle to obtain an approximation to the stress distribution. Unit thickness plate, Hooke's law material.

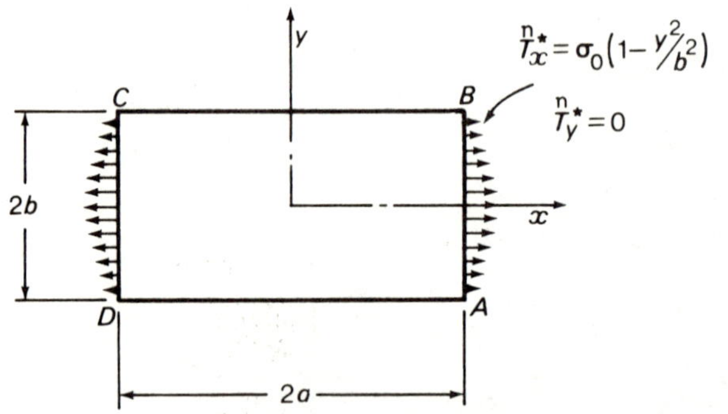

Fig. 7.4.2

To ensure satisfaction of internal equilibrium by the trial stresses, we work in terms of the Airy stress function. Then, we choose

$$\chi = \chi_0(x, y) + \sum_{i=1}^{N} c_i\chi_i(x, y) \tag{a}$$

where, if χ_0 satisfies the traction boundary conditions, the χ_i are chosen so as to yield zero tractions on the boundaries.

From Section 3.7, we know that prescribed tractions implies prescribed χ and $\partial\chi/\partial n$ so that in equation (a), we require χ_i and $\partial\chi_i/\partial n$ both to be zero along the edges. We can ensure this by assuming

$$\sum_{i=1}^{N} c_i\chi_i(x, y) = (a^2 - x^2)^2(b^2 - y^2)^2 \sum_{m,\, n = 1 \ldots} \sum a_{mn}x^{m-1}y^{n-1} \,. \tag{b}$$

Now consider the choice of $\chi_0(x, y)$. From equations (3.7.23) and (3.7.24), we require

$$\left.\frac{\partial\chi}{\partial y}\right|_{A,\, B} = \sigma_0\int\left(1 - \frac{y^2}{b^2}\right)\mathrm{d}y + \beta \,. \tag{c}$$

$$\left.\frac{\partial\chi}{\partial x}\right|_{A,\, B} = 0 + \alpha \,. \tag{d}$$

Integrating equation (c) with respect to y we have

$$\chi_{A,\,B} = \sigma_0\left(\frac{y^2}{2} - \frac{y^4}{12b^2}\right) + \alpha x + \beta y + \gamma \,. \tag{e}$$

The integration constants α, β and γ may be chosen arbitrarily here, since a linear function implies zero stresses. Choosing $\alpha = \beta = \gamma = 0$ renders $\chi_{A,B}$ symmetric in x and y which corresponds with the symmetry of the system in Fig. 7.4.2. Then, if we choose

$$\chi_0 = \frac{\sigma_0 y^2}{2}\left[1 - \frac{1}{6}\left(\frac{y}{b}\right)^2\right], \tag{f}$$

we see that it yields the required parabolic traction distribution along AB and CD and zero load along AD and BC.

As a first approximation, we choose a one parameter series in equation (a); then

$$\chi = \frac{\sigma_0 y^2}{2}\left[1 - \frac{1}{6}\left(\frac{y}{b}\right)^2\right] + c_1(a^2 - x^2)^2(b^2 - y^2)^2 \,. \tag{g}$$

Tractions are prescribed over the entire boundary of the plate so that the S_u of equation (6.5.16) is zero. Then

$$V^* = U^* \tag{h}$$

and for a Hooke's law material, the complementary energy principle becomes

$$\delta U(\chi) = 0 \,. \tag{j}$$

There are no natural boundary conditions here, so we can use the simpler expression for U, (see Section 6.6). That is

$$U = \frac{1}{2E}\iint\left(\frac{\partial^2\chi}{\partial x^2} + \frac{\partial^2\chi}{\partial y^2}\right)^2 \mathrm{d}x\mathrm{d}y \,. \tag{k}$$

Substituting from equation (g) in equation (k), we find $U = U(c_1)$ so that equation (j) requires

$$\frac{\partial U}{\partial c_1} = 0 \,. \tag{l}$$

Carrying out the integrations and differentiations,

$$c_1 = \sigma_0\,[9.1429a^4b^2 + 5.2245a^2b^4 + 9.1429b^6]^{-1} \,. \tag{m}$$

For a square plate $a = b$ and

$$c_1 = 0.0425\sigma_0/a^6 \,. \tag{n}$$

For a 'long' plate, $b \ll a$ and

$$c_1 = 0.1094\sigma_0/a^4b^2 \tag{p}$$

Corresponding to these two cases, the σ_x distributions are: for $a = b$

$$\sigma_x = \sigma_0\left(1 - \frac{y^2}{b^2}\right) - 0.1702\sigma_0\left(1 - \frac{x^2}{a^2}\right)^2\left(1 - \frac{3y^2}{b^2}\right). \tag{q}$$

For $a \gg b$

$$\sigma_x = \sigma_0\left(1 - \frac{y^2}{b^2}\right) - 0.1094\sigma_0\left(1 - \frac{x^2}{a^2}\right)^2\left(1 - \frac{3y^2}{b^2}\right). \qquad \text{(r)}$$

At section $x = 0$, these yield for $a = b$

$$\sigma_x = \sigma_0\left(0.8298 - 0.4894\frac{y^2}{b^2}\right). \qquad \text{(s)}$$

For $a \gg b$

$$\sigma_x = \sigma_0\left(0.8906 - 0.6719\frac{y^2}{b^2}\right). \qquad \text{(t)}$$

An improved solution was obtained by Timoshenko with

$$\chi = \frac{\sigma_0 y^2}{2}\left[1 - \frac{1}{6}\left(\frac{y}{b}\right)^2\right] + (a^2 - x^2)^2(b^2 - y^2)^2(c_1 + c_2x^2 + c_3y^2) . \qquad \text{(u)}$$

For $a = b$, he found

$$c_1 = 0.0404\sigma_0/a^6 , \quad c_2 = c_3 = 0.0117\sigma_0/a^8 . \qquad \text{(v)}$$

We obtain an improved solution by the Kantorovitch method in Example 7.4.3.

EXAMPLE 7.4.3

Solve Example 7.4.2, for a plate which is 'long' in the x direction, using the Kantorovitch method and demonstrate St. Venant's principle.

We assume a trial solution in the form

$$\chi = \frac{\sigma_0}{2}\left[y^2 - \frac{1}{6}\frac{y^4}{b^2}\right] + X(x)(b^2 - y^2)^2 \qquad \text{(a)}$$

where $X(x)$ is a function of x alone and is to be determined. As in Example 7.4.2, the first term satisfies the traction conditions prescribed along the edges so that the second term must yield zero tractions there. We see that the second term and its normal derivative do, indeed, yield zero traction along $y = \pm b$ [notice that $\partial/\partial n = \pm(\partial/\partial y)$ there]. It follows that $X(x)$ must satisfy $X = \mathrm{d}X/\mathrm{d}n = \pm(\mathrm{d}X/\mathrm{d}x) = 0$ along $x = \pm a$.

Since there are no natural boundary conditions to be satisfied, we may use the 'simplified' expression (k) of Example 7.4.2, for the complementary energy $U^* = U$. From equation (a)

$$\frac{\partial^2\chi}{\partial x^2} = X''(b^2 - y^2)^2 , \quad \frac{\partial^2\chi}{\partial y^2} = \sigma_0\left(1 - \frac{y^2}{b^2}\right) - 4X(b^2 - 3y^2) \qquad \text{(b)}$$

so that

$$U^* = \frac{1}{2E}\iint\{X''(b^2 - y^2)^2 + \sigma_0\left(1 - \frac{y^2}{b^2}\right) - 4X(b^2 - 3y^2)\}^2\,\mathrm{d}x\mathrm{d}y . \qquad \text{(c)}$$

Integrating first with respect to y and inserting the limits, we have

$$U^* = \frac{1}{2E}\int_{-a}^{a} \{0.812698b^9(X'')^2 + 1.066667\sigma_0^2 + 25.60b^5X^2$$

$$+ 1.828571\sigma_0 X'' - 4.266667\sigma_0 b^3 X - 4.876191b^7 X''\}\mathrm{d}x \quad \text{(d)}$$

or $$U^* = \int_{-a}^{a} F(X, X'')\mathrm{d}x \,. \quad \text{(e)}$$

From Section 4.9, condition $\delta U^* = 0$ for compatibility, yields the Euler-Lagrange equation

$$\frac{\partial F}{\partial x} - \frac{\mathrm{d}^2}{\mathrm{d}x^2}\left(\frac{\partial F}{\partial X''}\right) = 0 \,. \quad \text{(f)}$$

Substituting from equation (d) in equation (f), we find

$$\frac{\mathrm{d}^4 X}{\mathrm{d}x^4} - \frac{6}{b^2}\frac{\mathrm{d}^2 X}{\mathrm{d}x^2} + \frac{31.5}{b^4}X = 2.625\,\frac{\sigma_0}{b^6} \quad \text{(g)}$$

or $$X'''' + AX'' + BX = C \quad \text{(h)}$$

with A, B, C constants.

The solution to this equation consists of a complementary function and a particular integral. For the latter, using the D operator method, say, we have

$$X_{PI} = 0.0833\sigma_0/b^2 \,. \quad \text{(j)}$$

Trying a complementary function $X_{CF} = e^{mx}$, we find four possible values for m given by

$$m^4 + Am^2 + B = 0 \,. \quad \text{(k)}$$

The roots of this equation are complex. Using De Moivre's theorem, they are found to be

$$m_{1,2,3,4} = \pm\alpha \pm i\beta \,, \quad \text{(l)}$$

where $\alpha = 2.93471/b$, $\beta = 1.61632/b$, $i = \sqrt{-1}$. (m)

The general solution for X is then of the form

$$X = \left(C_1 \sin\frac{\beta x}{b} + C_2 \cos\frac{\beta x}{b}\right)\exp\left(\frac{\alpha x}{\beta}\right) + \left(C_3 \sin\frac{\beta x}{b} + C_4 \cos\frac{\beta x}{b}\right)\exp\left(-\frac{\alpha x}{\beta}\right)$$

$$+ 0.0833\,\frac{\sigma_0}{b^2} \,. \quad \text{(n)}$$

where exp (a) denotes e^a.

For convenience, we will choose the origin of coordinates at the centre of the left hand edge of the plate of Fig. 7.4.2. Then, since X is to remain finite for 'large' x, we conclude immediately that C_1 and C_2 must be zero. In order that the second term in equation (a) shall contribute zero traction along edge $x = 0$, we must have $X(0) = X'(0) = 0$. These conditions lead to

$$C_3 = -0.1513\,\sigma_0/b^2\,, \qquad C_4 = -0.8333\,\sigma_0/b^2\,. \tag{p}$$

The stress function (a) thus takes the form

$$\chi = \frac{\sigma_0}{2}\left[y^2 - \frac{1}{6}\frac{y^4}{b^2}\right] + \frac{\sigma_0}{12}\left[1 - \left(0.1513\sin\frac{\beta x}{b} + \cos\frac{\beta x}{b}\right)\exp\left(-\frac{\alpha x}{\beta}\right)\right] \times \left[1 - \frac{y^2}{b^2}\right]^2. \tag{q}$$

Corresponding to this, the longitudinal stress component σ_x in the plate is given by

$$\sigma_x = \sigma_0\left[1 - \frac{y^2}{b^2}\right] - \frac{\sigma_0}{3}\left[1 - \left(0.1513\sin\frac{\beta x}{b} + \cos\frac{\beta x}{b}\right)\exp\left(-\frac{\alpha x}{\beta}\right)\right] \times \left[1 - \frac{3y^2}{b^2}\right]. \tag{r}$$

When x is 'large', equation (r) reduces to

$$\sigma_x = \tfrac{2}{3}\sigma_0\,. \tag{s}$$

Evidently the longitudinal stress component has smoothed out to a constant value, which is just that required to balance the resultant action over the end, illustrating St. Venant's principle.

It is interesting to see just how rapidly the stress σ_x evens out in the plate. Taking $x/b = 0, 0.5, 1.0, 1.5, 2.0$ in equation (r), we obtain the results shown in Table 7.4.1.

Table 7.4.1

x/b	σ_x/σ_0
0.0	$1.0000 - y^2/b^2$
0.5	$0.8206 - 0.4619y^2/b^2$
1.0	$0.6970 - 0.0940y^2/b^2$
1.5	$0.6685 - 0.0054y^2/b^2$
2.0	$0.6656 + 0.0032y^2/b^2$

The table shows that σ_x evens out very rapidly indeed as x increases: at a distance of only three-quarters the plate width, it is virtually constant at the correct value. The reader should contrast these results with equation (t) of Example 7.4.2: the Kantorovitch solution is very much superior to the one term Rayleigh-Ritz one. In the latter case, the constraints imposed on the form of χ do not allow it to represent the uniform σ_x away from the ends of the long plate at all well. The reader should visualise the Kantorovitch and Rayleigh-Ritz forms of the χ surface to understand this. See also Example 7.3.2.

To take account of the parabolic distribution of traction along $x = a$, we can repeat the above analysis for the right hand end of the plate, choosing a coordinate 'x' penetrating back into the plate. If the plate is 'short', the conditions at $x = 0$ and $x = a$ interact and coefficients C_1 and C_2 are no longer zero; the conditions $X = X' = 0$ at $x = \pm a$ generate four simultaneous equations from which $C_1 \ldots\ldots C_4$ may be determined. A much greater computational effort is required now, but as we see from above, we can avoid the extra effort if the plate is not too short: the more easily obtained 'long' plate results will be adequate if the length is as little as twice the width. We must remember, of course, that the distribution of traction determines acceptable proportions for 'longness'. The more peaky the traction distribution, the further along the plate will the uneven stress distribution penetrate.

EXAMPLE 7.4.4

The so-called 'thick curved bar' is one for which the radial depth is of the same order of magnitude as the radius of curvature of its centre line as shown in Fig. 7.4.3. We suspect that the bending stress does not vary linearly as for a straight beam and we wish to provide an improved theory by the complementary energy principle. The material obeys Hooke's law.

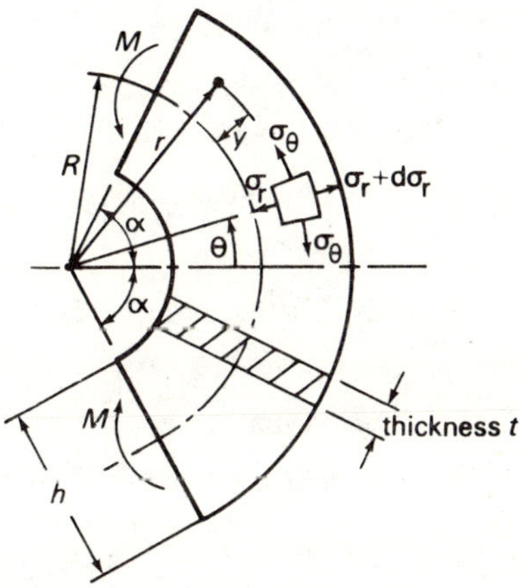

Fig. 7.4.3

From symmetry, we conclude that there is no shearing stress $\tau_{r\theta}$. Examining an element $rdrd\theta$ we see that, for radial equilibrium, a radial stress σ_r is required in addition to the bending stress σ_θ. However, for an approximate solution we shall tentatively assume that σ_r is very small compared with σ_θ, so that only the latter will contribute significantly to the complementary strain energy. As an improvement over the simple linear distribution of bending stress in straight bar, then, we assume

$$\sigma_\theta = A + By + Cy^2 , \quad \sigma_r = \text{negligible} . \tag{a}$$

Neglect of σ_r implies violation of equilibrium on a point by point basis, but, since pure bending is prescribed, we may satisfy equilibrium in an overall manner by requiring

$$\int_{-h/2}^{h/2} \sigma_\theta \, t\mathrm{d}y = 0, \tag{b}$$

and

$$\int_{-h/2}^{h/2} \sigma_\theta ty\mathrm{d}y = M. \tag{c}$$

Substituting from equations (a) in (b) and (c), we find that

$$A = -\frac{Ch^2}{12}, \tag{d}$$

and

$$B = \frac{12M}{th^3}, \tag{e}$$

so that

$$\sigma_\theta = \frac{12M}{th^3}y + \frac{Ch^2}{12}\left[\frac{12y^2}{h^2} - 1\right]. \tag{f}$$

Equation (f) represents a stress field which is *nominally* in equilibrium for all values of C. We now employ the complementary energy principle to determine the 'best' value of C for 'best' satisfaction of compatibility within the limitations we have set.

Tractions are prescribed over the entire surface of the bar so that $V^* = U^*$ and the complementary energy principle reduces to $\delta U^* = 0$.

Now

$$U^* = \int_{-\alpha}^{\alpha}\int_a^b \frac{\sigma_\theta^{\,2}}{2E} tr\mathrm{d}r\mathrm{d}\theta. \tag{g}$$

But $r = R + y$ so that $\mathrm{d}r = \mathrm{d}y$. Furthermore, σ_θ is independent of θ, so that

$$U^* = \frac{\alpha t}{E}\int_{-h/2}^{h/2} \sigma_\theta^{\,2}(R + y)\mathrm{d}y. \tag{h}$$

U^* is thus a function of the parameter C so that $\delta U^* = 0$ requires

$$\frac{\partial U}{\partial C} = 0 = \frac{2\alpha t}{E}\int_{-h/2}^{h/2} \sigma_\theta \frac{\partial \sigma_\theta}{\partial C}(R + y)\mathrm{d}y. \tag{j}$$

Substituting from equation (f) in (j), we find

$$C = -\frac{12M}{Rth^3}, \tag{k}$$

so that

$$\sigma_\theta = \frac{12My}{th^3}\left[1 - \frac{y}{R} + \frac{h^2}{12Ry}\right]. \tag{l}$$

We can write this as $\sigma_\theta = k\sigma_{\text{bend}}$ (m)

where σ_{bend} is the ordinary straight beam bending stress and k is a factor showing the effect of the curvature.

To test the adequacy of this solution, consider $R = h$, which represents a severely curved bar. In Table 7.4.2 we compare the exact solution, according to Timoshenko and Goodier, with our approximate one. The agreement is seen to be quite reasonable at $r = 3h/2$, but only fair at $r = h/2$.

Table 7.4.2

	$r = h/2$	$r = 3h/2$
$\left(\dfrac{th^2}{12M}\sigma_\theta\right)_{\text{ex}}$	-0.764	0.377
$\left(\dfrac{th^2}{12M}\sigma_\theta\right)_{\text{app}}$	-0.667	0.333
$\left(\dfrac{th^2}{12M}\sigma_\theta\right)_{\text{straight}}$	-0.500	0.500

The last figures show the straight beam result. The errors incurred in using these for this curved bar are seen to be very large indeed.

Problem 7.4

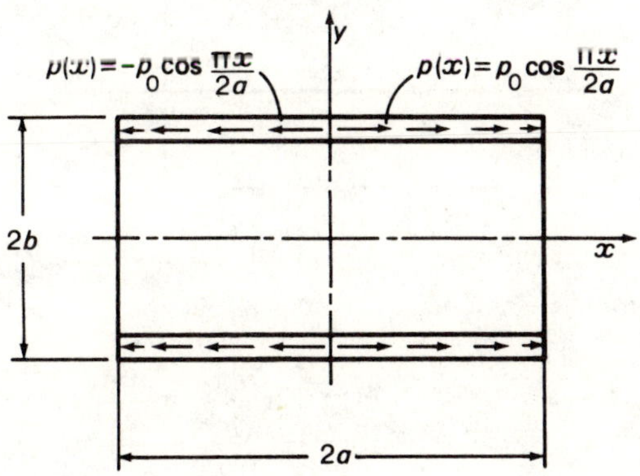

Fig. 7.4.4

7.4.1 A structural member consists of a thin rectangular sheet attached to two edge beams and is required, essentially, to support longitudinal distributed forces $p(x)$ per unit length of edge beam as shown in Fig. 7.4.4. Use the potential energy principle to describe, approximately, how the stresses are 'diffused' into the sheet. Assume a displacement pattern

$$u = u_0 \sin \frac{\pi x}{2a} \cosh \frac{k\pi y}{2b}, \qquad v = 0 \tag{a}$$

when $k = \sqrt{E/G}$ is a constant and u_0 is a parameter to be determined. The edge beams have a cross sectional area A and the sheet is t thick.

7.5 Plate bending problems

A simply supported rectangular plate carrying a transverse pressure presents one of the simplest problems of plate bending. We first use the potential energy principle to obtain the deflexion surface and examine the behaviour of such a rectangular plate in some detail.

EXAMPLE 7.5.1

Fig. 7.5.1 shows the plan form of a simply supported rectangular plate carrying a transverse pressure $p(x, y)$ over a rectangular patch of surface. Determine the response of the plate.

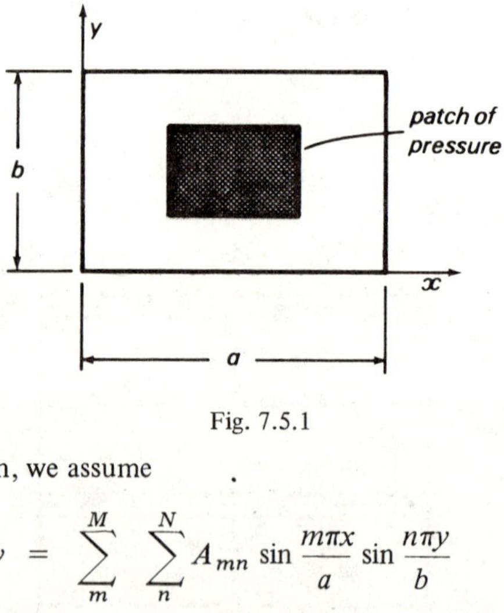

Fig. 7.5.1

To effect a solution, we assume

$$w = \sum_m^M \sum_n^N A_{mn} \sin \frac{m\pi x}{a} \sin \frac{n\pi y}{b} \tag{a}$$

from which

$$\frac{\partial^2 w}{\partial x^2} = -\sum_m \sum_n A_{mn} \left(\frac{m\pi}{a}\right)^2 \sin \frac{m\pi x}{a} \sin \frac{n\pi y}{b}, \tag{b}$$

$$\frac{\partial^2 w}{\partial y^2} = -\sum_m \sum_n A_{mn} \left(\frac{n\pi}{b}\right)^2 \sin \frac{m\pi x}{a} \sin \frac{n\pi y}{b}, \tag{c}$$

$$\frac{\partial^2 w}{\partial x \partial y} = \sum_m \sum_n A_{mn} \frac{\pi^2 mn}{ab} \cos \frac{m\pi x}{a} \cos \frac{n\pi y}{b} . \qquad \text{(d)}$$

We point out that not only does this series satisfy the geometric boundary conditions, which is essential for the potential energy method, but it also meets the mechanical condition of zero normal moment along the edges.

As we pointed out in Section 6.4, we may use a 'simplified' expression for the strain energy stored in a simply supported polygonal plate: then we can write

$$V = \frac{D}{2} \iint \left[\frac{\partial^2 w}{\partial x^2} + \frac{\partial^2 w}{\partial y^2} \right]^2 \mathrm{d}x\mathrm{d}y - \iint pw\mathrm{d}x\mathrm{d}y . \qquad \text{(e)}$$

Substituting from equations (a), (b), (c) in (e), carrying out the integrations would show that

$$V = V(A_{mn}\text{'s}) , \qquad \text{(f)}$$

so the A_{mn}'s constitute generalised coordinates.

For equilibrium $\delta V = 0$, or $\quad \dfrac{\partial V}{\partial A_{pq}} = 0 . \qquad$ (g)

That is,

$$0 = \frac{\partial V}{\partial A_{pq}} = D \iint [w_{xx} + w_{yy}] \left[\frac{\partial w_{xx}}{\partial A_{pq}} + \frac{\partial w_{yy}}{\partial A_{pq}} \right] \mathrm{d}x\mathrm{d}y$$

$$- \iint p \frac{\partial w}{\partial A_{pq}} \mathrm{d}x\mathrm{d}y , \qquad \text{(h)}$$

or
$$0 = D \int_0^b \int_0^a \Bigg\{ \Bigg[\sum\sum A_{mn} \left(\frac{m\pi}{a}\right)^2 \sin \frac{m\pi x}{a} \sin \frac{n\pi y}{b}$$

$$+ \sum\sum A_{mn} \left(\frac{n\pi}{b}\right)^2 \sin \frac{m\pi x}{a} \sin \frac{n\pi y}{b} \Bigg]$$

$$\times \left[\left(\frac{p\pi}{a}\right)^2 \sin \frac{p\pi x}{a} \sin \frac{q\pi y}{b} + \left(\frac{q\pi}{b}\right)^2 \sin \frac{p\pi x}{a} \sin \frac{q\pi y}{b} \right] \Bigg\} \mathrm{d}x\mathrm{d}y$$

$$- \iint_{\text{patch}} p \sin \frac{p\pi x}{a} \sin \frac{q\pi y}{b} \mathrm{d}x\mathrm{d}y , \qquad \text{(j)}$$

where $w_{xx} = \dfrac{\partial^2 w}{\partial x^2}$ etc.

Taking advantage of the orthogonality of the sets $\sin(m\pi x/a)$ and $\sin(n\pi y/b)$, and denoting the load term by I_{pq}, equation (j) yields

$$A_{pq} = \frac{4I_{pq}}{\pi^4 Dab[p^2/a^2 + q^2/b^2]^2} . \qquad \text{(k)}$$

As in Example 7.4.1, we obtain an explicit expression for A_{pq}, so that it is determined once and for all independently of how many terms are taken in the series.

To be explicit, we now suppose that $p(x, y)$ acts over the entire surface of the plate and is described by

$$p = p_0 \sin \frac{\pi x}{a} \sin \frac{\pi y}{b} . \tag{l}$$

Then,
$$I_{pq} = p_0 \int_0^a \sin \frac{p\pi x}{a} \sin \frac{\pi x}{a} \, dx \int_0^b \sin \frac{q\pi y}{b} \sin \frac{\pi y}{b} \, dy . \tag{m}$$

Evidently, I_{pq} is zero unless $p = q = 1$ and for this special load distribution,

$$A_{pq} = 0 , \quad \text{if } p \neq q \neq 1 ,$$

and
$$A_{11} = \frac{p_0 a^4 b^4}{\pi^4 D[a^2 + b^2]^2} , \tag{n}$$

so that
$$w = \frac{p_0 a^4 b^4}{\pi^4 D[a^2 + b^2]^2} \sin \frac{\pi x}{a} \sin \frac{\pi y}{b} . \tag{p}$$

It will be instructive now to see if the stress components $\sigma_z, \tau_{zx}, \tau_{zy}$, which were neglected in formulating the plate theory, do turn out to be small compared with σ_x, σ_y and τ_{xy}.† To fix ideas we let $a = b$.

First of all, we observe that σ_z is clearly of the order of p_0, denoted by $O(p_0)$, its value being $-p_0$ at $z = -h/2$.

Now from Section 6.4,

$$\sigma_x = \frac{-Ez}{1 - \nu^2} [w_{xx} + \nu w_{yy}] , \quad \sigma_y = \frac{-Ez}{1 - \nu^2} [w_{yy} + \nu w_{xy}] ,$$

and
$$\tau_{xy} = -\frac{E}{1 + \nu} z w_{xy} . \tag{q}$$

These three stress components will be greatest at the plate surface where $z = \pm h/2$. Then substituting from equation (p), we find that $\sigma_x, \sigma_y, \tau_{xy}$ are of the order of $p_0 a^2/h^2$

or
$$\sigma_x, \sigma_y, \tau_{xy} = O(p_0 a^2/h^2) . \tag{r}$$

Now for thin plates, $a \gg h$, so that since $\sigma_z = O(p_0)$, we see that it is negligible compared with $\sigma_x, \sigma_y, \tau_{xy}$.

To gain information about τ_{zx}, consider equilibrium in the x direction. Then

$$\frac{\partial \sigma_x}{\partial x} + \frac{\partial \tau_{xy}}{\partial y} + \frac{\partial \tau_{xz}}{\partial z} = 0 ,$$

† See Section 6.4.

when body forces are negligible. Substituting from equations (q), we find

$$-\frac{Ez}{(1-\nu^2)}\frac{\partial}{\partial x}(\nabla^2 w) + \frac{\partial \tau_{xz}}{\partial z} = 0 .$$

Integrating with respect to z,

$$\tau_{zx} = \frac{Ez}{2(1-\nu^2)}\frac{\partial}{\partial x}(\nabla^2 w) + f(x, y) .$$

$f(x, y)$ is an arbitrary function, but $\tau_{zx} = 0$ at $z = \pm h/2$ when only normal pressure is applied, so that

$$\tau_{zx} = -\frac{Eh^2}{8(1-\nu^2)}\frac{\partial}{\partial x}(\nabla^2 w)\left[1-\left(\frac{2z}{h}\right)^2\right] \tag{s}$$

or

$$\tau_{zx} = \frac{3}{2}Q_x\left[1-\left(\frac{2z}{h}\right)^2\right] . \tag{t}$$

Evidently τ_{zx} is greatest at the middle surface where $z = 0$. Substituting from equation (p) in (s)

$$\tau_{zx} = O\left(p_0\frac{a}{h}\right)$$

and similarly

$$\tau_{zy} = O\left(p_0\frac{a}{h}\right) . \tag{u}$$

Comparing equations (r) and (u), we find that τ_{zx} and τ_{zy} are engligible compared with σ_x, σ_y, τ_{xy}. Thus altogether we have given increased plausibility to the foundations of our theory for plates.

We now proceed to illustrate that there exist concentrated transverse reaction forces at the four corners of the plate. Following Section 6.4, we can calculate the effective shearing forces V_x and V_y to be

$$\left.\begin{aligned} V_x &= \left[\frac{\pi^3 DA_{11}(a^2+b^2)}{a^3b^2} + \frac{D(1-\nu)A_{11}\pi^3}{ab^2}\right]\cos\frac{\pi x}{a}\sin\frac{\pi y}{b} . \\ V_y &= \left[\frac{\pi^3 DA_{11}(a^2+b^2)}{a^2b^3} + \frac{D(1-\nu)A_{11}\pi^3}{a^2b}\right]\sin\frac{\pi x}{a}\cos\frac{\pi y}{b} . \end{aligned}\right\} \tag{v}$$

For the edges, we can calculate the resultant *upward* force, due to the distributed V_x and V_y, to be

$$R = \left|\int_0^a V_y\,\mathrm{d}x\right|_{y=0} - \left|\int_0^a V_y\,\mathrm{d}x\right|_{y=b} + \left|\int_0^b V_x\,\mathrm{d}y\right|_{x=0} - \left|\int_0^b V_x\,\mathrm{d}y\right|_{x=a}$$

or

$$R = \frac{4p_0ab}{\pi^2} + \frac{8(1-\nu)p_0}{\pi^4}\cdot\frac{a^3b^3}{(a^2+b^2)^2} . \tag{w}$$

The resultant downward force due to $p(x, y)$ is

$$P = \int_0^b \int_0^a p_0 \sin \frac{\pi x}{a} \sin \frac{\pi y}{b} \, dx dy$$

or

$$P = \frac{4p_0 ab}{\pi^2} . \tag{x}$$

Clearly P and R do not balance and, since all *distributed* transverse forces have been taken into account, there must be some concentrated ones left over. We show how to calculate these from the physical explanation of V_x and V_y given by Kelvin and Tait.

Fig. 7.5.2 shows a normal view of edge $x = a$, divided into overlaying elements of length Δy, with the couple M_{xy} replaced by equipollent pairs of equal, opposite forces at small distances Δy apart.†

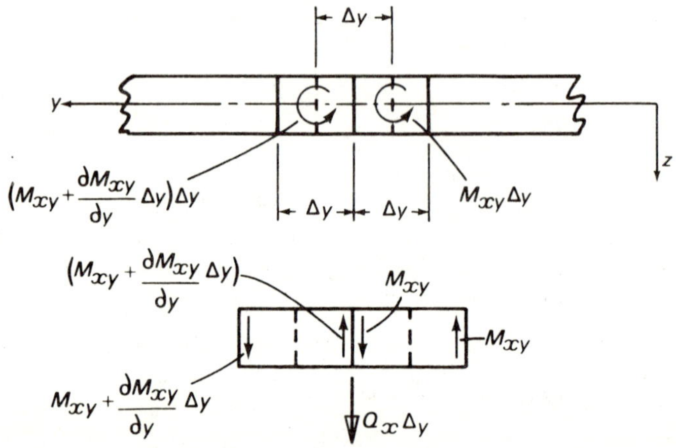

Fig. 7.5.2

For the dotted element, we see how the effective shearing force per unit length is

$$V_x = Q_x - \frac{\partial M_{xy}}{\partial y} .$$

If we now repeat this process as we travel to the left, we see that there is an unbalanced *downward* force M_{xy} remaining at the corner. Proceeding to the right, we end up with an unbalanced *upward* force M_{xy} at the corner. Following this argument through for all edges, we have the result summarised in Fig. 7.5.3.

† Since the plate thickness is small we may invoke St. Venant's principle to justify the claim that, in a very short distance in from the edge, the stress distribution is unaffected by this transformation.

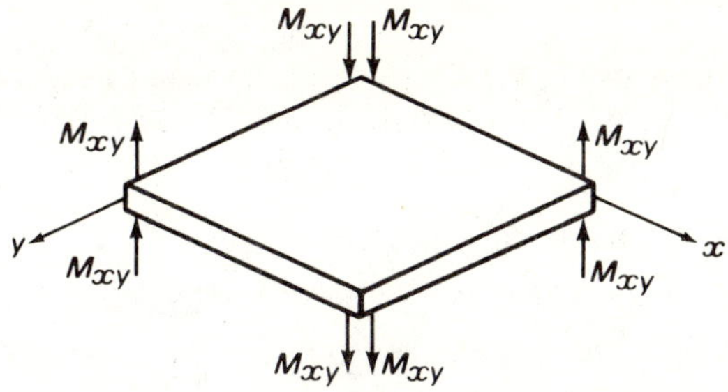

Fig. 7.5.3

Calculating M_{xy} for each corner for each edge and being very careful about the signs, we find the four concentrated forces shown in Fig. 7.5.4 with the notation

$$F_z = \frac{2(1-\nu)}{\pi^4}\frac{p_0 a^3 b^3}{(a^2+b^2)^2} . \tag{y}$$

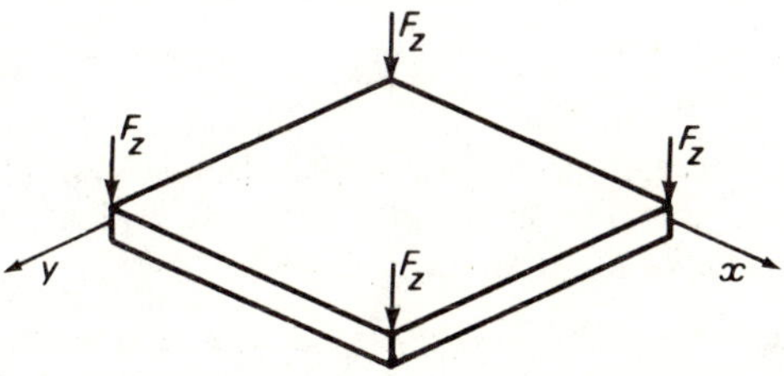

Fig. 7.5.4

Clearly, adding the contributions for all four corners, $4F_z$ is the additional downward force required to balance R of equation (w).

In conclusion we observe that by invoking the principle of super-position, these results may be applied to a simply supported plate supporting any load provided we represent it by a Fourier series

$$p(x, y) = \sum_{m=1}^{\infty}\sum_{n=1}^{\infty} p_{mn} \sin\frac{m\pi x}{a} \sin\frac{n\pi y}{b} . \tag{z}$$

Taking M and N to be infinite in equation (a) yields the exact solution.

The problem of the clamped rectangular plate cannot be treated as easily as the simply supported one, since $w(x, y)$ may no longer be expanded in an infinte trigonometric series. However, an approximate solution may readily be obtained by one of the direct methods. By way of illustration, we will employ Galerkin's technique.

EXAMPLE 7.5.2

The rectangular plate of Fig. 7.5.5 is clamped all round its edges and supports a uniformly distributed transverse load p_0. Determine the deflexion surface using Galerkin's method.

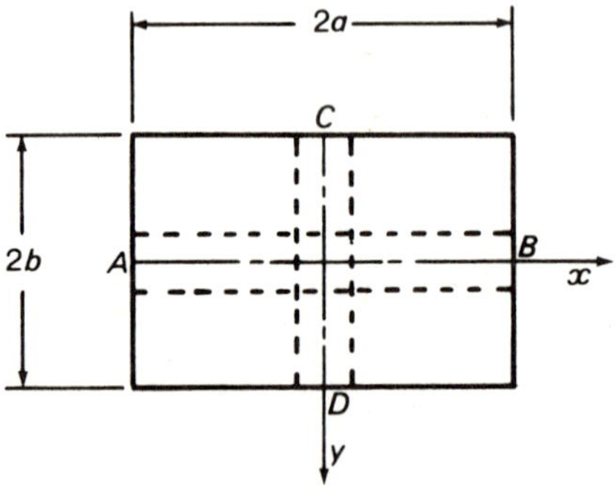

Fig. 7.5.5

In contrast to the preceeding example, we have labelled this plate to have sides $2a$ and $2b$. Equation (a) below shows why this is convenient, but the reader should notice that this change of notation has occurred.

To arrive at a reasonable trial function for the deflexion we may exercise some engineering judgement as follows. Isolating imaginary strips such as AB and CD, we can visualise them as deflecting rather like clamped-clamped beams subjected to uniformly distributed loads. Such problems are easily solved (or even more easily looked up in almost any book on strength of materials). In fact

$$w_{AB} = k_{AB}(x^2 - a^2)^2, \quad w_{CD} = k_{CD}(y^2 - b^2)^2, \tag{a}$$

where the k's depend on load and stiffness. Being guided by equation (a) we assume, as a first approximation,

$$w_1 = c_1\phi_1(x, y) = c_1(x^2 - a^2)^2(y^2 - b^2)^2 . \tag{b}$$

Clearly the prescribed geometric boundary conditions $w = \partial w/\partial n = 0$ are satisfied whilst there are no natural boundary conditions to worry about.

From Section 5.5, we see that determining c_1 by Galerkin's method requires us to form

$$\int_{-b}^{b}\int_{-a}^{a} [\nabla^4 w_1 - p_0/D]\phi_1(x, y)\mathrm{d}x\mathrm{d}y = 0 \tag{c}$$

or

$$\int_0^b\int_0^a \nabla^4 w_1\phi_1(x, y)\mathrm{d}x\mathrm{d}y = \int_0^b\int_0^a \frac{p_0}{D}\phi_1(x, y)\mathrm{d}x\mathrm{d}y . \tag{d}$$

From equation (b)

$$\nabla^4 w = 8c_1[3(y^2 - b^2)^2 + 4(3x^2 - a^2)(3y^2 - b^2) + 3(x^2 - a^2)^2].$$

The left hand side of equation (d) then becomes

$$\int_0^b \int_0^a 8c_1[3(y^2 - b^2)^2 + 4(3x^2 - a^2)(3y^2 - b^2) + 3(x^2 - a^2)^2] \times (x^2 - a^2)^2(y^2 - b^2)^2 \mathrm{d}x\mathrm{d}y. \quad \text{(f)}$$

Carrying out the integrations is straightforward but tedious. We observe symmetry: then, for some terms, the substitutions $x = a \sin\theta, y = b \sin\psi$ help and reduction formulae give the integrals immediately. In fact:

$$\text{L.H.S.} = 8c_1\left[\frac{3072}{4725}a^5b^9 + \frac{4096}{11025}a^7b^7 + \frac{3072}{4725}a^9b^5\right]. \quad \text{(g)}$$

The right hand side of equation (d) is

$$\text{R.H.S.} = \frac{p_0}{D}\int_0^a (x^2 - a^2)^2 \mathrm{d}x \int_0^b (y^2 - b^2)^2 \mathrm{d}y \quad \text{(h)}$$

or

$$= \frac{64}{225}\frac{p_0 a^5 b^5}{D}. \quad \text{(j)}$$

Then

$$c_1 = \frac{8}{225}\frac{p_0}{Db^4}\left[\frac{3072}{4725} + \frac{4096}{11025}\left(\frac{a}{b}\right)^2 + \frac{3072}{4725}\left(\frac{a}{b}\right)^4\right]^{-1}. \quad \text{(k)}$$

For a square plate side $2a = 2b$, we find the centre deflexion to be $0.0213 p_0a^4/D$ compared with $0.0202\, p_0a^4/D$ quoted by Timoshenko.

As a final example, we apply Kantorovitch's method to a plate with mixed boundary conditions.

EXAMPLE 7.5.3

The plate of Fig. 7.5.6 supports a transverse pressure $p(x, y) = \tilde{p}(x) \cdot \tilde{p}(y)$ over its entire surface. $\tilde{p}(x)$ and $\tilde{p}(y)$ are respectively functions of x and y alone. Use the Kantorovitch method to find the deflexion surface.

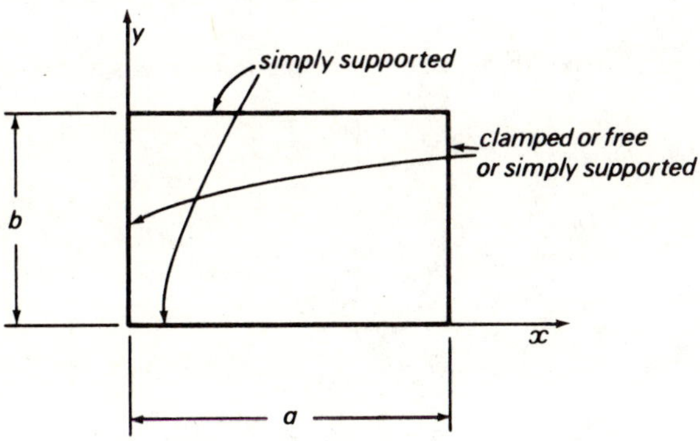

Fig. 7.5.6

We will assume a one term expansion

$$w = f(x)\phi(y) = f(x)\sin\frac{\pi y}{b} . \tag{a}$$

$$\text{Then } w_{xx} = f''\sin\frac{\pi y}{b}, \quad w_{yy} = -\frac{\pi^2}{b^2}f\sin\frac{\pi y}{b}, \quad w_{xy} = \frac{\pi}{b}f'\cos\frac{\pi y}{b} . \tag{b}$$

For the time being, we wish to have a free choice of boundary conditions along $x = 0$ and $x = a$. Then, although we have a polygonal plate, we must not use the 'simplified' expression for the strain energy stored. The starting point is thus

$$V = \frac{D}{2}\iint\{(\nabla^2 w)^2 - 2(1-\nu)[w_{xx}w_{yy} - (w_{xy})^2]\}\,\mathrm{d}x\mathrm{d}y - \iint pw\,\mathrm{d}x\mathrm{d}y \tag{c}$$

and

$$\delta V = D\iint\{\nabla^2 w\delta(\nabla^2 w) - (1-\nu)[w_{xx}\delta w_{yy} + w_{yy}\delta w_{xx} - 2w_{xy}\delta w_{xy}]\}\,\mathrm{d}x\mathrm{d}y$$
$$- \iint p\delta w\mathrm{d}x\mathrm{d}y . \tag{d}$$

$$\text{Now } \delta w_{xx} = \delta f''\sin\frac{\pi y}{b}, \quad \delta w_{yy} = -\frac{\pi^2}{b}\delta f\sin\frac{\pi y}{b}, \quad \delta w_{xy} = \frac{\pi}{b}\delta f'\cos\frac{\pi y}{b},$$

$$\delta(\nabla^2 w) = \left(\delta f'' - \frac{\pi^2}{b^2}\delta f\right)\sin\frac{\pi y}{b}, \quad \delta w = \delta f\sin\frac{\pi y}{b} . \tag{e}$$

Substituting from equation (e) in (d) and integrating first with respect to y, we find

$$\delta V = \frac{Db}{2}\int_0^a\left\{\left(f'' - \nu\frac{\pi^2}{b^2}f\right)\delta f'' + \left(\frac{\pi^4}{b^4}f - \nu\frac{\pi^2}{b^2}f''\right)\delta f + 2(1-\nu)\frac{\pi^2}{b^2}f'\delta f'\right\}\mathrm{d}x$$
$$- I_p\int_0^a\tilde{p}(x)\delta f\mathrm{d}x , \tag{f}$$

where

$$I_p = \int_0^b\tilde{p}(y)\sin\frac{\pi y}{b}\,\mathrm{d}y = \text{a definite number} . \tag{g}$$

Integrating by parts to obtain terms in δf alone under the integral sign

$$\delta V = \frac{Db}{2}\int_0^a\left\{f'''' - 2\frac{\pi^2}{b^2}f'' + \frac{\pi^4}{b^4}f - \frac{2I_p}{Db}\tilde{p}(x)\right\}\delta f\mathrm{d}x$$
$$+ \frac{Db}{2}\left|\left(f'' - \nu\frac{\pi^2}{b^2}f\right)\delta f'\right|_0^a - \frac{Db}{2}\left|\left(f''' - (2-\nu)\frac{\pi^2}{b^2}f'\right)\delta f\right|_0^a . \tag{h}$$

For equilibrium, $\delta V = 0$ and since δf is arbitrary over the plate surface, we have

$$\frac{\mathrm{d}^4 f}{\mathrm{d}x^4} - 2\frac{\pi^2}{b^2}\frac{\mathrm{d}^2 f}{\mathrm{d}x^2} + \frac{\pi^2}{b^2}f = \frac{2I_p}{Db}\tilde{p}(x) \tag{j}$$

as the governing differential equation for f. The boundary conditions are that at $x = 0$ and $x = a$,

$$\text{either} \qquad \frac{d^2 f}{dx^2} - \nu\frac{\pi^2}{b^2}f = 0 \quad \text{or} \quad \delta f' = 0 \tag{k}$$

$$\text{and either} \quad \frac{d^3 f}{dx^3} - (2 - \nu)\frac{\pi^2}{b^2}\frac{df}{dx} = 0 \quad \text{or} \quad \delta f = 0\,. \tag{l}$$

To proceed further, we will assume the plate is clamped along $x = 0$ and $x = a$ and that the transverse pressure is constant at p_0. Then,

$$I_p\widetilde{p}(x) = \frac{2p_0 b}{\pi}\,. \tag{m}$$

The solution to equation (j) consists of a complementary function and a particular integral. To obtain the former, assume $f = e^{mx}$, then

$$\frac{d^4 f}{dx^4} - 2\frac{\pi^2}{b^2}\frac{d^2 f}{dx^2} + \frac{\pi^4}{b^4}f = 0$$

$$\text{yields} \qquad m = \pm\frac{\pi}{b} \text{ and } \pm\frac{\pi}{b}\,.$$

That is, two pairs of equal roots so that

$$f_{CF} = (A + Bx)e^{m_1 x} + (C + Dx)e^{m_2 x}\,, \tag{n}$$

$$\text{where} \qquad m_1 = \frac{\pi}{b}\,, \quad m_2 = -\frac{\pi}{b}\,. \tag{p}$$

Using the operator D method, say,

$$f_{PI} = \frac{4p_0 b^4}{\pi^5 D} = -K\,, \text{ say}\,. \tag{q}$$

The general solution is thus

$$f = (A + Bx)e^{m_1 x} + (C + Dx)e^{m_2 x} + \frac{4p_0 b^4}{\pi^5 D}\,. \tag{r}$$

The values of A, B, C and D must be adjusted to suit the boundary conditions along $x = 0, x = a$. In our case we require $f = 0, f' = 0$ there so that

$$\left.\begin{aligned}
&A + C = K\,.\\
&e^{\pi a/b}A + ae^{\pi a/b}B + e^{-\pi a/b}C + ae^{-\pi a/b}D = K\,.\\
&\frac{\pi}{a}\left(\frac{a}{b}\right)A + B - \frac{\pi}{a}\left(\frac{a}{b}\right)C + D = 0\,.\\
&\frac{\pi}{a}\left(\frac{a}{b}\right)e^{\pi a/b}A + \left(1 + \frac{\pi a}{b}\right)e^{\pi a/b}B - \frac{\pi}{a}\left(\frac{a}{b}\right)e^{-\pi a/b}C + \left(1 - \frac{\pi a}{b}\right)e^{-\pi a/b}D = 0\,.
\end{aligned}\right\} \tag{s}$$

Equations (s) may be readily solved for A, aB, C, aD for a range of $r = a/b$ with the aid of a very simple computer programme. Indeed, for $a = b$

$$\left.\begin{array}{ll} A = 0.13949K\,, & B = 0.10231K/a\,, \\ C = 0.86051K\,, & D = 2.36744K/a\,. \end{array}\right\} \tag{t}$$

The centre deflection is given by

$$w_{\text{centre}} = f(a/2)\sin\pi/2\,, \tag{u}$$

or

$$w_{\text{centre}} = 1.962 \times 10^{-3}(p_0 a^4/D)\,. \tag{v}$$

This compares with a value of

$$w_{\text{centre}} = 1.914 \times 10^{-3}(p_0 a^4/D) \tag{w}$$

quoted by Timoshenko.

To treat various categories of load merely require changes in K whereas new boundary conditions require the setting up of new equations corresponding to (s).

Problems 7.5

7.5.1 The deflexion at a point (x, y) on a plate due to a unit point load at point (ξ, η) is called the **influence function** $a(x, y, \xi, \eta)$. For a simply supported rectangular plate, sides $a \times b$, assume a deflexion surface

$$w = \sum_{m=1}^{\infty}\sum_{n=1}^{\infty} a_{mn}\sin\frac{m\pi x}{a}\sin\frac{n\pi y}{b}\,. \tag{a}$$

Use the potential energy principle to show that $a(x, y, \xi, \eta) =$

$$\frac{4}{\pi^4 Dab}\sum_{m=1}^{\infty}\sum_{n=1}^{\infty}\left(\frac{1}{(m^2/a^2 + n^2/b^2)^2}\right)\sin\frac{m\pi x}{a}\sin\frac{n\pi y}{b}\sin\frac{m\pi\xi}{a}\sin\frac{n\pi\eta}{b}\,. \tag{b}$$

Notice that $a(x, y, \xi, \eta) = a(\xi, \eta, x, y)$ as we would expect from the recprocal theorem.

7.5.2 Treat the plate of Example 7.5.3 by the Kantorovitch method using the trial function

$$w = f(x)\left[1 - \left(\frac{2y}{b}\right)^2\right].$$

Place the origin of coordinates at the centre of one of the short sides.

7.5.3 When the deflexions of plates are not small compared with the thickness, the stretching of the middle surface becomes a significant factor in the analysis. To obtain an approximate solution for the large deflexion of a clamped circular plate, radius a, proceed as follows:

Assume that the deflexion w and the radial displacement u in the middle surface can be approximated by

$$w = w_0\left[1 - \left(\frac{r}{a}\right)^2\right]^2, \qquad u = r(a - r)(c_1 + c_2 r). \tag{a}$$

Then, accepting that the total strain energy stored in the plate is given by

$$U = U_b + U_d , \tag{b}$$

where U_b and U_d are the bending and direct strain energies respectively, show that the potential energy principle yields a cubic in w_0 which can be written

$$w_0 = \frac{pa^4}{64D}\left[1 + 0.488\left(\frac{w_0}{h}\right)^2\right]^{-1} . \tag{c}$$

$$\text{Accept that } U_b = \frac{D}{2}\int_0^r 2\pi\left[\left(\frac{\partial^2 w}{\partial r^2}\right)^2 + \frac{1}{r^2}\left(\frac{\partial w}{\partial r}\right)^2 + \frac{2\nu}{r}\frac{\partial w}{\partial r}\frac{\partial^2 w}{\partial r^2}\right] r\mathrm{d}r \tag{d}$$

$$\text{and} \qquad U_d = \frac{Eh^2}{2(1 - \nu^2)}\int_0^r 2\pi[\epsilon_r^2 + \epsilon_\theta^2 + 2\nu\epsilon_r\epsilon_\theta]\, r\mathrm{d}r , \tag{e}$$

$$\text{where} \qquad \epsilon_r = \frac{\mathrm{d}u}{\mathrm{d}r} + \frac{1}{2}\left(\frac{\mathrm{d}w}{\mathrm{d}r}\right)^2, \qquad \epsilon_\theta = \frac{u}{r} . \tag{f}$$

7.5.4 A simply supported variable thickness circular plate, radius a, carries a uniform pressure p. The bending stiffness is given by

$$D = D_0\left(1 - \frac{r}{2a}\right). \tag{a}$$

Assuming that the deflexion surface is approximated by

$$w = w_0\left[1 - \left(\frac{r}{a}\right)^2\right], \tag{b}$$

use the potential energy principle to show that

$$w_0 = \frac{3}{32}\,\frac{pa^4}{(1 + \nu)D_0} . \tag{c}$$

A mass is supported by a spring of stiffness k suspended from the centre of the above plate. Denoting the mass and plate centre deflexions by generalised coordinates q_1 and q_2 respectively, show that the stiffness matrix for the discretised assembly is

$$[K] = \begin{bmatrix} k & -k \\ -k & \left(\dfrac{16\pi(1 - \nu)D_0}{3a^2} + k\right) \end{bmatrix} \tag{d}$$

7.6 Further reading

Argyris, J.H. *and* Kelsey, S.	*Energy Theorems and Structural Analysis,* Butterworth (1960).
Bienzo, C.B. *and* Grammell, R.	*Engineering Dynamics,* Blackie (1955).
Duncan, W.J.	*Galerkin's Method in Mechanics and Differential Equations,* A.R.C. R. and M. No. 1798 (1937).
Dym, C.L. *and* Shames, I.H.	*Solid Mechanics: A Variational Approach,* McGraw Hill (1973).
Kantorovitch, L.V. *and* Krylov, V.I.	*Approximate Methods of Higher Analysis,* Noordhoff (1964).
Kelvin, W.T. *and* Tait, P.G.	*Elements of Natural Philosphy,* 2nd edition, Cambridge University Press (1885).
Langhaar, H.L.	*Energy Methods in Applied Mechanics,* John Wiley (1962).
Mansfield, E.H.	*The Bending and Stretching of Plates,* Pergamon (1964).
Sokolnikoff, I.S.	*The Mathematical Theory of Elasticity,* 2nd edition, McGraw Hill (1956).
Timoshenko, S. *and* Woinowsky-Kreiger, S.	*Theory of Plates and shells,* 2nd edition, McGraw Hill (1959).
Timoshenko, S. *and* Goodier, J.N.	*Theory of Elasticity,* 3rd edition, McGraw Hill (1970).
Williams, D.	*Introduction to the Theory of Aircraft Structures,* Arnold (1960).

Chapter 8

The Finite Element Method

8.1 Introduction

The examples of Chapter 7 illustrate the power of the direct methods of the variational calculus for obtaining approximate solutions to very difficult continuum problems. Central to these processes is the need to *assume* some admissible distribution of the wanted quantities in terms of functions defined throughout the region of the solid: the net effect is to transform the continuum with its infinite degrees of freedom, into a substitute finite degree of freedom system whose characteristics approximate those of the original model. Generating suitable approximating functions to satisfy the prescribed boundary conditions can be done fairly readily when the geometries of the solids of interest are reasonably simple: however, as the regions become more complex in shape, the task becomes increasingly difficult until, for most solids of practical interest, it becomes quite impossible. In the **finite element method,** these difficulties are avoided by the device of imagining the solid divided into a number of finite sized and conveniently shaped sub-regions or elements. The distribution of the wanted quantities is assumed to be relatively simple within these elements but by using a sufficient number of them, an acceptable representation of the overall real situation is obtained. The net effect is again to generate a finite degree of freedom substitute system: however, in contrast to the classical Ritz method, all the integrations required to define the appropriate functional must be evaluated in a piecewise manner from element to element.

The treatment of framed structures as assemblages of bars, having well understood individual behaviour, goes back a long way in the history of stress analysis. However, the advances made in the design of digital computers in the 1940's and 1950's gave a new impetus to the development of the subject. The newly available computing power allowed the solution of structural problems of great complexity and this, in turn, influenced the development of structural theory. Automatic computation demands highly systematic presentation and manipulation of data: the techniques of matrix algebra meet these requirements, and their embodiment in the so-called **matrix theory of structures** represents the beginning of the finite element method as a significant tool of engineering analysis. Initially, structural mechanics reasoning was used to extend these matrix methods to continuum problems and it was not until the early 1960's that the stiffness finite element was formulated in terms of the **principle of stationary total potential energy.** Recognition that the method could be built upon variational principles led to developments which would otherwise have been impossible: thus, a rational basis was provided for determining generalised nodal forces corresponding to distributed surface forces, body force, initial strains etc; formulations based on the complementary energy and Reissner principles appeared and applications to non-structural problems became possible.† More recently, the scope of the finite element technique has been widened even further by formulation in terms of the method weighted residuals; a discrete formulation may therefore be established whether the problem of interest is based on a variational principle or not.

The finite-element method has developed from being a refinement of structural theory to a powerful tool in mathematical physics with piece-wise Rayleigh-Ritz or Galerkin method interpretations. In this chapter, we aim to give a brief introduction to the subject as a natural follow-up to the earlier parts of the book; the reader wishing to pursue the topic in greater depth will find a starting point to a vast literature in the books and papers cited at the end of this chapter.

8.2 Outline of the finite element process

In applying piecewise methods to the approximate solution of continuum problems, the procedures are analogous to those described in Chapter 2. Thus, the region occupied by the solid is imagined as divided into a finite number of sub-regions by a mesh of surfaces. The inter-sections of these surfaces define a number of space curves which intersect at nodal points. Within each sub-region, or finite element, the distributions of the wanted functions (displacements and/ or stresses) are assumed in a suitable form with some associated parameters: the form of function is commonly a polynomial and the parameters may be identified with nodal values of the wanted functions and their derivatives. On the basis of these assumed distributions, an element property, such as strain energy or complementary strain energy, may be computed and a summation over all the elements yields the total strain energy (or complementary strain energy) stored in the solid. When the actions of the applied loads are taken into account, expressions for the total potential energy (or complementary energy) are obtained and an application of an appropriate variational principle yields sets of algebraic equations in which the nodal values of the wanted functions appear as the basic unkowns to be determined.

The process of solving a stress analysis problem by the finite element method can be conveniently summarised into five essential steps: for the so-called displacement or stiffness formulation, these are as follows:

(i) DEFINITION OF THE FINITE ELEMENT MESH. Depending on the problem at hand, it will be appropriate to discretise the continuum either into line, two dimensional or three dimensional sub-regions. The basis of the usual

† It is interesting to notice that as early as 1941 Courant showed, almost as an aside, how the torsion problem could be treated on the basis of a piece-wise Rayliegh Ritz method: his reasoning was essentially that used in the finite element method at present. Unfortunately this work did not appear in the main stream of engineering literature so that the practical significance of his approach was, apparently, not appreciated until some years later.

engineering theory of bars is that the deformations (and hence stresses) are characterised by the shape taken up by the centre line of the bar: the theory is thus one dimensional with bars represented by straight lines or curves. In Chapter 2, we saw that frames can be treated as assemblages of line elements: evidently an individual bar may also be treated as an assemblage of elements of length as shown in Fig. 8.2.1. Plane stress and plane strain situations are described by two dimensional theories so that finite elements for such problems must be planar figures: these may be straight sided or curvilinear triangles and quadrilaterals as shown in Fig. 8.2.2. The general three dimensional solid may be divided up into the tetrahedra of Fig. 8.2.3, or more general shapes, whilst if the solid enjoys axial symmetry in its geometry, the ring type elements of Fig. 8.2.4 may be used.

(ii) SELECTION OF DISPLACEMENT MODELS. The functions chosen to describe the (assumed) displacement pattern within a given element must meet certain criteria which will be described in Section 8.3 below. Usually the same form of function is used in all the elements of the discretisation, but it is not essential that this be so: if different types of elements are used in juxtaposition, it will be clear that these functions will need to be different in the different types of element if the criteria are to be met. The use of polynomial interpolating functions proves to be particularly convenient.

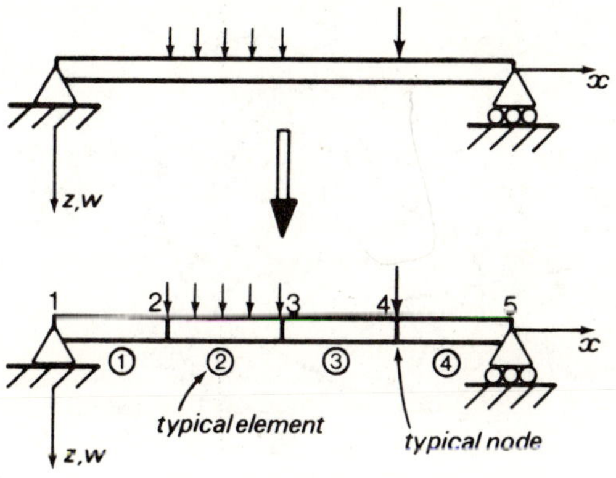

Fig. 8.2.1

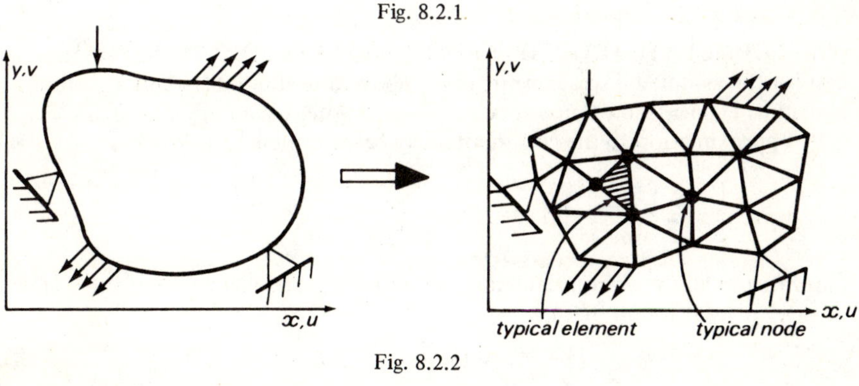

Fig. 8.2.2

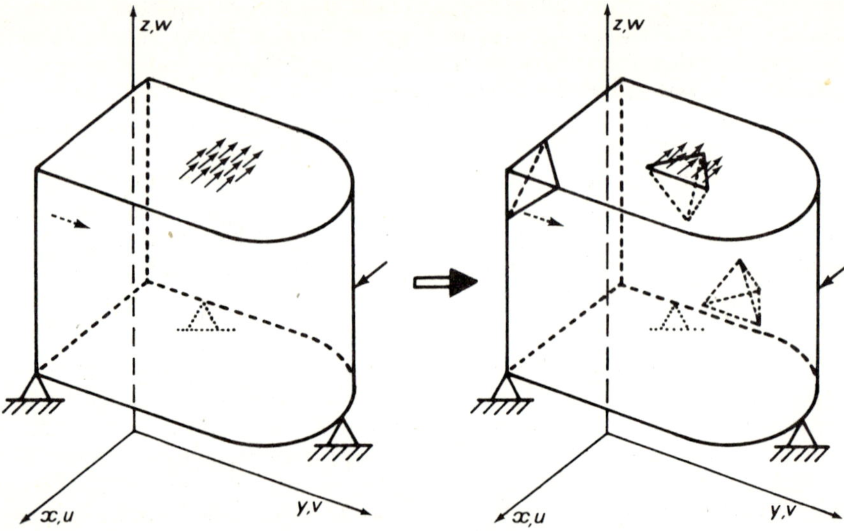

Fig. 8.2.3

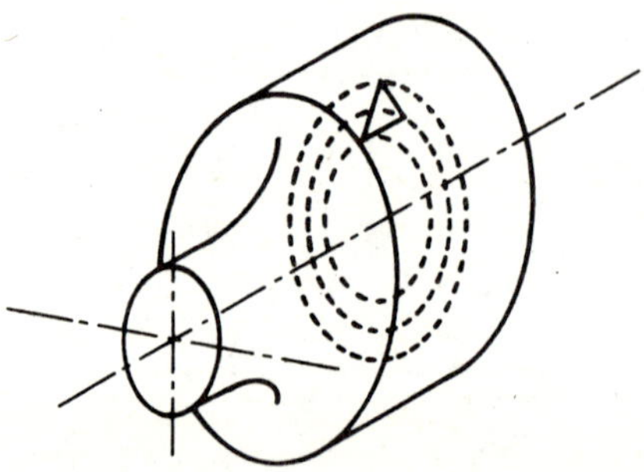

Fig. 8.2.4

(iii) FORMULATING THE DISCRETE STIFFNESS EQUATIONS. On the basis of the assumed displacements of (ii) above, the strain distribution within individual elements and consequently the total potential energy of the discrete approximation to the continuum may be determined from

$$V = U + \Omega = \sum_{\text{elements}} (U^e + \Omega^e) \tag{a}$$

Carrying out all the appropriate integrations (in a piecewise manner from element to element) it is found that

$$V = V(q_1, q_2 \ldots\ldots, q_n)\,. \tag{b}$$

Here, $q_1, q_2 \ldots\ldots, q_n$ are the generalised coordinates of the discretised system, usually they correspond to nodal displacements. For a linear elastic solid, U is a homogeneous quadratic function of the q's whilst for usual types of external forces, Ω is a linear function of the q's. Imposing the condition $\delta V = 0$ for equilibrium leads, in the manner typified in Chapters 1, 2 and 7, to the stiffness equations

$$[K]\{q\} = \{Q\} \; . \tag{c}$$

Here $[K]$ is a matrix of stiffness coefficients for the discrete system as a whole; it is synthesised from the element stiffness matrices in accordance with

$$U = \sum_e U^e \, .$$

$\{Q\}$ is a column vector of prescribed generalised forces corresponding to $\{q\}$.

(iv) SOLUTION OF THE STIFFNESS EQUATIONS. The solution of equations (c) is a standard procedure in matrix algebra; a number of equation solving routines, usually based on the Gaussian elimination process, being available at all computer centres. For the discretisation to yield an acceptable representation of the continuum system, the number of unknowns is very large so that solution cannot be contemplated without the aid of a high speed digital computer. $[K]$ is symmetric and the discretisation renders it sparsely populated, (that is, many of its elements are zero) and by suitable choice of nodal ordering, the non-zero elements of $[K]$ are clustered in a narrow band around its leading diagonal. The economical solution of all but the most trivial problems requires that these properties be fully exploited in drawing up computer programmes.

(v) DETERMINING ELEMENT STRAINS AND STRESSES. Once the nodal displacements have been determined, the element strains are readily calculated from the displacement model using strain-displacement relations typified in Section 3.2. The stresses are then obtainable by means of Hooke's law and, since they are proportional to derivatives of the displacements, they will not be as accurate as the displacements themselves. Equilibrium conditions are only satisfied in some average manner and not point by point: at particular stations within an element and on interelement boundaries, equilibrium conditions are violated.

The displacement method generates a substitute structural system which is stiffer than the true one which it approximates when compatible displacement fields are used. Increasing the fineness of the finite element mesh generates more degrees of freedom and the true stiffness is consequently approached from above.

8.3 Formulating a displacement model

In this chapter, we shall develop the finite element method by consideration of a series of increasingly sophisticated stressing situations. We shall see that once a suitable displacement pattern (or model) has been assumed to apply to a given element, the subsequent procedure, leading to the solution of the discretised problem, is standard. Indeed, we may recall that with the Rayleigh-Ritz method used in Chapter 7, the procedure was also perfectly standard once the displacement pattern was assumed. The choice of the displacement model is thus crucially important to the process and requires some attention.

Polynomial series are the most widely used approximating functions in finite element work. Their attraction stems mainly from (i) the ease with which various mathematical manipulations may be performed, (ii) the fact that truncating series at various levels yields recognisable degrees of approximation and (iii) that the criteria discussed below are usually readily met.

In a one dimensional situation (bar extension or flexure), the wanted function may be approximated by

$$u^e(x) \doteqdot u_n^{\,e}(x) = a_0 + a_1x + a_2x^2 + \ldots\ldots a_nx^n \;.$$

That is,

$$u_n^{\,e}(x) = [1\; x\; x^2 \ldots\ldots x^n] \begin{Bmatrix} a_0 \\ a_1 \\ \cdot \\ \cdot \\ a_n \end{Bmatrix}$$

or

$$u_n^{\,e}(x) = [F(x)]\,\{a\} \;. \tag{8.3.1}$$

Evidently, if the true solution is a polynomial of degree n, suitable adjustment of the a's allows $u_n^{\,e}(x)$ to represent $u^e(x)$ exactly, otherwise $u_n^{\,e}(x)$ can only be an approximation.†

In plane strain or plane stress situations, we are concerned with two components of displacement $u(x, y)$ and $v(x, y)$ in the x and y directions respectively. Then, for finite element purposes, we could assume

$$u^e(x, y) = a_0 + a_1x + a_2y + a_3x^2 + a_4xy + a_5y^2 + \ldots\ldots$$

$$v^e(x, y) = b_0 + b_1x + b_2y + b_3x^2 + b_4xy + b_5y^3 + \ldots\ldots$$

or

$$\begin{Bmatrix} u(x, y) \\ v(x, y) \end{Bmatrix}^e = \begin{bmatrix} [F(x, y)] & 0 \\ 0 & [F(x, y)] \end{bmatrix} \begin{Bmatrix} \{a\} \\ \{b\} \end{Bmatrix} . \tag{8.3.2}$$

Similarly, in three dimensions,

$$\begin{Bmatrix} u(x, y, z) \\ v(x, y, z) \\ w(x, y, z) \end{Bmatrix}^e = \begin{bmatrix} [G(x, y, z)] & 0 & 0 \\ 0 & [G(x, y, z)] & 0 \\ 0 & 0 & [G(x, y, z)] \end{bmatrix} \begin{Bmatrix} \{a\} \\ \{b\} \\ \{c\} \end{Bmatrix} .$$

In equations (8.3.2) and (8.3.3), we have used the same functional form for u and v and u, v and w respectively: this is usually the case although it is not essential.

† Fig. 7.2.5 illustrated how the true solution was approached by taking more terms in a series representing the bending moment diagram for a cantilever beam.

The approximating function chosen for the displacement model must be capable of reproducing certain features of the true displacements if convergence to the true solution† is to be obtained as more and more elements are used. To meet these requirements, the choice of a suitable polynomial is governed by the following considerations.

(i) NUMBER OF TERMS IN THE SERIES. By taking a finite number of terms in a power series to represent the element displacement field we are, in effect, forcing it to have only a finite number of degrees of freedom. To give physical significance to the coefficients in the series, we may express†† them in terms of the displacements at specified stations (nodes) in the element. To this end, we may use external nodes (that is, those on the element boundaries and which therefore are associated with more than one element), either alone or together with internal nodes (those associated with only one element). Thus if in the uniaxial element of Fig. 8.3.1 a two term series is used, evaluation of the two coefficients requires two nodal values of displacements, one external node at each end of the element. If a more refined element with a four term series is desired we may proceed in a variety of ways. Thus, we might: (a) add two interior nodes (3 and 4 in Fig. 8.3.1) and use the four nodal displacements as generalised coordinates; (b) retain only two nodes, 1 and 2, and treat two of the coefficients as nodeless parameters; (c) use the values of the displacements and their first derivatives at nodes 1 and 2 as nodal generalised coordinates.

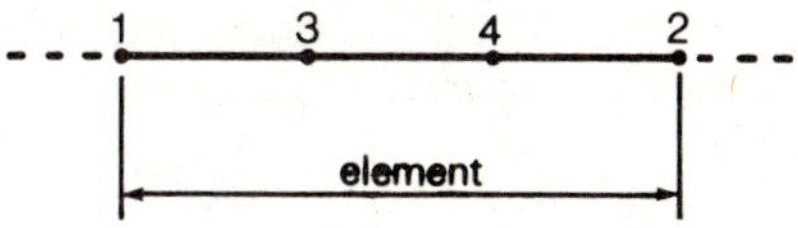

Fig. 8.3.1

In general, then, the total number of degrees of freedom of an element may comprise nodal displacements, rotations, strains etc., (the nodal degrees of freedom) plus any additional nodeless degrees of freedom chosen. The number of terms in the chosen series must clearly equal this *total* number of degrees of freedom.

(ii) COMPATIBILITY. The approximating function and certain of its derivatives must be continuous within the element and there must be compatibility between adjacent elements. The governing equations for the discretised system represent the stationary conditions for the potential energy functional: evidently, the approximating function must be differentiable at least to the extent called for in the functional for it to exist. Compatibility between adjacent elements requires that the chosen approximating function does not imply openings or

† The criteria will only ensure convergence in energy, a situation similar to that obtaining for the classical Rayleigh-Ritz process as pointed out in Chapter 5.

†† Notice that there is not a one to one correspondence between any one of the coefficients and only one of the nodal displacement components: the nodal displacements will be linear combinations of the coefficients.

overlappings between elements: this condition is met if the approximating function is such that there is at least continuity, between adjacent elements, of derivatives of order one less than that appearing in the functional. For example, we have seen that the strain energy in a bent beam involves the second derivative of the deflexion function: then, full compatibility between adjacent bending elements requires continuity of slope as well as deflexion.† We readily see that a polynomial of the third degree meets the requirements.

Elements meeting these conditions are said to be **conforming**. The conditions must be met to ensure that the solution converges in energy from below to the true one as a finer mesh used. As in example 7.2.1, we do not obtain a lower bound to the displacement at any chosen station; the displacements are too small only at the generalised coordinates.

(iii) COMPLETENESS. The simplest displacement pattern imaginable is a rigid body one for which there is no straining; next in simplicity is a state of constant strain: we see intuitively that the chosen approximating function must be capable of representing these two conditions. First, if the structure as a whole receives a rigid body motion, the nodal displacements must imply zero strain when substituted in the approximating function. Second, as a finer and finer mesh is used, the state of strain within an element becomes closer and closer to being constant: the approximating function must also be capable of accommodating this. These two conditions imply that a power series representation must begin with constant and linear terms.

Having set the scene with a general outline of the finite element process, we can best proceed by considering some specific applications: this we now do.

8.4 Displacement finite element formulation for bar extension

As a first application, we consider a tapered bar under a simple uniaxial loading system as shown in Fig. 8.4.1(a). The material obeys Hooke's law.

The first step is to define the finite element mesh. In this case, an element is merely a finite portion of length of the bar so that the discretisation is a linear sequence of elements interconnected at their end faces only as shown in Fig. 8.4.1 (b). The mesh defines elements and nodes and these have been numbered on a **global** or **system scale** as elements 1 to n and nodes 1 to $(n + 1)$. We know from the usual engineering (or mechanics of materials) theory of bars that, provided the taper is not too severe, the deformation is characterised by the axial displacement $u(x)$ of a cross section. Then, on a *global* scale we may label the nodal displacements as $u_1, u_2, \ldots\ldots u_{n+1}$, being quantities which characterise the deformation of the bar as a whole with the global or system reference frame xOy. When concentrating out attention on a typical element e, it proves con-

† Continuity of slope implies only a finite discontinuity in second derivative. Thus, we are able to state that $U = \sum_e U_e$ since integration for strain energy over an infinitely small region astride a common node will not yield any spurious contribution to U over and above each U_e.

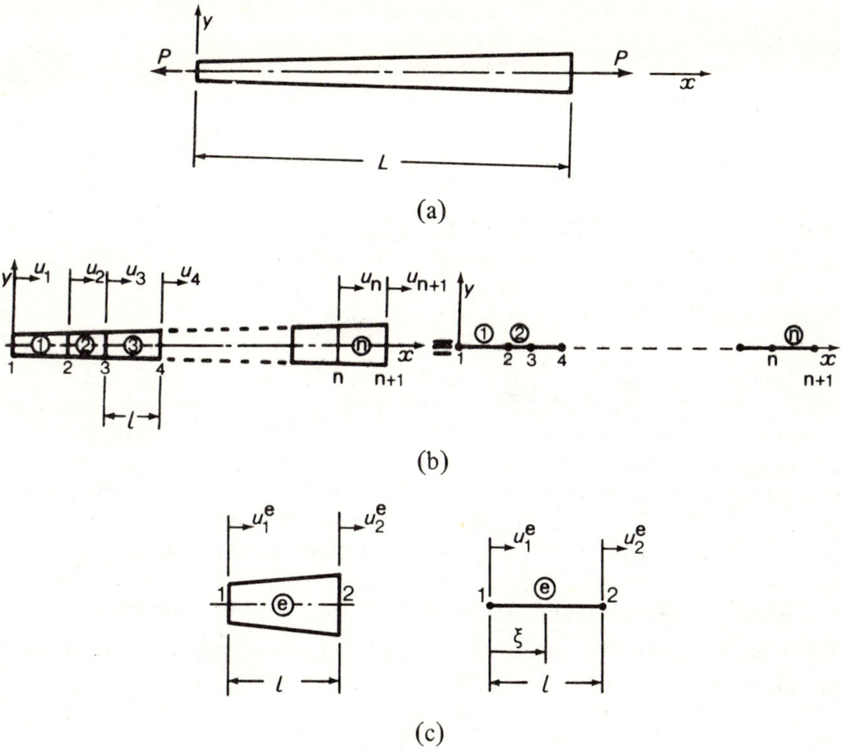

Fig. 8.4.1

venient to introduce a coordinate ξ local to it and to label its external nodes as 1 and 2 as shown in Fig. 8.4.1 (c). Then, on a local or element scale, deformation is characterised by the element nodal displacements u_1^e and u_2^e.

The next step is to select a displacement model for the element. Now, our knowledge of mechanics of materials tells us that the state of strain varies along the length of a tapered bar so that $u(x)$ will be a non-linear function of position in the bar. However, for the purposes of our approximate solution, we will *assume* that the displacement does, indeed, vary linearly within a typical element. The approximating function for u is thus

$$u^e(\xi) = a_0 + a_1\xi = [1\ \xi]\begin{Bmatrix} a_0 \\ a_1 \end{Bmatrix} = [f(\xi)]\{a\}\,. \qquad (8.4.1)$$

Referring to Fig. 8.4.1 (c), with $\xi = 0$ and $\xi = l$ in turn,

$$\begin{Bmatrix} u_1^e \\ u_2^e \end{Bmatrix} = \begin{bmatrix} 1 & 0 \\ 1 & l \end{bmatrix}\begin{Bmatrix} a_0 \\ a_1 \end{Bmatrix} \qquad (8.4.2)$$

or

$$\{u\}_0^e = [A]\{a\}\,. \qquad (8.4.3)$$

Solving for $\{a\}$ and substituting in equation (8.4.1) yields

$$u^e(\xi) = [f(\xi)]\,[A]^{-1}\,\{u\}_0^e = [N(\xi)]\,\{u\}_0^e\,.\dagger \tag{8.4.4}$$

In this particular case

$$[A]^{-1} = \begin{bmatrix} 1 & 0 \\ -1/l & 1/l \end{bmatrix}$$

so that

$$u^e(\xi) = [(1 - \xi/l)\;\xi/l]\begin{Bmatrix} u_1^e \\ u_2^e \end{Bmatrix} \tag{8.4.5}$$

or

$$u^e(\xi) = [N_1(\xi)\,N_2(\xi)]\begin{Bmatrix} u_1^e \\ u_2^e \end{Bmatrix} \tag{8.4.6}$$

We observe that $N_1(\xi_1) = 1,\ N_1(\xi_2) = 0,\ N_2(\xi_1) = 0,\ N_2(\xi_2) = 1 \ldots\ldots$ or $N_i(\xi_j) = \delta_{ij}$ = Kronecker δ: the element displacement field is thus represented by a linear interpolation between nodal displacements so that the $N_i(\xi)$ are interpolation polynomials. In the finite element literature $N(\xi)$ is often referred to as a **shape function** and we may readily confirm that it meets the criteria of Section 8.3. To avoid confusion, the reader should take special note that u_1^e and u_2^e are *element* nodal displacements as distinct from $u_1, u_2, \ldots\ldots u_{n+1}$ which are *global* or *system* nodal displacements. In due course,we shall have cause to relate these two sets of quantities in a systematic manner.

To begin the third step in the process, we obtain element strains and strain energy in terms of element nodal displacements.

Now,

$$\epsilon = \frac{\mathrm{d}u^e}{\mathrm{d}\xi} = \left[\frac{\mathrm{d}N}{\mathrm{d}\xi}\right]\{u\}_0^e = [B]\,\{u\}_0^e \tag{8.4.7}$$

so that in this case,

$$\epsilon = [-1/l \quad 1/l]\begin{Bmatrix} u_1^e \\ u_2^e \end{Bmatrix}\,. \tag{8.4.8}$$

Since Hooke's law obtains, $\sigma = E\epsilon$, but anticipating later work, we will generalise our notation slightly and write

$$\{\sigma\} = [D]\,\{\epsilon\}\,\dagger\dagger. \tag{8.4.9}$$

The strain energy stored in a typical element is (Section 2.2)

$$U^e = \int_e \frac{1}{2} E\epsilon^2\,\mathrm{d\,vol} = \int_e \frac{1}{2}\{\epsilon\}^t[D]\,\{\epsilon\}\,\mathrm{d\,vol}\,. \tag{8.4.10}$$

Substituting from equation (8.4.7),

$$U^e = \frac{1}{2}\{u\}_0^{e\,t}\left(\int_e [B]^t\,[D]\,[B]\,\mathrm{d\,vol}\right)\{u\}_0^e\,. \tag{8.4.11}$$

† The reader should note that u^e is the displacement field within the element whereas $\{u\}_0^e$ is a vector of element nodal displacements.

†† In later examples, we shall see that in general $\{\sigma\}$ and $\{\epsilon\}$ are column listings of stress and strain components; correspondingly, $[D]$ is a symmetric matrix of elastic constants. In equation (8.4.9) they are all of order 1 × 1.

Evaluating the triple matrix product and carrying out the integration in equation (8.4.11) yields a matrix of constant coefficients so that

$$U^e = \tfrac{1}{2}\{u\}_0^{et}[k]^e\{u\}_0^e \, . \tag{8.4.12}$$

This is the typical quadratic form for the strain energy stored in finite degree of freedom systems such as we met in Chapters 1 and 2. The matrix

$$[k]^e = \int_e [B]^t[D]\,[B]\,\mathrm{d\,vol} \tag{8.4.13}$$

is the **element stiffness matrix.** It is symmetric and, corresponding to the displacement model chosen here, it is of order 2×2.

Continuing with the third step of the overall process, we determine the total strain energy stored in the bar from

$$U = \sum_e U^e \, . \tag{8.4.14}$$

This summation may, with advantage, be expressed in matrix notation as follows. Denote a list of all the element nodal displacements by

$$\{\tilde{u}\} = \{\{u\}_0^e\} = \begin{Bmatrix} \{u\}_0^1 \\ \{u\}_0^2 \\ \cdot \\ \cdot \\ \cdot \\ \{u\}_0^n \end{Bmatrix} \tag{8.4.15}$$

Now display all the element stiffness matrices in

$$[\tilde{k}] = \begin{bmatrix} [k]^1 & 0 & 0 \ldots\ldots & 0 \\ 0 & [k]^2 & 0 \ldots\ldots & 0 \\ \vdots & \vdots & & \vdots \\ 0 & 0 & \ldots\ldots & [k]^n \end{bmatrix} . \tag{8.4.16}$$

Then, equation (8.4.14) can be written as

$$U = \tfrac{1}{2}\{\tilde{u}\}^t[\tilde{k}]\,\{\tilde{u}\} \, . \tag{8.4.17}$$

Now the elements of the vector $\{u\}_0^e$ are not completely independent. Referring to Figs 8.4.1 (b) and (c), we see that $u_2^{\,1} = u_1^{\,2} = u_2, u_2^{\,2} = u_1^{\,3} = u_3$ etc. We can collect all these **compatibility** statements together in matrix form as

$$\begin{Bmatrix} \{u\}_0^1 \\ \{u\}_0^2 \\ \cdot \\ \cdot \\ \cdot \\ \{u\}_0^n \end{Bmatrix} = \begin{bmatrix} [C]^1 \\ [C]^2 \\ \cdot \\ \cdot \\ \cdot \\ [C]^n \end{bmatrix} \begin{Bmatrix} u_1 \\ u_2 \\ \cdot \\ \cdot \\ \cdot \\ u_{n+1} \end{Bmatrix} \tag{8.4.18}$$

or

$$\{\tilde{u}\} = [C]\,\{u\} \, . \tag{8.4.19}$$

$[C]$ is a **compatibility** or **connexion matrix** which expresses the necessary compatibility between the locally and globally measured displacements. We readily see that in every row of $[C]$ there is only one non-zero element and that its value is unity.†

Equation (8.4.17) can now be expressed in terms of $\{u\}$, the vector of system generalised coordinates. That is,

$$U = \tfrac{1}{2}\{u\}^t[C]^t[\widetilde{k}]\,[C]\{u\} \tag{8.4.20}$$

or

$$U = \tfrac{1}{2}\{u\}^t[K]\{u\}\ , \tag{8.4.21}$$

where $[K]$ is the **assembled** or **system** stiffness matrix of the discrete approximation to the bar. It is synthesised from the individual element matrices by the process

$$[K] = [C]^t[\widetilde{k}]\,[C]\ . \tag{8.4.22}$$

We see immediately that since $[\widetilde{k}]$ is symmetric the process generates a symmetric $[K]$.

In practice the assembly, which requires a computer in any but the most trivial problems, is not carried out by literally executing the triple matrix product in equation (8.4.22): to do so would make for very inefficient computing. We shall see in Example 8.4.1 that, since $[C]$ is Boolean, the process has an extremely simple interpretation which allows easy coding in a programme.

Before we can finally establish the stiffness equilibrium equations, we must determine the potential energy of the applied loads. In Fig. 8.4.1, the only loads are P applied at nodes 1 and $(n + 1)$ so that

$$\Omega = -(-Pu_1) - Pu_{n+1}\ . \tag{8.4.23}$$

To provide a systematic presentation, however, denote $X_1, X_2, \ldots\ldots X_{n+1}$ as externally applied forces corresponding to $u_1, u_2, \ldots\ldots u_{n+1}$. Then, we can write

$$\Omega = -\{u\}^t\{X\} \tag{8.4.24}$$

where we recognise that in the present case,

$$[X_1\ X_2 \ldots\ldots X_n\ X_{n+1}]^t = [-P\ 0 \ldots\ldots 0\ P]^t.$$

The total potential energy in the discrete system is thus

$$V = \tfrac{1}{2}\{u\}^t[K]\{u\} - \{u\}^t\{X\}\ . \tag{8.4.26}$$

For equilibrium $\delta V = 0$ so that

$$\{\delta u\}^t([K]\{u\} - \{X\}) = 0\ .\dagger\dagger \tag{8.4.27}$$

But the δu are arbitrary so that $$[K]\{u\} = \{X\}\ . \tag{8.4.28}$$

† The reader should notice that $[C]$ is not merely the unity matrix $[I]$ since the non-zero elements do not necessarily appear on the leading diagonal. It is called a **Boolean matrix.**

†† Note that $\delta(\{u\}^t[K]\{u\}) = \{\delta u\}^t[K]\{u\} + \{u\}^t[K]\{\delta u\}$ which, since each term is a scalar and $[K]$ is symmetric, becomes $2\{\delta u\}^t[K]\{u\}$.

These are the required stiffness equilibrium equations for the 'assembled' approximate system; they have precisely the same general form as those derived for the discrete systems of Chapters 1 and 2.

Before proceeding to some generalisations, we will treat a specific problem.

EXAMPLE 8.4.1

The bar of Fig. 8.4.2 (a) is of constant thickness and linearly varying width. It is subjected to a variety of boundary conditions. Using only two *uniform* elements, as in Fig. 8.4.2 (b), and a linear displacement model, determine the displacements and stresses at the nodes. Compare with an analytic solution.

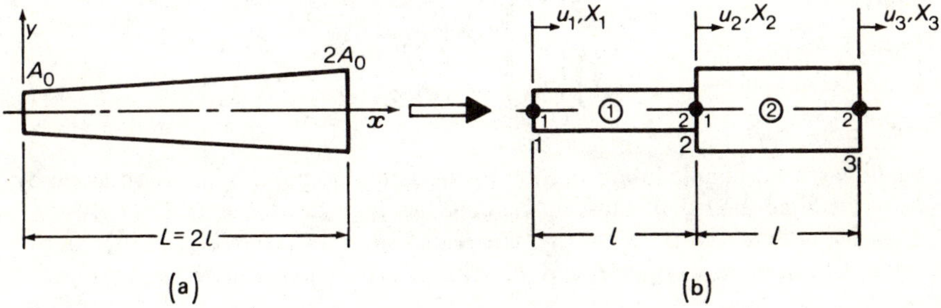

Fig. 8.4.2

Using equation (8.4.13) it is not especially difficult to determine $[k]^e$ for a tapered element. However, it will be illuminating to carry out the rather more gross idealisation of a uniform element decided upon here. Then, with the linear model of equation (8.4.6),

$$[k]^e = \frac{E}{l^2}\int_0^l \begin{Bmatrix} -1 \\ 1 \end{Bmatrix} [-1 \quad 1] A\,dx = \frac{EA}{l}\begin{bmatrix} 1 & -1 \\ -1 & 1 \end{bmatrix} . \qquad \text{(a)}$$

Using an average area for each element: for element 1, $A = 5/4A_0$, for element 2, $A = 7/4A_0$.

For compatibility, we must have

$$\begin{Bmatrix} u_1^1 \\ u_2^1 \\ u_1^2 \\ u_2^2 \end{Bmatrix} = \begin{bmatrix} 1 & 0 & 0 \\ 0 & 1 & 0 \\ 0 & 1 & 0 \\ 0 & 0 & 1 \end{bmatrix} \begin{Bmatrix} u_1 \\ u_2 \\ u_3 \end{Bmatrix} . \qquad \text{(b)}$$

With so few degrees of freedom, it is reasonable and instructive to use the formal assembly process of equation (8.4.22). Then

$$[K] = \begin{bmatrix} 1 & 0 & 0 & 0 \\ 0 & 1 & 1 & 0 \\ 0 & 0 & 0 & 1 \end{bmatrix} \begin{bmatrix} k_{11}{}^1 & k_{12}{}^1 & 0 & 0 \\ k_{21}{}^1 & k_{22}{}^1 & 0 & 0 \\ 0 & 0 & k_{11}{}^2 & k_{12}{}^2 \\ 0 & 0 & k_{21}{}^2 & k_{22}{}^2 \end{bmatrix} \begin{bmatrix} 1 & 0 & 0 \\ 0 & 1 & 0 \\ 0 & 1 & 0 \\ 0 & 0 & 1 \end{bmatrix} \tag{c}$$

or

$$[K] = \begin{bmatrix} k_{11}{}^1 & k_{12}{}^1 & 0 \\ k_{21}{}^1 & (k_{22}{}^1 + k_{11}{}^2) & k_{12}{}^2 \\ 0 & k_{21}{}^2 & k_{22}{}^2 \end{bmatrix}. \tag{d}$$

The assembled stiffness equilibrium equations are thus

$$\begin{bmatrix} k_{11}{}^1 & k_{12}{}^1 & 0 \\ k_{21}{}^1 & (k_{22}{}^1 + k_{11}{}^2) & k_{12}{}^2 \\ 0 & k_{21}{}^2 & k_{22}{}^2 \end{bmatrix} \begin{Bmatrix} u_1 \\ u_2 \\ u_3 \end{Bmatrix} = \begin{bmatrix} K_{11} & K_{12} & K_{13} \\ K_{21} & K_{22} & K_{23} \\ K_{31} & K_{32} & K_{33} \end{bmatrix} \begin{Bmatrix} u_1 \\ u_2 \\ u_3 \end{Bmatrix} = \begin{Bmatrix} X_1 \\ X_2 \\ X_3 \end{Bmatrix}. \tag{e}$$

We now see the simple interpretation of the assembly process. K_{ij} is obtained by merely adding the $k_{rs}{}^e$ of those elements providing a direct connexion between u_i and u_j with $u_r^e = u_i$, $u_s^e = u_j$. The reader should observe that in $k_{rs}{}^e$, the subscripts are *local* to the element whilst in K_{ij}, the subscripts are *global* ones.

Numerically, we have

$$\frac{EA_0}{4l} \begin{bmatrix} 5 & -5 & 0 \\ -5 & 12 & -7 \\ 0 & -7 & 7 \end{bmatrix} \begin{Bmatrix} u_1 \\ u_2 \\ u_3 \end{Bmatrix} = \begin{Bmatrix} X_1 \\ X_2 \\ X_3 \end{Bmatrix} \tag{f}$$

We notice that $[k]^e$ and $[K]$ are symmetric as we would expect.

First set of boundary conditions

Boundary tractions prescribed as $X_1 = -P$, $X_3 = P$ whilst $X_2 = 0$. Substituting these in the right hand side of equation (f), we find that we cannot solve for the u's since $[K]$ is singular (det $[K] = 0$). The physical explanation is that the boundary conditions provide no constraint against rigid body motion; equations (f) thus hold for an infinite number of $\{u\}$'s differing by only a rigid body translation. If we set any one of the u's to a prescribed value, say zero, a solution can be effected as described below.

Second set of boundary conditions

Boundary displacements prescribed as $u_1 = u_1{}^*$, $u_3 = u_3{}^*$: We thus seek u_2 and 'reactions' X_1 and X_3 observing that $X_2 = 0$.

To solve equations (f) for this case, it is convenient to re-arrange them so that the known and unknown displacements and forces are grouped together. Then

$$\frac{EA_0}{4l}\left[\begin{array}{c|cc} 12 & -5 & -7 \\ \hline -5 & 5 & 0 \\ -7 & 0 & 7 \end{array}\right]\left\{\begin{array}{c} u_2 \\ \hline u_1{}^* \\ u_3{}^* \end{array}\right\} = \left\{\begin{array}{c} 0 \\ \hline X_1 \\ X_3 \end{array}\right\} . \qquad (g)$$

Having *partitioned* the matrices in this way, the solution may be expressed in a formal matrix fashion but we defer this procedure until later and merely solve equation (g) directly to yield.

$$u_2 = 0.417u_1{}^* + 0.583u_3{}^* , \qquad (h)$$

$$X_1 = 0.729EA_0(u_1{}^* - u_3{}^*)/l , \qquad (j)$$

$$X_2 = -0.729EA_0(u_1{}^* - u_3{}^*)/l . \qquad (k)$$

With $X_1 = -X_2$, overall equilibrium is satisfied as would be expected.

According to our approximation, the strain is constant within an element and is given by

$$\epsilon = [B]\{u\}_0^e = \frac{1}{l}[-1 \quad 1]\{u\}_0^e . \qquad (l)$$

The stress is then $\sigma = E\epsilon$ and,

for element 1, $\sigma = 0.583E(u_3{}^* - u_1{}^*)/l$, (m)

for element 2, $\sigma = 0.417E(u_3{}^* - u_1{}^*)/l$. (n)

We see that according to our procedure, we may attribute either stress in equation (m) or (n) to a section at node 2; the difference is a measure of the imperfection in the solution. If we average results (m) and (n), we may hope to improve the solution at node 2. Indeed,

$$(\sigma_{\text{ave}})_{\textbf{node 2}} = 0.5E(u_3{}^* - u_1{}^*)/l . \qquad (p)$$

This problem may be readily solved by the usual mechanics of materials approach, giving

$$u(L/2) = 0.415u_1{}^* + 0.585u_3{}^* , \qquad (q)$$

$$\sigma(L/2) = 0.481E(u_3{}^* - u_1{}^*)/l . \qquad (r)$$

This shows remarkably good agreement with the finite element solution. However, at the ends of the bar,

$$\sigma(0) = 0.722E(u_3{}^* - u_1{}^*)/l, \qquad \sigma(L) = 0.361E(u_3{}^* - u_1)/l .$$

Agreement is not so good here but it can be improved by taking more elements.

Third set of boundary conditions

We prescribe $u_1 = 0$, $X_3 = P$ with $X_2 = 0$ again so that X_1 is an unkown reaction.

Equation (f) now becomes

$$\frac{EA_0}{4l}\left[\begin{array}{c|cc} 5 & -5 & 0 \\ \hline -5 & 12 & -7 \\ 0 & -7 & -7 \end{array}\right]\left\{\begin{array}{c} 0 \\ \hline u_2 \\ u_3 \end{array}\right\} = \left\{\begin{array}{c} X_1 \\ \hline 0 \\ P \end{array}\right\}. \tag{t}$$

Again we may partition the matrices as shown by the dotted lines. Solving for u_2 and u_3

$$u_2 = \frac{0.800Pl}{EA_0}, \qquad u_3 = \frac{1.371Pl}{EA_0} \tag{u}$$

which yields

$$X_1 = -P \tag{v}$$

to satisfy overall equilibrium.

Continuing with the bar extension problem, we now introduce some useful generalisations. Thus, suppose the bar experiences some initial strain ϵ_0, which may be due to temperature rise, a distributed axial force $\bar{X}$/unit volume and a concentrated axial force $\bar{\bar{X}}$ acting at some position in the bar which does not coincide with a node. We proceed to show that the net effect of these actions is to augment any nodal forces $\{X\}$ which may already be prescribed.

The influence of any initial strain appears in the stress strain law. (Equilibrium conditions are independent of such strains and the usual strain-displacement relations hold provided total strains are used.) Then, if ϵ is the total strain,

$$\sigma = E(\epsilon - \epsilon_0)$$

or, more generally,

$$\{\sigma\} = [D](\{\epsilon\} - \{\epsilon_0\}). \tag{8.4.29}$$

The strain energy stored in an element is then

$$U^e = \frac{1}{2}\int_e (\{\epsilon\}^t - \{\epsilon_0\}^t)[D](\{\epsilon\} - \{\epsilon_0\})\mathrm{d\,vol}$$

or

$$U^e = \frac{1}{2}\int_e \{\epsilon\}^t[D]\{\epsilon\}\mathrm{d\,vol} - \int_e \{\epsilon\}^t[D]\{\epsilon_0\}\,\mathrm{d\,vol}$$

$$+ \frac{1}{2}\int_e \{\epsilon_0\}^t[D]\{\epsilon_0\}\,\mathrm{d\,vol}\,.$$

But $\{\epsilon\} = [B]\{u\}_0^e$ so that

$$U^e = \frac{1}{2}\{u\}_0^{et}[k]^e\{u\}_0^e - \{u\}_0^{et}\{X_0\} + \frac{1}{2}\int_e \{\epsilon_0\}^t[D]\{\epsilon_0\}\,\mathrm{d\,vol}\,, \tag{8.4.30}$$

where

$$\{X_0\} = \int_e [B]^t[D]\{\epsilon_0\}\,\mathrm{d\,vol}\,. \tag{8.4.31}$$

The potential energy of the distributed and intermediate concentrated load associated with an element is given by

$$\Omega^e = -\int_e u^e(\xi)\bar{X}(\xi)\mathrm{d\,vol} - u^e(\xi_i)\bar{\bar{X}}_i \tag{8.4.32}$$

or
$$\Omega^e = -\{u\}_0^{e\,t}\{\bar{X}\}^e - \{u\}_0^{e\,t}\{\bar{\bar{X}}\}^e \tag{8.4.33}$$

where
$$\{\bar{X}\}^e = \int_e [N(\xi)]^t \bar{X}(\xi)\,\mathrm{d}\text{vol}\,, \qquad \{\bar{\bar{X}}\}^e = [N(\xi_i)]^t \bar{\bar{X}}_i\,. \tag{8.4.34}$$

Defining
$$\{\widetilde{X}_0\} = \begin{Bmatrix} \{X_0\}^1 \\ \{X_0\}^2 \\ \cdot \\ \cdot \\ \cdot \\ \cdot \\ \cdot \end{Bmatrix}$$

with similar expressions for $\{\widetilde{\bar{X}}\}$ and $\{\widetilde{\bar{\bar{X}}}\}$, and recalling the procedure implicit in equations (8.4.14) to (8.4.26), we can write

$$V = \tfrac{1}{2}\{u\}^t[K]\{u\} - \{u\}^t(\{X_0\} + \{\bar{X}\} + \{\bar{\bar{X}}\}) \tag{8.4.35}$$

where
$$\{X_0\} = [C]^t\{\widetilde{X}_0\} \text{ etc}\,. \tag{8.4.36}$$

Taking the variation of equation (8.4.34) and equating to zero, we again obtain a stiffness equilibrium equation

$$[K]\{u\} = \{X\}_{\text{tot}} = \{P\}, \text{ say}\,. \tag{8.4.37}$$

Here, $\{X\}_{\text{tot}}$ is a vector of total nodal forces synthesised from the initial strain effects etc., in accordance with equations (8.4.32) etc.†

Again, since $[C]$ is Boolean, the synthesis process is easy; thus, $(X_i)_{\text{tot}}$ is obtained merely by adding each $X_{0s}{}^e$ etc., for each element having a local coordinate $u_{0s}{}^e$ coinciding with global u_i.

Example 8.4.1 showed the need to partition the matrix equation when some of the displacements are prescribed. To generalise the discussion, consider the equation partitioned such that

$$\begin{bmatrix} [K_{\alpha\alpha}] & [K_{\alpha\beta}] \\ [K_{\beta\alpha}] & [K_{\beta\beta}] \end{bmatrix} \begin{Bmatrix} \{u_\alpha\} \\ \{u_\beta\} \end{Bmatrix} = \begin{Bmatrix} \{P_\alpha\} \\ \{P_\beta\} \end{Bmatrix}\,. \tag{8.4.38}$$

Here, $\{u_\alpha\}$ is a vector of prescribed displacements so that $\{P_\alpha\}$ is a vector of unkown 'reactions'. The prescribed nodal forces are $\{P_\beta\}$ with $\{u_\beta\}$ the corresponding unknown displacements. From equation (8.4.38), we have

$$[K_{\alpha\alpha}]\{u_\alpha\} + [K_{\alpha\beta}]\{u_\beta\} = \{P_\alpha\}\,. \tag{8.4.39}$$

$$[K_{\beta\alpha}]\{u_\alpha\} + [K_{\beta\beta}]\{u_\beta\} = \{P_\beta\}\,. \tag{8.4.40}$$

Equation (8.4.40) yields

$$\{u_\beta\} = [K_{\beta\beta}]^{-1}(\{P_\beta\} - [K_{\beta\alpha}]\{u_\alpha\})\,. \tag{8.4.41}$$

† Notice that the distributed loads etc. have been attributed to the nodes on a basis of work; in general such nodal forces are slightly different from those which would be obtained merely by 'lumping' the distributed actions to the nodes in a statical manner. The former procedure yields nodal forces which are proper generalised forces in accordance with Section 1.5.

Substituting in equation (8.4.39), we have

$$\{P_\alpha\} = ([K_{\alpha\alpha}] - [K_{\alpha\beta}][K_{\beta\beta}]^{-1}[K_{\beta\alpha}])\{u_\alpha\} + [K_{\beta\beta}]^{-1}\{P_\beta\}. \quad (8.4.42)$$

In the special case where the structure is mounted on rigid supports, $\{u_\alpha\} = \{0\}$, then

$$\{u_\beta\} = [K_{\beta\beta}]^{-1}\{P_\beta\}. \quad (8.4.43)$$

Now, $[K_{\beta\beta}]$ is non-singular so that $\{K_{\beta\beta}]^{-1}$ can be determined and $\{u_\beta\}$ found. The above manipulation shows us that to solve $[K]\{u\} = \{P\}$, when certain of the displacements are prescribed to be zero, it is only necessary to delete the rows and columns of $[K]$ corresponding to the zero u's to generate a reduced stiffness matrix $[K_{\beta\beta}]$. Example 8.4.1 showed this process for a small problem.

The procedure which we have described is quite standard for the displacement solution of any linear elastic stressing problem and, as we shall see in the remainder of the chapter, all that changes from situation to situation is the size and detailed content of the various matrices.

Problems 8.4

8.4.1 The line element of Fig. 8.4.3 has an internal node 3 in addition to the external nodes 1 and 2. Show that an axial displacement model $u = a_0 + a_1\xi + a_2\xi^2$ can be written in the form $u = [N]\{u\}_0^e$ where

$$[N] = [N_1 N_2 N_3] = \left[\left(1 - 3\frac{\xi}{l} + 2\frac{\xi^2}{l^2}\right)\left(-\frac{\xi}{l} + 2\frac{\xi^2}{l^2}\right)\left(\frac{4\xi}{l} - \frac{4\xi^2}{l^2}\right)\right]$$

and $\{u\}_0^{et} = [u_1\ u_2\ u_3]$.

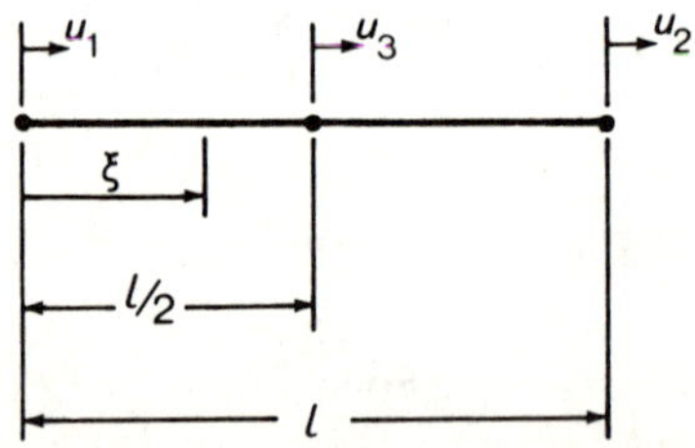

Fig. 8.4.3

Confirm that $N_i(\xi_j) = \delta_{ij}$ = Kronecker delta and sketch each $N_i(\xi)$.

8.4.2 Accepting the displacement model of Problem 8.4.1, show that the stiffness equilibrium equation for the element is

$$\frac{EA}{3l}\left[\begin{array}{cc|c} 7 & 1 & -8 \\ 1 & 7 & -8 \\ \hline -8 & -8 & 16 \end{array}\right]\left\{\begin{array}{c} u_1 \\ u_2 \\ \hline u_3 \end{array}\right\} = \left\{\begin{array}{c} P_1 \\ P_2 \\ \hline P_3 \end{array}\right\} \quad (a)$$

where $\{P\}$ is a vector of generalised nodal forces.

8.4.3 Node 3 of the element in Fig. 8.4.3 is not associated with any adjoining elements in an assembly. To reduce the total number of unkowns in a formulation, equation (a) of Problem 8.4.2 may be partitioned as shown and u_3 written in terms of u_1 and u_2. Show that this leads to an *effective* stiffness equilibrium equation

$$\frac{EA}{l}\begin{bmatrix} 1 & -1 \\ -1 & 1 \end{bmatrix}\begin{Bmatrix} u_1 \\ u_2 \end{Bmatrix} = \begin{Bmatrix} P_1 + P_3/2 \\ P_2 + P_3/2 \end{Bmatrix} .$$

This process of eliminating internal degrees of freedom is called **static condensation.**

8.4.4 Generate a stiffness matrix for a *tapered* element to suit the bar of example 8.4.1.

8.4.5 A rod rotates about an axis normal to its centre line. Deduce a load vector for the inertia load on an element assuming the displacement model of equation (8.4.5).

8.5 Displacement finite element formulation for bar flexure

According to the engineering theory of bar bending, the deformation is characterised by the deflexion curve $v(x)$ taken up by the centre line of the bar: the problem is thus one dimensional and a finite element comprises a line element. We saw in Section 2.4 that the strain energy functional involves $v''(x)$ so that according to the criteria of Section 8.3, an approximating polynomial $v^e(\xi)$ for the element of Fig. 8.5.1 must be at least of degree three to provide continuity of $v^e(\xi)$ and $v'^e(\xi)$ at the nodes. That is

$$v^e(\xi) = a_0 + a_1\xi + a_2\xi^2 + a_3\xi^3 = [f(\xi)]\{a\} . \qquad (8.5.1)$$

The element must therefore have at least four degrees of freedom.

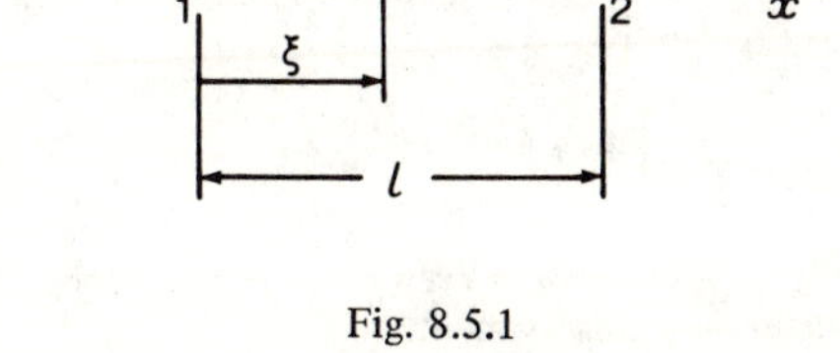

Fig. 8.5.1

Proceeding as before, we list the nodal displacements and slopes (the element generalised coordinates) as a vector

$$\{u\}_0^{et} = \left[v_1 \left(\frac{\mathrm{d}v}{\mathrm{d}\xi}\right)_1 v_2 \left(\frac{\mathrm{d}v}{\mathrm{d}\xi}\right)_2\right] = [v_1\ \theta_1\ v_2\ \theta_2] . \qquad (8.5.2)$$

Then with $\xi = 0$ and $\xi = l$ in turn in equation (8.5.1),

$$\begin{Bmatrix} v_1 \\ \theta_1 \\ v_2 \\ \theta_2 \end{Bmatrix} = \begin{bmatrix} 1 & 0 & 0 & 0 \\ 0 & 1 & 0 & 0 \\ 1 & l & l^2 & l^3 \\ 0 & l & 2l & 3l^2 \end{bmatrix} \begin{Bmatrix} a_0 \\ a_1 \\ a_2 \\ a_3 \end{Bmatrix} . \qquad (8.5.3)$$

or
$$\{u\}_0^e = [A]\{a\} . \qquad (8.5.4)$$

Equation (8.5.1) can now be written

$$v^e(\xi) = [f(\xi)]\,[A]^{-1}\{u\}_0^e = [N(\xi)]\{u\}_0^e . \qquad (8.5.5)$$

Determining $[A]^{-1}$ is clearly a little more involved than in Example 8.4.1 but it is still not too unreasonable by hand. However, if we observe that $[N(\xi)]\{u\}_0^e$ is an interpolating polynomial involving nodal function and derivative values we could, alternatively, use interpolation theory to generate $[N(\xi)]$ directly.† Indeed, $[N(\xi)]\{u\}_0^e$ is the **Hermite** or **osculating interpolation polynomial** for $v(\xi)$. In fact,

$$[N(\xi)] = \left[\left(1 - 3\frac{\xi^2}{l^2} + 2\frac{\xi^3}{l^3}\right)\left(\xi - 2\frac{\xi^2}{l} + \frac{\xi^3}{l^2}\right)\left(3\frac{\xi^2}{l^2} - 2\frac{\xi^3}{l^3}\right)\left(-\frac{\xi^2}{l} + \frac{\xi^3}{l^2}\right)\right] \qquad (8.5.6)$$

or $[N(\xi)] = [N_1(\xi)\,N_2(\xi)\,N_3(\xi)\,N_4(\xi)]$. (8.5.7)

where $N_i(\xi_j) = \delta_{ij}$ = Kronecker delta.

To proceed, we need $v''^e(\xi)$ and by regarding this quantity as a generalised strain, we may continue with our established notation. That is,

$$v''^e(\xi) = [N''(\xi)]\{u\}_0^e = [B(\xi)]\{u\}_0^e . \qquad (8.5.8)$$

The strain energy stored in an element is

$$U^e = \frac{1}{2}\int_0^l EI(v'')^2\,\mathrm{d}x \qquad (8.5.9)$$

so that, denoting $EI = [D]$, we find

$$U^e = \frac{1}{2}\{u\}_0^{et}\left(\int_0^l [B]^t[D]\,[B]\,\mathrm{d}x\right)\{u\}_0^e \qquad (8.5.10)$$

or
$$U^e = \frac{1}{2}\{u\}_0^{et}[k]^e\{u\}_0^e . \qquad (8.5.11)$$

† There are situations where $[A]^{-1}$ cannot even be determined; the use of interpolation polynomials then avoids some embarassment.

This is precisely the same format as for bar extension, but now $[k]^e$ is a 4 × 4 matrix. Carrying out the matrix multiplication and the integration in equation (8.5.10), we have, for a uniform element,

$$[k]^e = \frac{EI}{l^3}\begin{bmatrix} 12 & 6l & -12 & 6l \\ 6l & 4l^2 & -6l & 2l^2 \\ -12 & -6l & 12 & -6l \\ 6l & 2l^2 & -6l & 4l^2 \end{bmatrix} . \tag{8.5.12}$$

Supposing that the bar supports a distributed load $p(x)$ per unit length and concentrated loads P_i and couples M_i at a number of points along the span, then we may define the finite mesh to give nodes at the concentrated actions as in Fig. 8.5.2.

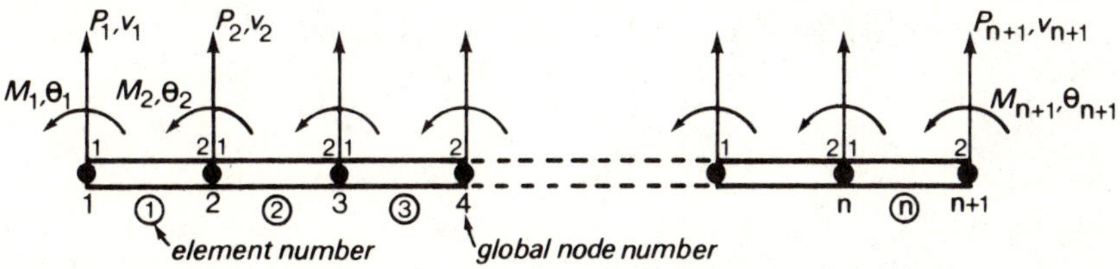

Fig. 8.5.2

The potential energy of the external loads is then

$$\Omega = -\sum_e\left(\int_e p(\xi)v(\xi)\mathrm{d}\xi\right) - P_1v_1 - M_1\theta_1 \ldots P_{n+1}v_{n+1} - M_{n+1}\theta_{n+1} . \tag{8.5.13}$$

Incorporating displacement model (8.5.5) and displaying $P_1, M_1, P_2, M_2 \ldots\ldots$, $v_1, \theta_1, v_2, \theta_2 \ldots\ldots$ as vectors $\{P_c\}$ and $\{u\}$ respectively, we have

$$\Omega = -\sum_e\left(\int_0^l \{u\}_0^{et}\,[N(\xi)]^t p(\xi)\mathrm{d}\xi\right) - \{u\}^t\{P_c\}$$

or

$$\Omega = -\sum_e\left(\{u\}_0^{et}\{P_d\}_0^e\right) - \{u\}^t\{P_c\} \tag{8.5.14}$$

where

$$\{P_d\}_0^e = \int_0^l [N(\xi)^t p(\xi)\mathrm{d}\xi . \tag{8.5.15}$$

For compatibility, we have $v_1^{\,1} = v_1, \theta_1^{\,1} = \theta_1, v_2^{\,1} = v_2, \theta_2^{\,1} = \theta_2$, $v_1^{\,2} = v_2, \theta_1^{\,2} = \theta_2, v_2^{\,2} = v_3, \theta_2^{\,2} = \theta_3$ etc., or using a compatibility matrix as in equation (8.4.19),

$$\{\tilde{u}\} = [C]\{u\} . \tag{8.5.16}$$

The total potential energy is thus

$$V = \tfrac{1}{2}\{u\}^t[K]\{u\} - \{u\}^t(\{P_d\} + \{P_c\}) \qquad (8.5.17)$$

with $[K]$ and $\{P_d\}$ generated in the manner implicit in equations (8.4.14) etc. For equilibrium $\delta V = 0$ and once more we have

$$[K]\{u\} = \{P\} \qquad (8.5.18)$$

with

$$\{P\} = \{P_d\} + \{P_c\} \ . \qquad (8.5.19)$$

So far, the beam has not been located in space: if, now, certain of the u's are prescribed, the reduced version of equation (8.5.18) can be solved for the remainder and the bending moment diagram determined from $M = EIv''$.

EXAMPLE 8.5.1

A propped cantilever beam is loaded as shown in Fig. 8.5.3(a). Using the two element representation of Fig. 8.5.3 (b), deduce the stiffness equilibrium equations.

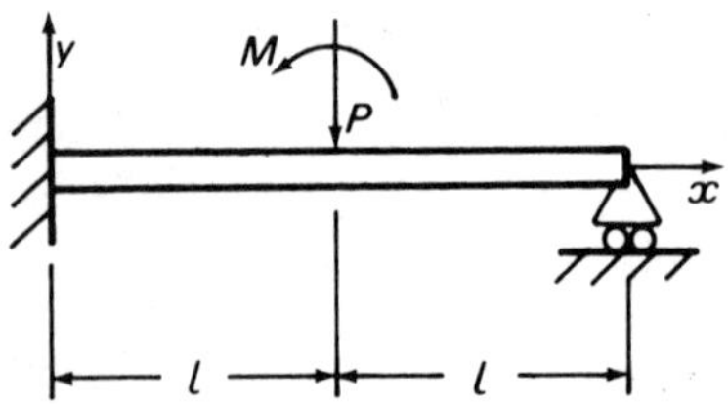

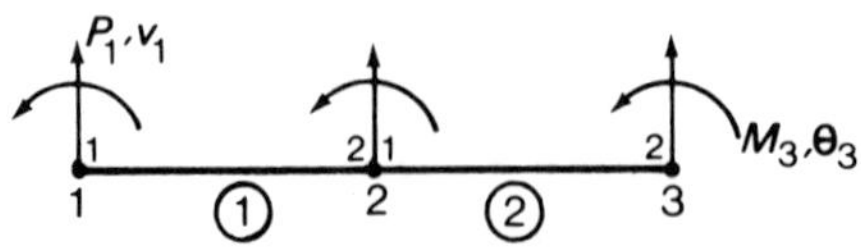

Fig. 8.5.3

For compatibility between element and system displacements, $\{\tilde{u}\} = [C]\{u\}$, or

$$\begin{Bmatrix} v_1{}^1 \\ \theta_1{}^1 \\ v_2{}^1 \\ \theta_2{}^1 \\ v_1{}^2 \\ \theta_1{}^2 \\ v_2{}^2 \\ \theta_2{}^2 \end{Bmatrix} = \begin{bmatrix} 1 & 0 & 0 & 0 & 0 & 0 \\ 0 & 1 & 0 & 0 & 0 & 0 \\ 0 & 0 & 1 & 0 & 0 & 0 \\ 0 & 0 & 0 & 1 & 0 & 0 \\ 0 & 0 & 1 & 0 & 0 & 0 \\ 0 & 0 & 0 & 1 & 0 & 0 \\ 0 & 0 & 0 & 0 & 1 & 0 \\ 0 & 0 & 0 & 0 & 0 & 1 \end{bmatrix} \begin{Bmatrix} v_1 \\ \theta_1 \\ v_2 \\ \theta_2 \\ v_3 \\ \theta_3 \end{Bmatrix} . \qquad (a)$$

Carrying out the assembly process $[K] = [C]^t[\widetilde{k}][C]$ etc., we find the equilibrium equations to be

$$\begin{bmatrix} k_{11}^1 & k_{12}^1 & k_{13}^1 & k_{14}^1 & 0 & 0 \\ k_{21}^1 & k_{22}^1 & k_{23}^1 & k_{24}^1 & 0 & 0 \\ k_{31}^1 & k_{32}^1 & (k_{33}^1 + k_{11}^2) & (k_{34}^1 + k_{12}^2) & k_{13}^2 & k_{14}^2 \\ k_{41}^1 & k_{42}^1 & (k_{43}^1 + k_{21}^2) & (k_{44}^1 + k_{22}^2) & k_{23}^2 & k_{24}^2 \\ 0 & 0 & k_{31}^2 & k_{32}^2 & k_{33}^2 & k_{34}^2 \\ 0 & 0 & k_{41}^2 & k_{42}^2 & k_{43}^2 & k_{44}^2 \end{bmatrix} \begin{Bmatrix} v_1 \\ \theta_1 \\ v_2 \\ \theta_2 \\ v_3 \\ \theta_3 \end{Bmatrix} = \begin{Bmatrix} P_1 \\ M_1 \\ P_2 \\ M_2 \\ P_3 \\ M_3 \end{Bmatrix} \tag{b}$$

or

$$[K]\{u\} = \{P\} . \tag{c}$$

The dotted boxes show the characteristic form of the assembly for beam finite elements. The reader may readily verify that with an increased number of elements, the system matrix may be generated from the element matrices in accordance with the general pattern of Fig. 8.5.4. Each of the solid boxes represent an element stiffness matrix and the overlap shows the coupling between them.

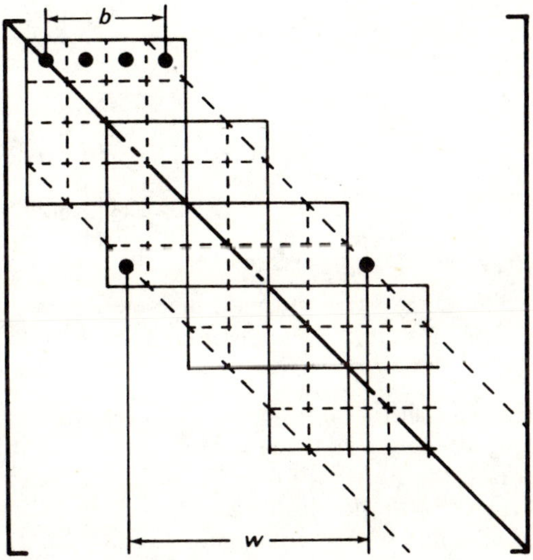

Fig. 8.5.4

The clustering of the non-zero numbers of $[K]$ in a narrow band about its leading diagonal is a characteristic feature of finite element work; it does not become apparent, however, unless a fairly large number of degrees of freedom is used. w, in Fig. 8.5.4, is called the **bandwidth** of the matrix and $b = (w + 1)/2$ is its **semi-bandwidth.** For the four degree of freedom bending element here, w is constant at a value of 7; for the two degree of freedom extensional element of Section 8.4, $w = 3$. For efficient use of a computer, we need only store and operate on the semi-band since $[K]$ is always symmetric.

Equation (c) cannot be solved as it stands since the system has not been located in space. Fig. 8.5.3 shows that, in fact, $v_1 = \theta_1 = v_3 = 0$; P_1, M_1, P_3 are thus unknown reactions. Deleting the rows and columns corresponding to v_1, θ_1 and v_3 in $[K]$, we obtain the reduced stiffness matrix as described in Section 8.4. The remaining quantities v_2, θ_2, θ_3 thus constitute the generalised coordinates for the system with $-P$, M, 0 being the corresponding generalised forces. Then, we have

$$\begin{bmatrix} (k_{33}{}^1 + k_{11}{}^2) & (k_{34}{}^1 + k_{12}{}^2) & k_{14}{}^2 \\ (k_{43}{}^1 + k_{21}{}^2) & (k_{44}{}^1 + k_{22}{}^2) & k_{24}{}^2 \\ k_{41}{}^2 & k_{42}{}^2 & k_{44}{}^2 \end{bmatrix} \begin{Bmatrix} v_2 \\ \theta_2 \\ \theta_3 \end{Bmatrix} = \begin{Bmatrix} -P \\ M \\ 0 \end{Bmatrix} \tag{d}$$

or

$$[K]\{q\} = \{Q\} . \tag{e}$$

Accepting result (8.5.12) for a uniform beam element, we finally have

$$\frac{EI}{l^3}\begin{bmatrix} 24 & 0 & 6l \\ 0 & 8l^2 & 2l^2 \\ 6l & 2l^2 & 4l^2 \end{bmatrix} \begin{Bmatrix} q_1 \\ q_2 \\ q_3 \end{Bmatrix} = \begin{Bmatrix} -P \\ M \\ 0 \end{Bmatrix}^{\dagger}, \tag{f}$$

where $q_1 = v_2$, $q_2 = \theta_2$, $q_3 = \theta_3$.

Inverting $[K]$,

$$\begin{Bmatrix} q_1 \\ q_2 \\ q_3 \end{Bmatrix} = \frac{l}{96EI}\begin{bmatrix} 7l^2 & 3l & -12l \\ 3l & 15 & -12 \\ -12l & -12 & 48 \end{bmatrix} \begin{Bmatrix} -P \\ M \\ 0 \end{Bmatrix} \tag{g}$$

Having computed the values of $\{q\}$, we may complete the solution by determining the reactions P_1, M_1, P_3 and the boundary moments and shearing forces in each element.

The reader should recall that for uniform beams supporting concentrated actions alone, the governingdifferential equation is $EIv'''' = 0$ so that $v = a_0 + a_1 x + a_2 x^2 + a_2 x^3$ is, in fact, the 'exact' shape of the deflexion curve. Solution (g) thus corresponds exactly with the usual engineering theory. Indeed, from beam theory, the deflexion at mid span of a propped cantilever length $2l$ supporting a mid span point load P alone is

$$v = -\frac{7Pl^3}{96EI} . \tag{h}$$

This corresponds with equation (g) when $M = 0$.

† The reader may notice that by lisiting only the non-zero u's the reduced stiffness matrix may be generated directly by inspection. However, a computer code must be more systematic than implied by this hand assembly process.

Problems 8.5

8.5.1 The system of Fig. 8.5.5 is to be used as a vehicle for demonstrating how elements having a suitable mixture of characteristics can be accommodated systematically in a finite element formulation. Using three beam elements, each having the stiffness matrix of equation (8.5.12), and observing that

$$U = \sum U_{\text{beams}} + \sum U_{\text{springs}}, \tag{a}$$

deduce the system stiffness matrix.

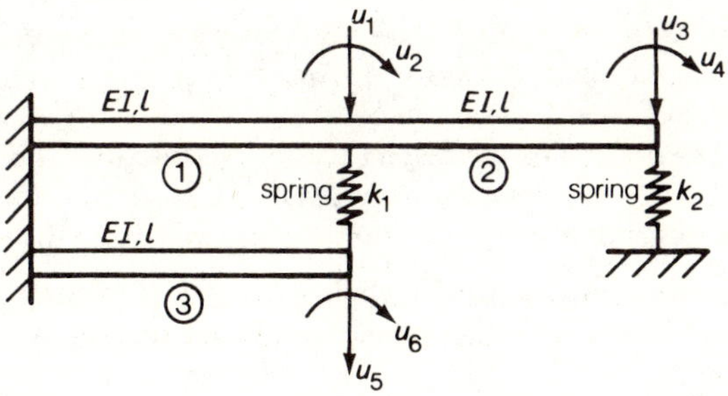

Fig. 8.5.5

Ans:

$$[K] = \frac{EI}{l^3}\begin{bmatrix} \left(24 + \frac{k_1 l^3}{EI}\right) & 0 & -12 & 6l & -\frac{k_1 l^3}{EI} & 0 \\ & 8l^2 & -6l & 2l^2 & 0 & 0 \\ & & \left(12 + \frac{k_2 l^3}{EI}\right) & -6l & 0 & 0 \\ & \text{symm} & & 4l^2 & 0 & 0 \\ & & & & \left(12 + \frac{k_1 l^3}{EI}\right) & -6l \\ & & & & & \left(4l^2 + \frac{k_1 l^3}{EI}\right) \end{bmatrix}$$

8.5.2 Deduce the load vector for a beam element of length l supporting a uniformly distributed transverse load p/unit length. Assume the displacement model of equation (8.5.6) and calculate the potential energy of p.

Ans: $$\{P\}_0^e = \left[\frac{1}{2}pl \; \frac{1}{12}pl^2 \; \frac{1}{2}pl \; -\frac{1}{12}pl^2\right]^t .$$

(Notice that if we merely 'shared out' the load statically, we would miss the couple terms $1/12pl^2$ and $-1/12pl^2$ which the work calculation has shown to be necessary. $\{P\}_0^e$ thus comprises proper generalised forces.)

8.6 Linear elastic frames

We have already pointed out that the usual treatment of frames as assemblages of bars subjected to axial load, bending and torsion is essentially a finite element process. When we discussed such systems in Chapter 2, we were primarily concerned with the utility of the various energy principles and whilst we saw that stiffness and flexibility matrices did, in fact, arise quite naturally, the simplicity of the frames considered made it unnecessary to adopt a formal and generally systematic approach to the solution. Clearly as the systems of interest become more complex a systematic approach becomes imperative and the matrix formulations of Sections 8.4 and 8.5 provide the framework within which to work: once the element stiffness matrices have been generated, the procedure is exactly as before. Our main reason for treating frames here is to illuminate further the notion of local and system coordinates and their relationships and to give a brief presentation of the matrix theory of structures. For our purpose it will be adequate to consider only plane frames; a fuller account will be found in texts on structural analysis.

The description of a plane frame as a whole is most conveniently executed with reference to some **global** or **system set** of coordinates such as the xOy set of Fig. 8.6.1.

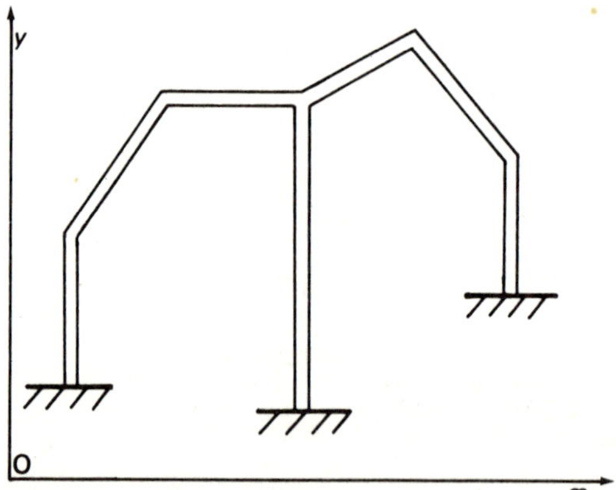

Fig. 8.6.1

On the other hand, the description of an individual element can be most conveniently carried out in relation to a set of **element** or **local coordinates** appropriate to it. In general, a typical element of a frame will be inclined to the global coordinates and the relationship between them and the element or local coordinates is not as simple as in Sections 8.4 and 8.5. To facilitate the solution process, we find it desirable to

(i) determine the element stiffness matrix with reference to convienient local coordinates;

(ii) by means of a coordinate transformation, relate the element stiffness matrix to the global coordinates;

(iii) use the transformed element matrices in the assembly process.

A typical inclined member will be bent and stretched so that its ends will be displaced as shown in Fig. 8.6.2.

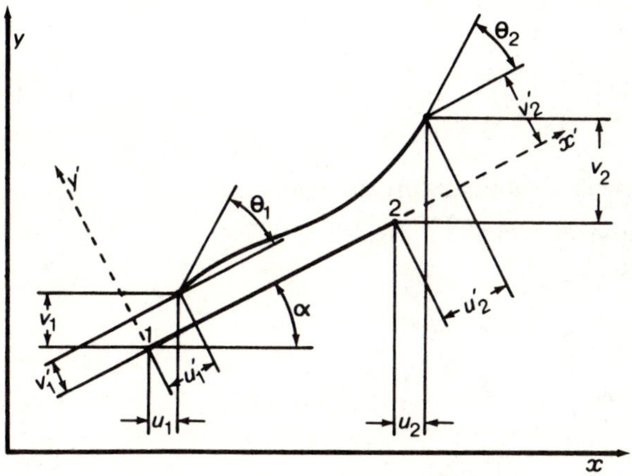

Fig. 8.6.2

In many situations, the coupling between axial and bending actions is negligible so that the stiffness matrix with respect to the $x'y'$ axes can, for a uniform element, be readily synthesised from equation (a) of Example 8.4.1 and equation (8.5.12) to give

$$\begin{bmatrix} EA/l & 0 & 0 & -EA/l & 0 & 0 \\ 0 & 12EI/l^3 & 6EI/l^2 & 0 & -12EI/l^3 & 6EI/l^2 \\ 0 & 6EI/l^2 & 4EI/l & 0 & -6EI/l^2 & 2EI/l \\ -EA/l & 0 & 0 & EA/l & 0 & 0 \\ 0 & -12EI/l^3 & -6EI/l^2 & 0 & 12EI/l^3 & -6EI/l^2 \\ 0 & 6EI/l^2 & 2EI/l & 0 & -6EI/l^2 & 4EI/l \end{bmatrix} \begin{Bmatrix} u_1' \\ v_1' \\ \theta_1' \\ u_2' \\ v_2' \\ \theta_2' \end{Bmatrix} = \begin{Bmatrix} X_1' \\ Y_1' \\ M_1' \\ X_2' \\ Y_2' \\ M_2' \end{Bmatrix}$$

or

$$[k']^e\{u'\}_0^e = \{P'\}_0^e \,. \tag{8.6.2}$$

Now we can certainly compute the total strain energy in the frame to be

$$U = \sum_e U^e = \sum_e \frac{1}{2}\{u'\}_0^{et}\,[k']^e\{u'\}_0^e \,. \tag{8.6.3}$$

But, imposing compatibility between the ends of meeting elements is not very convenient with local coordinates having different directions for different elements. However, from Fig. 8.6.2, we can write the geometric relations

$$\begin{Bmatrix} u_1' \\ v_1' \\ \theta_1' \\ u_2' \\ v_2' \\ \theta_2' \end{Bmatrix} = \begin{bmatrix} \cos\alpha & \sin\alpha & 0 & 0 & 0 & 0 \\ -\sin\alpha & \cos\alpha & 0 & 0 & 0 & 0 \\ 0 & 0 & 1 & 0 & 0 & 0 \\ 0 & 0 & 0 & \cos\alpha & \sin\alpha & 0 \\ 0 & 0 & 0 & -\sin\alpha & \cos\alpha & 0 \\ 0 & 0 & 0 & 0 & 0 & 1 \end{bmatrix} \begin{Bmatrix} u_1 \\ v_1 \\ \theta_1 \\ u_2 \\ v_2 \\ \theta_2 \end{Bmatrix} \tag{8.6.4}$$

or

$$\{u'\}_0^e = [T]\{u\}_0^e \tag{8.6.5}$$

where $[T]$ is a coordinate transformation matrix.

Substituting from equation (8.6.5) in (8.6.3)

$$U = \sum_e \frac{1}{2}\{u\}_0^{et}[T]^t[k']^e[T]\{u\}_0^e$$

or

$$U = \sum_e \frac{1}{2}\{u\}^{et}[k]^e\{u\}_0^e\,, \tag{8.6.6}$$

where

$$[k]^e = [T]^t[k']^e[T]\,. \tag{8.6.7}$$

By means of transformation (8.6.7), all the element stiffness matrices can be expressed with respect to the one set of global coordinates. Thus, provided this transformation is carried out first, the compatibility matrix $[C]$ is again Boolean and the assembly process is precisely the same as that followed in Sections 8.4 and 8.5. The assembled stiffness matrix will again be sparsely populated but, with the possibility of several members meeting at a given node in a frame, the bandedness of $[K]$ will be very much dependent on the nodal ordering. In fact, if d is the greatest difference in global node numbers for a single element and f is the number of degrees of freedom per node,

$$b = (d + 1)f\,. \tag{8.6.8}$$

For efficient computation, the node numbering must be carried out judiciously in order that b shall be as small as possible.

The potential energy of the externally applied nodal forces and couples can be determined directly as $\{u\}^t\{P_c\}$: distributed actions or initial strains can be treated in a manner similar to those employed in Sections 8.4 and 8.5 but incorporating the coordinate transformation matrix $[T]$. Finally, the condition $\delta V = 0$ yields the overall stiffness equilibrium equations

$$[K]\{u\} = \{P\}\,. \tag{8.6.9}$$

EXAMPLE 8.6.1

Generate the overall stiffness matrix for the pinjointed frame of Fig. 8.6.3 (a).

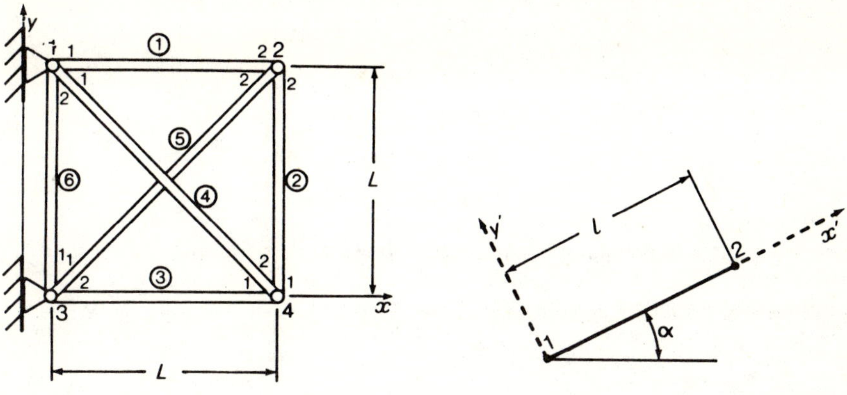

(a) (b)

Fig. 8.6.3

The stiffness matrix for a typical member with respect to local coordinates, Fig. 8.6.3 (b), is from equation (a) of Example 8.4.1,

$$[k']^e = \frac{EA}{l}\begin{bmatrix} 1 & -1 \\ -1 & 1 \end{bmatrix} . \tag{a}$$

The coordinate transformation matrix $[T]$ is defined by

$$\begin{Bmatrix} u_1' \\ v_1' \\ u_2' \\ v_2' \end{Bmatrix} = \begin{bmatrix} c & s & 0 & 0 \\ -s & c & 0 & 0 \\ 0 & 0 & c & s \\ 0 & 0 & -s & c \end{bmatrix} \begin{Bmatrix} u_1 \\ v_1 \\ u_2 \\ v_2 \end{Bmatrix} \tag{b}$$

or

$$\{u'\}_0^e = [T]\{u\}_0^e . \tag{c}$$

Here, $c = \cos\alpha$ and $s = \sin\alpha$.

Now $[k']^e$ has the form of equation (a) since an axial member is considered to have no transverse stiffness. Before we can execute transformation (8.6.7) however, we rewrite equation (a) as

$$[k']^e = \frac{EA}{l}\begin{bmatrix} 1 & 0 & -1 & 0 \\ 0 & 0 & 0 & 0 \\ -1 & 0 & 1 & 0 \\ 0 & 0 & 0 & 0 \end{bmatrix} \tag{d}$$

in order that $[k']^e$ shall be conformable with $[T]$.

Then, following equation (8.6.7),

$$[k]^e = \frac{EA}{l}\begin{bmatrix} c^2 & sc & -c^2 & -sc \\ sc & s^2 & -sc & -s^2 \\ -c^2 & -sc & c^2 & sc \\ -sc & -s^2 & sc & s^2 \end{bmatrix}. \tag{e}$$

For element 1, $\alpha = 0$ and equation (e) reduces to equation (d).

For elements 2 and 6, observing the element nodal ordering, $\alpha = 90°$ so that

$$[k]^2 = \frac{EA}{L}\begin{bmatrix} 0 & 0 & 0 & 0 \\ 0 & 1 & 0 & -1 \\ 0 & 0 & 0 & 0 \\ 0 & -1 & 0 & 1 \end{bmatrix} \tag{f}$$

For element 3, $\alpha = 180°$ so that

$$[k]^3 = \frac{EA}{L}\begin{bmatrix} 1 & 0 & -1 & 0 \\ 0 & 0 & 0 & 0 \\ -1 & 0 & 1 & 0 \\ 0 & 0 & 0 & 0 \end{bmatrix}. \tag{g}$$

For elements 4 and 5,

$$[k]^4 = \frac{\sqrt{2}EA}{4L}\begin{bmatrix} 1 & -1 & -1 & 1 \\ -1 & 1 & 1 & -1 \\ -1 & 1 & 1 & -1 \\ 1 & -1 & -1 & 1 \end{bmatrix},$$

$$[k]^5 = \frac{\sqrt{2}EA}{4L}\begin{bmatrix} 1 & 1 & -1 & 1 \\ 1 & 1 & -1 & -1 \\ -1 & -1 & 1 & 1 \\ -1 & -1 & 1 & 1 \end{bmatrix}. \tag{h}$$

The assembly process of equation (8.4.22) was found to have a very simple interpretation, illustrated in Fig. 8.5.4, when the $[C]$ matrix is Boolean. Applying the $[T]$ matrices to the elements ensures this favourable situation and the system matrix $[K]$ for the frame of Fig. 8.6.3 may be determined by inspection to be

$$[K] = \frac{EA}{L}\begin{bmatrix} 1+\frac{\sqrt{2}}{4} & -\frac{\sqrt{2}}{4} & -1 & 0 & 0 & 0 & -\frac{\sqrt{2}}{4} & \frac{\sqrt{2}}{4} \\ -\frac{\sqrt{2}}{4} & 0 & 0 & 0 & 0 & -1 & \frac{\sqrt{2}}{4} & -\frac{\sqrt{2}}{4} \\ -1 & 0 & 1+\frac{\sqrt{2}}{4} & \frac{\sqrt{2}}{4} & -\frac{\sqrt{2}}{4} & -\frac{\sqrt{2}}{4} & 0 & 0 \\ 0 & 0 & \frac{\sqrt{2}}{4} & 1+\frac{\sqrt{2}}{4} & -\frac{\sqrt{2}}{4} & -\frac{\sqrt{2}}{4} & 0 & -1 \\ 0 & 0 & -\frac{\sqrt{2}}{4} & -\frac{\sqrt{2}}{4} & 1+\frac{\sqrt{2}}{4} & \frac{\sqrt{2}}{4} & -1 & 0 \\ 0 & -1 & -\frac{\sqrt{2}}{4} & -\frac{\sqrt{2}}{4} & \frac{\sqrt{2}}{4} & 1+\frac{\sqrt{2}}{4} & 0 & 0 \\ -\frac{\sqrt{2}}{4} & \frac{\sqrt{2}}{4} & 0 & 0 & -1 & 0 & 1+\frac{\sqrt{2}}{4} & -\frac{\sqrt{2}}{4} \\ \frac{\sqrt{2}}{4} & -\frac{\sqrt{2}}{4} & 0 & -1 & 0 & 0 & -\frac{\sqrt{2}}{4} & 1+\frac{\sqrt{2}}{4} \end{bmatrix}. \quad \text{(j)}$$

At this stage, $[K]$ is singular since the constraints on the structure have not been imposed. If we suppose $u_1 = v_1 - u_3 = v_3 = 0$, deleting the rows and columns corresponding to these quantities in equation (j) yields the reduced matrix and allows solution for u_2, v_2, u_4, v_4 in accordance with equations (8.4.38) etc. In fact, the matrix equation of equilibrium is

$$\frac{EA}{L}\begin{bmatrix} 1+\frac{\sqrt{2}}{4} & \frac{\sqrt{2}}{4} & 0 & 0 \\ \frac{\sqrt{2}}{4} & 1+\frac{\sqrt{2}}{4} & 0 & -1 \\ 0 & 0 & 1+\frac{\sqrt{2}}{4} & -\frac{\sqrt{2}}{4} \\ 0 & -1 & -\frac{\sqrt{2}}{4} & 1+\frac{\sqrt{2}}{4} \end{bmatrix}\begin{Bmatrix} u_2 \\ v_2 \\ u_4 \\ v_4 \end{Bmatrix} = \begin{Bmatrix} X_2 \\ Y_2 \\ X_4 \\ Y_4 \end{Bmatrix} \quad \text{(k)}$$

This is precisely the solution found by the same principle but using less systematic means in Example 2.6.6.

Problems 8.6

8.6.1 Use the formal procedures of Section 8.6 to establish the matrix equilibrium equation for the frame of Fig. 8.6.4.

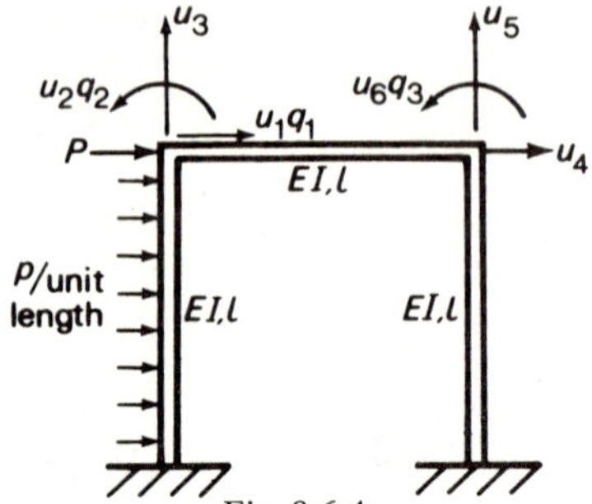

Fig. 8.6.4

Ans:

$$\begin{bmatrix} \left(\frac{12EI}{l^3}+\frac{EA}{l}\right) & \frac{6EI}{l^2} & \frac{6EI}{l^2} & \left(-\frac{EA}{l}\right) & 0 & 0 \\ & \left(\frac{12EI}{l^3}+\frac{EA}{l}\right) & \frac{6EI}{l^2} & 0 & \frac{-12EI}{l^3} & \frac{6EI}{l^2} \\ & & \frac{8EI}{l} & 0 & -\frac{6EI}{l^2} & \frac{2EI}{l} \\ & & & \left(\frac{12EI}{l^3}+\frac{EA}{l}\right) & 0 & \frac{6EI}{l} \\ & \text{symm} & & & \left(\frac{12EI}{l^3}+EA\right) & \frac{-6EI}{l^2} \\ & & & & & \frac{8EI}{l} \end{bmatrix}$$

$$\times \begin{Bmatrix} u_1 \\ u_2 \\ u_3 \\ u_4 \\ u_5 \\ u_6 \end{Bmatrix} = \begin{Bmatrix} p+\frac{1}{2}pl \\ 0 \\ \frac{1}{12}pl^2 \\ 0 \\ 0 \\ 0 \end{Bmatrix}$$

Confirm that, if bar extensions are negligible, the stiffness matrix with respect to q_1, q_2, q_3 is precisely that of Example 2.7.4.

8.7 Plane stress and plane strain problems

Comparing Figs. 8.2.1 or 8.6.1, and Fig. 8.2.2, we see that there is a significant new feature affecting the choice of displacement model for a finite element treatment of a two dimensional continuum as compared with the treatment of a bar or frame. Now, adjacent elements are in contact along a common interface as well as at nodes so that to meet the criteria of Section 8.3, displacement models must be devised to ensure continuity of displacements across an interface as well as at the nodes. To illustrate, we will consider the simplest two dimensional element, the **constant strain triangle.**†

Fig. 8.7.1 (a) shows a portion of a triangular mesh whilst a typical element is isolated in Fig. 8.7.1 (b).

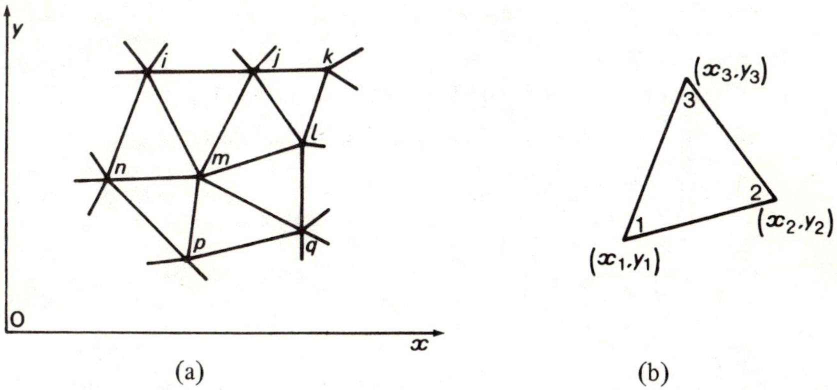

Fig. 8.7.1

Assuming that displacements vary linearly within the element, the x and y component are given by

$$\begin{aligned} u^e(x, y) &= a_0 + a_1 x + a_2 y , \\ v^e(x, y) &= a_3 + a_4 x + a_5 y . \end{aligned} \tag{8.7.1}$$

That is,

$$\begin{Bmatrix} u \\ v \end{Bmatrix}^e = \begin{bmatrix} 1 & x & y & 0 & 0 & 0 \\ 0 & 0 & 0 & 1 & x & y \end{bmatrix} \begin{Bmatrix} a_0 \\ \vdots \\ a_5 \end{Bmatrix} \tag{8.7.2}$$

or

$$\{u\}^e = [f(x, y)]\{a\} . \tag{8.7.3}$$

This displacement model utilises six coefficients so that six degrees of freedom are implied. With a node at each apex, each having two degrees of freedom, the three node triangle of Fig. 8.7.1 (b) fits the situation.

† A brief discussion of more elaborate elements is given in Section 8.9.

Proceeding as before, the a's are related to the nodal u's and v's by

$$\begin{Bmatrix} u_1 \\ v_1 \\ u_2 \\ v_2 \\ u_3 \\ v_3 \end{Bmatrix} = \begin{bmatrix} 1 & x_1 & y_1 & 0 & 0 & 0 \\ 0 & 0 & 0 & 1 & x_1 & y_1 \\ 1 & x_2 & y_2 & 0 & 0 & 0 \\ 0 & 0 & 0 & 1 & x_2 & y_2 \\ 1 & x_3 & y_3 & 0 & 0 & 0 \\ 0 & 0 & 0 & 1 & x_3 & y_3 \end{bmatrix} \begin{Bmatrix} a_0 \\ a_1 \\ a_2 \\ a_3 \\ a_4 \\ a_5 \end{Bmatrix} \tag{8.7.4}$$

or

$$\{u\}_0^e = [A]\{a\} . \tag{8.7.4a}$$

Solving for the a's,

$$\{a\} = [A]^{-1}\{u\}_0^e \tag{8.7.5}$$

Substituting from equation (8.7.5) in (8.7.3),

$$\{u\}^e = [f(x, y)]\,[A]^{-1}\{u\}_0^e = [N(x, y)]\,\{u\}_0^e\,\dagger \tag{8.7.6}$$

Here, $[N(x, y)] = \Big[[N(x, y)]_1\,[N(x, y)]_2\,[N(x, y)]_3\Big]$, where

$$[N(x, y)]_1 = \frac{1}{2\Delta}(a_1 + b_1x + c_1y)[I] = \frac{1}{2\Delta}N_1(x, y)[I] \text{ etc} \tag{8.7.7}$$

and

$$\left.\begin{aligned} a_1 &= x_2y_3 - x_3y_2 , \\ b_1 &= y_2 - y_3 , \\ c_1 &= x_3 - x_2 . \end{aligned}\right\} \tag{8.7.8}$$

The other constants $a_2, b_2 \ldots\ldots c_3$ are given by equations (8.7.8) merely by cyclic change of the subscripts. $[I]$ is the unit matrix and Δ is the area of the element given by

$$\Delta = \frac{1}{2}\begin{vmatrix} 1 & 1 & 1 \\ x_1 & x_2 & x_3 \\ y_1 & y_2 & y_3 \end{vmatrix} . \tag{8.7.9}$$

The reader may readily confirm that

$$N_i(x_jy_j) = \delta_{ij} , \tag{8.7.10}$$

where δ_{ij} is the Kronecker delta: $[N(x, y)]$ is thus seen to be a linear interpolation function matrix. Further, the displacements along any one side of an element depend only on those at the nodes at each end of that side so that by requiring the nodal displacements of adjacent elements to match in the usual compatibility statement through a $[C]$ matrix, compatibility of displacements across the common interface is also assured. Indeed, this displacement model meets all the criteria of Section 8.3.

† The reader should distinguish between $\{u\}^e$ and $\{u\}_0^e$. The former denotes the displacement field within the element and is of order 2×1, whilst the latter is a vector of element nodal displacements of order 6×1.

The strain components implied by the assumed displacements are given by

$$\begin{Bmatrix} \epsilon_x \\ \epsilon_y \\ \gamma_{xy} \end{Bmatrix} = \begin{Bmatrix} \dfrac{\partial u}{\partial x} \\ \dfrac{\partial v}{\partial y} \\ \dfrac{\partial u}{\partial y} + \dfrac{\partial v}{\partial x} \end{Bmatrix} = \begin{bmatrix} \dfrac{\partial}{\partial x} & 0 \\ 0 & \dfrac{\partial}{\partial y} \\ \dfrac{\partial}{\partial y} & \dfrac{\partial}{\partial x} \end{bmatrix} \begin{Bmatrix} u \\ v \end{Bmatrix}^e \tag{8.7.11}$$

or

$$\{\epsilon\} = [\partial]\{u\}^e. \tag{8.7.12}$$

Noting equation (8.7.6), we have

$$\{\epsilon\} = [\partial]\,[N(x, y)]\,\{u\}_0^e = [B]\{u\}_0^e\,. \tag{8.7.13}$$

For the linear displacement model chosen here, $[B]$ turns out to be a matrix of constants and hence the description 'constant strain element' in this case. In fact,

$$\{\epsilon\} = \frac{1}{2\Delta}\begin{bmatrix} b_1 & 0 & b_2 & 0 & b_3 & 0 \\ 0 & c_1 & 0 & c_2 & 0 & c_3 \\ c_1 & b_1 & c_2 & b_2 & c_3 & b_3 \end{bmatrix}\{u\}_0^e\,. \tag{8.7.14}$$

Hooke's law has the form $\{\sigma\} = [D]\{\epsilon\}$, (8.7.15)

with $\{\sigma\} = [\sigma_x \sigma_y \tau_{xy}]^t$, whilst

$$[D] = \frac{E}{1-\nu^2}\begin{bmatrix} 1 & \nu & 0 \\ \nu & 1 & 0 \\ 0 & 0 & (1-\nu)/2 \end{bmatrix} \text{ for plane stress}, \tag{8.7.16}$$

$$\text{or } [D] = \frac{E}{(1+\nu)(1-2\nu)}\begin{bmatrix} (1-\nu) & \nu & 0 \\ \nu & (1-\nu) & 0 \\ 0 & 0 & (1-2\nu)/2 \end{bmatrix} \text{ for plane strain}. \tag{8.7.17}$$

The strain energy stored in an element of thickness t is given by

$$U^e = \frac{1}{2}\iint_e \{\epsilon\}^t[D]\{\epsilon\}t\,\mathrm{d}x\mathrm{d}y\,.\dagger \tag{8.7.18}$$

Substituting from equation (8.7.13),

$$U^e = \frac{1}{2}\{u\}_0^{et}\left(\iint_e [B]^t[D]\,[B]\,t\,\mathrm{d}x\mathrm{d}y\right)\{u\}_0^e \tag{8.7.19}$$

or

$$U^e = \frac{1}{2}\{u\}_0^{et}\,[k]^e\{u\}_0^e\,. \tag{8.7.20}$$

† In the plane strain situation a slice of unit thickness may be assumed.

which is precisely the same form as before. Here, $[k]^e$ is a 6×6 matrix corresponding to the element's six degrees of reedom.

The loading may, typically, consist of a body force $\{F_b\} = [F_x F_y]^t$ per unit volume and a surface traction $\{T_s\} = [T_x T_y]^t$ per unit surface. Then for an element,

$$-\Omega^e_{\text{body}} = \iint_e \{u\}^{et}\{F_b\}\, t\mathrm{d}x\mathrm{d}y = \{u\}^{et}_0 \iint_e [N]^t\{F_b\}\, t\mathrm{d}x\mathrm{d}y$$

or
$$\Omega^e_{\text{body}} = -\{u\}^{et}_0\, \{F_b\}^e\,. \tag{8.7.21}$$

$$-\Omega^e_{\text{surface}} = \int_{S_T} \{u\}^{et}_{S_T}\, \{T_s\}\, t\mathrm{d}S = \{u\}^{et}_0 \left(\int_{S_T} [N]^t\{T_s\}\, t\mathrm{d}S\right)$$

or
$$\Omega^e_{\text{surface}} = -\{u\}^{et}_0\, \{F_s\}^e\,. \tag{8.7.22}$$

Here, the integration only extends over that portion of the element surface on which the surface traction is prescribed. The total potential energy of the system is then

$$V = \sum_e U^e + \sum_e \Omega^e_{\text{body}} + \sum_{e'} \Omega^e_S \tag{8.7.23}$$

where e' represents only those elements at the surface of the solid (edge of the region) where tractions are prescribed. Once again, the $[C]$ matrix allows a formal coding of the assembly process for stiffnesses and forces and $\delta V = 0$ yields the stiffness equilibrium equations

$$[K]\{u\} = \{P\}\,. \tag{8.7.24}$$

Once the u's have been determined, we return to equations (8.7.13) and (8.7.15) to yield the element strains and stresses.

EXAMPLE 8.7.1

Determine the stiffness matrix for the plane stress element of Fig. 8.7.2. Thickness t is constant.

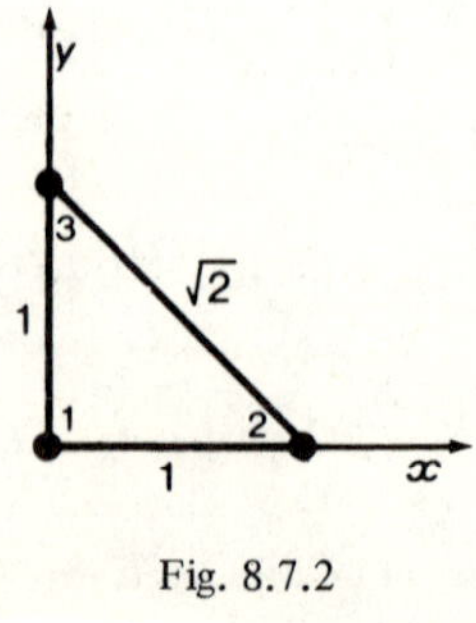

Fig. 8.7.2

Area of element $$\Delta = \tfrac{1}{2}. \tag{a}$$

$$b_1 = y_2 - y_3 = -1; \quad b_2 = y_3 - y_1 = 1; \quad b_3 = y_1 - y_2 = 0. \tag{b}$$

$$c_1 = x_3 - x_2 = -1; \quad c_2 = x_1 - x_3 = 0; \quad c_3 = x_2 - x_1 = 1. \tag{c}$$

Then from equation (8.7.14),

$$[B] = \begin{bmatrix} -1 & 0 & 1 & 0 & 0 & 0 \\ 0 & -1 & 0 & 0 & 0 & 1 \\ -1 & -1 & 0 & 1 & 1 & 0 \end{bmatrix}. \tag{d}$$

Since $[B]$ and t are constant, the integration in equation (8.7.19) is trivial and

$$[k]^e = \Delta t [B]^t [D] [B] \tag{e}$$

so that

$$[k]^e = \frac{Et}{2(1-\nu^2)} \begin{bmatrix} \frac{3-\nu}{2} & \frac{1+\nu}{2} & -1 & -\frac{1-\nu}{2} & -\frac{1-\nu}{2} & -\nu \\ & \frac{3-\nu}{2} & -\nu & -\frac{1-\nu}{2} & -\frac{1-\nu}{2} & -1 \\ & & 1 & 0 & 0 & \nu \\ & \text{symm} & & \frac{1-\nu}{2} & \frac{1-\nu}{2} & 0 \\ & & & & \frac{1-\nu}{2} & 0 \\ & & & & & 1 \end{bmatrix}. \tag{f}$$

EXAMPLE 8.7.2

Nodes 1 and 3 of the element in Fig. 8.7.3 lie on the edgeof a plate where tractions are prescribed. If the tractions consist of a pressure and upward shear varying linearly from p_1 to p_3 and s_1 to s_3 respectively, per unit surface between nodes 1 and 3, determine the element load vector.

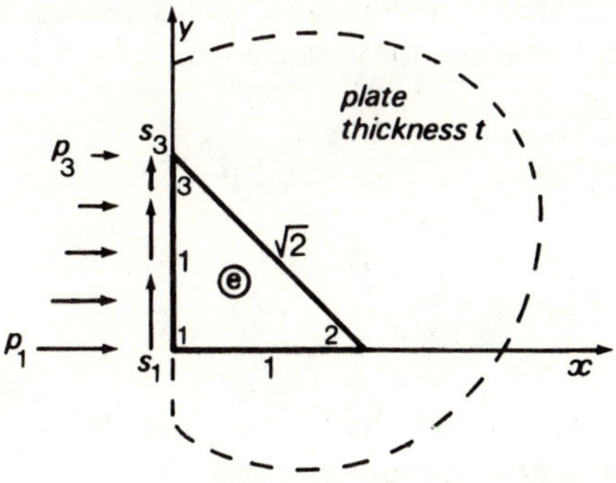

Fig. 8.7.3

From equation (8.7.22)

$$\{F_s\}^e = \left|\left(\int_0^1 [N]^t \{T_s\}\, t\mathrm{d}y\right)\right|_{x=0} \tag{a}$$

Using equation (8.7.7) with $\{T_s\} = [p\ s]^t$,

$$[N]^t\{T_s\} = \frac{1}{2\Delta}\begin{bmatrix} N_1\,[I] \\ N_2\,[I] \\ N_3\,[I] \end{bmatrix}\begin{Bmatrix} p \\ s \end{Bmatrix} = \frac{1}{2\Delta}\begin{Bmatrix} N_1 p \\ N_1 s \\ N_2 p \\ N_2 s \\ N_3 p \\ N_3 s \end{Bmatrix}. \tag{b}$$

Now, for $x = 0, N_i = a_i + c_i y$. Further, for this element $2\Delta = 1, a_1 = 1$, $a_2 = 0, a_3 = 0,\ c_1 = -1, c_2 = 0,\ c_3 = 1$. Carrying out the multiplications in equation (b), we find from equation (a) that

$$\{F_s\}^e = \int_0^1 \begin{Bmatrix} p_1 + (p_3 - 2p_1)y - (p_3 - p_1)y^2 \\ s_1 + (s_3 - 2s_1)y - (s_3 - s_1)y^2 \\ 0 \\ 0 \\ p_1 y + (p_2 - p_1)y^2 \\ s_1 y + (s_3 - s_1)y^2 \end{Bmatrix} \mathrm{d}y\,. \tag{c}$$

Evaluating the integrals,

$$\{F_s\}^e = [(\tfrac{1}{3}p_1 + \tfrac{1}{6}p_3)\ (\tfrac{1}{3}s_1 + \tfrac{1}{6}s_3)\ 0\ 0\ (\tfrac{1}{6}p_1 + \tfrac{1}{3}p_3)\ (\tfrac{1}{6}s_1 + \tfrac{1}{3}s_3)]^t. \tag{d}$$

We observe that the components of load vector attributable to node 2 have automatically turned out to be zero; we would expect this since node 2 is an internal one. It so happens in this problem that the load vector determined on the above basis of work corresponds exactly to that which would be obtained by sharing the traction statically between nodes 1 and 3. This is so because the assumed element displacement pattern is linear; in general, the results of the two procedures are different.

Problems 8.7

8.7.1 Generate the stiffness matrix for the plane stress element of Fig. 8.7.4 assuming the displacement model of equations (8.7.6, 7, 8, 9). Thickness t = constant.

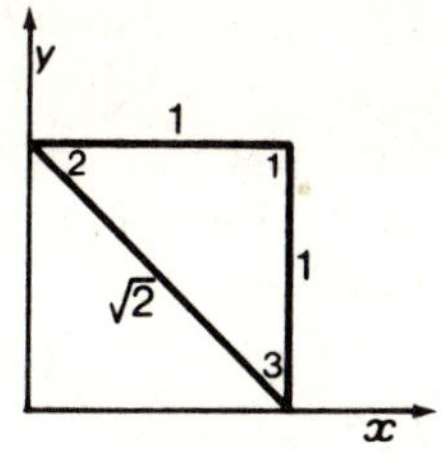

Fig. 8.7.4

Ans:

$$[k]^e = \frac{Et}{2(1-\nu^2)} \begin{bmatrix} \frac{3-\nu}{2} & \frac{1+\nu}{2} & -1 & -\frac{1-\nu}{2} & -\frac{1-\nu}{2} & -\nu \\ & \frac{3-\nu}{2} & -\nu & -\frac{1-\nu}{2} & -\frac{1-\nu}{2} & -1 \\ & & 1 & 0 & 0 & \nu \\ & & & \frac{1-\nu}{2} & \frac{1-\nu}{2} & 0 \\ \text{symm} & & & & \frac{1-\nu}{2} & 0 \\ & & & & & 1 \end{bmatrix}$$

8.7.2 Using the element stiffness matrices of Example 8.7.1 and Problem 8.7.1, generate the stiffness matrix for the system of Fig. 8.7.5. Impose the appropriate boundary condition in order that symmetry shall be exploited so that the solution applies to the first quadrant. Plate thickness is 1 mm whilst Young's modulus and Poisson's ratio are $E = 207$ kN/mm^2, $\nu = 0.3$.

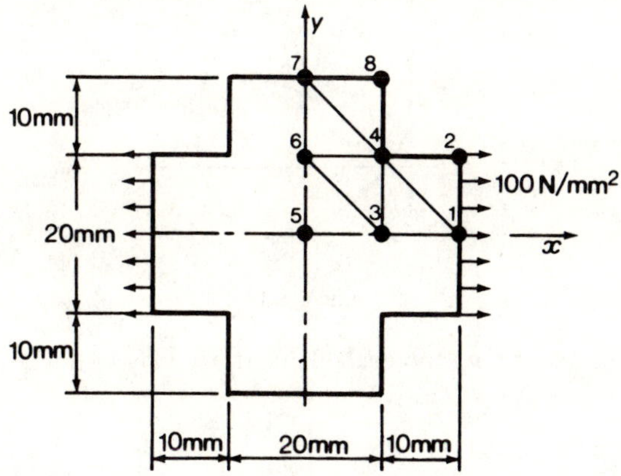

Fig. 8.7.5

8.7.3 Deduce the load vector for the system of Problem 8.7.2.

Ans: $\{P_{x_1}\ P_{y_1}\ P_{x_2}\ P_{y_2} \ldots\ldots P_{y_8}\}^t = \{\tfrac{1}{2}p\ 0\ \tfrac{1}{2}p\ 0\ 0\ 0 \ldots\ldots 0\}^t$.

8.7.4 The element of Example 8.7.1 is subjected to an initial strain $\{\epsilon_0\}$. Deduce the appropriate load vector if $\{\epsilon_0\}^t = \alpha T[1\ \ 1\ \ 0]$, corresponding to a thermal strain.

Ans: $$\{F_{\epsilon_0}\}^t = \frac{tE\alpha T}{2(1-\nu)}[-1\ \ -1\ \ 1\ \ 0\ \ 0\ \ 1] .$$

(α is the temperature coefficient of linear expansion, T is the temperature rise.)

8.8 The torsion problem

As a further application of the finite element technique to two dimensional continua, we will treat the problem of torsion of a prismatic bar. In Section 5.6 we saw that the problem may be formulated in terms of a warping function or a stress function; a variational treatment of the former problem must be based on the principle of potential energy whilst the latter comes from the complementary energy principle. To provide a contrast with the procedures described earlier in this chapter, we will use a complementary energy formulation.

From Section 6.7,

$$V^* = \frac{l}{2G}\iint\left[\left(\frac{\partial\phi}{\partial x}\right)^2 + \left(\frac{\partial\phi}{\partial y}\right)^2 - 4G\alpha\phi\right]\mathrm{d}x\mathrm{d}y \qquad (8.8.1)$$

and $\phi = 0$ on the boundary of a simply connected region.

The cross section of the bar may be divided into triangular elements as shown in Fig. 8.8.1.

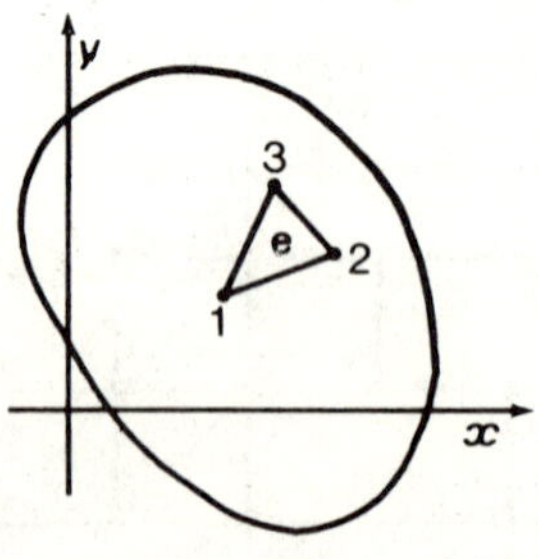

Fig. 8.8.1

Assuming ϕ varies linearly within each element, we may write

$$\phi^e = [1\ \ x\ \ y]\begin{Bmatrix} a_0 \\ a_1 \\ a_2 \end{Bmatrix} = [f(x, y)]\{a\} . \qquad (8.8.2)$$

Then, considering each node in turn,

$$\begin{Bmatrix} \phi_1 \\ \phi_2 \\ \phi_3 \end{Bmatrix}^e = \begin{bmatrix} 1 & x_1 & y_1 \\ 1 & x_2 & y_2 \\ 1 & x_3 & y_3 \end{bmatrix} \begin{Bmatrix} a_0 \\ a_1 \\ a_2 \end{Bmatrix} , \tag{8.8.3}$$

or

$$\{\phi\}_0^e = [A]\{a\} , \tag{8.8.4}$$

so that

$$\phi^e = [f(x, y)]\,[A]^{-1}\{\phi\}_0^e = [N_1(x, y)\,N_2(x, y)\,N_3(x, y)]\{\phi\}_0^e$$
$$= [N(x, y)]\{\phi\}_0^e \,. \tag{8.8.5}$$

The functions $N_i(x, y)$ have precisely the same form as that defined by equations (8.7.7) etc.

To utilise equation (8.8.1), we form

$$\frac{\partial \phi^e}{\partial x} = \left[\frac{\partial N}{\partial x}\right]\{\phi\}_0^e = [B_x]\{\phi\}_0^e \,,$$
$$\frac{\partial \phi^e}{\partial y} = \left[\frac{\partial N}{\partial y}\right]\{\phi\}_0^e = [B_y]\{\phi\}_0^e \,. \tag{8.8.6}$$

Then, for a typical element

$$V^{*e} = \frac{l}{2G}\iint_e\Big(\{\phi\}_0^{et}[B_x]^t[B_x]\{\phi\}_0^e + \{\phi\}_0^{et}[B_y]^t[B_y]\{\phi\}_0^e$$
$$- 4G\alpha\{\phi\}_0^{et}[N]^t\Big)\,\mathrm{d}x\mathrm{d}y \,, \tag{8.8.7}$$

or

$$\frac{V^{*e}}{l} = \frac{1}{2}\{\phi\}_0^{et}\left(\frac{1}{G}\iint_e([B_x]^t[B_x] + [B_y]^t[B_y])\mathrm{d}x\mathrm{d}y\right)\{\phi\}_0^e$$
$$- 2\alpha\{\phi\}_0^{et}\left(\iint_e [N]^t\mathrm{d}x\mathrm{d}y\right) , \tag{8.8.8}$$

or

$$\frac{V^{*e}}{l} = \frac{1}{2}\{\phi\}_0^{et}[h]^e\{\phi\}_0^e - \{\phi\}_0^{et}\{u\}^e \,. \tag{8.8.9}$$

With ϕ^e varying linearly as supposed above,

$$h_{ij}{}^e = \frac{1}{4G\Delta}(b_i b_j + c_i c_j), \quad u_i{}^e = -\frac{2\alpha\Delta}{3} \tag{8.8.10}$$

Denoting $\{\phi\}$ as the vector of global nodal ϕ's, we have, for continuity in ϕ

$$\{\tilde{\phi}\} = \{\{\phi\}_0^e\} = [C]\{\phi\} \,, \tag{8.8.11}$$

where $[C]$ is a Boolean continuity matrix. Denoting also

$$\{\tilde{u}\} = \{\{u\}^e\} \,, \tag{8.8.12}$$

$$[\widetilde{h}] = \begin{bmatrix} [h]^1 & 0 & 0\ldots\ldots \\ 0 & [h]^2 & 0\ldots\ldots \\ \cdot & \cdot & \cdot \\ \cdot & \cdot & \cdot \\ \cdot & \cdot & \cdot \\ \cdot & \cdot & \cdot \end{bmatrix}, \qquad (8.8.13)$$

the total complementary energy for the bar becomes

$$\frac{V^*}{l} = \sum_e \frac{V^{*e}}{l} = \frac{1}{2}\{\widetilde{\phi}\}^t[\widetilde{h}]\{\widetilde{\phi}\}^t - \{\widetilde{\phi}\}^t\{\widetilde{u}\}$$

$$= \frac{1}{2}\{\phi\}^t[C]^t[\widetilde{h}][C]\{\phi\} - \{\phi\}^t[C]\{\widetilde{u}\},$$

or

$$\frac{V^*}{l} = \frac{1}{2}\{\phi\}^t[H]\{\phi\} - \{\phi\}^t\{u\}. \qquad (8.8.14)$$

For compatibility, $\delta V^* = 0$, so that

$$[H]\{\phi\} = \{u\}. \qquad (8.8.15)$$

The H_{ij} are obtained by merely adding together the h_{rs}^e for those elements providing a connexion between ϕ_i ($\equiv\phi_r^e$) and ϕ_j ($\equiv\phi_s^e$). u_i is obtained by adding together u_r^e for each element containing ϕ_i ($\equiv\phi_r^e$). The procedure is thus completely analogous to that followed in the displacement formulation of the earlier sections in this chapter.

As it appears in equation (8.8.15), $[H]$, which is of the nature of a generalised flexibility matrix, is singular. Before solution, the boundary conditions on $\{\phi\}$ must be imposed and since $\phi = 0$ on the boundary it is only necessary to delete the rows and columns of $[H]$ corresponding to the boundary nodes, to produce the non-singular coefficient matrix.† Once the nodal ϕ's have been determined, the stress components and torsion constant are obtained by means of equations (3.6.14) and (3.6.19).

EXAMPLE 8.8.1

Determine the matrix equation of compatibility for the torsion of square section bar of Fig. 8.8.2 when only four elements are used.

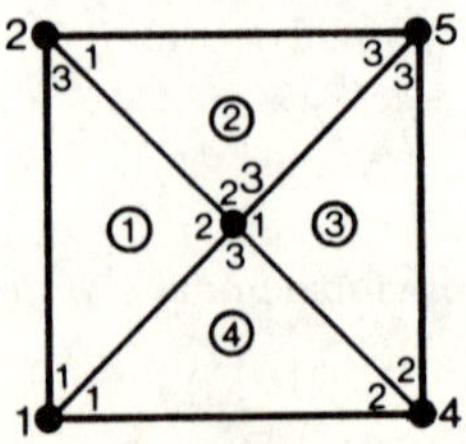

Fig. 8.8.2

† See the discussion relating to equation (8.4.38) etc.

Following the procedure described in the text, by inspection:

$$\begin{bmatrix} (h_{11}{}^1+h_{11}{}^4) & h_{13}{}^1 & (h_{12}{}^1+h_{13}{}^4) & h_{12}{}^4 & 0 \\ h_{31}{}^1 & (h_{33}{}^1+h_{11}{}^2) & (h_{32}{}^1+h_{12}{}^2) & 0 & h_{13}{}^2 \\ (h_{21}{}^1+h_{31}{}^4) & (h_{23}{}^1+h_{21}{}^2) & \begin{pmatrix} h_{22}{}^1+h_{22}{}^2 \\ + \\ h_{11}{}^3+h_{33}{}^4 \end{pmatrix} & (h_{12}{}^3+h_{32}{}^4) & (h_{23}{}^2+h_{13}{}^3) \\ h_{21}{}^4 & 0 & (h_{21}{}^3+h_{23}{}^4) & (h_{22}{}^3+h_{22}{}^4) & h_{23}{}^3 \\ 0 & h_{31}{}^2 & (h_{32}{}^2+h_{31}{}^3) & h_{32}{}^3 & (h_{33}{}^2+h_{33}{}^3) \end{bmatrix} \begin{Bmatrix} \phi_1 \\ \phi_2 \\ \phi_3 \\ \phi_4 \\ \phi_5 \end{Bmatrix} = \frac{\alpha}{6} \begin{Bmatrix} 2 \\ 2 \\ 4 \\ 2 \\ 2 \end{Bmatrix} \quad \text{(a)}$$

Observing that $\phi_1 = \phi_2 = \phi_5 = \phi_4 = 0$, the reduced coefficient matrix consists of only the element in the third row and column for equation (a). That is,

$$(h_{22}{}^1 + h_{22}{}^2 + h_{11}{}^3 + h_{33}{}^4)\phi_3 = \frac{4\alpha}{6} . \quad \text{(b)}$$

Of course, if we only ever generated these equations by hand, we could list the non-zero ϕ's and form the reduced matrix directly by inspection. However, as we have previously pointed out, in serious practical work there are so many unknowns that the systematic process described here must be coded for handling in a computer.

Problems 8.8

8.8.1 Determine the coefficient matrices for the elements of Fig. 8.8.3 (a) and (b) on the basis of linear interpolation for the Prandtl stress function.

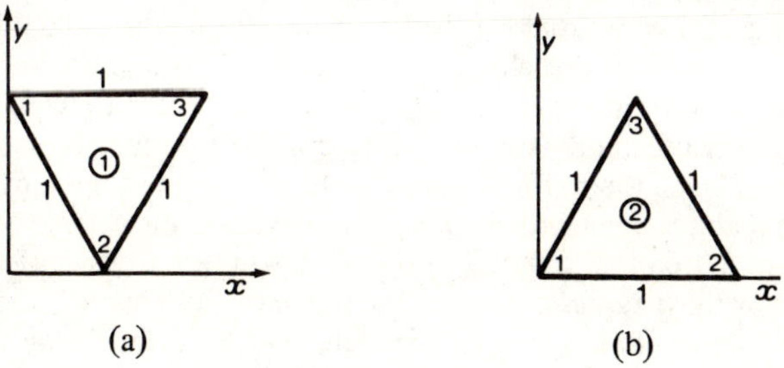

Fig. 8.8.3

Ans: $$[h]^1 = \frac{\sqrt{3}}{6G}\begin{bmatrix} 2 & -1 & -1 \\ -1 & 2 & -1 \\ -1 & -1 & 2 \end{bmatrix} = [h]^2 .$$

8.8.2 Use the results of Problem 8.8.1 to deduce the matrix compatibility equation for the torsion of the section of Fig. 8.8.4.

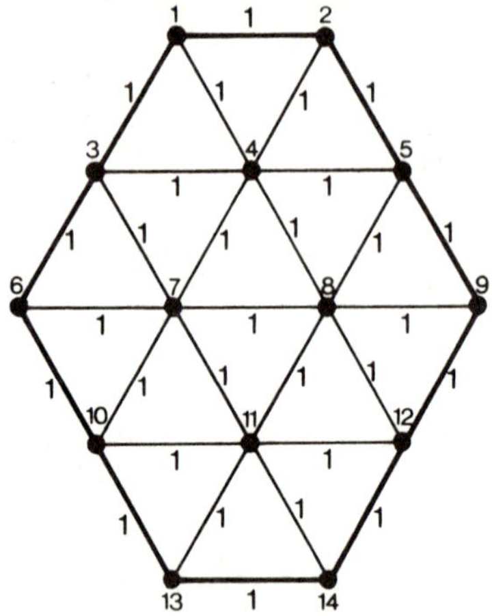

Fig. 8.8.4

Ans:

$$\begin{bmatrix} -6 & 1 & 1 & 0 \\ 1 & -6 & 1 & 1 \\ 1 & 1 & -6 & 1 \\ 0 & 1 & 1 & -6 \end{bmatrix} \begin{Bmatrix} \phi_4 \\ \phi_7 \\ \phi_8 \\ \phi_{11} \end{Bmatrix} = -3G\alpha \begin{Bmatrix} 1 \\ 1 \\ 1 \\ 1 \end{Bmatrix}$$

8.9 Higher order elements

In the earlier sections of this chapter, we have been concerned with acquiring an understanding of the essentials of the finite element method. To this end, it was possible to work with simple shaped elements within which the wanted quantities were assumed to vary in a relatively simple manner. Whilst these basic techniques have been widely used and have yielded acceptable solutions to many practically important problems, their effectiveness has been improved upon by a number of developments and refinements. Increased sophistication of element design and problem formulation figure significantly amongst these advances and whilst a full account of these topics is beyond the scope of this chapter, some reference to them is desirable for the sake of completeness. Accordingly we shall, in this section, briefly discuss the refined design of elements and, in Section 8.10, give a short account of some generalisations of problem formulation.

In the treatment of bar extension, plane stress/strain and torsion problems, we saw that the appropriate functionals were defined and solutions could be effected using a simple linear model for the wanted quantity. In contrast, the functional for beam bending involved the second derivative of the deflexion so that it was mandatory to use a higher order displacement model to ensure continuity of beam slopes (first derivatives) at the nodes: evidently, use of this displacement model for bar extension would ensure continuity of nodal strains and stresses and, for a given number of unknowns (degrees of freedom), we might expect better accuracy in the overall solution than that afforded by the linear model. We may follow this argument in other situations: thus, in say, the plane stress case, we would expect to need a large number of constant strain triangle elements to obtain acceptable accuracy in regions of high stress gradients whereas if higher order displacement fields were used, fewer elements would suffice. However, if the element shape is still triangular, say, fewer elements implies less accurate matching with curved portions of solid boundaries so that one is led to consider elements with curved sides. To summarise, then, one can visualize alternative discretisations as

(i) simple element shapes with simple displacement/stress fields;
(ii) simple element shapes with higher order displacement/stress fields;
(iii) complicated element shapes with higher order fields.

We have already discussed the first alternative and now move on to the second.

In Sections 8.4, 8.5, 8.7, 8.8, some manipulation showed that the displacement fields were expressible in terms of interpolation functions: evidently, if these could be generated directly, the inversion of matrix $[A]$ would be avoided and the whole treatment rather more elegant. In books on numerical analysis, a general form for an interpolation polynomial in one variable is given as

$$u_{pq} = \sum_{j=1}^{p} \sum_{i=0}^{q} N_{ij}(x)\, u^i(x_j) \; . \tag{8.9.1}$$

Here, $u(x)$ is the function to be approximated by u_{pq}, the $N_{ij}(x)$ are polynomials and the superscript i denotes the i'th derivative of $u(x)$ computed at x_j. By choosing $N_{ij}(x)$, p, q and x_j in various ways, various families of interpolating polynomials are obtained. When only function values and no derivatives are involved, we have

$$u_p = \sum_{j=1}^{p} N_j(x)u(x_j) \, . \tag{8.9.2}$$

Choosing

$$N_j = \frac{(x-x_1)(x-x_2)\,.\,.\,.\,.\,.\,(x-x_{j-1})(x-x_{j+1})\,.\,.\,.\,.\,.\,(x-x_p)}{(x_j-x_1)(x_j-x_2)\,.\,.\,.\,.\,.\,(x_j-x_{j-1})(x_j-x_{j+1})\,.\,.\,.\,.\,.\,(x_j-x_p)} \tag{8.9.3}$$

we observe that $N_j(x_i) = \delta_{ij}$ = the Kronecker delta. Such functions are **Lagrangian interpolation polynomials** and equation (8.9.2) then becomes a Lagrangian interpolation formula. If, for example, $p = 2$, we have the simple linear interpolation

$$u_2(x) = \frac{x - x_2}{x_1 - x_2}\, u(x_1) + \frac{x - x_1}{x_2 - x_1}\, u(x_2) \; . \tag{8.9.4}$$

As a refinement, choose $p = 2$, for two points $x_1 = 0$ and $x_2 = l$, $q = 1$, so that function and first derivative values are used, and

$$\left.\begin{aligned} N_{01}(x) &= 1 - \frac{3x^2}{l^2} + \frac{2x^3}{l^3}, \quad & N_{02}(x) &= \frac{3x^2}{l^2} - \frac{2x^3}{l^3}, \\ N_{11}(x) &= x - \frac{2x^2}{l} + \frac{x^3}{l^2}, \quad & N_{12}(x) &= -\frac{x^2}{l} + \frac{x^3}{l^2}. \end{aligned}\right\} \tag{8.9.5}$$

Equation (8.9.1) then becomes

$$\begin{aligned} u(x) = &\left(1 - \frac{3x^2}{l^2} + \frac{2x^3}{l^3}\right)u(0) + \left(x - \frac{2x^2}{l} + \frac{x^3}{l^2}\right)u'(0) \\ &+ \left(\frac{3x^2}{l^2} - \frac{2x^3}{l^3}\right)u(l) + \left(-\frac{x^2}{l} + \frac{x^3}{l^2}\right)u'(l), \end{aligned} \tag{8.9.6}$$

or $$u(x) = [N(x)]\{u\}_0 \tag{8.9.7}$$

with $\{u\}_0^t = [u(0)\, u'(0)\, u(l)\, u'(l)]$. This is called **Hermite interpolation** and we recognise equation (8.9.6) as corresponding to the beam displacement model of Section 8.5.† Again we observe that $N_i(x_j) = \delta_{ij}$, the Kronecker delta.

The formulations and integrations involved in generating coefficient matrices for more refined elements are much facilitated by the use of **natural coordinates.** These coordinates are local to an element and they allow the location of a point within an element to be described by dimensionless numbers which never exceed unit value. As a first example, consider the location of point P in the one dimensional element of Fig. 8.9.1.

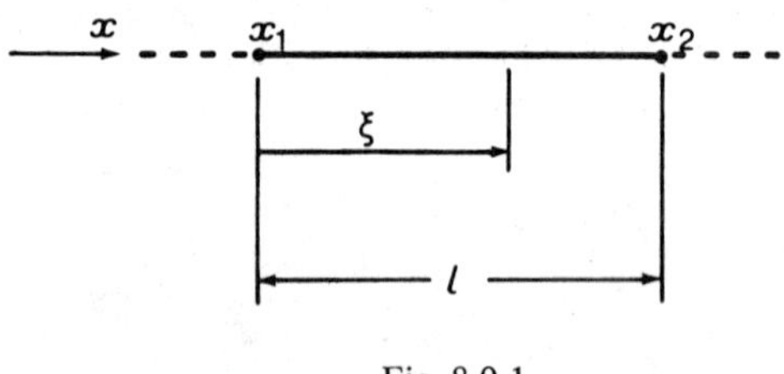

Fig. 8.9.1

The ends of the element are at global coordinates x_1 and x_2. Local coordinate ξ, varying from 0 to l at nodes 1 and 2 respectively, may be introduced but $\bar{\xi} = \xi/l$ varying from 0 to 1 is better. Introducing the dimensionless quantities

$$\bar{\xi}_1 = \left(1 - \frac{\xi}{l}\right), \quad \bar{\xi}_2 = \frac{\xi}{l}, \tag{8.9.8}$$

† Hermite polynomials are obtained from Lagrange polynomials according to

$$l_{0j}(x) = [1 - 2(x - x_j)l_j'(x_j)]\, l_j^2(x),$$

$$l_{1j}(x) = (x - x_j)\, l_j^2(x).$$

point P has coordinates $(\bar{\xi}_1, \bar{\xi}_2)$ so that nodes 1 and 2 will have coordinates (1, 0) and (0, 1) respectively. This symmetric representation is seen to be attractive in the context of interpolation. From equation (8.9.8), it follows that

$$\bar{\xi}_1 + \bar{\xi}_2 = 1 . \qquad (8.9.9)$$

The quantities $\bar{\xi}_1$ and $\bar{\xi}_2$ comprise a set of natural coordinates for the element and they are related to the global coordinates by

$$x = \bar{\xi}_1 x_1 + \bar{\xi}_2 x_2 . \qquad (8.9.10)$$

The linear displacement model for the axial element of Section 8.4 can now be written in alternative forms. First,

$$u(\xi) = \left[\left(1 - \frac{\xi}{l}\right) \frac{\xi}{l}\right]\begin{Bmatrix} u_1 \\ u_2 \end{Bmatrix}^e = [N_1(\xi)\, N_2(\xi)]\,\{u\}_0^e \qquad (8.9.11)$$

or, in view of equation (8.9.8),

$$u(\bar{\xi}_1, \bar{\xi}_2) = [N_1(\bar{\xi}_1, \bar{\xi}_2)\, N_2(\bar{\xi}_1, \bar{\xi}_2)]\,\{u\}_0^e . \qquad (8.9.12)$$

In this case, $N_1(\bar{\xi}_1, \bar{\xi}_2)$ and $N_2(\bar{\xi}_1, \bar{\xi}_2)$ have the particularly simple form given by equation (8.9.8), so that

$$u(\bar{\xi}_1, \bar{\xi}_2) = \bar{\xi}_1 u_1 + \bar{\xi}_2 u_2 . \qquad (8.9.13)$$

We observe the similarity between equations (8.9.10) and (8.9.13), a matter to which we shall return, and that $N_1 + N_2 = 1$.

As a second example, we reconsider the quadratic displacement model of Problem 8.4.1. Using Lagrangian interpolation with three points 1, 3, 2 at $\xi = 0, l/2, l$, respectively, equation (8.9.3) gives

$$\begin{aligned} N_1 &= \frac{(\xi - \xi_2)(\xi - \xi_3)}{(\xi_1 - \xi_2)(\xi_1 - \xi_3)} = 1 - \frac{3\xi}{l} + \frac{2\xi^2}{l^2} , \\ N_2 &= \frac{(\xi - \xi_1)(\xi - \xi_3)}{(\xi_2 - \xi_1)(\xi_2 - \xi_3)} = -\frac{\xi}{l} + \frac{2\xi^2}{l^2} , \\ N_3 &= \frac{(\xi - \xi_1)(\xi - \xi_2)}{(\xi_3 - \xi_1)(\xi_3 - \xi_2)} = \frac{4\xi}{l} - \frac{4\xi^2}{l^2} . \end{aligned} \qquad (8.9.14)$$

Rewritten in terms of the natural coordinates of equation (8.9.8), we have

$$u(\bar{\xi}_1, \bar{\xi}_2) = [\bar{\xi}_1(2\bar{\xi}_1 - 1)\ \bar{\xi}_2(2\bar{\xi}_2 - 1)\ 4\bar{\xi}_1 \bar{\xi}_2]\,\{u\}_0^e . \qquad (8.9.15)$$

Natural coordinates for points in the triangle of Fig. 8.9.2 are readily generated using vector algebra.

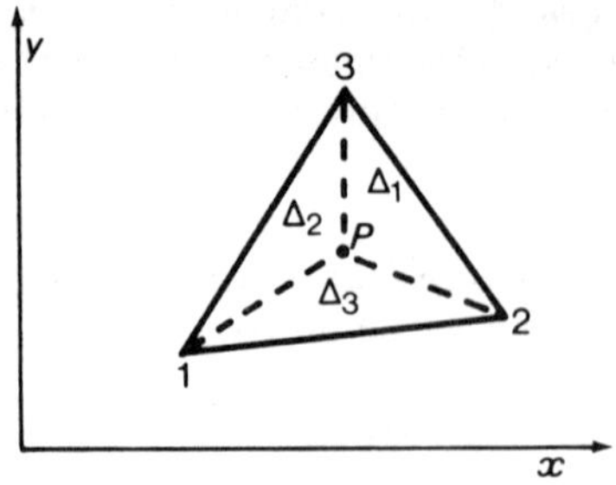

Fig. 8.9.2

With the triangle located in a global system xOy, the vertices 1, 2,3 and point P are defined by position vectors $\mathbf{r}_1, \mathbf{r}_2, \mathbf{r}_3$ and $\mathbf{r}$ respectively so that P is joined to the vertices by vectors $(\mathbf{r} - \mathbf{r}_1), (\mathbf{r} - \mathbf{r}_2), (\mathbf{r} - \mathbf{r}_3)$. But these three are linearly dependent so that

$$\alpha_1(\mathbf{r} - \mathbf{r}_1) + \alpha_2(\mathbf{r} - \mathbf{r}_2) + \alpha_3(\mathbf{r} - \mathbf{r}_3) = 0 \qquad (8.9.16)$$

where the α's are scalars. Rearranging,

$$\mathbf{r} = \frac{\alpha_1}{\alpha_1 + \alpha_2 + \alpha_3}\mathbf{r}_1 + \frac{\alpha_2}{\alpha_1 + \alpha_2 + \alpha_3}\mathbf{r}_2 + \frac{\alpha_3}{\alpha_1 + \alpha_2 + \alpha_3}\mathbf{r}_3 \qquad (8.9.16a)$$

or $\mathbf{r} = \bar{\xi}_1\mathbf{r}_1 + \bar{\xi}_2\mathbf{r}_2 + \bar{\xi}_3\mathbf{r}_3$ (8.9.16b)

Adjusting the values of the $\bar{\xi}$'s locates point P relative to the vertices; in particular, the vertices have coordinates (1, 0, 0), (0, 1, 0) and (0, 0, 1). The $\bar{\xi}$'s evidently comprise natural coordinates for the triangle with sides 23, 31, 12 corresponding to $\bar{\xi}_1 = 0, \bar{\xi}_2 = 0, \bar{\xi}_3 = 0$ respectively. Expanding equation (8.9.16b) into scalar form,

$$\begin{aligned} x &= \bar{\xi}_1 x_1 + \bar{\xi}_2 x_2 + \bar{\xi}_3 x_3 , \\ y &= \bar{\xi}_1 y_1 + \bar{\xi}_2 y_2 + \bar{\xi}_3 y_3 , \end{aligned} \qquad (8.9.17)$$

or

$$\begin{Bmatrix} x \\ y \end{Bmatrix} = \begin{bmatrix} \bar{\xi}_1 & 0 & \bar{\xi}_2 & 0 & \bar{\xi}_3 & 0 \\ 0 & \bar{\xi}_1 & 0 & \bar{\xi}_2 & 0 & \bar{\xi}_3 \end{bmatrix} \left\{\begin{Bmatrix} x \\ y \end{Bmatrix}\right\}_0^e . \qquad (8.9.18)$$

Observing that $\bar{\xi}_1 + \bar{\xi}_2 + \bar{\xi}_3 = 1$, (9.8.19)

we may solve for the $\bar{\xi}$'s to yield

$$\begin{Bmatrix} \bar{\xi}_1 \\ \bar{\xi}_2 \\ \bar{\xi}_3 \end{Bmatrix} = \frac{1}{2\Delta}\begin{bmatrix} a_1 & b_1 & c_1 \\ a_2 & b_2 & c_2 \\ a_3 & b_3 & c_3 \end{bmatrix}\begin{Bmatrix} 1 \\ x \\ y \end{Bmatrix} . \qquad (8.9.20)$$

Here, Δ is the area of the triangle whilst $a_1 \ldots\ldots c_3$ are precisely the quantities defined in equations (8.7.8). That is

$$a_1 = x_2y_3 - x_3y_2\,, \quad b_1 = y_2 - y_3\,, \quad c_1 = x_3 - x_2$$

with the other coefficients obtained by cyclic change of the subscripts. A little manipulation shows that the $\bar{\xi}_i = \Delta_i/\Delta =$ area ratios in Fig. 8.9.2 and so they are also called **area coordinates.**

Using interpolation functions, the two dimensional linear displacement model of Section 8.7 can be written directly as

$$\begin{Bmatrix} u(\bar{\xi}_i) \\ v(\bar{\xi}_i) \end{Bmatrix} = \begin{bmatrix} N_1(\bar{\xi}_i) & 0 & N_2(\bar{\xi}_i) & 0 & N_3(\bar{\xi}_i) & 0 \\ 0 & N_1(\bar{\xi}_i) & 0 & N_2(\bar{\xi}_i) & 0 & N_3(\bar{\xi}_i) \end{bmatrix} \{u\}_0^e\,, \tag{8.9.21}$$

where $N_r[\bar{\xi}_1(s)\bar{\xi}_2(s)\bar{\xi}_3(s)] = \delta_{rs}$. But in a manner completely analogous to that for the axial element above, we see that $N_1 = \bar{\xi}_1$, $N_2 = \bar{\xi}_2$, $N_3 = \bar{\xi}_3$ and

$$\{u(\bar{\xi}_i)\}^e = [\bar{\xi}_1\,[I]\;\; \bar{\xi}_2\,[I]\;\; \bar{\xi}_3\,[I]\,]\{u\}_0^e\,, \tag{8.9.22}$$

which, in view of equations (8.9.20), corresponds precisely with equations (8.7.6) and (8.7.7). We observe the similarity between equations (8.9.18) and (8.9.22). The reader is asked to consider quadratic interpolation for a six node triangle in Problem 8.9.1.

We have observed that equation (8.9.10), (8.9.13) and (8.9.18), (8.9.22) are of exactly the same form so that the geometry of and the displacement field in the element are described in terms of the same parameters. Elements enjoying this property as referred to as **isoparametric**†; when higher order displacement models are contemplated, the concept allows elements with curved sides to be generated very conveniently. Equations such as

$$\begin{Bmatrix} x \\ y \end{Bmatrix}^e = [N(\bar{\xi}_i)] \left\{ \begin{Bmatrix} x \\ y \end{Bmatrix} \right\}_0^e \tag{8.9.23}$$

can be regarded as a mapping of figures from the natural coordinate domain into the global system, equation (8.9.18) maps a triangle into a triangle, whilst $N_1 = \bar{\xi}_1(2\bar{\xi}_1 - 1)$, $N_2 = \bar{\xi}_2(2\bar{\xi}_2 - 1)$, $N_3 = \bar{\xi}_3(2\bar{\xi}_3 - 1)$, $N_4 = 4\bar{\xi}_1\bar{\xi}_2$, $N_5 = 4\bar{\xi}_2\bar{\xi}_3$, $N_6 = 4\bar{\xi}_3\bar{\xi}_1$ generate the sides of the curvilinear triangle of Fig. 8.9.3.

The reader will find it instructive to sketch these transformations and confirm that curves 142, 253 and 361 are parabolic in the xOy plane: if, say, points 1, 4, 2 are collinear, curve 142 degenerates into a straight line. Elements can be generated in this fashion to satisfy the continuity and completeness requirements of Section 8.3.

† Elements for which the displacement model is of lower order than the geometric one are called **superparameteric** whilst those with converse descriptions are called **subparametric**.

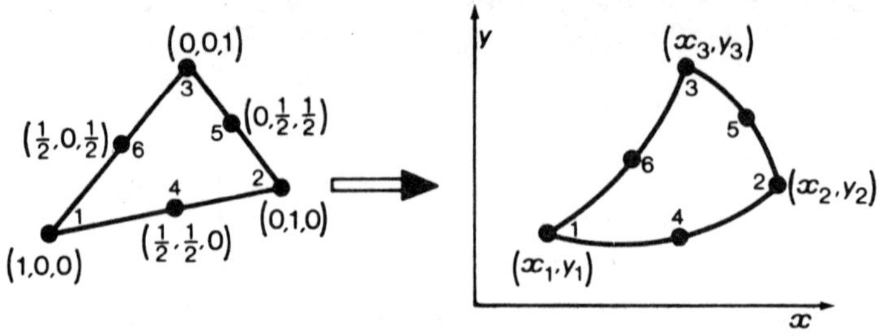

Fig. 8.9.3

Generating the element stiffness matrix requires the performace of certain differentiations and integrations and to effect these, we need to determine transformations between differential coefficients and integrals expressed in terms of natural and global coordinates. Thus, for a plane element

$$[k] = \iint_{\Delta} t[B]^t[D][B]\,\mathrm{d}x\mathrm{d}y\,, \tag{8.9.24}$$

where
$$[B] = [\partial][N(x,y)] = \begin{bmatrix} \dfrac{\partial}{\partial x} & 0 \\ 0 & \dfrac{\partial}{\partial y} \\ \dfrac{\partial}{\partial y} & \dfrac{\partial}{\partial x} \end{bmatrix} [N(x,y)]\,. \tag{8.9.25}$$

But $[N]$ is most conveniently expressed in terms of natural coordinates so that we need to invoke the chain rule of differentiation:

$$\frac{\partial}{\partial x} = \sum_i \frac{\partial \bar{\xi}_i}{\partial x}\frac{\partial}{\partial \bar{\xi}_i}\,, \quad \frac{\partial}{\partial y} = \sum_i \frac{\partial \bar{\xi}_i}{\partial y}\frac{\partial}{\partial \bar{\xi}_i}\,. \tag{8.9.26}$$

In the case of the constant strain triangle, equation (8.9.20) holds so that $\bar{\xi}_i = (1/2\Delta)(a_i + b_i x + c_i y)$.

Then
$$\frac{\partial \bar{\xi}_i}{\partial x} = \frac{b_i}{2\Delta}\,, \quad \frac{\partial \bar{\xi}_i}{\partial y} = \frac{c_i}{2\Delta} \tag{8.9.27}$$

and
$$\frac{\partial}{\partial x} = \sum_i \frac{b_i}{2\Delta}\frac{\partial}{\partial \bar{\xi}_i}\,, \quad \frac{\partial}{\partial y} = \sum_i \frac{c_i}{2\Delta}\frac{\partial}{\partial \bar{\xi}_i}\,. \tag{8.9.28}$$

More generally,

$$\frac{\partial}{\partial \bar{\xi}_i} = \begin{bmatrix} \dfrac{\partial x}{\partial \bar{\xi}_i} & \dfrac{\partial y}{\partial \bar{\xi}_i} \\ \vdots & \vdots \\ \vdots & \vdots \end{bmatrix} \begin{Bmatrix} \dfrac{\partial}{\partial x} \\ \dfrac{\partial}{\partial y} \end{Bmatrix} = [J] \begin{Bmatrix} \dfrac{\partial}{\partial x} \\ \dfrac{\partial}{\partial y} \end{Bmatrix}, \tag{8.9.29}$$

where
$$\frac{\partial x}{\partial \bar{\xi}_i} = \begin{bmatrix} \dfrac{\partial N_1}{\partial \xi_1} & \dfrac{\partial N_2}{\partial \xi_2} & \cdots\cdots \end{bmatrix} \begin{Bmatrix} x \\ y \end{Bmatrix}_0^e \text{ from above}\,.$$

In general, the integration of equation (8.9.24) must be carried out numerically and processes such as Newton-Cotes and Gaussian quadrature for doing this are described in books on numerical analysis.† Additionally, the following expressions are valuable

$$\int_l \bar{\xi}_1^{\,p}\,\bar{\xi}_2^{\,q}\,\mathrm{d}x = \frac{p!q!}{(p+q+1)!}\,l \text{ in one dimension} , \tag{8.9.30}$$

$$\iint_\Delta \bar{\xi}_1^{\,p}\,\bar{\xi}_2^{\,q}\,\bar{\xi}_3^{\,r}\,\mathrm{d}x\mathrm{d}y = \frac{p!q!r!}{(p+q+r+2)!}\,2\Delta \text{ in the plane.} \tag{8.9.31}$$

For further details, including applications to three dimensional situations, the reader should turn to the reading list.

EXAMPLE 8.9.1

Deduce the $[B]$ matrix for the constant strain triangle

$$[B] = \frac{1}{2\Delta}\begin{bmatrix} \sum b_i\dfrac{\partial}{\partial\bar{\xi}_i} & 0 \\ 0 & \sum c_i\dfrac{\partial}{\partial\bar{\xi}_i} \\ \sum c_i\dfrac{\partial}{\partial\bar{\xi}_i} & \sum b_i\dfrac{\partial}{\partial\bar{\xi}_i} \end{bmatrix}\begin{bmatrix} \bar{\xi}_1 & 0 & \bar{\xi}_2 & 0 & \bar{\xi}_3 & 0 \\ 0 & \bar{\xi}_1 & 0 & \bar{\xi}_2 & 0 & \bar{\xi}_3 \end{bmatrix}$$

$$\text{i.e.}\quad [B] = \frac{1}{2\Delta}\begin{bmatrix} b_1 & 0 & b_2 & 0 & b_3 & 0 \\ 0 & c_1 & 0 & c_2 & 0 & c_3 \\ c_1 & b_1 & c_2 & b_2 & c_3 & b_3 \end{bmatrix}$$

which corresponds exactly with equation (8.7.14).

Problems 8.9

8.9.1 Show that the functions below define a quadratic interpolation displacement field for the six node triangle element of Fig. 8.9.3. Confirm that continuity of displacements across element boundaries is assured. Sketch N_1 and N_4, as representative of the rest, and confirm that $N_1 + N_2 + \ldots . . N_6 = 1$.

$$N_1 = \bar{\xi}_1(2\bar{\xi}_1 - 1), \quad N_2 = \bar{\xi}_2(2\bar{\xi}_2 - 1), \quad N_3 = \bar{\xi}_3(2\bar{\xi}_3 - 1)$$

$$N_4 = 4\bar{\xi}_1\bar{\xi}_2, \quad N_5 = 4\bar{\xi}_2\bar{\xi}_3, \quad N_6 = 4\bar{\xi}_3\bar{\xi}_1 .$$

† Indeed, useful information is given in Zienkiewicz's book cited in the reading list.

8.9.2 If the nodal displacements for Problem 8.9.1 are listed as $\{u\}_0^{et} = [u_1 \ldots\ldots u_6 \;\; v_1 \ldots\ldots v_6]$ show that the corresponding $[B]$ matrix can be written as

$$[B] = \begin{bmatrix} [B]_1 & [0] \\ [0] & [B]_2 \\ [B]_2 & [B]_1 \end{bmatrix} \tag{a}$$

where

$$[B]_1 = \frac{1}{2\Delta}[b_1(4\overline{\xi}_1 - 1) \;\; b_2(4\overline{\xi}_2 - 1) \;\; b_3(4\overline{\xi}_3 - 1) \;\; 4(b_1\overline{\xi}_2 + b_2\overline{\xi}_1)$$
$$4(b_2\overline{\xi}_3 + b_3\overline{\xi}_2) \;\; 4(b_3\overline{\xi}_1 + b_1\overline{\xi}_3)]$$
$$[B]_2 = \frac{1}{2\Delta}[c_1(4\overline{\xi}_1 - 1) \ldots\ldots 4(c_3\overline{\xi}_1 + c_1\overline{\xi}_3)] \tag{b}$$

Confirm that the element stiffness matrix is of order 12×12.

8.10 Hybrid and mixed formulations

In a finite element method based on the potential energy principle, the assumed displacement field must satisfy certain continuity conditions, whilst the stress field in a complementary energy formulation must satisfy equilibrium requirements. The generation of displacement and stress models to meet these conditions is fairly straight forward in many situations, but there are a number of problem areas in which it is difficult and this has provided the motivation for the development of a number of sophisticated alternative methods. Certain of these derive from modifications to the basic variational principles which can be achieved by means of the method of Lagrange multipliers. These advanced methods can be regarded as a culmination of the work described in the earlier sections of this book, so that a brief discussion of them provides a natural conclusion to this last chapter.

Hybrid and mixed methods are those in which independent displacement and stress fields are utilised simultaneously. Thus, in the hybrid equilibrium approach, an equilibrium stress field is defined in terms of generalised parameters within the elements whilst a displacement field, ensuring inter-element continuity, is defined along the element boundaries in terms of nodal displacements. Application of a modified complementary energy principle yields flexibility equations but these are transformed into stiffness relations before the element contributions are assembled. In the mixed method, stress and displacement fields are defined independently within the elements, and applications of the Reissner principle generates mixed equations which are also transformed into stiffness relations prior to assembly and solution. Pian and Tong† have presented a logical classification of the various finite element approaches; here we present, in

† See the paper cited in the list of further reading.

their notation, an outline of what they described as stiffness methods in illustration of the more advanced techniques which have been developed.

There are a number of finite element methods in which the primary unknowns are displacements but only one of these begins from the basis of an assumed displacement field within the elements: application of the potential energy principle leads to stiffness equilibrium equations and the procedure is that which has occupied the majority of this chapter. A second approach, deriving from the complementary energy principle, utilises an assumed equilibrium stress field and is called, simply, the equilibrium method. Assuming compatible displacement functions along inter-element boundaries, simultaneously with taking an assumed equilibrium stress field within the element, leads to the hybrid equilibrium method deriving from a modified complementary energy principle. A fourth approach comprises the mixed method in which independent stress and displacement fields are assumed within the elements and Reissner's principle is used. It is instructive to present all these together to gain some insight into the relationships between them.

We have noted above that both the equilibrium and hybrid equilibrium methods derive from the principle of stationary total complementary energy, $\delta V^* = 0$, where

$$V^* = U^* - \int_{S_u} \overset{n}{T}_i u_i^* \mathrm{d}S .\dagger \tag{8.10.1}$$

The stress field must satisfy equilibrium in the interior of each element, continuity of traction across inter-element boundaries and equilibrium with applied surface tractions where the latter are prescribed. For our purposes here, it is convenient to account for the second of these conditions by means of the Lagrange multiplier method and to generate a modified complementary energy functional as follows.

Supposing elements '1' and '2' to have a common segment of surface S_c, then we require

$$[\overset{n}{T}_i(S)]_1 + [\overset{n}{T}_i(S)]_2 = 0, \quad i = 1, 2, 3 \tag{8.10.2}$$

over S_c. But these are equations of constraint so that they may be introduced into equation (8.10.1) by means of a Lagrange multiplier term

$$L = \int_{S_c} \lambda_i(S) \{ [\overset{n}{T}_i(S)]_1 + [\overset{n}{T}_i(S)]_2 \} \mathrm{d}S \tag{8.10.3}$$

or

$$L = \left(\int_{S_c} \lambda_i \overset{n}{T}_i \mathrm{d}S \right)_1 + \left(\int_{S_c} \lambda_i \overset{n}{T}_i \mathrm{d}S \right)_2 . \tag{8.10.4}$$

When all the inter-element boundaries have been taken into account, the modified complementary energy functional becomes

$$\bar{V}^* = \sum_e \left(\int_{V_e} U_e^* \,\mathrm{d\,vol} - \int_{S_e} \lambda_i \overset{n}{T}_i \mathrm{d}S - \int_{S_{u_e}} \overset{n}{T}_i u_i^* \mathrm{d}S \right) . \tag{8.10.5}$$

† See section 6.8.

The summation is carried out over all the elements 'e' having volumes V_e and surfaces S_e. S_{u_e} is that part of S_e on which displacements are prescribed. Taking variations of V^*, we find that the λ_i are equal to u_i, the displacements along the inter-element boundary, so that equation (8.10.5) can be re-written as

$$\bar{V}^* = \sum_e \left(\int_{V_e} U_e^* \, \mathrm{d\,vol} - \int_{S_e} \overset{n}{T}_i u_i \mathrm{d}S - \int_{S_{u_e}} \overset{n}{T}_i u_i^* \mathrm{d}S \right) . \qquad (8.10.6)$$

Both the equilibrium and hybrid equilibrium methods may be derived from equation (8.10.6), but they are distinguished from each other by the manner in which the inter-element surface integrals are treated. We consider the equilibrium method first.

The stress field may be written in the form

$$\{\sigma\} = [P]\{\beta\} + [P_F]\{\beta_F\} . \qquad (8.10.7)$$

Here, $\{\beta\}$ is a vector of unknown generalised stress parameters with $[P]\{\beta\}$ satisfying internal equilibrium in the absence of body forces. The second term is a particular solution in which the stresses satisfy equilibrium with both body forces and any prescribed surface traction so that $\{\beta_F\}$ is a vector of known parameters. Now the surface tractions can be expressed in terms of the stresses by means of Cauchy's formula $\overset{n}{T}_i = \sigma_{ji} n_j$; this can be expressed in matrix form and, taking equation (8.10.7) into account, we can write

$$\{T\} = [R]\{\beta\} + [R_F]\{\beta_F\} . \qquad (8.10.8)$$

Values of the surface tractions $\overset{n}{T}_i$ at selected stations on the boundary may be chosen as generalised forces $\{Q\}$; $\{T\}$ may be expressed in terms of $\{Q\}$ by means of an interpolation function $[\phi]$ in

$$\{T\} = [\phi]\{Q\} . \qquad (8.10.9)$$

Eliminating $\{T\}$ between equations (8.10.8) and (8.10.9) yields

$$\{Q\} = [G]^t\{\beta\} + [G_F]^t\{\beta_F\} , \qquad (8.10.10)$$

where $[G]^t = [\phi]^{-1}[R]$ etc.

Element generalised displacements $\{q\}$, corresponding to $\{Q\}$, may be defined on an equivalent work basis by

$$\{Q\}^t\{q\} = \int_{S_e} \overset{n}{T}_i u_i \mathrm{d}S = \{Q\}^t \int_{S_e} [\phi]^t \{u(S)\} \mathrm{d}S \qquad (8.10.11)$$

or

$$\{q\} = \int_{S_e} [\phi]^t \{u(S)\} \mathrm{d}S . \qquad (8.10.12)$$

The q's are thus weighted integrals, over the element surface, of the boundary displacements so that inter-element compatibility is only satisfied in some average way.

Substituting from equations (8.10.7), (8.10.10) and (8.10.11) in (8.10.6),

$$\bar{V}^* = \sum_e (\tfrac{1}{2}\{\beta\}^t[H]\{\beta\} + \{\beta\}^t[H_F]\{\beta_F\} + \tfrac{1}{2}\{\beta_F\}^t[H_{FF}]\{\beta_F\} - (\{\beta\}^t[G] + \{\beta_F\}^t[G_F])\{q\}), \tag{8.10.13}$$

where,

$$\left.\begin{aligned} [H] &= \int_{V_e} [P]^t[D]^{-1}[P]\,\mathrm{d\,vol}, \\ [H_F] &= \int_{V_e} [P]^t[D]^{-1}[P_F]\,\mathrm{d\,vol}, \\ [H_{FF}] &= \int_{V_e} [P_F]^t[D]^{-1}[P_F]\,\mathrm{d\,vol}, \\ \{\epsilon\} &= [D]^{-1}\{\sigma\}. \end{aligned}\right\} \tag{8.10.14}$$

Taking variations with respect to $\{\beta\}$ for a typical element yields

$$[H]\{\beta\} + [H_F]\{\beta_F\} - [G]\{q\} = 0. \tag{8.10.15}$$

Solving for $\{\beta\}$ in terms of $\{q\}$ and substituting back into equation (8.10.13) gives $\bar{V}^*$ in terms of the q's.†

That is $$\bar{V}^* = \sum_e (\tfrac{1}{2}\{q\}^t[k]\{q\} - \{\bar{Q}\}^t\{q\} + C_e), \tag{8.10.16}$$

$$\left.\begin{aligned} \text{where } [k] &= [G]^t[H]^{-1}[G] = \text{an element stiffness matrix}, \\ \{Q\} &= [G]^t[H]^{-1}[H_F]\{\beta_F\} - \{\beta_F\}^t[G_F], \\ C_e &= \tfrac{1}{2}\{\beta_F\}^t[H_F]^t[H]^{-1}[H_F]\{\beta_F\} + \tfrac{1}{2}\{\beta_F\}^t[H_{FF}]\{\beta_F\} \end{aligned}\right\} . \tag{8.10.17}$$

Taking variations of equation (8.10.16) with respect to the q's yields a stiffness equation from which the generalised displacements may be found.

In the hybrid equilibrium approach, the element boundary displacements $\{u(S)\}$ are expressed in terms of nodal displacements $\{q\}$ using interpolation functions so that

$$\{u(S)\} = [L]\{q\}.^{\dagger\dagger} \tag{8.10.18}$$

This is independent of the choice of stress field and represents an important difference from the equilibrium approach in which the q's and the stress field are related through equation (8.10.12). Proceeding as before, equations (8.10.16) and (8.10.17) are obtained but now,

$$[G] = \int_{S_n} [R]^t[L]\,\mathrm{d}S, \quad [G_F] = \int_{S_n} [R_F]^t[L]\,\mathrm{d}S \tag{8.10.19}$$

† Pian and Tong point out that if the β's are to be determinable in terms of the q's, then the total number of stress modes must be greater than the number of generalised displacements by at least the number of rigid body modes for the structure.

†† Since the interpolating functions $[L]$ are applied to individual boundary segments, it is relatively easy to construct them to ensure that inter-element compatibility is maintained.

and

$$\{\bar{Q}\} = [G]^t[H]^{-1}[H_F]\{\beta_F\} - \{\beta_F\}[G_F] + \int_{S_{T_e}} \{T\}^t[L]\,dS. \quad (8.10.20)$$

Once more, the procedure leads to stiffness equations from which displacements may be found. A hybrid displacement method may be developed by assuming displacement fields independently for each element and introducing the constraint of inter-element compatibility through Lagrange multipliers. The basis of this approach is the potential energy principle and the multipliers can be interpreted as inter-element tractions.

We saw in Section 6.8 that a stationary value for the Reissner functional yields the equilibrium and stress-strain-displacement relations as Euler equations whilst all boundary conditions are natural ones. An approximate solution can thus imply approximate satisfaction, not only of the equilibrium and stress-displacement conditions in the interior of the solid, but also of boundary displacements on S_u and boundary tractions on S_T: this represents a considerable relaxation of the conditions to be met in the potential energy or complementary energy approaches. In the finite element context, independent stress and displacement fields are assumed within the elements but in view of the mixed term $\frac{1}{2}\sigma_{ij}(u_{i,j} + u_{j,i})$, the only one involving derivatives in the Reissner functional (6.8.13), it is necessary to examine inter-element continuity of stresses and displacements. For finite element work, equation (6.8.13) can be written

$$V_R = \sum_e \left\{ \int_{V_e} [\tfrac{1}{2}\sigma_{ij}(u_{i,j} + u_{j,i}) - F_i u_i - U_0{}^*(\sigma_{ij})]\,d\text{vol} \right.$$
$$\left. - A_n - \int_{S_{T_e}} \overset{n}{T}_i{}^* u_i\,dS - \int_{S_{u_e}} \overset{n}{T}_i(u_i - u_i{}^*)\,dS \right\} \quad (8.10.21)$$

where A_n is a non-zero term when the u_i are discontinuous across inter-element boundaries. To simplify matters for the remainder of this discussion, suppose the chosen fields are such that $A_n = 0$ and that the prescribed displacements are met on S_u.

Let the stress field, which need not be equilibriant, be assumed in the form

$$\{\sigma\} = [P_\sigma]\{\beta\}, \quad (8.10.22)$$

where $[P_\sigma]$ is a function of position and $\{\beta\}$ is a vector of generalised stress parameters. Let the displacement field be

$$\{u\} = [P_u]\{q\} \quad (8.10.23)$$

where $[P_u]$ is an interpolation function and the q's are nodal displacements. Equation (8.10.21) then becomes

$$V_R = \sum_e (-\tfrac{1}{2}\{\beta\}^t[H]\{\beta\} + \{\beta\}^t[G]\{q\} - \{Q\}^t\{q\}) \quad (8.10.24)$$

where
$$[H] = \int_{V_e} [P_\sigma]^t [D]^{-1} [P] \, \mathrm{d}\,\mathrm{vol} \,,$$

$$[G] = \int_{V_e} [P_\sigma]^t [P_u] \, \mathrm{d}\,\mathrm{vol} \,, \tag{8.10.25}$$

$$Q = \int_{V_e} [P_u]^t \{F\} \mathrm{d}\,\mathrm{vol} + \int_{S_{T_e}} [P_u(S)]^t \{T\} \mathrm{d}S \,.$$

Considering the variation with respect to $\{\beta\}$ and $\{q\}$ of the contribution of a typical element to V_R, we have

$$\begin{aligned} \delta V_R^e &= -\{\delta\beta\}^t [H] \{\beta\} + \{\delta\beta\}^t [G] \{q\} + \{\beta\}^t [G] \{\delta q\} - \{Q\}^t \{\delta q\} \\ &= 0 \,. \end{aligned} \tag{8.10.26}$$

Since $\{\delta\beta\}$ and $\{\delta q\}$ are arbitrary, this yields a pair of matrix equations

$$\begin{aligned} -[H]\{\beta\} + [G]\{q\} &= \{0\} \,, \\ [G]^t \{\beta\} &= \{Q\} \,. \end{aligned} \tag{8.10.27}$$

Solving the first equation for the β's and substituting into the second, we have

$$[k]\{q\} = \{Q\} \,, \tag{8.10.28}$$

where $[k] = [G]^t [H]^{-1} [G]$ = an element stiffness matrix . (8.10.29)

Once again, sophisticated reasoning has led to a stiffness formulation from which nodal displacements may be found.

In design work, one is commonly interested in stresses rather than displacements so that the equilibrium, hybrid equilibrium and mixed formulations hold promise of a better accuracy for the stresses than does the displacement method for a given effort. The differentiation of approximate displacements necessary to yield stresses in the latter approach is essentially an inaccurate process and so represents a disadvantage. However, it is still the case that the displacement method is the most widely used in structural mechanics and the adoption of the sophisticated elements described in Section 8.9 enhances its value. It provides an upper bound to the element stiffness whilst the equilibrium approach provides a lower bound. Mixed and hybrid formulations yield stiffnesses somewhere between these two so that, on balance, one might expect then to yield a better solution for a given effort.

The classification of finite element methods has been admirably summarised by Pian and Tong. Table 8.10.1 is reproduced, with minor changes, from their paper by kind permission of the *International Journal of Numerical Methods in Engineering.*

TABLE 8.10.1 Classification of Finite Element Methods in Solid Mechanics (after Pian and Tong)

Model	*Variational Principle*	*Assumed Inside Each Element*	*Along Inter-element Boundary*	*Unknowns in Final Equations*
Compatible	Minimum Potential Energy	Smooth Displacement Distribution	Continuous Displacement	Nodal Displacements
Equilibrium	Minimum Complementary Energy	Smooth and Equilibrating Stress Distribution	Equilibrium of Boundary Tractions	(a) Generalised Nodal Displacements (b) Stress Function Parameters
Hybrid (I)	Modified Complementary Energy	Smooth and Equilibrating Stress Distribution	Assumed Compatible Displacements	Nodal Displacements
Hybrid (II)	Modified Potential Energy	Smooth Displacement Distribution	Assumed Equilibrating Boundary Tractions	Boundary Redundant Forces
Mixed	Reissner's Variational Principle	Smooth Displacement and Stress Distribution	Various Combinations of Boundary Displacements and Tractions	Various Combinations of Boundary Displacements and Traction Parameters

EXAMPLE 8.10.1

Use the hybrid equilibrium method to generate the stiffness matrix for the beam element of Fig. 8.10.1. Assume $M = \beta_1 + \beta_2 x$ for the 'stress' field, where M is the bending moment.

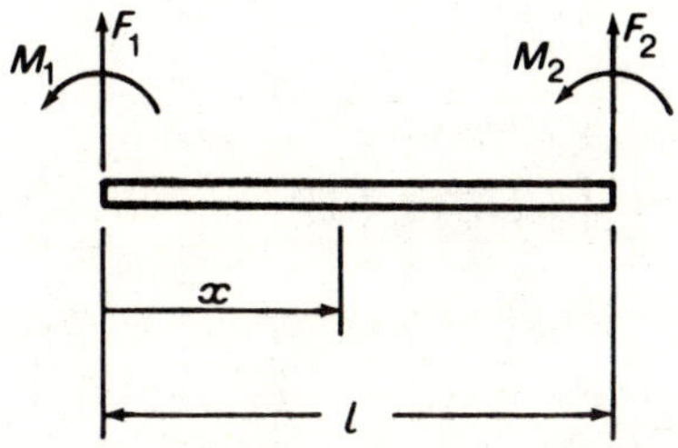

Fig. 8.10.1

The usual convention for positive bending moment M and shearing force F is shown in Fig. 8.10.2 from which equilibrium requires

$$\frac{\mathrm{d}M}{\mathrm{d}x} = F .$$

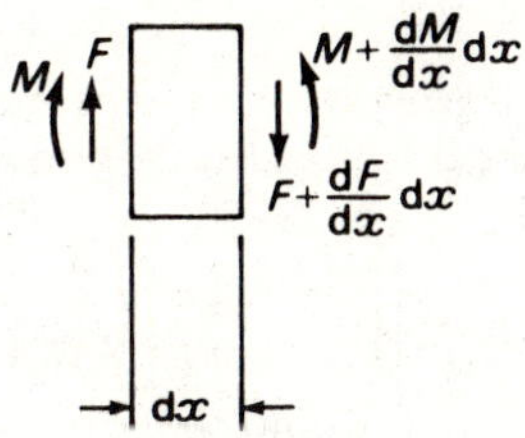

Fig. 8.10.2

Since

$$M = [P]\{\beta\} = [1 \quad x]\begin{Bmatrix}\beta_1\\ \beta_2\end{Bmatrix}, \tag{a}$$

$$F = [P']\{\beta\} = [0 \quad 1]\begin{Bmatrix}\beta_1\\ \beta_2\end{Bmatrix}. \tag{b}$$

The boundary traction $\{T\}$ of equation (8.10.8) is then

$$T = \begin{Bmatrix}F_1\\ M_1\\ F_2\\ M_2\end{Bmatrix} \quad \begin{bmatrix}0 & -1\\ 1 & 0\\ 0 & 1\\ -1 & -l\end{bmatrix}\begin{Bmatrix}\beta_1\\ \beta_2\end{Bmatrix} = [R]\{\beta\} . \tag{c}$$

Denoting $[T] = [\phi]\{Q\}$, $\{\phi\} = [I] =$ unity matrix, so that $[G]^t = [R]$,

$$[H] = \int_0^l \frac{1}{EI}\begin{Bmatrix}1\\ x\end{Bmatrix}[1 \quad x]\,dx = \frac{l}{6EI}\begin{bmatrix}6 & 3l\\ 3l & 2l^2\end{bmatrix} \tag{d}$$

and

$$[H]^{-1} = \frac{EI}{l^3}\begin{vmatrix}4l^2 & -6l\\ -6l & 12\end{vmatrix} \tag{e}$$

The element stiffness matrix is given by

$$[k] = [G]^t[H]^{-1}[G] \tag{f}$$

or

$$[k] = \frac{EI}{l^3}\begin{bmatrix}12 & 6l & -12 & 6l\\ 6l & 4l^2 & -6l & 2l^2\\ -12 & -6l & 12 & -6l\\ 6l & 2l^2 & -6l & 4l^2\end{bmatrix}$$

which is precisely that obtained by the displacement method in equation (8.5.12).

EXAMPLE 8.10.2

Use the mixed formulation to generate the stiffness matrix for a beam element assuming an interpolation model for the bending moment as

$$M = \left[\left(1 - \frac{\xi}{l}\right)\frac{\xi}{l}\right]\begin{Bmatrix}M_1\\ M_2\end{Bmatrix} \tag{a}$$

and equation (8.5.6) for the displacement.

Observing that the Reissner functional for a beam is given in equation (b) of Example 6.8.1,

$$[H] = \int_{V_e}[P_\sigma]^t[D]^{-1}[P_\sigma]\,\mathrm{d\,vol}$$

becomes

$$[H] = \int_0^l \frac{1}{EI}\begin{Bmatrix}\left(1 - \frac{\xi}{l}\right)\\ \frac{\xi}{l}\end{Bmatrix}\left[\left(1 - \frac{\xi}{l}\right)\frac{\xi}{l}\right]\mathrm{d}\xi\,. \tag{b}$$

That is

$$[H] = \frac{l}{6EI}\begin{bmatrix}2 & 1\\ 1 & 2\end{bmatrix} \tag{c}$$

and

$$[H]^{-1} = \frac{2EI}{l}\begin{bmatrix}2 & -1\\ -1 & 2\end{bmatrix}. \tag{d}$$

With the assumed displacement field,

$$[G] = \int_{V_e} [P_\sigma]^t [P_u] \, \mathrm{d}\,\mathrm{vol}$$

$$= \int_0^l \begin{Bmatrix} \left(1 - \dfrac{\xi}{l}\right) \\ \dfrac{\xi}{l} \end{Bmatrix} \left[\left(-\frac{6}{l^2} + \frac{12\xi}{l^3}\right)\left(-\frac{4}{l} + \frac{6\xi}{l^2}\right)\left(\frac{6}{l^2} - \frac{12\xi}{l^3}\right)\left(-\frac{2}{l} + \frac{6\xi}{l^2}\right)\right] \mathrm{d}\xi \, .$$

$$\text{or } [G] = \frac{1}{l}\begin{bmatrix} -1 & -l & 1 & 0 \\ 1 & 0 & -1 & l \end{bmatrix} . \tag{e}$$

The reader may readily confirm that

$$[k] = [G]^t [H]^{-1} [G]$$

leads to the usual beam element stiffness matrix. We would expect this result, since the assumed bending moment distribution is consistent with the assumed displacement which is itself the basis of the previous potential energy formulation.

Problems 8.10

8.10.1 Use the mixed formulation to generate a beam element stiffness matrix assuming a linear interpolation model for both the bending moment and the displacement.

8.10.2 Verify that the following stress model does indeed satisfy equilibrium for the plane stress rectangular internal element of Fig. 8.10.3. Body forces are zero.

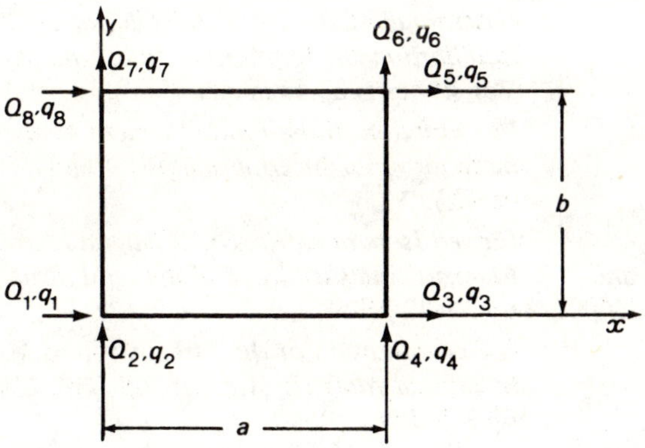

Fig. 8.10.3

$$\begin{Bmatrix} \sigma_x \\ \sigma_y \\ \tau_{xy} \end{Bmatrix} = \begin{bmatrix} 1 & y & 0 & 0 & 0 & x & 0 & y^2 & 0 & x^2 \\ 0 & 0 & 1 & x & 0 & 0 & y & 0 & x^2 & y^2 \\ 0 & 0 & 0 & 0 & 1 & -y & -x & 0 & 0 & -2xy \end{bmatrix} \begin{Bmatrix} \beta_1 \\ \beta_2 \\ \cdot \\ \cdot \\ \cdot \\ \beta_{10} \end{Bmatrix}$$

Generate the $[R]$ matrix of equation 8.10.8. Assuming a linear interpolation along the boundary segments, generate the $[\phi]$ matrix of equation (8.10.9) and the $[L]$ matrix of equation (8.10.18).

Without carrying out all the detailed manipulations, determine the order of the $[k]$ matrix according to a hybrid equilibrium formulation.

8.11 Further reading

The literature on finite-element methods is vast so that only a few representative papers and books are cited here.

Argyris, J.H. *Recent Advances in Matrix Methods of Structural Analysis,* Pergamon Press (1964).

Clough, R.W. *The Finite Element Method in Plane Stress Analysis: Proc. 2nd A.S.C.E. Conf. on Electronic Computation,* (1960).

Courant, R. *Variational Methods for the Solution of Problems of Equilibrium and Vibrations, Bull. Am. Math. Soc.* Vol. **49** pp 1–23, (January 1943).

Desai, C.S. *and* Abel, J.F. *Introduction to the Finite Element Method: a numerical method for engineering analysis,* Van Nostrand Reinhold (1972).

Ergatoudis, J., Irons, B.H. *and* Zienkiewiez, O.C. *Curved Isoparametric, Quadrilateral Elements for Finite Element Analysis, Int. J. Solids and Structures,* Vol. **4** pp 31–42 (1968).

Jones, R.E. *A Generalisation of the Direct-stiffness Method of Structural Analysis, A.I.A.A. Jnl.* Vol. **2**, No. 5, pp 821–826 (1964).

Martin, H.C. *and* Carey, G.F. *Introduction to Finite Element Analysis: Theory and Application,* McGraw Hill (1973).

Norrie, D.H. *and* de Vries, G. — *The Finite Element Method,* Academic Press (1973)

Pian, T.H.H. *and* Tong, Pin — *Basis of Finite Element Methods for Solid Continua, Int. J. Num. Methods in Eng.* Vol. **1** pp 1–28 (1969).

Severn, R.T. *and* Taylor, P.R. — *The Finite Element Method for Flexure of Slabs when Stress Distributions are Assumed, Proc. Inst. Civ. Engrs.* Vol. 34, pp 153–170 (1966).

Tottenham, H. *and* Brebbia, C. — *Finite Element Techniques in Structural Mechanics,* Southampton University Press (1970).

Turner, M.J., Clough, R.W., Martin, H.C. *and* Tupp, L.J. — *Stiffness and Deflexion Analysis of Complex Structures, Jnl. Aero. Sci.* Vol. **23**, No. 9, pp 805–824 (1956).

Veubeke, F. de — *Matrix Methods of Structural Analysis,* Pergamon Press (1964).

Zienkiewiez, O.C. — *The Finite Element Method in Engineering Science,* McGraw Hill (1971).

Appendix

A.1 Introduction

Some mathematical ideas, which are useful for the remainder of the text, are collected together in this Appendix. Clearly it is neither appropriate nor practical for the presentation to be exhaustive so that we merely summarise certain information to provide a convenient reference adequate for our purpose: for a more extensive coverage, it is necessary to refer to the various specialised texts quoted in the Reading List at the end of this chapter.

A.2 Indicial notation and the summation convention

In discussing problems in continuum mechanics, a considerable amount of mathematical manipulation is involved so that, if the usual x, y, z notation is used, it is often all too easy to lose track of the main threads of an argument in a mass of algebra. The use of the indicial notation helps to overcome such difficulties by a clarity of expression which assists the various thought processes involved and highlights any regularity of pattern in analytic results. Whilst the benefits of the notation are not so evident in the theoretical description of particular situations, they re-appear when problems are formulated in terms suitable for handling in a modern computer.

Suppose we have a set of n quantities $x_1, x_2 \ldots\ldots x_n$ within a particular class, then an alternative description could be 'the set x_i, $i = 1, 2 \ldots\ldots n$'. Provided the circumstances allow us to take n for granted without confusion, we can write, even more briefly, 'the quantities x_i'. We can also take x_i to denote any typical one of the quantities; the appropriate sense will be clear from the context. The symbol i is an **index** and may be a superscript or a subscript.

As a first example, instead of describing a point P in space as having coordinates x, y, z, we could use x_1, x_2, x_3. Then, we could speak of point $P(x_i)$; $x_i, i = 1, 2, 3$ are the coordinates of P.

Many times in this text, we calculate the work done by forces $F_1, F_2 \ldots\ldots F_n$ acting through corresponding displacements† $u_1, u_2 \ldots\ldots u_n$ to be

$$W = F_1 u_1 + F_2 u_2 + \ldots\ldots F_n u_n = \sum_{i=1}^{n} F_i u_i \, . \qquad \text{(A.2.1)}$$

† A displacement u_i *corresponds* to a force F_i if it is measured at the *point of application* of F_i and *in the direction* of F_i.

Observing that the subscript i is repeated in the term on the extreme right hand side in equation (A.2.1), we can simplify the presentation further. Thus, as a **summation convention**, we can take it that the repetition of an index in a term implies a summation with respect to that index over its range with no need to attach the symbol Σ. The **range** of an index i is the set of values $1, 2, 3 \ldots\ldots n$ which the index can take. An index that is summed over is called a **dummy index** and the symbol used for it is immaterial. Thus, we can write equation (A.2.1) in the alternative forms

$$W = F_i u_i = F_j u_j = F_p u_p = F_q u_q \qquad \text{(A.2.2)}$$

provided the ranges of i, j, p, q are the same. A **free index** is one that is not summed over. Thus, suppose we have n linear relations

$$\begin{aligned} a_{11}x_1 + a_{12}x_2 + \ldots\ldots a_{1n}x_n &= y_1\,, \\ a_{21}x_1 + a_{22}x_2 + \ldots\ldots a_{2n}x_n &= y_2\,, \\ a_{n_1}x_1 + a_{n2}x_2 + \ldots\ldots a_{nn}x_n &= y_n\,. \end{aligned} \qquad \text{(A.2.3)}$$

Then a typical y_i is related to the x's by the i'th equation

$$y_i = a_{i1}x_1 + a_{i2}x_2 + \ldots\ldots a_{in}x_n = a_{ij}x_j\,, \qquad \text{(A.2.4)}$$

where we have observed the summation convention over the subscripts j. Here i is a free index, and it identifies which equation (and so which y) is temporarily of interest. Alternatively we may speak of the **set** of equations $y_i = a_{ij}x_j$; the context of the discussion makes it clear whether we are indeed referring to the set or the single quantity y_i.

To express the sum of squares in indicial notation, we must write

$$S = x_1^2 + x_2^2 + \ldots\ldots x_n^2 = x_i x_i\,, \quad i = 1, 2 \ldots\ldots n\,. \qquad \text{(A.2.5)}$$

The **Kronecker delta** is an important quantity defined by

$$\delta_{ij} = \begin{cases} 0 & \text{if } i \neq j\,, \\ 1 & \text{if } i = j\,. \end{cases} \qquad \text{(A.2.6)}$$

Occasionally, it may be useful to use this symbol to write results such as (A.2.5), slightly differently as

$$S = \delta_{ij} x_i x_j\,. \qquad \text{(A.2.7)}$$

Here, both i and j are repeated and a double summation is implied. The reader should verify, by direct expansion over $i, j = 1, 2, 3$, that equations (A.2.5) and (A.2.7) do agree and that it is immaterial whether summation is carried out first on i and then on j or *vice-versa.*

A second important quantity is the **permutation symbol** defined by

$$e_{ijk} = \begin{cases} 1 \text{ if } i, j, k \text{ form one of the sequences } 1, 2, 3 \text{ or } 2, 3, 1 \text{ or } 3, 1, 2\,, \\ 0 \text{ if the values of any two indices are the same}\,, \\ -1 \text{ if } i, j, k \text{ form one of the sequences } 2, 1, 3 \text{ or } 1, 3, 2 \text{ or } 3, 2, 1. \end{cases} \quad \text{(A.2.8)}$$

The Kronecker delta and the permutation symbol are related by the so-called $e - \delta$ identity:

$$e_{ijk}e_{ist} = \delta_{js}\delta_{kt} - \delta_{jt}\delta_{ks}\,. \quad \text{(A.2.9)}$$

When a function $F(x_1, x_2 \ldots\ldots x_n)$ is to be differentiated with respect to x_i, we may write

$$\frac{\partial F}{\partial x_i} = F_{,i}\,. \quad \text{(A.2.10)}$$

The comma notation $_{,i}$ denotes differentiation with respect to x_i and not with respect to i. Recalling the summation convention, the differential of F may be written

$$\mathrm{d}F = \frac{\partial F}{\partial x_i}\mathrm{d}x_i = F_{,i}\mathrm{d}x_i \quad \text{(A.2.11)}$$

In Chapter 5, we see that for equilibrium in the x_1 direction in the interior of a continuum, the components of stress and body force are related by

$$\frac{\partial \sigma_{11}}{\partial x_1} + \frac{\partial \sigma_{21}}{\partial x_2} + \frac{\partial \sigma_{31}}{\partial x_3} + F_1 = 0\,. \quad \text{(A.2.12)}$$

Similar expressions hold for the x_2 and x_3 directions so that we can write, compactly,

$$\sigma_{ji,j} + F_i = 0\,, \quad i, j = 1, 2, 3\,. \quad \text{(A.2.13)}$$

Problems A.2

1. Show that

$$\text{(i) } \delta_{ii} = 3\,; \quad \text{(ii) } \delta_{ij}\delta_{ij} = 3\,; \quad \text{(iii) } \delta_{ij}\delta_{jk} = \delta_{ik}\,;$$
$$\text{(iv) } e_{ijk}A_jA_k = 0; \quad \text{(v) } e_{ijk}e_{jki} = 6\,.$$

A.3 Scalars, Vectors and Cartesian Tensors

In elementary mechanics, we meet entities called scalars and vectors; in continuum mechanics, tensors arise. Here, we summarise some information relating to these quantities.

Scalars are entities which require only a single number for their specification at a point. Thus, the density of a body is described by a function of position ρ which has the same value for a given point irrespective of the choice of coordinate axes. On the other hand, quantities such as forces **F** or displacements **u** have both magnitude and direction and so can be represented as *directed line segments.* Such quantities are **vectors** and must be specified relative to some reference frame. In printed work, vectors are usually symbolised by bold type, **u**, **F**, **v** etc. and their magnitudes printed as $|\mathbf{u}|$, $|\mathbf{F}|$ or merely u, F etc. In handwritten work $\vec{u}, \vec{F}$, $|\vec{u}|$, $|\vec{F}|$, are commonly used and, as we shall see below, the indicial notation is also especially convenient.

The following definitions are fundamental to vector algebra:

(i) Two vectors **A** and **B** are considered equal when they have the same magnitude and direction irrespective of their initial point.† A **unit** vector has unit magnitude whilst the **null** or **zero** vector has zero magnitude.

(ii) When vector **B** is equal in magnitude but opposite in direction to vector **A**, then $\mathbf{B} = -\mathbf{A}$.

(iii) The sum or resultant of two vectors **A** and **B** is a third vector **C** obtained by the parallelogram law as shown in Fig. A.3.1.

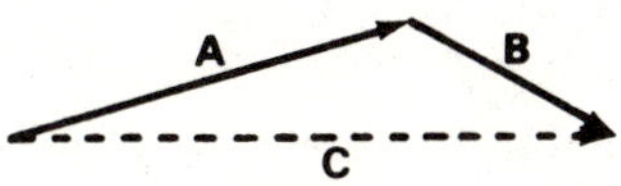

Fig. A.3.1

Vector addition is commutative and associative so that we may write

$$\mathbf{D} = \mathbf{A}+\mathbf{B}+\mathbf{C} = \mathbf{C}+\mathbf{B}+\mathbf{A} = (\mathbf{C}+\mathbf{B})+\mathbf{A} = \mathbf{C}+(\mathbf{A}+\mathbf{B}) \text{ etc.} \quad \text{(A.3.1)}$$

In view of (ii) above $$\mathbf{C} = \mathbf{A}-\mathbf{B} = \mathbf{A}+(-\mathbf{B}) \quad \text{(A.3.2)}$$

and $$\mathbf{A}-\mathbf{A} = \mathbf{A}+(-\mathbf{A}) = \mathbf{O} = \text{null vector .} \quad \text{(A.3.3)}$$

(iv) The product of vector **A** by a scalar m is a new vector $m\mathbf{A}$ having a magnitude $|m|$ times that of **A** and having a direction the same as or opposite to **A** according as m is positive or negative

$$m\mathbf{A} = \mathbf{A}m \quad \text{(A.3.4)}$$

$$(m+n)\mathbf{A} = m\mathbf{A}+n\mathbf{A} \quad \text{(A.3.5)}$$

$$(mn)\mathbf{A} = m(n\mathbf{A}) = n(m)\mathbf{A} \quad \text{(A.3.6)}$$

If **a** is a vector of unit magnitude in the direction of **A**, then

$$\mathbf{A} = A\mathbf{a} \ . \quad \text{(A.3.7)}$$

† The initial point of a vector is that point from which it issues. When an arrow is used to indicate a vector, its tail end is at the initial point, its length represents the magnitude to some scale, its direction gives the vector's direction and the arrow head gives the sense.

Expressing vectors in terms of unit vectors is often convenient. In particular, if **i**, **j**, **k** are unit vectors along the positive x, y, z cartesian axes respectively

$$\mathbf{A} = A_x\mathbf{i} + A_y\mathbf{j} + A_z\mathbf{k} \qquad \text{(A.3.8)}$$

where A_x, A_y, A_z are the (scalar) components of **A** as shown in Fig. A.3.2.

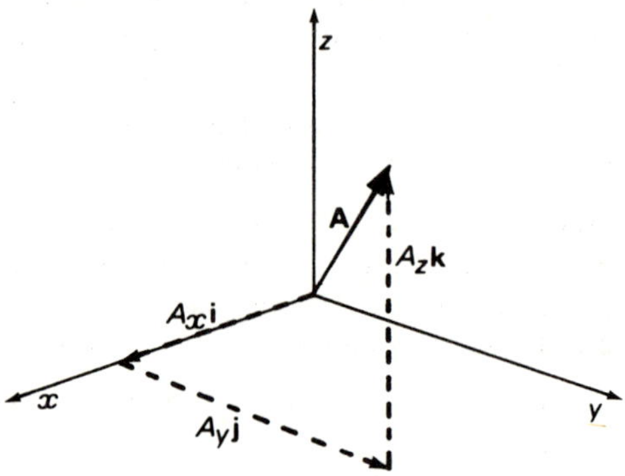

Fig. A.3.2

It follows that

$$\mathbf{A} + \mathbf{B} = (A_x + B_x)\mathbf{i} + (A_y + B_y)\mathbf{j} + (A_z + B_z)\mathbf{k} \text{ etc.} \qquad \text{(A.3.9)}$$

From Fig. A.3.2 we also see that the magnitude of **A** is

$$|\mathbf{A}| = A = \sqrt{A_x^2 + A_y^2 + A_z^2}\,. \qquad \text{(A.3.10)}$$

(v) The **dot,** or **scalar,** or **inner product** of two vectors **A** and **B** is defined by

$$\mathbf{A} \cdot \mathbf{B} = AB\cos\theta\,, \qquad 0 \leqslant \theta \leqslant \pi\,, \qquad \text{(A.3.11)}$$

where θ is the angle between the vectors. We see that $\mathbf{A} \cdot \mathbf{B}$ is a scalar quantity; it is instructive to write it as

$$\mathbf{A}\cdot\mathbf{B} = (\text{magnitude of } \mathbf{A}) \cdot (\text{magnitude of component } \mathbf{B} \text{ along } \mathbf{A})\,. \qquad \text{(A.3.12)}$$

The dot product has a particular significance in mechanics. Thus if **B** is a force and **A** is the displacement of the point of application of **B**, then $\mathbf{A} \cdot \mathbf{B}$ is the work done by **B.**

Some useful properties of the dot product are

$$\mathbf{A} \cdot \mathbf{B} = \mathbf{B} \cdot \mathbf{A}\,. \qquad \text{(A.3.13)}$$

$$\mathbf{A} \cdot (\mathbf{B} + \mathbf{C}) = \mathbf{A} \cdot \mathbf{B} + \mathbf{A} \cdot \mathbf{C}\,. \qquad \text{(A.3.14)}$$

$$\mathbf{i}\cdot\mathbf{i} = \mathbf{j}\cdot\mathbf{j} = \mathbf{k}\cdot\mathbf{k} = 1\,; \qquad \mathbf{i}\cdot\mathbf{j} = \mathbf{j}\cdot\mathbf{k} = \mathbf{k}\cdot\mathbf{i} = 0\,. \qquad \text{(A.3.15)}$$

If $\mathbf{A} \cdot \mathbf{B} = 0$ and neither **A** nor **B** is null, then they are mutually perpendicular. If **b** is a unit vector, then

$$\mathbf{A} \cdot \mathbf{b} = A \cos \theta = \text{component of } \mathbf{A} \text{ in direction of } \mathbf{b}\,. \qquad \text{(A.3.16)}$$

From the above results, we can write alternatively

$$\mathbf{A} \cdot \mathbf{B} = A_x B_x + A_y B_y + A_z B_z\,. \qquad \text{(A.3.17)}$$

When using the indicial notation, we may denote unit vectors $\mathbf{e}_i$ along the x_i axes and write

$$\mathbf{A} = A_i \mathbf{e}_i\,. \qquad \text{(A.3.18)}$$

Then

$$\mathbf{A} \cdot \mathbf{B} = A_i B_i\,. \qquad \text{(A.3.19)}$$

Thus, instead of using bold type, we may write vector A_i or B_i etc. The context will make it clear whether we are referring to vector A_i or merely the ith component of **A**.

(vi) The **cross** or **vector product** of **A** and **B** is a new vector **C** defined by

$$\mathbf{C} = \mathbf{A} \times \mathbf{B} = AB \sin \theta\, \mathbf{c} \qquad 0 \leqslant \theta \leqslant \pi\,. \qquad \text{(A.3.20)}$$

c is a unit vector perpendicular to both **A** and **B** with **A**, **B** and **C** forming a right handed set as shown in Fig. A.3.3(a).

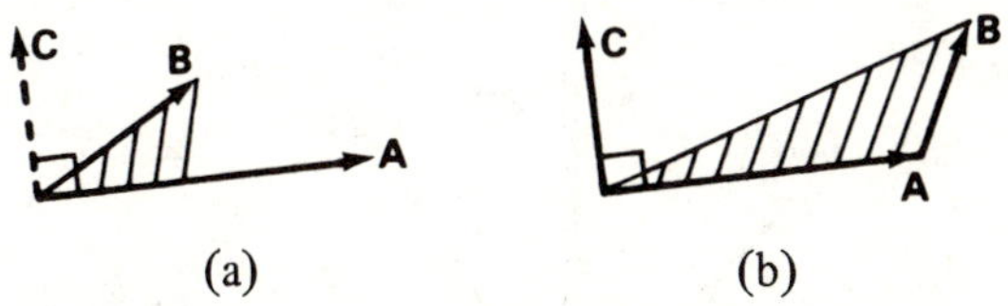

Fig. A.3.3.

The cross product also has a particular significance in mechanics: if **A** is a radius vector from a point O to the initial point of a force vector **B**, then $\mathbf{A} \times \mathbf{B}$ is the moment of the force about point O [Fig. A.3.3.(b)]. $|\mathbf{C}| = AB \sin \theta$.

Some useful properties of the cross product are

$$\mathbf{A} \times \mathbf{B} = -\mathbf{B} \times \mathbf{A}\,. \qquad \text{(A.3.21)}$$

$$\mathbf{A} \times (\mathbf{B} + \mathbf{C}) = \mathbf{A} \times \mathbf{B} + \mathbf{A} \times \mathbf{C}\,. \qquad \text{(A.3.22)}$$

$$\mathbf{i} \times \mathbf{i} = \mathbf{j} \times \mathbf{j} = \mathbf{k} \times \mathbf{k} = 0\,; \quad \mathbf{i} \times \mathbf{j} = \mathbf{k},\ \mathbf{j} \times \mathbf{k} = \mathbf{i},\ \mathbf{k} \times \mathbf{i} = \mathbf{j}\,. \qquad \text{(A.3.23)}$$

An alternative expression for the cross product is

$$\mathbf{A} \times \mathbf{B} = \begin{vmatrix} \mathbf{i} & \mathbf{j} & \mathbf{k} \\ A_x & A_y & A_z \\ B_x & B_y & B_z \end{vmatrix} \qquad \text{(A.3.24)}$$

or, on expanding the determinant,

$$\mathbf{A} \times \mathbf{B} = (A_y B_z - A_z B_y)\mathbf{i} - (A_x B_z - A_z B_x)\mathbf{j} + (A_x B_y - A_y B_x)\mathbf{k} . \quad \text{(A.3.25)}$$

Using the indicial notation,

$$\mathbf{A} \times \mathbf{B} = C_i \mathbf{e}_i \quad \text{(A.3.26)}$$

where

$$C_i = e_{ijk} A_j B_k \quad \text{(A.3.27)}$$

is the *i*th component of **C**.

So far, our discussion of vectors has been largely geometric in character; an alternative analytic description is available which has scope for generalisation to tensors. Central to this approach is the influence of coordinate transformations on the description of these quantities so that, as a first step, we examine the transformation of coordinates under rotation of axes. In this connexion, we shall find it convenient to use the indicial notation.

Suppose we have a description of some physical situation relative to an 'old' set of axes x_i and that we wish to change to some 'new' set $x_p{}'$ by rotation as shown in Fig. A.3.4.

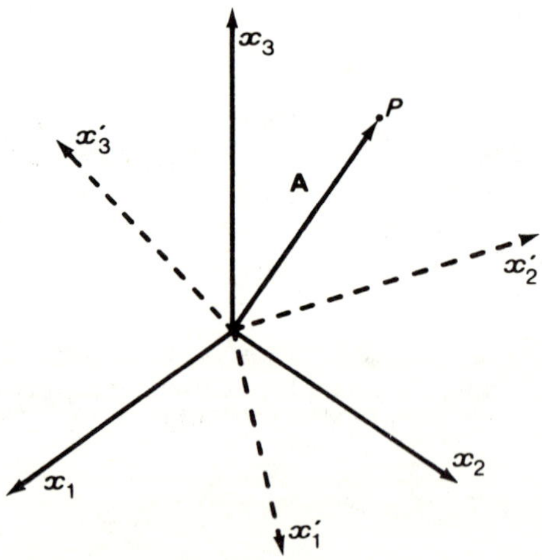

Fig. A.3.4

Denote the cosine of the angle between 'new' $x_p{}'$ and 'old' x_i by $l_{pi} = \cos(x_p{}' x_i)$. If point *P*, defined by radius vector **A**, is deemed fixed whatever set of axes is used, then

$$\mathbf{A} = x_i \mathbf{e}_i = x_p{}' \mathbf{e}_p{}' . \quad \text{(A.3.28)}$$

Here, $\mathbf{e}_i$ and $\mathbf{e}_p{}'$ are unit vectors along the 'old' and 'new' axes respectively whilst $x_i = (x_i - 0)$ and $x_p{}' = (x_p{}' - 0)$ are the components of **A** along the 'old' and 'new' axes. Using the dot product, we can write the projection (component) of **A** along x_p' as

$$A_p' = x_p' = \mathbf{A} \cdot \mathbf{e}_p' = A_i e_{p_i}' . \qquad \text{(A.3.29)}$$

Here, A_i and e_{p_i}' are the components of **A** and $\mathbf{e}_p'$ along the 'old' axes. But since $\mathbf{e}_p'$ is a unit vector, $\mathbf{e}_{p_i}' = \cos(x_p' x_i) = l_{p_i}$. Furthermore $A_i = x_i$ so that

$$x_p' = l_{p_i} x_i \qquad \text{(A.3.30)}$$

and

$$A_p' = l_{p_i} A_i \qquad \text{(A.3.31)}$$

Expressed in words, equation (A.3.31) states that the projection of **A** onto a 'new' direction is the sum of the projections of its 'old' components $\mathbf{A}_i$ onto the new direction. The reader should prove that in going from 'new' to 'old'

$$A_j = l_{pj} A_p' . \qquad \text{(A.3.32)}$$

We now have an analytic definition of a vector: a quantity is a vector if it requires three scalar numbers A_i to specify it with respect to rectangular axes x_i and if these numbers transform according equations (A.3.31) or (A.3.32) under rotation of axes.

An important property of the direction cosines is deduced very conveniently as follows. Recalling from Section A.2, the definition of the Kronecker delta, we can write equation (A.3.32) as

$$A_j = \delta_{ij} A_i = l_{pj} A_p' . \qquad \text{(A.3.33)}$$

Substituting from equation (A.3.31)

$$\delta_{ij} A_i = l_{pj} l_{pi} A_i$$

or

$$(\delta_{ij} - l_{pj} l_{pi}) A_i = 0 .$$

But, A_i is an arbitrary vector so that

$$l_{pi} l_{pj} = \delta_{ij} . \qquad \text{(A.3.34)}$$

A coordinate transformation enjoying this property is said to be **orthogonal**.†

We now define a second order tensor. Suppose two vectors **A** and **B** have components A_i, B_j, A_p', B_q' with respect to 'old' and 'new' coordinates x_i and x_p'. Let us form nine quantities T_{ij} defined by

$$T_{ij} = A_i B_j . \qquad \text{(A.3.35)}$$

These could be displayed as a two dimension array or matrix as described in Section A4. We may similarly form nine new quantities.

$$T_{pq}' = A_p' B_q' . \qquad \text{(A.3.36)}$$

Then, in view of equation (A.3.31),

$$T_{pq}' = l_{pi} l_{qj} A_i B_j ,$$

or

$$T_{pq}' = l_{pi} l_{qj} T_{ij} . \qquad \text{(A.3.37)}$$

† Compare with Problem A.4.1. The matrix of direction cosines, $[l]$, is orthogonal so that $[l]^t = [l]^{-1}$.

Entities which require nine components for their specification and the components of which transform according to the nine equations (A.3.37) are called **second order tensors.** Since here we are only concerned with the transformation from one set of Cartesian axes to another, we are defining **Cartesian tensors.** When curvilinear coordinates are used the transformations are not a simple as in equations (A.3.30) and we have cause to distinguish between **co-variant** and **contravariant tensors**: for Cartesian tensors, such a distinction is not necessary. Several entities of importance in engineering are second order tensors; the states of stress and strain being examples of particular interest to us in this text.

We are now in a position to make the generalised statement alluded to above. If, on rotation of axes,

(i) a quantity $\phi(x_i)$ transforms to $\phi'(x_p')$ according to

$$\phi = \phi' ,$$

it is a **zero order tensor** or **scalar.**

(ii) a quanty $A_i(x_j)$ transforms to $A_p'(x_q')$, $i, j, p, q = 1, 2, 3$, according to

$$A_p' = l_{pi} A_i ,$$

it is a **first order tensor** or **vector.**

(iii) a quantity $T_{ij}(x_r)$ transforms to $T_{pq}'(x_s')$, $i, j, p, q, r, s = 1, 2, 3$, according to

$$T_{pq}' = l_{pi} l_{qj} T_{ij}$$

it is a **second order tensor.**

The extension to an nth order tensor is immediate, the definition being embodied in

$$T_{pqr}' \ldots\ldots = l_{pi} l_{qj} l_{rk} \ldots\ldots T_{ijk} \ldots\ldots \qquad \text{(A.3.38)}$$

It is useful to observe that the number of free indices corresponds exactly to the order of the tensor. Thus, a scalar has no free subscripts, a vector has one etc.

When tensors are functions of position, we usually have cause to differentiate or integrate them with respect to the space variables x_i. We first show that differentiating an nth order tensor leads to an $(n + 1)$th order tensor. According to the comma notation of section A.2,

$$T_{pqr}' \ldots\ldots ,_s = \frac{\partial}{\partial x_s'} (T_{pqr}' \ldots\ldots)$$

$$= \frac{\partial}{\partial x_s'} (l_{pi} l_{qi} l_{rk} \ldots\ldots T_{ijk} \ldots\ldots)$$

$$= l_{pi} l_{qi} l_{rk} \ldots\ldots \frac{\partial T_{ijk} \ldots\ldots}{\partial x_s'} .$$

but $$\frac{\partial T_{ijk}}{\partial x_s'} = \frac{\partial T_{ijk}}{\partial x_l} \frac{\partial x_l}{\partial x_s'}$$

and $$\frac{\partial x_l}{\partial x_s'} = l_{sl} = \cos(x_s' x_l)$$

so that $$T_{pqr\ldots\ldots,s}' = l_{pi} l_{qj} l_{rk} \ldots l_{sl} T_{ijk\ldots\ldots,l} \qquad \text{(A.3.39)}$$

which is the transformation defining an $(n + 1)$th order tensor. We may continue similarly to higher order derivatives.

Certain useful differential operators are met with in vector analysis. Thus, if $\phi(x_i)$ is a *scalar* function of position, its gradient is a *vector*

$$\text{grad}\,\phi = \nabla\phi = \frac{\partial\phi}{\partial x}\mathbf{i} + \frac{\partial\phi}{\partial y}\mathbf{j} + \frac{\partial\phi}{\partial z}\mathbf{k} = \phi_{,i}\mathbf{e}_i\,. \qquad \text{(A.3.40)}$$

$$\nabla \equiv \mathbf{e}_i \frac{\partial}{\partial x_i} \qquad \text{(A.3.41)}$$

is a vector operator called *del.*

The **divergence** of a *vector* function $\mathbf{A}(x_i)$ is a *scalar* quantity

$$\text{div}\,\mathbf{A} = \nabla\cdot\mathbf{A} = \frac{\partial A_x}{\partial x} + \frac{\partial A_y}{\partial y} + \frac{\partial A_z}{\partial z} = A_{i,i}\,. \qquad \text{(A.3.42)}$$

The **curl** of a *vector* **A** is a *vector* quantity

$$\text{curl}\,\mathbf{A} = \nabla\times\mathbf{A} = \begin{vmatrix} \mathbf{i} & \mathbf{j} & \mathbf{k} \\ \dfrac{\partial}{\partial x} & \dfrac{\partial}{\partial y} & \dfrac{\partial}{\partial z} \\ A_x & A_y & A_z \end{vmatrix} = \begin{vmatrix} \mathbf{e}_1 & \mathbf{e}_2 & \mathbf{e}_3 \\ \dfrac{\partial}{\partial x_1} & \dfrac{\partial}{\partial x_2} & \dfrac{\partial}{\partial x_3} \\ A_1 & A_2 & A_3 \end{vmatrix}$$

or $$\nabla\times\mathbf{A} = e_{ijk} A_{k,j} \mathbf{e}_i\,. \qquad \text{(A.3.44)}$$

The Laplace operator ∇^2 is required time again and we may write

$$\nabla^2\phi = \nabla\cdot(\nabla\phi) = \phi_{,ii}\,. \qquad \text{(A.3.45)}$$

Just as in elementary calculus we find it convenient to use the formula for integration by parts and similar devices, so in the higher calculus of continuum mechanics do we find certain integral transformation theorems helpful. In particular, taking advantage of these theorems eases the developments of Chapters 6 and 7. We first consider **Gauss's theorem.**

Suppose that an nth tensor $T_{ijk\ldots\ldots}$ is defined within a region V of space bounded by a closed surface S. We assume that S is smooth (or at least, piecewise smooth) so that tangent planes turn continuously (or piecewise continuously). We wish to find an alternative expression for

$$I = \int_V \frac{\partial}{\partial x_r}(T_{ijk\ldots\ldots})\,\mathrm{d\,vol}\,. \qquad \text{(A.3.46)}$$

In Fig. A.3.5, $\mathrm{d\,vol} = \mathrm{d}x_1\,\mathrm{d}x_2\,\mathrm{d}x_3$.

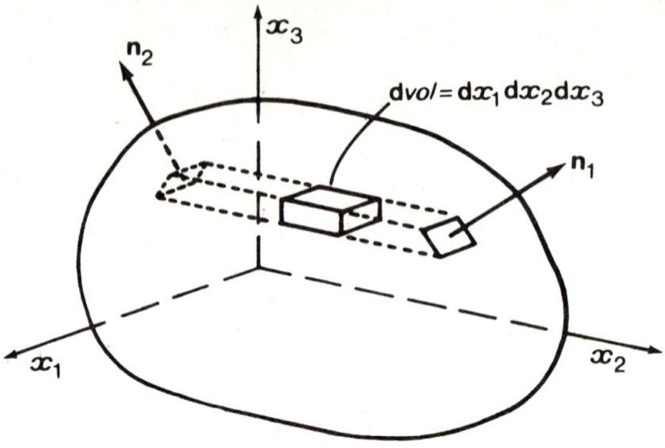

Fig. A.3.5

To proceed, we consider, typically, $r = 2$ and integrate first with respect to x_2. Then,

$$I = \iiint \frac{\partial}{\partial x_2}(T_{ijk\,.....})\mathrm{d}x_1\,\mathrm{d}x_2\,\mathrm{d}x_3$$

$$= \left|\iint T_{ijk\,.....}\;\mathrm{d}x_1\,\mathrm{d}x_3\right|_1 - \left|\iint T_{ijk\,.....}\;\mathrm{d}x_1\,\mathrm{d}x_3\right|_2 .$$

At ends 1 and 2 respectively of the prismatic element in Fig. A.3.5, $(\mathrm{d}x_1\,\mathrm{d}x_3)_1 = |n_2|_1\,\mathrm{d}S$ and $(\mathrm{d}x_1\,\mathrm{d}x_3)_2 = -|n_2|_2\,\mathrm{d}S$ where $\mathrm{d}S$ is an element of the surface S; $\mathbf{n}_1$ and $\mathbf{n}_2$ are unit vector normals so that $n_i = \cos(\mathbf{n}x_i)$.†

Then, $\displaystyle\int_V \frac{\partial}{\partial x_2}(T_{ijk\,.....})\mathrm{d}\,\mathrm{vol} = \int_S (|T_{ijk\,.....}n_2\,\mathrm{d}S|_1 + |T_{ijk\,.....}n_2\,\mathrm{d}S|_2)$.

Accumulating the contributions from all such infinitesimal prisms, we find

$$\int_V \frac{\partial}{\partial x_2}(T_{ijk\,.....})\mathrm{d}\,\mathrm{vol} = \int_S T_{ijk\,.....}\,n_2\,\mathrm{d}S .$$

Repeating the process for $r = 1$ and 3 we have, in general,

$$\int_V \frac{\partial}{\partial x_r}(T_{ijk\,.....})\mathrm{d}\,\mathrm{vol} = \int_S T_{ijk\,.....}\,n_r\mathrm{d}S , \quad r = 1, 2, 3 . \tag{A.3.47}$$

This is Gauss's theorem.

Depending on the order of $T_{ijk\,.....}$, we can write equation (A.3.47) in various forms. Thus, if $T_{ijk\,.....}$ is a scalar ϕ

$$\int_V \phi_{,r}\,\mathrm{d}\,\mathrm{vol} = \int_S \phi n_r \mathrm{d}S \tag{A.3.48}$$

† Whilst we have assumed here that a line parallel to Ox_2 cuts the surface in only two points, result (A.3.47) holds more generally.

or, since $\phi_{,r}\mathbf{e}_r = \nabla\phi$ and $n_r\mathbf{e}_r\mathrm{d}S = \mathbf{n}\mathrm{d}S = \mathrm{d}\mathbf{S}$,

$$\int_V \nabla\phi \mathrm{d}\,\mathrm{vol} = \int_S \phi \mathrm{d}\mathbf{S} \,. \tag{A.3.49}$$

If $T_{ijk\,.....}$ is a vector $\mathbf{A}_i$, we have

$$\int_V A_{i,r}\mathrm{d}\mathrm{vol} = \int_S A_i n_r \mathrm{d}S \,. \tag{A.3.50}$$

Choosing $r = i$ and observing that $A_{i,i} = \mathrm{div}\,\mathbf{A}$ and $A_i n_i \mathrm{d}S = \mathbf{A}\cdot\mathbf{n}\mathrm{d}S = \mathbf{A}\mathrm{d}\mathbf{S}$, we have the **divergence theorem** or **Green's theorem in space**

$$\int_V \mathrm{div}\,\mathbf{A}\,\mathrm{d}\,\mathrm{vol} = \int_S \mathbf{A}\mathrm{d}\mathbf{S} \,. \tag{A.3.51}$$

Equation (A.3.50) is a convenient starting point for deriving a formula for integration by parts. Thus, suppose $A_i = uv$, the product of two functions, then since

$$\frac{\partial A_i}{\partial x_r} = u\frac{\partial v}{\partial x_r} + v\frac{\partial u}{\partial x_r} \,,$$

we have immediately

$$\int_V u\frac{\partial v}{\partial x_r}\mathrm{d}\,\mathrm{vol} = \int_S uvn_r\mathrm{d}S - \int_V v\frac{\partial u}{\partial x_r}\mathrm{d}\,\mathrm{vol} \,. \tag{A.3.52}$$

Returning to equation (A.3.50) and specialising to plane stiuations with $i = r$,

$$\int_S \left(\frac{\partial A_1}{\partial x_1} + \frac{\partial A_2}{\partial x_2}\right)\mathrm{d}S = \int_{\mathscr{C}} (A_1 n_1 + A_2 n_2)\mathrm{d}C \,.$$

If we denote $A_1 = N(x, y)$, $A_2 = -M(x, y)$ and observe that $\mathrm{d}S = \mathrm{d}x\mathrm{d}y$, $n_1\mathrm{d}C' = \mathrm{d}y$, $-n_2\mathrm{d}C' = \mathrm{d}x$, Fig. A.3.6, then

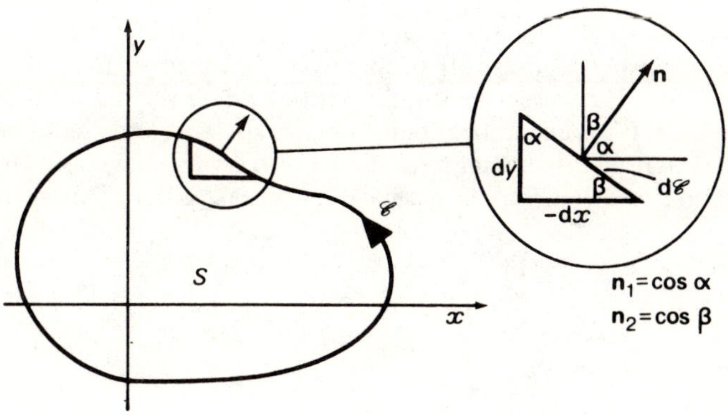

Fig. A.3.6

$$\iint\left(\frac{\partial N}{\partial x} - \frac{\partial M}{\partial y}\right)\mathrm{d}x\mathrm{d}y = \int_{\mathscr{C}} [M\mathrm{d}x + N\mathrm{d}y] \ . \qquad \text{(A.3.53)}$$

This is **Green's theorem in the plane.** We see that if $(\partial N/\partial x) = (\partial M/\partial y)$, the integral around the closed path $\mathscr{C}$ is zero. Under those circumstances $M\mathrm{d}x + N\mathrm{d}y$ is said to be an **exact differential.**

Problems A.3

1. Three forces of 20N each act in the xOy plane. Show that their resultant is $20(\sqrt{3} - 1)$N in the y direction if they act at 60°, 120° and 270° respectively to the x axis.

2. An object is moved round in a closed path in a constant vector force field **F**. Show that the total work done is zero.

3. A rigid body rotates about an axis through O at an angular speed ω. If a force **F** acts at point **r** show that its instantaneous rate of doing work (power) is $P = \mathbf{F} \cdot \omega \times \mathbf{r}$.

4. Using definitions (A.3.19) and (A.3.27) together with the $e \sim \delta$ identity (A.2.9), show that

$$\mathbf{A} \times (\mathbf{B} \times \mathbf{C}) = (\mathbf{A} \cdot \mathbf{C})\mathbf{B} - (\mathbf{A} \cdot \mathbf{B})\mathbf{C} \ .$$

5. If $T_{ijk \ldots\ldots}$ is an nth order tensor, prove that $T_{ijk \ldots\ldots, rs}$ is an (n+ 2)th order tensor.

6. Hooke's law states that $\sigma_{ij} = C_{ijkl}\epsilon_{kl}$. Show that the elastic constants C_{ijkl} are the components of a fourth order tensor.

7. Use Green's theorem in the plane to show that twice the area bounded by a simple closed curve $\mathscr{C}$ is

$$2A = \int_{\mathscr{C}} (x\mathrm{d}y - y\mathrm{d}x) \, .$$

A.4 Determinants and matrices

We have seen in the text that determining the characteristics of mechanical systems having many degrees of freedom involves the manipulation of large amounts of data. The systematic handling of this data is very much facilitated by the use of matrices and in this connexion, we also encounter determinants. It is important to realise that determinants are not the same things as matrices but, under certain circumstances, it is meaningful to speak of the determinant of a matrix.

The symbol

$$\det a = |a| = \begin{vmatrix} a_{11} & a_{12} & a_{13} & \ldots\ldots a_{1n} \\ a_{21} & a_{22} & & \\ \cdot & & & \\ \cdot & & & \\ \cdot & & & \\ a_{n1} & a_{n2} & \ldots\ldots\ldots\ldots & a_{nn} \end{vmatrix}$$

represents an nth order determinant where a_{ij} is the element lying in the ith row and jth column of the determinant. The number of rows equals the number of columns and the parallel lines on either side of the array indicates that a sequence of operations is to be performed to lead to a *value* for the determinant. For example, if $|a|$ is of order 2×2, then we define

$$\begin{vmatrix} a_{11} & a_{12} \\ a_{21} & a_{22} \end{vmatrix} = a_{11}a_{22} - a_{12}a_{21} . \tag{A.4.1}$$

Laplace's expansion provides the means of extending this procedure to the evaluation of an nth order determinant as described below.

The **minor** M_{pq} of element a_{pq} of $|a|$ is defined as the sub-determinant obtained after row p and column q have been deleted from $|a|$. For example, if

$$|a| = \begin{vmatrix} a_{11} & a_{12} & a_{13} \\ a_{21} & a_{22} & a_{23} \\ a_{31} & a_{32} & a_{33} \end{vmatrix}$$

then

$$M_{11} = \begin{vmatrix} a_{22} & a_{23} \\ a_{32} & a_{33} \end{vmatrix}, \quad M_{23} = \begin{vmatrix} a_{11} & a_{12} \\ a_{31} & a_{32} \end{vmatrix}, \quad M_{32} = \begin{vmatrix} a_{11} & a_{13} \\ a_{21} & a_{23} \end{vmatrix} \text{ etc.}$$

The **cofactor** A_{pq} of element a_{pq} of $|a|$ is defined as a signed minor. That is,

$$A_{pq} = (-1)^{(p+q)}M_{pq} . \tag{A.4.2}$$

In this example, $A_{11} = (-1)^{(1+1)}M_{11} = M_{11}$, $A_{23} = (-1)^{(2+3)}M_{23} = -M_{23}$ etc.

We can now write down the rule for expanding (evaluating) an nth order determinant. We can proceed in either one of two ways:

(i) expand by row: select any row i of $|a|$, then

$$|a| = \sum_{j=1}^{n} a_{ij}A_{ij} . \tag{A.4.3}$$

If A_{ij} is itself greater than 2×2, then it must be expanded in an analogous way and so on. When we finally obtain cofactors which are 2×2, rule (A.4.1) is applied.

(ii) expand by column: select any column j of $|a|$ then

$$|a| = \sum_{i=1}^{n} a_{ij}A_{ij} . \tag{A.4.4}$$

Using rule (A.4.3) for row 1 below, we find that

$$|a| = \begin{vmatrix} 1 & 2 & 3 \\ 4 & 8 & 12 \\ 7 & 8 & 9 \end{vmatrix} = 1 . \begin{vmatrix} 8 & 12 \\ 8 & 9 \end{vmatrix} - 2 . \begin{vmatrix} 4 & 12 \\ 7 & 9 \end{vmatrix} + 3 . \begin{vmatrix} 4 & 8 \\ 7 & 8 \end{vmatrix} .$$

or $|a| = -24 + 96 - 72 = 0$.

The reader should confirm that expanding on, say, column 2 yields the same result.

Determinants have a number of useful properties, and by taking advantage of them computations can often be simplified considerably. We will not list them all here, but we will state one which has special significance for us:

> If at least one row (column) of a determinant is a linear combination of the other rows (columns), then the determinant is zero.

In the above example, row 2 is four times row 1 and, as we saw, $|a| = 0$.

A number of quantities displayed as a rectangular array comprise a **matrix.** If the array has m rows and n columns, the matrix is of order $m \times n$ as, for example, in

$$[a] = \begin{bmatrix} a_{11} & a_{12} \cdots\cdots a_{1n} \\ a_{21} & \cdots\cdots \\ \vdots & \\ a_{m1} \cdots\cdots\cdots & a_{mn} \end{bmatrix}.$$

A matrix does not have a *value* in the way that a determinant has. We shall have cause to *operate upon* matrices, but determinants are operators themselves; consequently, we can conceive of the determinant of a square matrix but not of a rectangular one. The reader should observe the notation $[a]$ for a matrix and $|a|$ for a determinant.

Square matrices have an equal number of rows and columns and have a special significance for us in stress analysis. The **leading diagonal** of a square matrix extends from the top left element to the bottom right element, the matrix is **symmetric** if $a_{ij} = a_{ji}$ for all i and j. A **diagonal matrix** is square with non-zero elements on the leading diagonal only; the **unit** or **identity matrix** $[I]$ has elements of unit value on the leading diagonal and zeroes in all the off diagonal positions so that $a_{ij} = \delta_{ij}$ = the Kronecker delta. The determinant of $[a]$, $\det [a] = |a|$. A scalar is a square matrix of order 1×1.

A column matrix, having n rows and *one* column, is also called a **column vector** and often denoted, as in this text, by braces $\{a\}$. A row matrix (or vector) has one row and n columns and is denoted by $\lfloor a \rfloor$.

The transpose $[a]^t$ of a matrix $[a]$, square or rectangular, is obtained by interchanging the rows for the columns, thus:–

If $\quad [a] = \begin{bmatrix} a_{11} & a_{12} & a_{13} \\ a_{21} & a_{22} & a_{23} \end{bmatrix}, \quad$ then $\quad [a]^t = \begin{bmatrix} a_{11} & a_{21} \\ a_{12} & a_{22} \\ a_{13} & a_{23} \end{bmatrix}.$

From this definition, we see that if $[k]$ is a square symmetric matrix, $[k]^t = [k]$.

Two matrices $[a]$ and $[b]$ are equal only if

(i) they are of the same order;

(ii) each element a_{ij} of $[a]$ is equal to the corresponding element b_{ij} of $[b]$.

From these two statements we see that if any two elements of $[a]$ are interchanged, a new matrix $[a']$ is formed which is not equal to $[b]$.

Matrices of the same order can be added or subtracted. Thus if

$$[d] = [a] \pm [b] \pm [c]$$

then $d_{ij} = a_{ij} \pm b_{ij} \pm c_{ij}$ for every i and j.

Multiplying a matrix $[a]$ by a scalar α gives $\alpha[a]$ which means that every element a_{ij} becomes αa_{ij} for all i and j. Two or more matrices may be multiplied together; a matrix product is defined as follows.

A matrix $[a]$ of order $(m \times p)$ pre-multiplying matrix $[b]$ of order $(p \times n)$ yields a matrix $[c]$ of order $(m \times n)$ denoted by

$$[c] = [a]\,[b].$$

Each c_{ij} is obtained by multiplying rows of $[a]$ into columns of $[b]$ so that

$$c_{ij} = \sum_{k=1}^{p} a_{ik}b_{kj} \text{ for all } i \text{ and } j. \qquad \text{(A.4.5)}$$

We see from this definition that the product can be formed only if the number of columns of $[a]$ is the same as the number of rows in $[b]$. Such matrices are said to be **conformable.** If $[a]$ and $[b]$ are of order $(m \times n)$ and $(n \times m)$ respectively, it is possible for $[a]$ to **pre**-multiply $[b]$ as in $[a]\,[b]$ and for $[a]$ to **post**-multiply $[b]$ as in $[b]\,[a]$. However, it is important to realise that, in general, $[a]\,[b] \neq [b]\,[a]$. For example

if $$[a] = \begin{bmatrix} 1 & 2 \\ 3 & 4 \end{bmatrix}, \quad [b] = \begin{bmatrix} 5 & 6 \\ 7 & 8 \end{bmatrix},$$

then $$[a]\,[b] = \begin{bmatrix} 19 & 22 \\ 43 & 50 \end{bmatrix}, \quad [b]\,[a] = \begin{bmatrix} 23 & 34 \\ 31 & 46 \end{bmatrix}.$$

Matrix manipulations follow the usual rules of scalar algebra *except* that multiplication is not commutative as we saw above. In particular, the associative law of multiplication does hold so that

$$[a]\,[b]\,[c] = ([a]\,[b])[c] = [a]([b]\,[c])\,. \qquad \text{(A.4.6)}$$

It is necessary that the pairs of matrices involved in an intermediate multiplication be conformable.

If $[a]$ is a square matrix, then we write

$$[a]\,[a] \,\ldots\ldots\, [a] = [a]^n \qquad \text{(A.4.7)}$$

when $[a]$ is multiplied by itself n times.

The reader should choose a simple example to demonstrate that for any matrix $[b]$,

$$[b]\,[I] \;=\; [I]\,[b] \;=\; [b]\,. \tag{A.4.8}$$

The order of $[I]$ must be chosen so that it is conformable for either pre- or post-multiplication as required.

Using the basic rule (A.4.5), it may be shown that the transpose of a matrix product is equal to the product of the individual transposes in reverse order.

If $$[d] \;=\; [a]\,[b]\,[c]\,, \quad \text{then} \quad [d]^t \;=\; [c]^t\,[b]^t\,[a]^t\,. \tag{A.4.9}$$

Computations are often facilitated by **partitioning** matrices. Thus, we may write

$$[a] \;=\; \left[\begin{array}{ccc|c} a_{11} & a_{12} & a_{13} & a_{14} \\ a_{21} & a_{22} & a_{23} & a_{24} \\ \hline a_{31} & a_{32} & a_{33} & a_{34} \end{array}\right] \;=\; \begin{bmatrix} [P]\,[Q] \\ [R]\,[S] \end{bmatrix}$$

The dotted lines define sub-matrices of $[a]$: we may operate in terms of these sub-matrices. For example,

$$[B] \;=\; \left[\begin{array}{cc|cc} 1 & 2 & 0 & 0 \\ 3 & 4 & 0 & 0 \\ \hline 0 & 0 & 5 & 6 \\ 0 & 0 & 7 & 8 \end{array}\right] \left[\begin{array}{cc|cc} 1 & 0 & 1 & 0 \\ 0 & 1 & 0 & 1 \\ \hline 1 & 0 & 1 & 0 \\ 0 & 1 & 0 & 1 \end{array}\right] \;=\; \begin{bmatrix} [P_1]\,[Q_1] \\ [R_1]\,[S_1] \end{bmatrix} \begin{bmatrix} [I]\,[I] \\ [I]\,[I] \end{bmatrix}.$$

Treating $[P_1]$ etc., as elements in a reduced matrix, we follow the rules of matrix multiplication to give

$$[B] \;=\; \begin{bmatrix} ([P_1] \;+\; [Q_1]) & ([P_1] \;+\; [Q_1]) \\ ([R_1] \;+\; [S_1]) & ([R_1] \;+\; [S_1]) \end{bmatrix},$$

where we have noted $[P_1]\,[I] \;=\; [P_1]$ etc.

Since in this example $[Q_1] \;=\; [R_1] \;=\; [0]$, a **null matrix**, the manipulation is much simplified and

$$[B] \;=\; \begin{bmatrix} 1 & 2 & 1 & 2 \\ 3 & 4 & 3 & 4 \\ 5 & 6 & 5 & 6 \\ 7 & 8 & 7 & 8 \end{bmatrix},$$

as may be confirmed by direct multiplication.

The **inverse** or **reciprocal** of a square matrix $[a]$ is written as $[a]^{-1}$ and defined by

$$[a]^{-1}\,[a] \;=\; [a]\,[a]^{-1} \;=\; [I]\,. \tag{A.4.10}$$

The reader should notice particularly that this definition does *not* imply $[a]^{-1}$ is equal to $1/[a]$. In fact, defining the **adjoint**, adj of $[a]$ to be

$$\text{adj}[a] = \begin{bmatrix} [A_{11}] & [A_{21}] & [A_{31}] \ldots\ldots & [A_{n1}] \\ [A_{12}] & [A_{22}] \ldots\ldots\ldots\ldots & & \\ \vdots & \vdots & & \\ [A_{1n}] & [A_{2n}] \ldots\ldots\ldots\ldots & & [A_{nn}] \end{bmatrix} \tag{A.4.11}$$

where $[A_{ij}]$ is the cofactor of a_{ij}.

$$[a]^{-1} = \frac{\text{adj}[a]}{|a|} . \tag{A.4.12}$$

The reader should note the order of the subscripts in adj$[a]$. Equation (A.4.12) shows that $[a]^{-1}$ does not exist if $|a| = 0$, when this condition obtains, the matrix is said to be **singular**. Evidently only non-singular matrices have inverses.

There is a useful rule for determining the inverse of a product. Thus, if $[a] = [b]\,[c]\,[d]$, then

$$[a]^{-1} = [d]^{-1}\,[c]^{-1}\,[b]^{-1} . \tag{A.4.13}$$

Time and again, in this text, sets of equations arise in the form

$$\begin{aligned} a_{11}x_1 + a_{12}x_2 + \ldots\ldots a_{1n}x_n &= y_1 , \\ a_{21}x_1 + a_{22}x_2 + \ldots\ldots \quad &= y_2 , \\ \vdots \qquad\qquad\qquad &\quad \vdots \\ a_{n1}x_1 + \ldots\ldots\ldots\ldots a_{nn}x_n &= y_n \end{aligned}$$

From the rules of matrix multiplication, we can write these as

$$[a]\{x\} = \{y\} . \tag{A.4.14}$$

With $\{y\}$ prescribed, the unknown $\{x\}$ is given by pre-multiplying equation (A.4.14) by $[a]^{-1}$. That is

$$[a]^{-1}\,[a]\{x\} = [a]^{-1}\{y\} .$$

Invoking equation (A.4.10), we see that

$$\{x\} = [a]^{-1}\{y\} . \tag{A.4.15}$$

Solution of equations (A.4.14) thus involves determining the inverse of the coefficient matrix, For efficiency in computing practice, the inverse matrix is not determined literally according to definition (A.4.12); various processes, based essentially on **Gaussian elimination**, are used. The procedure may be illustrated by a simple example. Thus, suppose we wish to solve for the x's in

$$\begin{bmatrix} 1 & -3 & 2 \\ 2 & -5 & 7 \\ 1 & -2 & 4 \end{bmatrix} \begin{Bmatrix} x_1 \\ x_2 \\ x_3 \end{Bmatrix} = \begin{Bmatrix} 4 \\ -3 \\ 2 \end{Bmatrix}.$$

If we subtract twice the first equation from the second, and subtract the first equation from the third, we have

$$\begin{bmatrix} 1 & -3 & 2 \\ 0 & 1 & 3 \\ 0 & 1 & 2 \end{bmatrix} \begin{Bmatrix} x_1 \\ x_2 \\ x_3 \end{Bmatrix} = \begin{Bmatrix} 4 \\ -11 \\ -2 \end{Bmatrix}.$$

If in this new system we subtract the second equation from the third, we have

$$\begin{bmatrix} 1 & -3 & 2 \\ 0 & 1 & 3 \\ 0 & 0 & -1 \end{bmatrix} \begin{Bmatrix} x_1 \\ x_2 \\ x_3 \end{Bmatrix} = \begin{Bmatrix} 4 \\ -11 \\ 9 \end{Bmatrix}.$$

That is, we have transformed the original coefficient matrix into an **upper triangular matrix**. We may now solve for x_3 from the third equation, substitute back into the second to yield x_2 and finally substitute the newly found x_3 and x_2 back in the first equation to yield x_1. Indeed, we find

$$x_3 = -9, \quad x_2 = 16, \quad x_1 = 70.$$

The process corresponds to a triangular decomposition of the coefficient matrix $[a]$ into a product of upper and a lower triangular matrices. That is $[a] \to [l]\,[u]$, where $[l]$ is a lower triangular matrix with unit values on the leading diagonal and $[u]$ is an upper triangular matrix of the form generated in the example. We thus have

$$[a]\{x\} = [l]\,[u]\{x\} = \{y\}. \tag{A.4.16}$$

Denoting $[u]\{x\} = \{z\}$, we have

$$[l]\{z\} = \{y\} \tag{A.4.17}$$

as the forward reduction and

$$[u]\{x\} = \{z\} \tag{A.4.18}$$

as the back substitution.

In most of the situations discussed in this book, the coefficient matrix $[a]$ is symmetric and positive definite so that **Cholesky's method** may be used.† Now, the triangular decomposition is specialised to $[a] = [l]\,[l]^t$ where $[l]$ is a lower triangular matrix the diagonal terms of which are not necessarily unity. The process is now $[a]\{x\} = [l]\,[l]^t\{x\} = \{y\}$: denoting $[l]^t\{x\} = \{z\}$,

$$[l]\{z\} = \{y\} \tag{A.4.19}$$

† The coefficient matrix generated by Reissner's principle does not enjoy positive definiteness.

is forward reduction and $$[l]^t\{x\} = \{z\} \tag{A.4.20}$$
is back substitution. To solve a system of n equations by the Cholesky scheme requires $n^3/6 + 3n^2/2 + 7n/3$ operations: Gaussian elimination can require four times this number whilst **Kramer's rule** [essentially equation (A.4.15)] requires $n!$ multiplications for evaluating the determinant alone.

Gaussian elimination and its variants comprise a class of **direct methods** of solution. Alternative methods, such as the **Gauss-Seidel** one, are **iterative** or **relaxation methods.** Here, we begin from an initial trial vector $\{x\}^0$ and generate a sequence of vectors $\{x\}^s$ which, hopefully, converges to the correct solution $\{x\}$. Indeed, since for very large sets of equations the round-off errors generated in the elimination methods may be unacceptably large, iterative methods may be used to refine such solutions. Computer centres have ready prepared packages for carrying out this kind of work so that the stress analyst need not develop his own equation solving routines; he may use these packages at will and concentrate his energies on formulating and solving his engineering problems.

Problems A.4

1. Determine the quantities $[k]\{u\}$, $\{u\}^t[k]$, $\{u\}^t([k]\{u\})$, and $(\{u\}^t[k])\{u\}$ when
$$[k] = \begin{bmatrix} 1 & 1 & 1 \\ 1 & 2 & 2 \\ 1 & 2 & 3 \end{bmatrix}, \quad \{x\} = \begin{Bmatrix} 1 \\ 2 \\ 1 \end{Bmatrix}.$$

2. Show that the product of a matrix and its transpose is always a square symmetric matrix.

3. If forces $F_1, F_2 \ldots\ldots F_n$ act through corresponding displacements u_1, $u_2 \ldots\ldots u_n$, demonstrate that the work done is
$$W = \{u\}^t\{F\} = \{F\}^t\{u\}.$$

4. The configuration of a system of discrete members is defined by coordinates $\{u\}$ at which forces $\{F\}$ act. Geometric constraints are introduced such that
$$\{u\} = [C]\{q\},$$
where $\{q\}$ is a vector of generalised coordinates and $[C]$ a matrix of constant coefficients. Show that the generalised forces are obtained from
$$\{Q\} = [C]^t\{F\}.$$
Recall that if virtual displacements were applied, then the corresponding virtual work would be invariant with respect to a change of coordinates. (Chapter 1).

5. Referring to Fig. A.4.1, show that rotation of axes leads to
$$\begin{Bmatrix} x' \\ y' \end{Bmatrix} = \begin{bmatrix} \cos\alpha & \sin\alpha \\ -\sin\alpha & \cos\alpha \end{bmatrix} \begin{Bmatrix} x \\ y \end{Bmatrix}$$
or $$\{x'\} = [t]\{x\}.$$

Show that $[t]$ is **orthogonal**, that is $[t]^{-1} = [t]^t$ so that $[t][t]^t = [I]$. [Compare with equation (A.3.34) for the three dimensional case.]

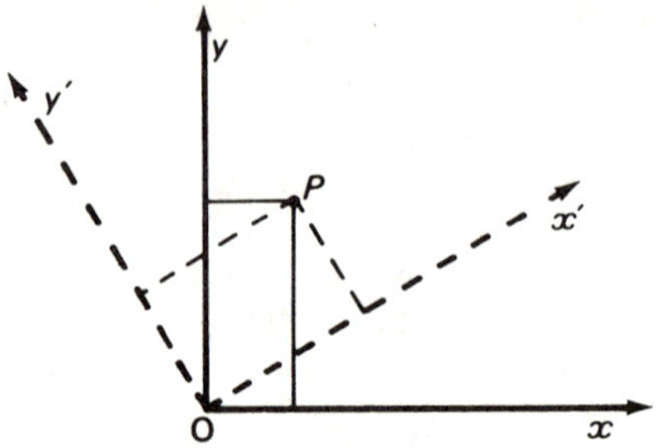

Fig. A.4.1

A.5 Quadratic forms

An expression of the type

$$Q = \sum_{i=1}^{n} \sum_{j=1}^{n} a_{ij} x_i x_j = \{x\}^t [a] \{x\} , \qquad \text{(A.5.1)}$$

where $[a]$ is a symmetric matrix, is called a **quadratic form** in the variables x_i, $i = 1, 2, \ldots .. n$. We have seen that the strain energy stored in a finite degree of freedom linear elastic system is described by this form: it may also be shown that the kinetic energy of such a system is a quadratic form in the velocities. Forms such as (A.5.1) thus have considerable significance in the energy methods of stress analysis so that it will be useful to summarise some of their properties.

If we restricted ourselves to coefficients a_{ij} and variables x_i which are all real, then Q is real. When all the x_i are zero, then so is Q. Consider now the range of possible values of Q as the x_i take independent real values except all simultaneously being zero. One of the following conditions will occur:

(i) $Q > 0$, then the quadratic form is said to be positive definite;
(ii) $Q \geqslant 0$, the form is positive semi-definite;
(iii) $Q \leqslant 0$, the form is negative semi-definite;
(iv) $Q < 0$, the form is negative definite;
(v) $\infty < Q < \infty$, the form is indefinite.

Examples of quadratic forms of the above types are:

$x_1^2 + x_2^2$	Positive definite.
$(x_1 + x_2)^2 = x_1^2 + 2x_1x + x_2^2$	Positive semi-definite.
$-(x_1 + x_2)^2 = -x_1^2 - 2x^1x^2 - x_2^2$	Negative semi-definite.
$-x_1^2 - x_2^2$	Negative definite.
x_1x_2	Indefinite.

The type of quadratic form is determined by the matrix of coefficients or, more particularly, by its **rank.** If $[a]$ is of order n and $|a| \neq 0$, then its rank is said to be n. If however $|a| = 0$, we consider the determinants of sub-matrices obtained by deleting rows and columns of $[a]$. Then, if at least one non-singular $r \times r$ matrix can be determined in this way, with all larger sub-matrices being singular, $[a]$ is said to be of rank r.† Evidently, if $r < n$, $[a]$ is singular and so is the quadratic form.

Suppose the x_i are expressible in terms of new variables y_i by the non-singular†† linear transformation

$$x_i = \sum_{j=1}^{n} t_{ij} y_j \, , \qquad i = 1, 2 \ldots\ldots n \tag{A.5.2}$$

or

$$\{x\} = [t]\{y\} \, . \tag{A.5.3}$$

Then, the quadratic form becomes

$$Q = \{y\}^t [t]^t [a] [t] \{y\} = \{y\}^t [b] \{y\} \, . \tag{A.5.6}$$

Now, it can be shown that it is always possible to choose some $[t]$ such that $[b]$ is diagonal with the special property that

$$Q = y_1^2 + y_2^2 + \ldots\ldots y_{\mathrm{I}}^2 - y_{\mathrm{I}+1}^2 - \ldots\ldots - y_r^2 \, . \tag{A.5.7}$$

Here, r is the rank of $[a]$ and I is the **index** of the form; I is determined uniquely by $[a]$. Equation (A.5.7) is said to be the **canonical form** of the quadratic form.

The type of quadratic form is determined by r and I as follows.

(i) If $\mathrm{I} = n$, Q is positive definite.
(ii) If $\mathrm{I} = r < n$, Q is positive semi-definite.
(iii) If $\mathrm{I} = 0, r < n$, Q is negative semi-definite.
(iv) If $\mathrm{I} = 0, r = n$, Q is negative definite.
(v) If $0 < \mathrm{I} < r$, Q is indefinite.

There are direct methods for determining index I without explicitly considering the transformation into sums and differences of squares. In particular, if we wish to ascertain whether or not some matrix is positive definite, we may use a theorem which states that when $[a]$ is positive definite all its principal minors are positive.

Because the type of quadratic form is determined by the coefficient matrix $[a]$, it is common to speak of positive definite, positive semi-definite etc., matrices. The stiffness and flexibility matrices met with in this book are positive definite whilst the coefficient matrix generated by Reissner's principle is not.

† This implies that r is the maximum number of linearly independent rows or columns in $[a]$.

†† The transformation is non-singular if $|t| \neq 0$ so that $\{y\} = [t]^{-1}\{x\}$ can be determined.

Problems A.5

1. Determine the class to which the following quadratic forms belong.

(i) $x_1^2 + 4x_2^2 + 12x_3^2 + 4x_1x_2 + 12x_2x_3 - 2x_3x_4 + 6x_1x_3 - 4x_1x_4 + 6x_2x_4$.

(ii) $u_1^2 + u_2^2 + \frac{1}{2}u_3^2 + \frac{1}{2}u_4^2 - u_3u_4$.

(iii) $10x_1^2 + 7x_2^2 + 5x_3^2 + 12x_4^2 - x_1x_2 - 2x_1x_3 - 3x_3x_4$.

(iv) $5v_1^2 + 5v_2^2 + v_3^2 - 8v_1v_2 - 4v_2v_3 + 4v_3v_1$.

Further Reading

Ayres, F.	*Schaum's Outline of Theory and Problems of Matrices,* McGraw Hill (1974).
Bickley, W.G. *and* Thompson, R.S.H.G.	*Matrices: Meaning and Manipulation,* The English Universities Press (1964).
Chorlton, F.	*Vector and Tensor Methods,* Ellis Horwood, Publisher (1976).
Courant, R.	*Differential and Integral Calculus, Vol II,* Blackie (1954).
Frazer, R.A., Duncan, W.J. *and* Collar, A.R.	*Elementary Matrices,* Cambridge University Press (1955).
Jaeger, L.G.	*Cartesian Tensors in Engineering Science,* Pergamon Press (1966).
National Physical Laboratory (NPL)	*Modern Computing Methods (Notes on Applied Science No. 16*) H.M.S.O. (1961).
Pipes, L.A. *and* Hovanessian, S.A.	*Matrix Computer Methods in Engineering,* John Wiley (1969).
Sokolnikoff, I.S.	*Tensor Analysis: Theory and Applications,* John Wiley (1951).
Speigel, M.R.	*Vector Analysis,* Schaum Publishing Co. (1957).

Index